Polymer Tribology

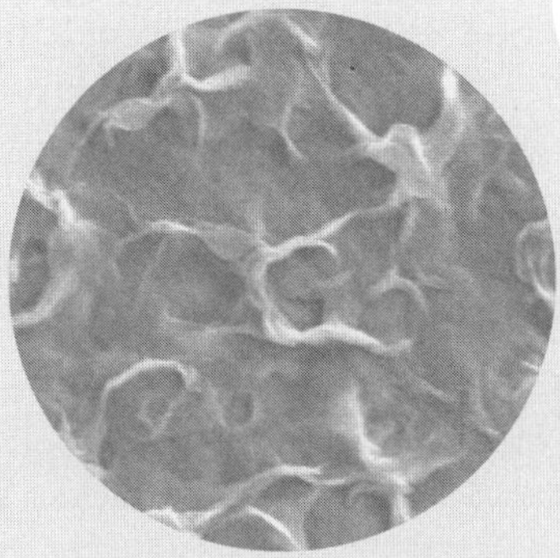

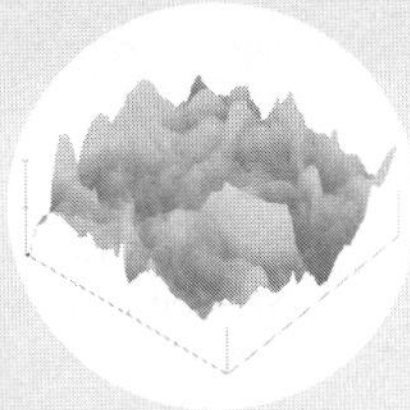

Polymer Tribology

Editors

Sujeet K Sinha

National University of Singapore, Singapore

Brian J Briscoe

Imperial College London, UK

Imperial College Press

Published by

Imperial College Press
57 Shelton Street
Covent Garden
London WC2H 9HE

Distributed by

World Scientific Publishing Co. Pte. Ltd.
5 Toh Tuck Link, Singapore 596224
USA office: 27 Warren Street, Suite 401-402, Hackensack, NJ 07601
UK office: 57 Shelton Street, Covent Garden, London WC2H 9HE

British Library Cataloguing-in-Publication Data
A catalogue record for this book is available from the British Library.

POLYMER TRIBOLOGY

ISBN-13 978-1-84816-202-0
ISBN-10 1-84816-202-2

Printed in Singapore by B & JO Enterprise

EDITORIAL

POLYMER TRIBOLOGY — A PREFACE

Tribology is defined as the branch of study of interacting solids. In more technical terms, it is the science and technology of two contacting solid surfaces in relative motion that produces effects such as friction, wear, adhesion, stiction, surface fatigue and so on. These simple terms obviously entail a host of complex surface mechanical and materials issues. Tribology is a very materials-related discipline where each material may produce entirely different phenomenon during contact sliding and a "trail of events" that will decide its performance in terms of the measured quantities such as the coefficient of friction, wear rate and the durability. The fact that tribology is an interfacial phenomenon, it also greatly depends upon the surrounding (environmental) factors such as the presence of contamination, oxide layer on the solid surfaces, thermal heat and lubricants, in addition to the intrinsic chemical, physical and mechanical properties of the two interacting bodies and operational conditions of contact pressure and relative speed. There is another factor that may change the tribological performance of two bodies and that is the presence of loose debris materials that is ejected from either of the interacting bodies and this is known as the "third-body".

The above serves as a broad definition of tribology and presents the complexity that is faced while dealing with the design of any contacting mechanical system where relative motion, intentional or unintentional, is experienced. With this background, tribology of polymers and their different modifications presents further complexity as polymers are easily influenced by the operating conditions and the prevailing environment. However, tribology of polymers is also a fascinating area because polymers can be modified by various chemical and physical means to suit a particular application. In addition, they are available in liquid, solid or semi-solid forms at room temperature and thus provide opportunities of their use as solid or liquid lubricants, low friction and wear resistant bulk solid or as films.

Although research activities on polymer tribology started almost at the same time when tribology was gaining importance as recognized mechanical and materials field in engineering, important works really started with the works on rubbers and elastomers. This was because of the importance of rubbers and elastomers in the development of modern automotive industry. These materials served as tyres and brake pads which were all subjected to severe sliding and loading conditions coupled with the onslaught of external factors such as the adverse road conditions, dirt and water. Earliest reference that can be found on the friction studies on rubber is by Roth, Driscoll and Holt [1]. The concern for the tribological (mainly friction and wear) performance of rubbers led to the classical work by Schallamach by needles and spherical balls under scratching or sliding conditions [2]. His works have remained the hallmark of many subsequent works on rubbers and elastomers. Further adaptations of these works can be found in later researches.

Tribological work on non-elastomeric polymers can be traced back at least to the publication by Shooter and Thomas [3] in 1949 in which they investigated friction coefficients for polymer sliding on polymer and on steel for Teflon (PTFE), poly(ethylene), poly(styrene) and Perspex (PMMA). They found that the area of contact was proportional to the normal load, thus verifying Bowden and Tabor's proposal for metals. This work established that the value of friction coefficient of PTFE and PE were very low (low interfacial shear strength), a result that led to many tribological applications of these two linear thermoplastics. By late 1950s, many other polymers were also recognized for their value as tribological materials and work had begun on understanding the fundamental mechanisms of friction and wear of many polymers of the day. Research on polymer tribology intensified in 1960s and 1970s and many of the current understandings in these subjects can be traced back to the classical works by Tabor's group [4-6] at Cavendish Laboratory, University of Cambridge, UK. Works by Tanaka [7, 8], Lancaster [9], Steijn [10], Grosch [11] and by the Russian group are also notable [12, 13] on PTFE and many other thermoplastics. These early works found many new tribological applications of thermoplastics in areas such as biology and space technology. One notable example is the application of ultra-high molecular weight poly(ethylene) (UHMWPE) as load bearing acetabular cup in hip prostheses [14, 15] that was used almost accidently after a failed attempt with PTFE by John Charnley in 1962. UHMWPE is still the most popular material for hip/knee prosthetic joints. Similar tribological applications of polymers in engineering are abound.

The successful applications of bulk polymers in engineering led to the work of using polymers as composites where modifications to properties could be achieved by adding fillers such as strong fibres, lubricating solid particles and even metallic particle and fibres. Polymer composites have been very successful in finding many engineering applications as tribological materials for gears, bearing cage, shoe soles, automotive brakes and so on. Important early works on polymer composites were conducted by groups led by Briscoe [16, 17], Bahadur [18] and Rhee [19]. The popularity of the field of tribology of polymer composites led to the compilations of many edited works and a notable early work among them was the edited book by Friedrich [20] in 1986. Polymer composites have been so successful in tribology that in today's time, probably it would be hard to find a polymer for tribological application which is not in composite or even hybrid form. This work has further extended to the application of nano-sized fillers such as nano-clay, carbon nanotubes, nano-sized ceramic particles (Si_3N_4) and so on, with further improvements in the friction and/or wear performances [21, 22]. Presently, composites are used as bulk or as films for various applications where surface friction and wear are important design requirements.

In line with the developments in the area of polymer tribology, as briefly outlined above, this book was commissioned to present the achievements in a manner that incorporates all facets of the tribology of polymers, their modifications and most importantly, their applications. Each selected topic has been written by experts in the field. Though not initially intended, this book has naturally followed its division into three major parts: bulk polymers, reinforced polymers and polymer films. These are the three important forms of polymeric solids that a user may find during the course of engineering design and materials selection or in tribological research involving polymers. Part I of this book deals with polymers that have not been modified with any strengthening agent for bulk mechanical properties. Chapters in this part are on the fundamental studies of adhesion, friction, hardness and wear of some important thermoplastics. This part also includes chapters on the application of pristine thermoplastics for biomedical and Microsystems, and, a chapter on the tribology of rubbers, all in un-reinforced form. Part II deals with the popular subject of tribology of reinforced polymers, also called polymer composites. As many of the polymers are inferior in load carrying capacity when compared to other engineering materials such as metals, it is useful to add fillers to increase the strength and thus reduce wear. This trend started with the use of fillers such as glass fibre, carbon fibre, aramid fibre and a host of particulate fillers in polymer matrices such as epoxy, phenolics and many thermoplastics. Fillers are also added to polymers to modify their friction property and thus wear performance; addition of PTFE particle comes under this class of materials. These trends in the area of tribology of polymer composites have been captured by several chapters in this book. There are many applications of polymer composites and hence we have included chapters on the tribological applications in gears, brake and other automotive components. Currently, there is a new trend on the tribological application of polymer films and coatings. Polymer coatings, in pure or blended (composite) form, can be applied to almost all materials by chemical, physical or even mechanical bonding or interlocking. There are many advantages of having polymer coatings on substrates such as metals, Si or even some polymers which are very poor tribological materials. Use of solid or partly-lubricated polymer films would avoid the need for liquid lubrication if the conditions are well-controlled. Therefore, Part III of this book is dedicated to some works on polymer films. This includes chapters on the mechanical property characterizations and several applications such as in Microsystems for achieving low friction and low wear performance. A very novel approach in this field is the use of polymer brushes that have shown excellent tribological properties and hence we have included one chapter on polymer brushes as well.

Overall, we are very hopeful that this book provides an excellent spectrum of results and analysis on the tribology of polymeric solids in their various forms and manifestations. It has been a privilege to collaborate with leading researchers from around the world for this book and we are very grateful to them for providing us their chapters as presented in this book. The meticulous hard work by the authors of the chapters made our editing job relatively easier and any error still found in this book is entirely due to our incapability.

The Editors
Sujeet K. Sinha, *National University of Singapore*
Brian J. Briscoe, *Imperial College, London, UK*

References

[1] F. L. Roth, R. L. Driscoll and W. L. Holt, *J. Res. Natl. Bur. Stand*, 28 (1942).
[2] A. Schallamach, *J. Polym. Sci.*, 9(5), 1952.
[3] K. V. Shooter and O. H. Thomas, *Research*, 2 (1949) p. 533.
[4] S. C. Cohen and D. Tabor, *Proc. R. Soc. London*, **A291** (1966).
[5] K. C. Ludema and D. Tabor, *Wear*, 9 (1966).
[6] C. M. Pooley and D. Tabor, *Proc. R. Soc. London*, **A 329** (1972).
[7] K. Tanaka, Y. Uchiyama and S. Toyooka, *Wear*, 23 (1973).
[8] K. Tanaka and T. Miyata, *Wear*, 41 (1977).
[9] J. K. Lancaster, *Tribol. Conv.*, (Inst. Mech. Eng., London, 1969), p. 100.
[10] R. P. Steijn, *Wear*, 12 (1968).
[11] K. A. Grosch, *Proc. R. Soc. London*, **A274** (1963).
[12] S. N. Ratner, I. I. Farberoua, O. V. Radyukeuich and E. G. Lure, *Sov. Plast.*, 7 (1964).
[13] S. B. Ratner and E. G. Lure, *Dokl. Akad. Nauk SSSR*, 166(4), (1966).
[14] D. Dowson, in *Polymer Surfaces*, Ed. D. T. Clark and J. Feast, (John Wiley and Sons, New York, 1978), p. 400.
[15] H. McKellop, I. C. Clarke, K. L. Markolf and H. C. Amstutz, *J. Biomed. Mat. Res.*, 12 (1978).
[16] B. J. Briscoe, A. K. Pogosian and D. Tabor, *Wear*, 27 (1974).
[17] B. J. Briscoe, M. D. Steward and A. Groszek, *Wear*, 42 (1977).
[18] M. K. Kar and S. Bahadur, *Wear*, 30 (1974).
[19] T. Liu, and S. K. Rhee, in *Wear of Materials 1977*, (The ASME, New York, 1977) p. 552.
[20] K. Friedrich, (Ed), *Friction and Wear of Polymer Composites*, (Elsevier, Amsterdam, 1986).
[21] Q. Wang, J. Xu, W. Shen and W. Liu, *Wear*, 196 (1996), p. 82.
[22] M. Z. Rong, M. Q. Zhang, H. Liu, H. M. Zeng, B. Wetzel and K. Friedrich, *Ind. Lubr. Tribol.*, 53 (2001), p. 72.

ACKNOWLEDGEMENTS

An edited work such as this one would not have been possible without the generous help from the following individuals. First and foremost, we would like to thank the authors of the chapters who put in painstaking effort in completed their chapters within a tight schedule. Many of the authors also served as the peer-reviewers for the chapters submitted for publication.

We are very grateful to Dr. N. Satyanarayana and Mr. Myo Minn for going through the manuscript and suggesting many changes as well as for compiling the contents and index pages of the book and for formatting many of the chapters.

Finally, we would like to thank our wives and children for bearing the added work stress during the compilation of this book. Without their love and great support, this work would not have been completed.

CONTENTS

PART I
Bulk Polymers

CHAPTER 1

ADHESION AND FRICTION OF POLYMERS

Nikolai K. Myshkin and Alexander V. Kovalev

Department of Tribology, Metal-Polymer Research Institute
Belarus National Academy of Science
Kirov St. 32A, Gomel, Belarus
E-mail: nkmyshkin@mail.ru

Surface phenomena and the structure of polymers affect essentially their tribological characteristics, the adhesion of polymers to solids being one of the dominating factors. The latter is explained, in particular, by the adsorption (chemisorption) of the functional groups, variations in the polymer crystallinity and morphology, constrained molecular mobility, and catalytic effects of the additives on the reactions occurring in the contact zone. It is shown how load, sliding velocity, and temperature affect friction. Different modes of wear of polymers and friction transfer are considered.

1. Introduction

Friction is a common phenomenon in life and industry, which is governed by the processes prevailing on the surface layers of sliding bodies. The simple idea used in studies of friction is that there are two main components of friction, namely, adhesion and deformation. Such approach is correct for all materials including polymers. The behavior of polymers has distinguishing features, some of which are described by Briscoe.[1,2] The main concept of the tribology of polymers consists of three basic elements involved in friction:[3-5]

(i) adhesive junctions, their type and strength;
(ii) shear and rupture of the rubbing materials in the contact;
(iii) real contact area.

The deformation component of friction results from the resistance of the polymer to "ploughing" by the asperities of the harder counterface. Polymer surface asperities experience elastic, plastic, and visco-elastic deformation depending on the material properties. The adhesion component stems from the adhesive junctions formed on the spots of real contact between the mated surfaces. The adhesion component of friction for polymers is believed to exceed much the deformation. Special consideration is needed for transfer films, being the key factor, which determine the tribological behavior of polymers and polymer composites.

2. Polymers and Polymer Composites

It is known that the size of linear macromolecules of polymers is very big and the main feature of the polymer structure is that macromolecules consist of the rigid segments which can rotate thus providing the flexibility of the molecular chains. Another feature of polymers is that strong chemical forces link the atoms in a polymer chain whereas the intermolecular forces which are significantly weaker, link the chains. The structural features of polymers and the possibility of changing their properties within a wide range provide a variety of the tribological applications of polymers and polymer composites.

The application of the different fillers gives an opportunity of improving the tribological behavior of polymers.[6] For example; the reinforcement with short fibres (glass or carbon) is used most often to increase the mechanical strength, hence, the load-bearing capacity of polymer composites. Solid lubricants such as polytetrafluoroethylene (PTFE), graphite, molybdenum disulphide added to polymers affect significantly the formation of the transfer films on the counterface and decrease the friction coefficient.[7-9]

In recent years, owing to rapid advances in nanotechnologies[10], polymer nanocomposites being a polymer matrix filled with the particles 100 nm and smaller in size, have become more and more common. Commonly used nanofillers in plastics are the carbon materials (fullerene and its derivatives), layered clayey minerals, and nanoparticles of metals or their organic and inorganic compounds. Information is available on the polymer/polymer nanoblends in which the dispersed phase is broken up into the nanometer fragments within the polymer matrix.[11]

3. Mechanical Behavior of Polymer-Based Materials

The mechanical behavior of polymers is governed by the combination of elasticity and viscosity. At small deformation, polymers behave as the Hook elastic body ($\sigma = E\varepsilon$, where σ and ε are the stress and strain, and E is the modulus of elasticity) modeled with a spring and Newtonian fluid ($\sigma = \eta d\varepsilon/dt$, where η is the viscosity and t is the time); the latter is presented by a damper. The combination of these elements gives a simple description of visco-elasticity (see Table 1).

In general, the constitutive law for polymers is written as the ordinary differential equation with the constant coefficients:

$$a_0\sigma + a_1\frac{d\sigma}{dt} + \ldots + a_m\frac{d^m\sigma}{dt^m} = b_0\varepsilon + b_1\frac{d\varepsilon}{dt} + \ldots + b_n\frac{d^n\varepsilon}{dt^n}, \qquad (1)$$

where $a_0,\ldots, a_m, b_0,\ldots, b_n$ are the constant coefficients governing the mechanical behavior.

4. Adhesion of Polymers

The adhesive bond is formed in two stages. The first stage includes the movement of the polymer molecules towards the solid surface and their certain orientation on it. This process is promoted by the increase in the temperature and the pressure as well as the transition of one of the phases into the liquid state due to its dissolution or melting. If the polymer surface and the counterface are thoroughly cleaned the contact becomes stronger.

The second stage of adhesion involves the direct interaction of the polymer and the solid surface which can be caused by different forces. The covalent forces become predominant at the distances between atoms and molecules not exceeding 0.5 nm; ionic and van der Waals forces are effective at farther distances[12,13] (approximately from 1 to 100 nm). The interatomic interaction of the contacting phases completes the adhesion process that corresponds to the minimal interphase energy.

Table 1. Simplest models of viscoelasticity.

№	Model	Equation	Relaxation (ε=ε$_0$)	Creep (σ=σ$_0$)
1	Maxwell 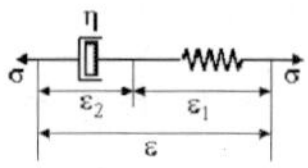	$\frac{d\sigma}{dt}+\frac{1}{\tau}\sigma=E\frac{d\varepsilon}{dt}$	$\sigma=E\varepsilon_0\exp\left(-\frac{E}{\eta}t\right)$ 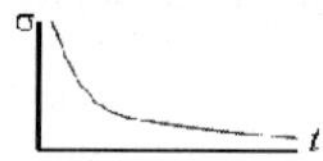	$\varepsilon=\frac{\sigma_0}{\eta}t+\varepsilon_0$ 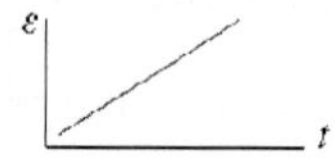
2	Kelvin 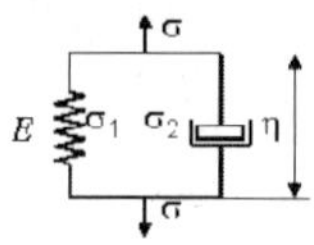	$\frac{d\varepsilon}{dt}+\frac{1}{\tau}\varepsilon=\frac{1}{\eta}\sigma$		$\varepsilon=\frac{\sigma_0}{E}\left(1-\exp\left(-\frac{E}{\eta}t\right)\right)$ 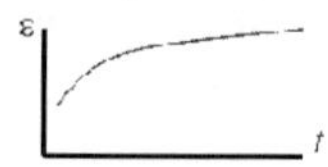
3	3-element 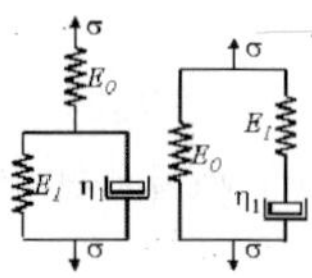	$\frac{d\sigma}{dt}+\frac{E_1}{\eta_1}\sigma=$ $=(E_0+E_1)\frac{d\varepsilon}{dt}+\frac{E_0E_1}{\eta_1}\varepsilon$	$\sigma=\sigma_\infty+E_1\varepsilon_0\exp\left(-\frac{E_1}{\eta_1}t\right)$ 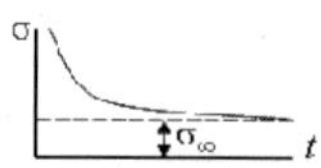	$\varepsilon=\varepsilon_\infty-\frac{\sigma_0}{E_1}\exp\left(-\frac{E_1}{\eta_1}t\right)$ 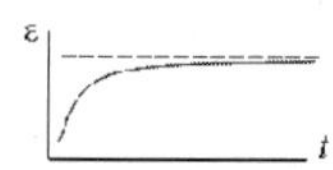
4	3-element 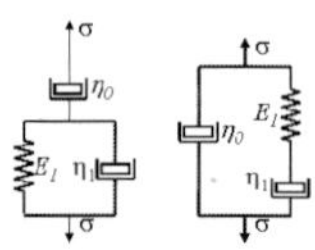	$\frac{d\sigma}{dt}+\frac{E_1}{\eta_1}\sigma=$ $=\eta_0\frac{d^2\varepsilon}{dt^2}+E_1\left(1+\frac{\eta_0}{\eta_1}\right)\frac{d\varepsilon}{dt}$	$\sigma=E_1\varepsilon_0\exp\left(-\frac{E_1}{\eta_1}t\right)$ 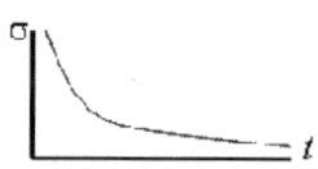	$\varepsilon=\frac{\sigma_0}{\eta_0}t+\frac{\sigma_0}{E_1}\left(1-\exp\left(-\frac{E_1}{\eta_1}t\right)\right)$ 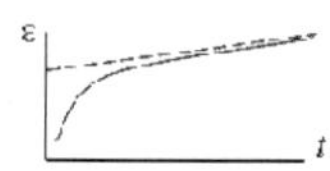

The bonds formed between the contact surfaces are followed by junctions appearing on real contact spots. The formation and rupture of the junctions govern the adhesion component of friction. The simple model of junction formation has been proposed by Bowden and Tabor.[3] For the majority of polymers van der Waals and hydrogen bonds are typical.[14,15] The hydrogen bond exists at a very short distance in the polymers containing the chemical groups where the hydrogen atom is linked with an electronegative atom.

The junctions sheared under the applied tangential force result in the friction force. In general, the interfacial junctions are influenced by the nature of the mated surfaces as well as by the surface chemistry and the stresses in the surface layers.

If the interfacial bonding is stronger then the cohesive strength of the weaker material, the latter is fractured and the polymer transfer takes place. Otherwise, fracture occurs at the interface. According to well accepted phenomenon, in polymers the surface forces and the forces acting between the polymer chains are nearly equal and fracture often occurs in the bulk of polymers. But, it was observed for metal-polymer contact that the metal is transferred to the polymer surface under certain conditions.[14,15]

Tsukruk[16-18] *et al.* described the formation of stable molecular lubrication films from the functionalized tri-block copolymer with poly[styrene-b-(ethylene-co-butylene)-b-styrene (SEBS). This polymer was chemically deposited to the silicon oxide surface covered with the interfacial epoxy-terminated monolayer. The authors optimized the grafting density and thickness of the SEBS films to assure the development of organized structure within the molecularly thick (~10 nm) films. The key components of this microstructure, which, the authors believe, are critical for obtaining the superior interfacial properties, are the 2D net of interconnected glassy polystyrene nano-domains reinforcing the resin matrix and the dense chemical grafting of the resin matrix to the substrate. These SEBS films with optimal microstructure show tribological properties far exceeding those of other molecular coatings and self-assembled monolayers. They possess a very low friction coefficient, modest adhesion, and the superior wear stability compared with the non-structured or non-reinforced organic lubricating coatings.

4.1. *Thermodynamic Approach (Specific Surface Energy)*

When two solids approach one another it is just the interactions of their molecular fields that result in the attraction forces. Strong adhesion requires deformation at contact points to be plastic since in elastic contact the stored potential energy can rupture the junction when the contact is unloaded. The thermodynamic work of adhesion between solids 1 and 2 equal to the work of reversible adhesion rupture is determined by the Dupre formula:

$$\gamma = \gamma_1 + \gamma_2 - \gamma_{12} , \qquad (2)$$

where γ_1 and γ_2 are the energies required to form the unit surfaces of solids 1 and 2 (their free energy) and γ_{12} is the excessive or interphase energy.

When discussing the surface phenomena it is necessary to distinguish the work required to form a unit surface and the work required to expand the available surface S by the same area. This means that the surface energy and surface tension are different characteristics.[19] The surface energy γ_S and surface tension σ_S are related by the known expression:

$$\gamma_S = \sigma_S + S\frac{\partial \sigma_S}{\partial S} . \qquad (3)$$

In general, the difference between σ_S and γ_S can be considerable for liquids and negligible for crystalline solids if the experiment is performed at elevated temperatures and lasts for a long time.

4.2. *Contact Adhesion*

Several models have been developed to describe the contact adhesion. The Johnson-Kendall-Roberts (JKR) model[20,21] sometimes referred to as the model of contact mechanics and the Derjaguin-Muller-Toporov (DMT) model[22] are best known. The comparative analysis of the models[23] shows that the JKR model is applied to the bodies of micrometer and greater sizes having the properties of polymers, whereas the DMT model is valid for bodies of nanometer sizes having the properties of metals. Therefore, we briefly discuss below the JKR model.

The JKR model is based on the assumption on an infinitely small radius of effect of surface forces, i.e. it is assumed that interactions occur only within the contact area. Elastic contact between a sphere of radius R and half-space is analyzed with consideration of van der Waals forces which compress the mated bodies together in addition to the applied load. The energy of molecular interactions is $W_m = -\pi a^2\gamma$. The contact stiffness resists the action of the forces. Using energy balance the equations for the main contact parameters are derived based on the combination of the Hertzian pressure distribution (loading) and Boussinesq distribution (unloading). Such combination yields the compression in the middle of the contact and an infinite tensile stress at its edges.

The formula for calculating the radius of adhesive contact is

$$a^3 = \frac{R}{K}\left(P + 3\pi R\gamma + \sqrt{6\pi RP\gamma + (3\pi R\gamma)^2}\right), \tag{4}$$

where $K = \frac{4}{3}\frac{1-\nu^2}{E}$ is the elastic constant, E is Young's modulus, and P is the normal load.

Therefore, it is apparent that without adhesion ($\gamma = 0$) the Hertz equation is obtained while if $\gamma > 0$ the contact area always exceeds that of the Hertzian contact area under the same normal load P. We note that when the contact is fully unloaded ($P = 0$) it does not disappear but remains finite with the radius

$$a = \left(6\pi R^2\gamma/K\right)^{1/3}. \tag{5}$$

Only the application of a tensile (negative) load can reduce this radius, and then the contacting surfaces would separate at the least load corresponding to the conversion of the radicand in Eq. (4) into zero

$$P_{pull-off} = -\frac{3}{2}\pi R\gamma. \tag{6}$$

This circumstance is the specific feature of the JKR model.

Derjaguin was the first (1934) to formulate and solve approximately the problem of the effect of elastic contact deformation on adhesion. Later on Derjaguin, Muller, and Toporov developed it further in their works resulting in creating the DMT model describing the contact of an elastic sphere with a rigid half-space. This model is based on the following two postulates: surface forces do not change the deformed profile of the sphere and it remains Hertzian; the attraction force acts outside the contact circle pressing the bodies together with the contact region being under compression by the stresses distributed according to Hertz.

Equilibrium is reached when the deformation is sufficient for the elastic response (the force of elastic restoration of the sphere) F_e to counterbalance the joint effect of the applied external load P and the forces of molecular attraction F_s:

$$F_e = P + F_s \,. \tag{7}$$

Let us assume that attraction is represented by the Lennard–Jones potential. The profile of the deformed sphere outside the contact area for a given radius r is known to be described by the equation

$$z(r,a) = \frac{1}{\pi R}\left[a\left(r^2 - a^2\right)^{1/2} - \left(2a^2 - r^2\right)\arctan\left(r^2/a^2 - 1\right)^{1/2} \right] \tag{8}$$

to which the equilibrium state z_0 (the clearance within the contact site) should be added. Then the molecular attraction force is calculated by the direct integration:

$$F_s = 2\pi \int_a^\infty p(z+z_0)rdr \,.$$

The calculation of this integral is rather difficult yet the number of approximate formulas have been published which facilitate the use of the model in question. In particular, there is a simple Maugis relation between the load and approach obtained for the conditions of the DMT model[24]:

$$\frac{P}{P_c} = \frac{1}{\sqrt{3}}\left(\frac{\delta}{\delta_c}\right)^{3/2} - \frac{4}{3}, \tag{9}$$

where δ is elastic displacement, $P_c = \frac{3}{2}\pi\gamma R, \quad \delta_c = \left(\frac{P_c^2}{3K^2R}\right)^{1/3}$.

Analysis shows that each of these models is true for the certain combinations of physical-mechanical and geometric characteristics of the bodies (see Fig. 1). The DMT theory is applicable to the materials for which the point with the coordinates (K, γ) lies below the corresponding

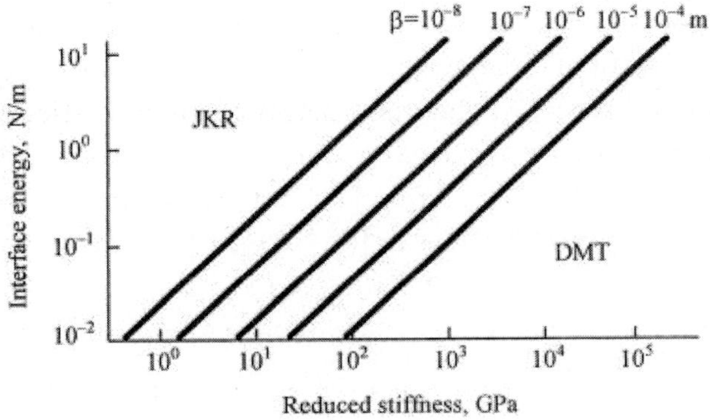

Fig. 1. Domains of parameters (K, γ) where either of two models (JKR or DMT) is correct.

line plotted at the constant radius of asperity tip β. Here K is the reduced stiffness of the contacting materials and γ is the interface energy.

4.3. *Contact of Rough Surfaces*

When two surfaces approach each other their opposing asperities with the maximum heights come into contact forming the individual spots. The total area of these spots is known as the real contact area (RCA).

When simulating the real contact in the case of plastics, the temperature and sliding velocity should be taken into account[3,25-27].

Let us consider the contact of rigid rough body and a smooth elastic half-space.[27,28] The half-space is taken to be homogeneous and isotropic with the Young's modulus E, Poisson's ratio ν, and the thermal expansion coefficient α. The problem for a heated sphere of the radius R was solved by Barber[29] under the following assumptions: the temperature of the heated sphere is T while outside the contact the surface of the half-space is free of the mechanical and thermal loads.

The desired solution, i.e. the relationship between the total load P, circular contact radius a, and approach of the sphere δ is sought as a superposition of the solutions for the problems considering mechanical loading and heating.

Following the Greenwood-Williamson approach[30], we can be able to find the total area of all contact spots (RCA) and the total load. The non-dimensional equations for the normal distribution of the asperity heights are written as follows:

$$A_r / A_a = \pi R D \sigma_r \frac{1}{\sqrt{2\pi}} \int_h^{\infty} (\xi - h) \exp\left[-\xi^2 / 2\right] d\xi , \tag{10}$$

$$\begin{aligned} \frac{P}{A_a} \frac{(1-\nu^2)}{E} &= \frac{4}{3} R^{1/2} D \sigma_r^{3/2} \frac{1}{\sqrt{2\pi}} \int_h^{\infty} (\xi - h)^{3/2} \exp\left[-\xi^2 / 2\right] d\xi \\ &+ \frac{2\alpha T}{\pi} (1+\nu) R D \sigma_r \frac{1}{\sqrt{2\pi}} \int_h^{\infty} (\xi - h) \exp\left[-\xi^2 / 2\right] d\xi . \end{aligned} \tag{11}$$

Here A_a is the apparent contact area, A_r is the real contact area, D is the surface density of asperities, σ_r is the root- mean -square roughness, α is the bandwidth parameter, ξ is the non-dimensional height of asperity, and h is the non-dimensional separation.

In a study by Petrokovets (1999)[28] it was shown that the RCA of two bodies with different temperatures becomes smaller when the temperature difference increases, and the RCA is always smaller than in the isothermal case. However, if the mechanical behavior of the material is sensitive to temperature changes, the above RCA decrease may be "hidden" by RCA rise due to the reduction in the mechanical characteristics of the material.

If the rheological behavior is governed by a single relaxation time, the simple exponential dependence describes its temperature-dependent modulus:

$$E = E_0 \exp(-\beta T) \tag{12}$$

where β is a constant having dimension of the reciprocal of the temperature and conventionally termed as the rheological parameter.

The calculation was performed under several loads with different values of β. Some results are presented in Fig. 2. It is seen that the temperature-dependent RCA may pass the minimum with increasing the temperature. The existence of this minimum and its value depend on the combination of the thermal (α) and mechanical (rheological) properties of the contacting materials. Increase in the rheological parameter β, i.e. the sharpening of the temperature dependence of the modulus, results in degeneration of the minimum (the ascending RCA ~ *E* curve) all other things being the same (see Fig. 2).

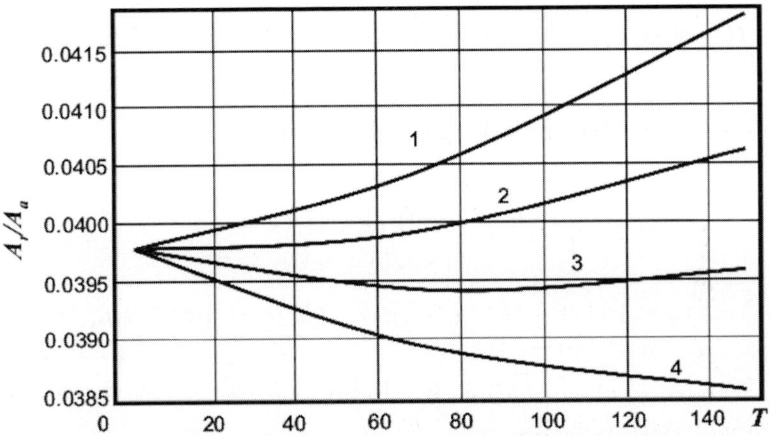

Fig. 2. Effect of the rheological parameter β on temperature-dependent real contact area A_r/A_a: 1– β = 0/0020; 2 – 0.0018, 3 – 0.0016; 4 – β = 0.0014.

4.4. *Contact of Rough Surfaces with Adhesion*

It is commonly known that the real surfaces are rough. Experiments show that the roughness affects the adhesion of rough surfaces to a large extent. As an example, contact between a sphere and a rough half-space is considered. The analysis is based on the Greenwood and Tripp model[31] where the load-approach relation is treated as in the case of a single asperity. It can be shown that the total force acting on the sphere is

$$P = 2\pi P_c D \int_0^{a^*} \int_0^{\infty} r \overline{P}_i (z-h) \Phi_\alpha(z) dz dr .$$

After the change of variables and transformations the force *P* becomes

$$P = P_c \frac{3\sqrt{2}}{8} \pi \Delta_c \mu_c \sqrt{\frac{\beta}{R}} I(h, \alpha, \Delta_c), \tag{13}$$

where $P_c = \frac{3}{2}\pi R\gamma$ is the adhesion of the smooth sphere of the radius R; $\mu = \frac{8}{3} D\sigma_r \sqrt{2R\beta}$ is the complex parameter of surface topography;

$$I(h,\alpha,\Delta_c) = \int_0^{a^*}\int_0^{\infty} \overline{P}_1(x)\Phi^*_\alpha(h + a_0 y + b_0 y^2 + \Delta_c x)dxdy$$

$$h = d/\sigma_r;\ a_0 = 1 - \zeta/c;\ b_0 = \frac{3}{2}(k - \frac{1}{2})\frac{\zeta}{c^3};$$

$$\zeta = \frac{\sqrt{\pi}}{4}\mu F_{3/2}(h,\alpha)\Gamma(k+1)/(k+\frac{1}{2});$$

$$a^* = \frac{3}{4}\sqrt{\pi}c\Gamma(k+\frac{3}{2})/\Gamma(k+2);\ F_{3/2}(h,\alpha) = \int_h^{\infty}\Phi^*_\alpha(\xi)d\xi.$$

Here $\Delta_c = \delta_c/\sigma_r$ is the adhesion parameter, and $\Phi^*_\alpha(\xi)$ is the Nayak distribution of asperity peaks.[32]

Thus, Eq. (13) allows the force compressing two rough surfaces to be estimated. It describes the situation where the contact is unloaded. After unloading the compressive force is zero whereas the separation remains non-zero and has the definite value h_0 which is found from the condition

$$I(h_0, \alpha, \Delta_c) = 0. \tag{14}$$

In order to detach the surfaces fully ($h_0 = 0$), the certain negative (tensile) force should be applied. Then due to adhesion the asperities with the initial height h_0 are lengthened by δ_c which is found from Eq. (14) where the lower limit of the integration is replaced with $-L$ ($L = \min(\Delta_c, h - h_0)$). The minimal value of P is the pull-off force (adhesion force). The calculations have indicated that the roughness initiates sharp decrease in the adhesion. For example, if the root-mean-square roughness σ_r increases twice the adhesion drops by about two orders of magnitude. This effect of the roughness cannot be explained by the reduction of the contact area. It can be attributed by the successive rupture of adhesive junctions at the elastic recovery of the highest asperities during the separation of the contacting surfaces. Some results of the calculations for the contact of polymers are in satisfactory

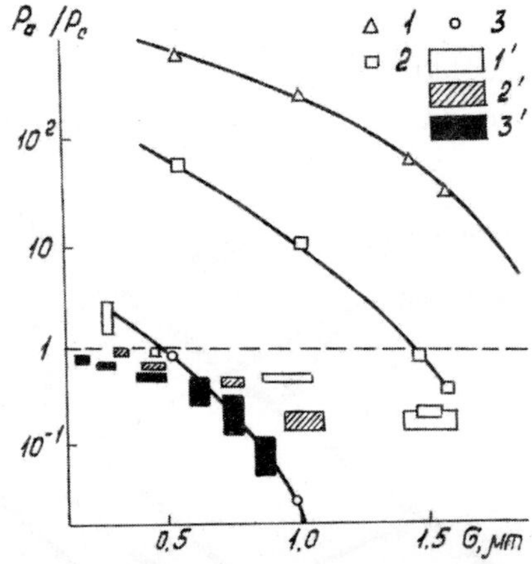

Fig. 3. Effect of roughness on adhesion: 1, 1' – K = 0.29 MPa, γ = 0.068 N/m, Δ_c = 1.99-6.49; 2, 2' – K = 0.91 MPa, γ = 0.040 N/m, Δ_c = 0.66-2.14; 3, 3' – K = 3.20 MPa, γ = 0.034 N/m, Δ_c = 0.26-0.84; 1-3 – calculation, 1'-3' – experiment.

agreement with the data obtained in the experiments of Fuller and Tabor[33] (see Fig. 3).

There is no question that the surface asperities are not smooth, i.e. they have smaller asperities of the nanoscale size which result from the molecular and supramolecular structure of polymers. In this case the RCA should be estimated based on the two-level model of Archard type. To take into account microroughness on large asperities the solution of Greenwood and Tripp[31] for the contact of two rough spheres is used as the governing equation for a single asperity. Analysis of the two-level model has shown that the highest asperities of the first level (roughness) are coming into contact and form the individual contact spots. Yet, contrary to traditional view, the spots are not continuous but multiple connected, that is, each spot consists of a set of smaller spots whose total area is conditionally named the "physical contact area". This area is less than the real contact area by an order of magnitude. Taking into account the height distribution of asperities, the dependence of dimensionless contact area on dimensionless load is obtained in parametric form:

$$\tilde{A} = 3^{1/3}\Delta_c I(h,\alpha,\Delta_c)$$
$$\tilde{p} = 3^{1/3}\Delta_c^{3/2} I(h,\alpha,\Delta_c).$$

On the other hand, for the two-level model (roughness plus subroughness) the calculation of the contact area revealed that it may increase several times due to the effect of the surface force (see Fig. 4).

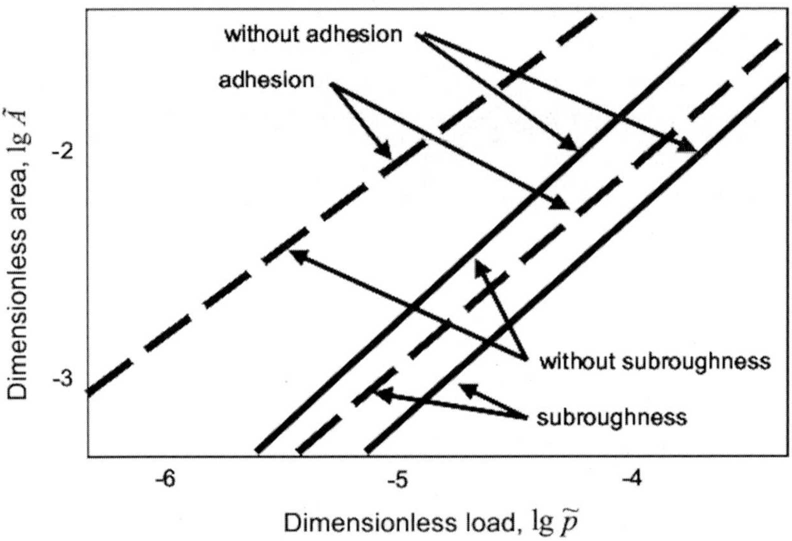

Fig. 4. Effect adhesion on discrete contact area.

The two-level model with adhesion is of interest for precision engineering. It can be developed in the same manner as in the above case. A rough sphere models the asperity of the first level. The load-contact radius relation for a single microasperity is assumed to result from the JKR or DMT theory. The calculation procedure is described elsewhere.[34,35]

4.5. *Measurement of Adhesion*

A modern instrument for measuring adhesion is the surface forces apparatus (SFA). The SFA allows one to measure directly the laws of forces in liquids and vapors.[36,37] It contains two crossed atomically smooth mica cylinders. One mica cylinder is mounted to a piezoelectric translator, through which the distance is adjusted. The other mica cylinder is mounted to a spring with the known and adjustable spring constant. The separation between the surfaces is measured optically using the multiple beam interference fringes. Knowing the position of one cylinder and the distance to the surface of the second cylinder the deflection of the spring and the force can be calculated.

One of the main problems arising when measuring molecular forces is that they increase rapidly with shortening the distance between the specimens under testing. Hence, the measurements should be carried out at a very low speed that can not be provided by the design of the common balance.

Derjaguin *et al.* proposed to solve the problem[13] by applying the principle of a feedback balance. Following this idea we have developed a new apparatus for studying surface forces – the contact adhesion meter[38] (see Fig. 5).

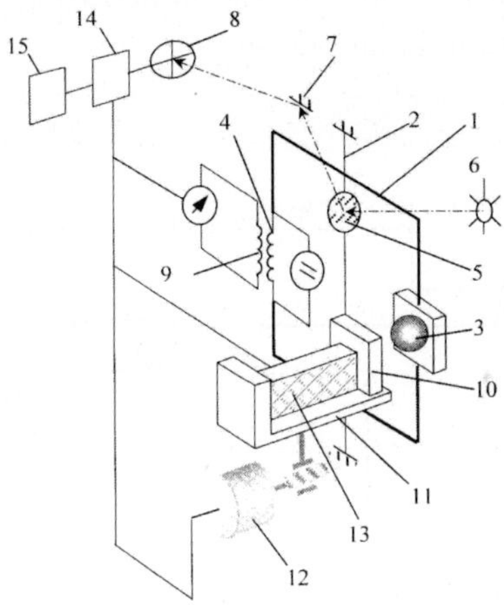

Fig. 5. Principal scheme of adhesion meter: 1 – frame; 2 – string 2; 3 – holder; 4 – movable coil; 5 – mirror; 6 – laser; 7 – expander of optical base; 8 – photodetector; 9 – coil; 10 – specimen; 11 – table; 12 – stepping motor 12; 13 – system of fine positioning controlled by piezodrive.

The device consists of a ball probe fixed on the arm of a highly sensitive electromagnetic balance with negative feedback. The dependence of adhesion forces on the distance between the ball probe and the sample is measured in both approach and pull-off modes. The interface forces which tend to rotate the arm with the ball are compensated by an electromagnet placed on the opposite arm of the balance. The measurement of the compensating current in the electromagnet allows us to determine the acting interface forces.

The contact adhesion meter (CAM) is capable of measuring the force interaction of surfaces in two regimes. When the surfaces are separated the rupture of bonds (pull-off force) is obtained. The whole process is shown in Fig. 6 which presents the force-distance curve.[39] In the Fig. 6 the point A shows the beginning of the interaction of the approaching surfaces, the portion AB corresponds to pure attraction without formation of the real mechanical contact between the solids. To the left from the point B the force interaction and elastic deformation occur simultaneously. At the point C the elastic force of resistance to penetration becomes dominating. The point D corresponds to the moment when the elastic force of resistance to penetration equals the adhesive force of the mutual attraction.

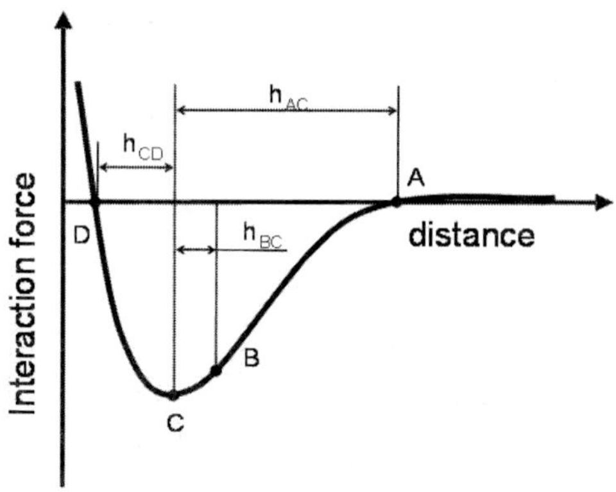

Fig. 6. Stages of contact formation and main points on force–distance curve: h_{AC} – effective radius of surface forces; h_{BC} – corresponds to tensile elastic deformation of surface; h_{CD} – corresponds to elastic mutual penetration with account for adhesive attraction between solids.

We have measured the surface energy of coatings on a silicon plate (crystal structure (111), homopolar semiconductor). The coatings were 3 nm thick octadecyltrichlorsilane (OTS), 8 nm thick poly[styrene-b-(ethylene-co-butylene)-b-styrene (SEBS), and 1 nm thick epoxysilane.

Also the organic SAMs (self-assembled monolayers) of $ODPO_4$ (octadecylphosphoric acid ester) and $DDPO_4$ (dodecylphosphoric acid ester) 2 nm thick with the titanium and titanium oxide underlayers were studied. The silicon and titanium balls with the radius of 1 and 1.5 mm, respectively, were used as probes.

These measurements[39,40] have demonstrated that the specific surface energy of the studied coatings depends on the coating and probe materials as well as the probe radius. It is 0.004 J/m^2 for OTS and epoxysilane (probe radius 1 mm) and 0.003 J/m^2 for epoxysilane and 0.002 J/m^2 for SEBS (probe radius 1.5 mm). The following results were obtained for SAMs: 0.007 J/m^2 for Si/TiO_x/$ODPO_4$, 0.011 for Si/Ti/$ODPO_4$, and 0.002 for Si/Ti/$DDPO_4$ (probe radius 1 mm), and 0.004 J/m^2 for $ODPO_4$ (probe radius 1.5 mm).

5. Friction of Polymer over the Hard Counterface

Several authors[4,41-49] have studied the state of art in polymer tribology. The hardness, plasticity index (which describes the deformation properties of rough surfaces), and elastic modulus were obtained for organic polymers (poly(methylmethacrylate), PMMA; poly(styrene), PS; poly(carbonate), PC; ultra-high molecular weight poly(ethylene), UHMWPE) using the contact compliance method.[50] The dependence of the imposed penetration depth, the maximum load, and the strain rate upon the hardness and elastic modulus was described;[51,52] typical penetration depths were in the range from about 10 nm to 10 μm while the applied loads were less than 300 mN.

It should be noted that almost without exception ploughing is accompanied by adhesion and under certain conditions it may result in microcutting, that is, an additional work should be done increasing the friction force.

There are other mechanisms of energy dissipation at deformation. For example, when a polymer with visco-elastic behavior slides over a hard rough surface energy dissipation is caused by the high hysteresis losses. This deformation component is known as friction due to the elastic hysteresis.[3,53] The energy may also be carried away, for example, with elastic waves generated at the interface and outgoing to the infinity, owing to the nucleation and development of microcracks in the material.[1,54]

The mechanical component results from the resistance of the softer material to "ploughing" by asperities of the harder one. The adhesion component comes from adhesive bonds formed between the surfaces in the friction contact. It is believed that for polymers the adhesion (molecular) component exceeds much the mechanical one.[15] This is explained by transfer films generated on the metal counterface. Special consideration is given to transfer films as the key factor which determines the tribological behavior of polymer materials.[4,7] The following factors affect considerably the friction force: the contact load, sliding velocity, and temperature. The effects are not independent, for example, depending on the contact load and velocity the contact temperature may vary considerably changing the friction mode.

5.1. *Effect of Load on Friction*

It is a common knowledge that the friction force is proportional to the normal load and experiments have shown that this law is valid for some polymers (see Table 2). The friction coefficient remains practically constant under the load from 10 to 100 N when a steel ball of the radius 6.35 mm slides over polytetrafluoroethylene (PTFE), polymethyl methacrylate (PMMA), polyvinylchloride (PVC), polyethylene (PE), and Nylon.[55] Other authors[56,57] have obtained similar results with the same materials as well as with some others, e.g. with PTFE, polytrifluorochloroethylene (PTFCE), PVC, polyvinyliden chloride (PVDC), and PE under the load from 2 to 15 N, with PTFE, PMMA, polystyrene (PS) and PE under the load from 10 to 40 N, and so on.

Outside this range, to the left and right, the proportionality between the friction force and applied load breaks down. Thus, it was shown that in the range of moderate loads 0.02 to 1 N the friction coefficient decreases with increasing the load.[58] Such a behavior may be explained by the elastic deformation of surface asperities. Of interest is the fact that similar behavior is characteristic of rubbers for which the elastic deformation is typical.[59,60]

Table 2. Effect of load on friction coefficient.

Author(s)	Materials and load	Graphical representation
Bowers, Clinton, Zisman[56]	2–15 N Steel–polymer (PTFE, PFCE, PVC, PVDC, PE)	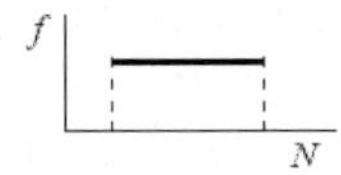
Shooter, Thomas[57]	10–40 N Steel–polymer (PTFE, PE, PMMA, PC)	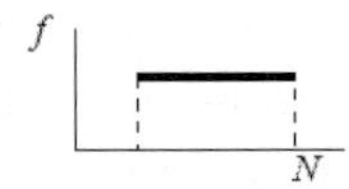
Shooter, Tabor[55]	10–100 N Steel–polymer (PTFE, PE, PMMA, PVC, Nylon)	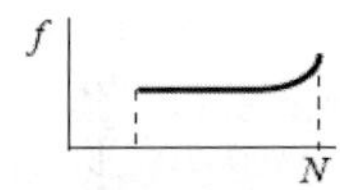
Rees[58]	Steel–polymer (PTFE, PE, Nylon)	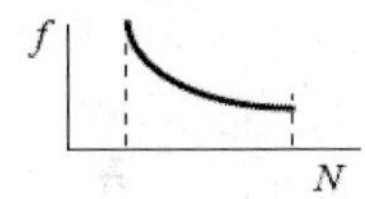
Bartenev[54], Schallamach[65]	Theory Steel–rubber	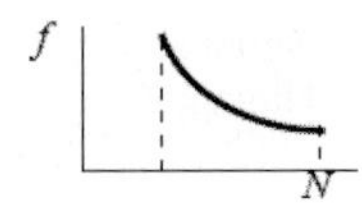
Kragelskii[4]	Theory Steel–rubber	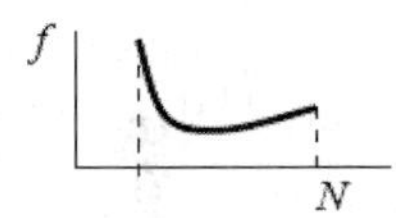

On the other side of the proportionality range, the friction coefficient increases with increasing the load. This is often explained by the plastic deformation of asperities in the contact. Thus, the friction of polymers as a function of the load varies in the manner which was described by Kragelskii.[4] That is, the friction coefficient passes the minimum which corresponds to transition from the elastic contact to plastic one.

We should bear in mind that the load can vary the temperature of visco-elastic transitions in polymers and thereby the mechanism of friction.

5.2. *Effect of Sliding Velocity on Friction*

Theoretically, the friction force should not depend on the sliding velocity. For polymers, however, this statement is only true if the temperature within the contact area grows but negligibly. Usually there is a complex dependence of the friction coefficient on the velocity explained by variations in the relaxation properties and physicochemical activity of macromolecules. When the test temperature of the polymer sample approaches that of glass-transition a strong dependence of the friction coefficient on the velocity is observed. At lower temperatures f does not practically depend on the velocity.[61]

It is difficult to separate the effects of the velocity and temperature on friction. Examples of great diversity of the available results are shown in Table 3.

Table 3. Effect of sliding velocity on friction coefficient.

Author(s)	Materials and sliding velocity	Graphical representation
Shooter, Thomas[57]	0.01–1.0 cm/s Steel–polymer (PTFE, PE, PMMA, PC)	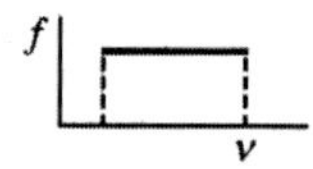
Milz, Sargent[66]	4 – 183 cm/s Polymer–polymer 1 – Nylon, 2 – PC	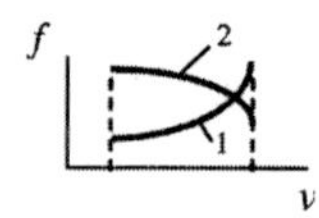
Fort[68]	10^{-5}–10 cm/s Steel–polymer (PTFE)	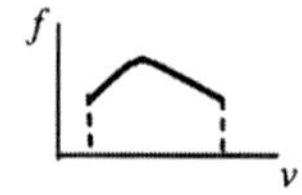
White[64]	0.1 – 10 cm/s Steel–polymer (1 – PTFE, 2 – Nylon)	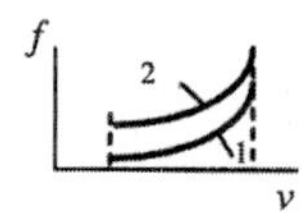
Flom, Porile[62,63]	1.1– 180 cm/s Steel–polymer (PTFE)	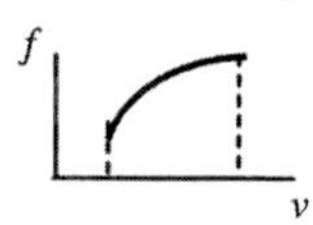
Oloffson, Gralben[60]	1.5 cm/s Polymer–polymer (fibers)	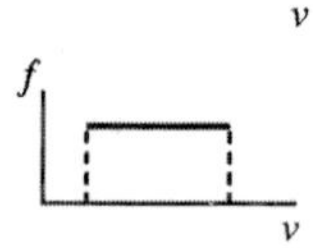
Bartenev and Lavrentiev[54], Schallamach[59]	Theory Steel–rubber	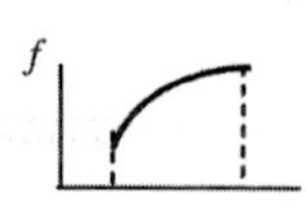

Velocity-independent friction occurs only within a limited range of velocities (0.01–1.0 cm/s) for PTFE, PE, PMMA, and PS[57] as well as for fiber-fiber contact.[60] But more complex relationships between the friction coefficient and sliding velocity are most often observed. Such relationships can be related to the visco-elastic behavior of polymers.

In the range of low velocities the viscous resistance in the contact zone increases with increasing the velocity. When the contact pressure is high, an abnormal viscous flow is observed which leads to a sharp rise in the viscosity due to an increase in the velocity.[62-64] Molecular-kinetic considerations lead one to the same dependence.[54,65]

In the range of high velocities, elastic behavior is prevalent in the contact zone and, as a result, the friction force depends only slightly on the velocity or it decreases with the velocity.[66,67] In addition, it should be borne in mind that the duration of contact is short at high velocities leading to further decrease in the friction force.

In the intermediate range of velocities all the above factors are in competition and a maxima appears on the friction force–sliding velocity curve whose position depends on the relaxation properties of the polymer.[86]

It should be recognized that the friction force–sliding velocity relationship depends essentially on the test temperature.[61] When the tests are conducted near the glass-transition temperature the sliding velocity has a pronounced effect on friction whereas at lower temperatures friction hardly depends on the sliding velocity.

5.3. *Effect of Temperature on Friction*

Polymers as visco-elastic materials are very sensitive to frictional heating. It is commonly known that heat generation at friction results from the deformation of the material on real contact spots. Another source of heat can be attributed to the formation and rupture of adhesive bonds. These processes are most probably energetically non-equivalent and the energy difference may cause the generation or absorption of heat.

Often it is believed that the effect of temperature on friction can be taken into account using the mechanical characteristics of polymers measured at certain temperatures. In support of this assumption, a correlation of the friction coefficient with the hardness and shear strength was found for some polymers.[61,68,69] Such correlation is valid only when the temperature has no effect on adhesion.[70] Some friction patterns as functions of temperature are presented in Table 4.

Table 4. The effect of temperature on friction coefficient.

Author(s)	Material	Temperature, test conditions	Graphical representation
Shooter, Thomas[57]	*1* – PS, *2* – PTFE	20–80°C Steel–polymer	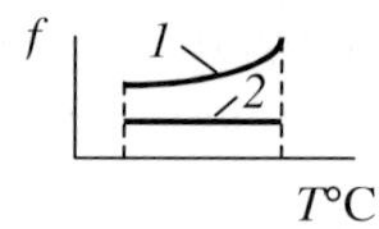
Ludema, Tabor[69]	*1*, *2* – PCTFE, *3* – PP	–50–+150 °C Steel–polymer 1 – v = 3.5×10^{-5} cm/s, 2 – v = 3.5×10^{-2} cm/s	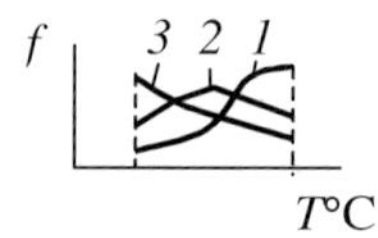
King, Tabor[70]	*1* – PE, *2* – PTFE	–40–+20 °C Steel–polymer	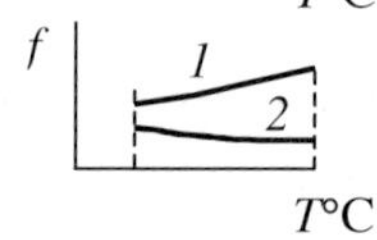
Schallamach[59]	Rubber	20–200 °C Steel–polymer	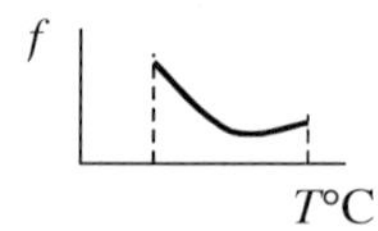

6. Wear of Polymers

The main mechanisms of wear for polymers are adhesion, abrasion, and fatigue.[4,7] Abrasive wear is caused by hard asperities on the counterface and/or hard particles that move over the polymer surface. This mode of wear occurs when the roughness is the determinative parameter in friction.

The adhesive mode of wear occurs in the sliding of a polymer over unlike surface, e.g. metal, when the strength of the adhesive bonds formed between the contacting materials can exceed the cohesive strength of the polymer. As a result, some part of the material is transferred onto the counterface and forms a transfer film; another part of the worn material is removed from the friction zone as the wear debris.

The fatigue wear of polymers is caused by the crack propagation at the repeated deformation of the material at friction. Fatigue results in the pitting, crack generation, and delamination. The wear debris are formed as a result of the growth and intersection of the small cracks on the polymer surface which are perpendicular to the sliding direction. The fatigue wear occurs after prolonged friction that may be important in the absence of adhesive wear when the counterface is smooth.

In general, three stages of wear can be distinguished:[71] running-in, steady regime and catastrophic wear. As applied to a metal/polymer friction system, the first stage implies the generation of a transfer film on the metal counterface. For most metal/polymer systems the tribological characteristics much depend on the properties of the transfer film, i.e. its nature, strength of bonding to the metal surface, the part of the counterface area covered with the transfer film, the severity of adhesive interaction on the polymer/transfer film interface. This has been supported experimentally and reported by several authors.[7,72,73]

As polymers adhere to hard surfaces through adsorption[74] it can be assumed that the adhesion component of the friction force depends on the adsorptivity of macromolecules. Immobilization of their functional groups and constraining the molecular mobility must affect strongly the tribological behavior of the polymer.

The friction of polymers over metals is greatly affected by tribochemical reactions occurring in the contact zone.[75] The tribochemical transformations in macromolecules follow the free-radical mechanism consisting of three main stages: initiation of macromolecules; growth of chains and their termination.

The direction and kinetics of chemical processes in polymers depend not only on the chemical structure of macromolecules but also on the arrangement of molecules, perfection and size of crystallites, the degree of molecular orientation, and molecular mobility. The structural heterogeneity in polymers leads to non-uniform distribution of additives and reagents in their bulk. In the partially crystalline polymers the low-molecular weight substances (oxygen, products of oxidation, inhibitors, plasticizers, dyes, fillers, etc.) tend to concentrate in the amorphous regions of the polymer; most reactive fragments of macromolecules are also localized there. The local concentrations of reagents may differ much from the average concentrations; consequently, the local rates of tribochemical reactions must be different from the average rates. Thus, the introduction of additives into the polymer that alters its physical structure unavoidably affects its tribological properties. The tribochemical transformations in macromolecules and wear of polymer can be affected by the composition of the counterface metal. Effective catalysts of chemical processes in polymers that follow the radical mechanisms are Pt, Pd, Rh, Mo, Ta, Cr, and Ti. No effect is produced by Au and Ag. Quite typical tribochemical processes are hydrogenation (catalyzed by alloys containing Ni, Co, Fe, Cu, Pt, and Re), dehydrogenation (catalyzed by oxides of alkali-earth, transition metals and rare-earth elements; sulphides, tellurides, stibides, arsenides, selenides of Mg, Ca, Zn, Cd, Cr, Ni, Mo, etc; borides, nitrides, carbides, silicides, phosphides of V, Ti, Cr, Mo, W, etc); oxidation (partial oxidation is catalyzed by oxides of V–VIII groups of metals e.g. Mo, Bi, Co, Fe, etc). The compounds resulted from interaction of the macromolecules and the products of tribochemical reactions have various chemical structures. Often, especially at friction in air, these may be compounds of the coordination type. The coordination compounds generated on a surface seem typical enough for metal/polymer systems as this process is observed in the static contact between a polymer and metal.[75]

6.1. *Abrasive Wear*

Abrasion displays scratches, gouges, and scoring marks on the worn surface; the debris produced by abrasion are frequently shaped as the fine cutting chips similar to those produced during machining, although at a much finer scale. Most of the models associated with abrasive wear incorporate geometric asperity descriptions, so that wear rates turn out to be quite dependent on the shape and apex angles of the abrasive points moving over the surface.

There are two distinct modes of deformation when an abrasive particle acts on the plastic material. The first mode is the plastic grooving often referring to as ploughing. It occurs when a prow is pushed ahead of the particle and material is continually displaced sideways to form the ridges adjacent to the developing groove. No material is removed from the surface. The second mode is named the cutting because it is similar to micromachining and all the material displaced by the particle is removed as a chip.

There is another approach to description of abrasive wear. Experiments have shown that the abrasive wear rate is in proportion to $1/\sigma_u\varepsilon_u$ where σ_u is the ultimate tensile stress and ε_u is the corresponding strain (see Fig. 7). The correlation was found by Lancaster and Ratner and is often referred to by their names.[76,77]

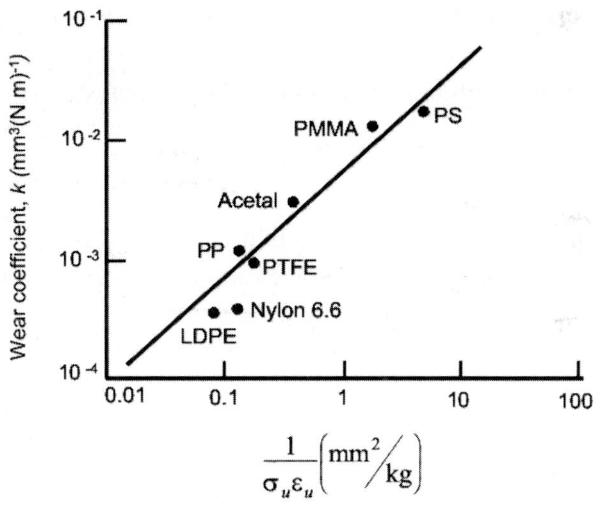

Fig. 7. Ratner-Lancaster correlation for abrasion: σ_u, ε_u are ultimate tensile stress and elongation (after Briscoe (1981)[43]).

In two-body abrasion some asperities produce ploughing, the rest make cutting depending on two controlling factors: the attack angle of the particle and the interfacial shear strength expressed as the ratio between the shear stress at the interface and the shear yield stress of plastically deformed material. In the case of three-body abrasion the free abrasive particle readily penetrates the polymer surface which begins to operate as an emery cloth that increases the wear of the counterface.

6.2. *Adhesive Wear and Friction Transfer*

Adhesive wear results from the shear of friction junctions. This wear mode involves the formation of adhesive junctions, their growth and rupture. Belyi *et al.* noted that the transfer of a polymer is the most important characteristic of the adhesive wear of polymer-based materials[15].

The phenomenon of friction transfer is observed for nearly all materials. The consequences of material transfer may be significantly distinct.[78-80] If the small particles of micrometer size are transferred from one surface to the other the wear rate varies only slightly. Under certain conditions the situations take place such that a thin film of the soft material is transferred onto the hard mating surface, for example, polymer on metal. If the transferred polymer film is carried away from the steel surface and is continuously formed the wear rate increases. In the case when the film is held in place, the friction occurs between the similar materials which may eventually result in seizure. Spreading of the polymer on steel gives rise to an abrupt jump of the friction force but the wear changes insignificantly.

It has been known that under certain conditions the hard material is transferred on the soft surface. For example, bronze is transferred on polymer. The transferred hard particles are embedded in the soft material and act as the abrasive which scratches the parent material.

Friction transfer attracted a lot of attention in the tribological community generating many new concepts and hypotheses.[81]

Polymers are most susceptible to the friction transfer when rubbing both against metals and polymers. As an illustration let us consider friction between PTFE and PE.[15] Experiments were carried on the wear tester with the block-on-ring geometry. It has been found that PTFE is transferred in the form of flakes of very small sizes during the initial period of friction. The thickness of the transferred film increases monotonically and then oscillates about a mean value; the magnitude and amplitude of the oscillations depend on the test conditions, especially on the load and sliding velocity (see Fig. 8).

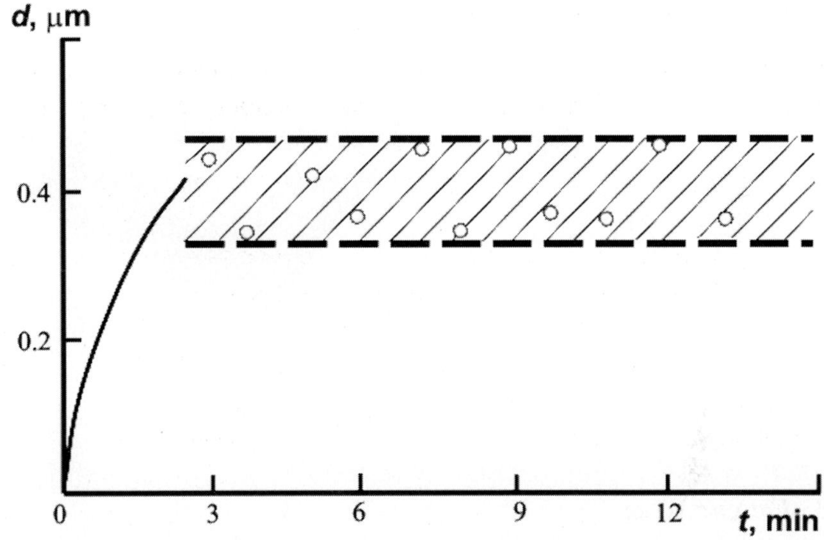

Fig. 8. Thickness of transferred film of PTFE as a function of friction duration (load = 0.05 MPa, sliding velocity = 0.35 m/s).

Self-lubricating additives such as PTFE are used to optimize the frictional properties of solids.[82] It is important that the lubricant is transferred to the surface of the contact material during sliding reducing the interfacial shear stress and friction coefficient.

Figure 9a shows the well-known schematic representation of the semi-crystalline nature of PTFE.[6] There are lamellae of crystalline layers with amorphous regions between them.

Figure 9b illustrates the AFM image of the PTFE surface after friction tests. Crystallites are clearly seen on the friction surface. They are disordered which proves their free motion occurred during friction. Therefore, the presence of free slices governs the reduction of the friction coefficient and frictional transfer for the materials filled with PTFE.

One more consequence of the polymer transfer is a change in the roughness of both surfaces in contact. The roughness of polymer surface undergo large variation during the unsteady wear until the steady wear is reached, while the roughness of the metal surface is modified due to the transfer of polymer.[83]

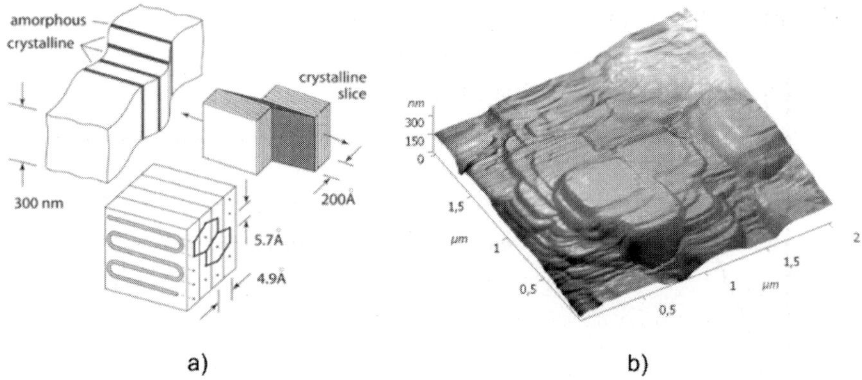

Fig. 9. a) Microstructure of PTFE; b) AFM-image of PTFE friction surface.

6.3. *Fatigue Wear*

The fatigue is known to be a change in the material state due to repeated stressing. Its characteristic feature is the accumulation of irreversible changes, which give rise to the generation, and the development of cracks. The similar process takes place during friction being named the friction fatigue. Unlike the bulk fatigue, it spans only the surface and sub-surface regions. It has been known that the fatigue cracks are initiated at the points where the maximum tangential stress or the tensile strain takes place. The theoretical and experimental studies show that under the contact loading the maximum tangential stress position is dependent on the friction coefficient.

With low friction coefficient, the point where the shear stress is maximal is located below the surface ($f < 0.3$). When the friction coefficient increases ($f > 0.3$), the point emerges on the surface. On the other hand, if a solid is subjected to a combination of the normal and tangential loading, the surface and sub-surface regions appear where the tensile strain and thereby the frictional heating occurs. Therefore, the cracks may be nucleated in surface and/or below it.[26,27]

The initiation of the fatigue cracks is assisted by the defects, which are responsible for stress concentration. These are the scratches, dents, marks and pits on the surface, and impurities, voids, cavities in sub-surface region. Both the surface and sub-surface cracks which open due to the repeated stressing will gradually grow, join, cross each other and meet the surface until the wear debris are detached after a certain number of the stressing cycles.

The fatigue wear rate is dependent on the numerous factors: physical, mechanical and chemical properties of solid surface, lubricant, environment, surface quality, temperature, etc. Although the friction force is decreased by lubrication and hence the tensile stress drops, the fatigue wear occurs and a number of cycles to the surface damage increases insignificantly.

6.4. *Frictional Behaviour of Polymer Materials with Nanofillers*

The reported data evidence a strong influence of nano-additives on the structure and properties of polymers. One of the most useful results achieved with the nano-dimensional fillers is increased mechanical properties (mechanical strength, modulus of elasticity) of polymer materials. This is especially pronounced with nanoclays and carbon nanomaterials.[84,85] It is clearly proved at relatively low contents of the nanofiller. One of the main reasons for this is strong adsorptive interaction of macromolecules with rather developed filler surface, and also polymer transition to some particular conditions with constricted molecular mobility. The increased cohesive strength of the polymer binder can result in the improved tribological properties of the composites such as the abrasion resistance and resistance to fatigue wear.

The nano-particle fillers can markedly raise the heat distortion temperature (HDT) of polymer materials. For example, in the case of Nylon 6, HDT under loading increases from 65°C for the pure polymer up to 152°C for the nanocomposite containing montmorillonite (MMT) – 14.7 wt %.[86] An increase in heat resistance is also typical of non-polar polymers. For PP, for example, HDT rises from 109°C up to 152°C at a clay content of 6 wt %.[87] As the high temperature may develop in the friction contact zone, a rise in HDT must be favourable for tribological characteristics.

It was found that clay added to polymer materials enhances their heat resistance as well as the thermal resistance including the resistance to thermal-oxidative degradation.[84] The nanoparticles prevent oxygen from diffusion into the polymer and escape of the volatile products of thermolysis. It was also noticed that prolonged heating of a nanocomposite can strengthen the barrier quality owing to the backing of a silicate in the surface layer of the sample and result in a networked structure.[88]

The inhibitive action of fullerenes and carbon nanomaterials with respect to thermal and thermal-oxidative degradation of polymer materials has been studied and is still under intensive study.[89,90] Fullerenes are electrophilic being the active acceptors of electrons. They tend to accept nucleophilic reagents and also hydrogen, free-radical and carbenoidal particles. Therefore, fullerenes present in the surface layer of rubber must influence the course of the tribochemical transformations and consequently the friction parameters.

At strong adsorptive interaction of nanofillers and macromolecules the molecular mobility becomes abruptly frozen in the polymer amorphous phase. The functional groups in macromolecules become blocked by the interaction with the filler surface.[84] As a result, the obstacles are created that prevent from adhesive interaction with the counterface, and the adhesion component of friction force decreases.

The presence of nano-particles within a friction contact may influence the processes of the mass transfer, the accumulation of static electricity, the heat transfer on individual sites of contacting surfaces, the creation of real contact areas, *etc*.

Thus, effects of nano-fillers on the tribological properties of polymers can alter both bulk and surface properties of the materials.

7. Conclusions

A widespread interest in plastics has grown in the mid twentieth century due to the features of their structure, specific mechanical behavior, and considerable possibility to change the polymer properties. But, creep behavior of polymers (strong dependence of their properties on temperature), low heat conductivity, and sensitivity to the environment often posed the numerous problems. The extensive studies over many years have developed the field of modern engineering in which the plastics can be applied as tribological materials, more commonly in the form of coatings and solid lubricants. The latter are used either in the pure form or as the composites or laminated structures.[91] The thin polymer films, e.g. self-assembled monolayers formed by the chemisorption and physisorption of organic molecules (polymers) are prospective boundary lubricants in the fast-growing area of the memory storage devices, the microelectromechanical systems and other precision mechanisms.[92,93]

Further progress in the field of friction and wear of polymers and their composites should be based on solving a number of the important problems which will allow us to establish more refined mechanisms which occur in the working surfaces of friction pairs. It appears important to study the structural changes at molecular level in the surface layers, and to investigate the tribo-chemical reactions. It is also important to find the methods of controlled regulation over the structure and frictional properties of polymers, based on the physical premises and concepts.

It is clear that the progress in engineering will provide many new opportunities in applications of plastics and therefore research in their mechanical and tribological behavior will be a challenging and fruitful field of science and technology.

Acknowledgments

Present work was partially funded under INTAS grant No 05-7940. This chapter was greatly inspired by the ideas of late Dr. M.I. Petrokovets (1937-2006).

References

1. B. J. Briscoe, in: *Friction and Wear of Polymer Composites*, Ed. F. Klaus (Elsevier, Amsterdam, 1986), p. 25.
2. B. J. Briscoe, in: *Fundamental of Friction: Macroscopic and Microscopic Processes*, Ed. I. L. Singer and H. M. Pollok (Kluwer Academic Publishers, Dordrecht, 1992), p. 167.

3. F. P. Bowden and D. Tabor, *Friction and Lubrication of Solids* (Clarendon Press, Oxford, 1964).
4. I. V. Kragelskii, *Friction and Wear* (Pergamon Press, Elmsford, 1982).
5. N. K. Myshkin, C. K. Kim and M. I. Petrokovets, *Introduction to Tribology* (CMG Publishers, Seoul, 1997).
6. L. S. Schadler, L. C. Brinson and W.G. Sawyer, *JOM*, **59**, 50 (2007).
7. S. Bahadur, *Wear*, **245**, 92 (2000).
8. S. Bahadur, and V. K. Polineni, *Wear*, **200**, 95, (1996).
9. A. A. Cenna, P. Dastoor, A. Beehag and N. V. Page, *J. Mater. Sci.*, **36**, 891 (2001).
10. H. S. Nalwa, *Encyclopedia of Nanoscience and Nanotechnology* (American Scientific Publishers, California, 2004).
11. Z. Zhang and K. Friedrich, in: *Polymer Composites – from Nano- to Macroscale* (Springer, Berlin 2005), p. 169.
12. K. W. Allen, *Phys. Technol.*, **19**, 234 (1988).
13. B. V. Deryagin, N. A. Krotova and V. P. Smilga, *Adhesion of Solids*, (Consultant Bureau, New York, 1978).
14. D. H. Buckley, Surface Effects in Adhesion, Friction, Wear, and Lubrication, (Elsevier, Amsterdam, 1981).
15. V. A. Bely, A. I. Sviridenok, M. I. Petrokovets and V. G. Savkin, *Friction and Wear in Polymer-Based Materials* (Pergamon Press, Oxford, 1982).
16. V. V. Tsukruk, H.-S. Ahn, D. Kim and A. Sidorenko, *Appl. Phys. Lett.*, **80**, 4825 (2002).
17. I. Luzinov, D. Julthongpiput, V. Gorbunov and V.V. Tsukruk, *Tribol. Int.*, **34**, 327 (2001).
18. M. Lemieux, S. Minko, D. Usov, M. Stamm, and V. V. Tsukruk, *Polymer Preprints*, **44**, 490 (2003).
19. Ya. E. Geguzin and N. N. Ovcharenko, *Sov. Phys. Usp.*, **5**, 129 (1962).
20. K. L. Johnson, K. Kendall and A. D. Roberts, *Proc. Roy. Soc.* **A324**, 301 (1971).
21. K. L. Johnson, *Contacts Mechanics,* (Cambridge University Press, Cambridge, 1987).
22. B. V. Deryagin, V. M. Muller and Yu. P. Toporov, *J. Colloid Interface Sci.*, **53**, 314 (1975).
23. K. L. Jonson and J. A. Greenwood, *J. Colloid Interf. Sci.*, **192**, 326 (1997).
24. D. Maugis, *J. Colloid Interf. Sci.*, **150**, 243 (1992).
25. I. M. Hutchings, Tribology: Friction and Wear of Engineering Materials (Edward Arnold, London, 1992).
26. Y. Yamaguchi, *Tribology of Plastic Materials* (Elsevier, Amsterdam, 1990).
27. N. K. Myshkin and M. I. Petrokovets, in *Mechanical Tribology. Materials, Characterization, and Applications*, Eds. G. E. Totten and H. Liang, (Marcel Dekker, New York and Basel, 2004).
28. M. I. Petrokovets, *Journal of Friction and Wear*, **20**, 1 (1999).
29. J. R. Barber, *Int. J. Mech. Sci.*, **15**, 813 (1973).

30. J. A. Greenwood and J. B. P. Williamson, *Proc. Roy. Soc.*, **A295**, 300 (1966).
31. J. A. Greenwood and J. H. Tripp, *ASME J. Appl. Mech.*, **34**, 153 (1967).
32. P. R. Nayak, ASME J. of Lubrication Technology, **93**, 398 (1971).
33. K. N. G. Fuller and D Tabor, *Proc. Roy. Soc.* **A345**, 327 (1975).
34. N. K. Myshkin, M. I. Petrokovets and S. A. Chizhik, *Tribol. Int.*, **31**, 79 (1998).
35. N. K. Myshkin, M. I. Petrokovets and SA Chizhik, *Tribol. Int.*, **32**, 379 (1999).
36. J. N. Israelachvili, *Intermolecular and Surface Forces* (Academic Press, London, 1991).
37. J. N. Israelachvili, *Surf. Sci. Rep.,* **14**, 109 (1992).
38. N. K. Myshkin, A. Ya. Grigoriev, A. M. Dubravin, O. Yu. Komkov, N. D. Spencer, and M. Tosatti, in *Proc. of 14th Int. Colloquium Tribology "Tribology and Lubrication Engineering"*, (Germany, Esslingen, 2004), p. 73.
39. N. K. Myshkin, A. V. Kovalev, I. N. Kovaleva and A. Ya. Grigoriev, in *Proc. of "Tribology of Surface Layers and Coatings"*, (Czech Republic, Prague, 2004), p. 3.
40. N. Myshkin, A. Kovalev, W. Scharff and M. Ignatiev, in *Proc. of 2nd Vienna International Conference "Micro- and Nano-Technology VIENNANO'07"*, (Austria, Vienna, 2007), p. 229.
41. N. K. Myshkin, A. V. Kovalev and W. Scharff, in *Proc. of 7th International Symposium INSYCONT'06*, (Poland, Krakow, 2006), p. 249.
42. V. E. Starzhynsky, A. M. Farberov, S. S. Pesetskii, S. A. Osipenko and V. A. Braginsky, *Precision Plastic Parts and Their Production Technology*, (Nauka i Tekhnika, Minsk, 1992) (in Russian).
43. B. J. Briscoe, *Tribology Int.,* **14**, 231, (1998).
44. G. Jintang, *Wear*, **245**, 100 (2000).
45. H. Unal, U. Sen and A. Mimaroglu, *Tribol. Int.*, **37**, 727 (2004).
46. H. Unal and A. Mimaroglu, *Mater. Des.*, **24**, 183 (2003).
47. Y. K. Chen, O. P. Modi, A. S. Mhay, A. Chrysanthou and J. M. O'Sullivan, *Wear*, **255**, 714 (2003).
48. C. J. Schwartz and S. Bahadur, *Wear*, **251**, 1532 (2001).
49. Y. M. Xu and B. G. Mellor, *Wear*, **251**, 1522 (2001).
50. B. J. Briscoe, L, Fiori, and E. Pelillo, *J. Phys. D: Appl. Phys.*, **31**, 2395 (1998).
51. H. Shulga, A. Kovalev, N. Myshkin, and V. V. Tsukruk, *European Polymer Journal*, **40**, 949 (2004).
52. A. Kovalev, H. Shulga, M. Lemieux, N. Myshkin, and V. V. Tsukruk, *J. Mater. Res.*, **19**, 716 (2004).
53. D. F. Moore, *The Friction and Lubrication of Elastomers,* (Pergamon Press, Oxford, 1972).
54. G. M. Bartenev and V. V. Lavrentev, *Friction and Wear of Polymers*, (Elsevier, Amsterdam, 1981).
55. K. Shooter and D. Tabor, *Proc. Roy. Soc.* **B65**, 661 (1952).
56. R. C. Bowers, W. C. Clinton, and W. A. Zisman, *Lubrication Eng.*, **5**, 204 (1953).
57. K. Shooter and R. H. Thomas, *Research*, **2**, 533 (1952).

58. B. L. Rees, *Research*, **10**, 331 (1957).
59. A. Schallamach, *Proc. Phys. Soc.* **B65**, 658 (1952).
60. B. Oloffson and N. Gralben, *Text. Res. J.*, **17**, 488 (1947).
61. G. V. Vinogradov, G. M. Bartenev, A. I. Elkin and V K Mikhaylov, *Wear,* **16**, 213 (1970).
62. D. G. Flom and N. T. Porile, *Nature*, **175**, 682 (1955).
63. D. G. Flom and N. T. Porile, *J. Appl. Phys.*, **26**, 1080 (1955).
64. N. S. White, *J. Res. Nat. Bur. Stand.*, **57**, 185 (1956).
65. A. Schallamach, *Proc. Phys. Soc.* **B66**, 1161 (1955).
66. W. C. Milz and L. E. Sargent, *Lubrication Eng.*, **11**, 313 (1955).
67. K Tanaka, *Wear*, **100**, 243 (1984).
68. T. Fort, *J. Phys. Chem.*, **66**, 1136 (1962).
69. K. C. Ludema and D. Tabor. *Wear*, **9**, 329 (1966).
70. R. T. King and D. Tabor. *Proc. Phys. Soc.* **B66**, 728 (1953).
71. I. V. Kragelskii, M. N. Dobychin and V. S. Kombalov, *Friction and Wear Calculation Methods*, (Pergamon Press, Oxford, 1982).
72. S. P. Gubin, Yu. A. Koksharov, G. B. Khomutov and G. Yu. Yurkov, *Russ. Chem. Rev.*, **74**, 489 (2005).
73. L. Yu and S. Bahadur, *Wear*, **214**, 245 (1998).
74. V. L. Vakula and L. M. Pritykin, *Physical Chemistry of Polymer Adhesion*, (Khimia, Moscow, 1984) (in Russian).
75. G. Heinicke, *Tribochemistry*, (Akademie-Verlag, Berlin, 1984).
76. J. K. Lancaster, *Proc. Instn. Mech. Engrs.*, **183**, 98 (1968).
77. S. B. Ratner, I. I. Farberova, O. V. Radyukevich and E. G. Lure, *Soviet Plastics*, vol. 7, 37 (1964) (in Russian).
78. K. R. Makinson and D. Tabor, *Proc. Roy. Soc.*, **A281**, 49 (1964).
79. A. I. Sviridenok, V. A. Bely, V. A. Smurugov, and V. G. Savkin, *Wear*, **25**, 301 (1973).
80. K. Tanaka, Y. Uchiyama and S. Toyooka, *Wear*, **23**, 153 (1973).
81. N. K. Myshkin, *Wear*, **245**, 116 (2000).
82. C. H. Surberg, E. Badisch and A. Pauschitz, in Proc. of 2nd Vienna International Conference "Micro- and Nano-Technology VIENNANO'07", (Vienna, Austria, 2007), p. 201.
83. V. K. Jain and S Bahadur, in *Proc. of International Conference on Wear of Materials*, (Dearborn, 1979), p. 581.
84. B. J. Briscoe and S. K. Sinha, in *Tribolofy of Polymeric Nanocomposites*, Ed. K. Friedrich and A. Schlarb (Elsevier, Amsterdam, 2008), p. 1.
85. B. Jurkowska, B. Jurkowski, P. Kamrowski, S. S. Pesetskii, V. N. Kowal, L. S. Pinchuk and Y. A. Olkhov, *J. Appl. Polym. Sci.*, **100**, 390 (2006).
86. Y. Kojima, A. Usuki, M. Kawasumi, A. Okada, T. Kuraucki and O. Kamigaito, *J. Polym. Sci. Part A. Polym. Chem.*, **31**, 983 (1993).

87. P. H. Nam, P. Maiti, M. Okamoto, T. Kotaka, M. Ohshima, A. Usuki, N. Hasegawa and H. Okamoto, *Polymer Engineering and Science*, **42**, 1907 (2002).
88. J. W. Gilman, T. Kashiwagi, E. P. Giannelis, E. Manias, S. Lomakin, J. D. Lichtenhan and P. Jones, in *Chemistry and Technology of Polymer Additives*, (Blackwell Science, Oxford, 1999), p. 249.
89. B. M. Ginzburg, L. A. Shibaev, O. F. Kireenko, A. A. Shepelevskii, E. Yu. Melenevskaya and V. L. Ugolkov, *Polym. Sci.* **A47**, 160 (2005).
90. B, Troitskii, A. Domrachev, L. V. Khokhlova and L. I. Anikina, *Polym. Sci.* **A43**, 964 (2001).
91. G. C. Pratt, in *Encyclopedia of Materials Science and Engineering*, Ed. M. B. Bever, (Pergamon Press, Oxford, 1986), p. 281.
92. V. V. Tsukruk, T. Nguen, M. Lemieux, J. Hazel, W. N. Weber, V. V. Shevchenko, N. Klimenko and E. Sheludko. in *Tribology Issues and Opportunities in MEMS*, Ed. B.Bhushan, (Kluwer Academic Publishers, Dodrecht, 1998), p. 607.
93. B. Bhushan. *Tribology and Mechanics of Magnetic Storage Devices*, (IEEE Press, Springer, New York, 1996).

CHAPTER 2

TRIBOPHYSICAL INTERPRETATION OF POLYMER SLIDING MECHANISMS

Pieter Samyn[1], Gustaaf Schoukens[1] and Jan Quintelier[2]*

[1]*Department of Textiles, Ghent University*
Technologiepark 907, B-9052 Zwijnaarde, Belgium
[2]*Department of Mechanical Construction and Production, Ghent University*
Sint-Pietersnieuwstraat 41, B-9000 Gent, Belgium
**E-mail: Jan.Quintelier@UGent.be*

This chapter gives an overview of micro-structural changes at polymer surfaces induced by friction, and the way they influence the sliding behaviour. Raman spectroscopy is presently applied on worn polymer surfaces giving quantitative and qualitative information about the chemical nature (structural units or additives), conformational order (trans-gauche molecular isomerism), state of order (amorphous, crystalline) and molecular orientation (alignment of polymer chain and side groups) after sliding. Depositions of internal lubricants and formation of a sliding film on the polymer surface are characterized. Degradation mechanisms due to chemical reactions, thermal effects or radical formation are illustrated for polyacetals, polyamides, polyesters and thermosetting or thermoplastic polyimides. Crystallization deteriorates the sliding stability while the formation of a rigid amorphous phase results in low friction and controllable wear rates. Other information is obtained from thermo-analytical analysis (DTA/TGA) of wear debris particles. Through the joint action of repeated loading and high temperature, the wear debris undergoes physical and chemical reactions since its generation. As a result, the properties of those small particles would not be identical to those of the bulk material and are characteristic for the wear process. For polyimides, hydrolysis is the origin of high friction at 100 to 180°C and lacking transfer film while imidization at 180 to 260°C promotes transfer film formation with low friction and stable wear rates.

1. Introduction

Polymer tribology is a multidisciplinary field of research that can be approached from mechanical or from tribophysical point of view. Many experimental works and theoretical models available in literature describe friction and wear mechanisms as mechanical effects, considering the formation of a transfer film by adhesion and mechanical interlocking[1]. According to early research, the formation of a friction film reduces the counterface roughness when debris particles accumulate between roughness asperities. The film thus increases the true contact area and therefore reduces localized stresses. Others found that the formation of a transfer film depends on adhesive forces[2]: wear rates generally decreased when the adhesion between the transfer material and the counterface improves, which could be attributed to a stable transfer layer.

From a tribophysical point of view, most energy consumed during sliding is transformed into heat, increasing the temperature on the rubbing surface. Chemical reactions such as polymerization, crystallization or irreversible degradation and finally melting of the polymer surface and/or the wear debris particles take place and further control the sliding stability, possibly leading to transfer or severe wear. The chemical reactivity at the polymer surface may become so high that physico-chemical interactions between the sliding surfaces or between the sliding surface and the environment are the origin for specific interfacial bonds. Some interactions are controlled by radical formation due to mechanical rupture of molecules on the sliding surface. It was reported that the friction and wear processes at high temperatures are rather controlled by pyrolysis[3]. Thermal and oxidative reactions cause degraded wear particles to incorporate in the transfer film and change its stability. The debris particles created at low temperatures often show stronger cohesion than debris that is partially degraded through formation at high sliding temperatures. Chemical changes at the sliding interface are frequently studied with techniques such as EDX, XPS and SIMS[4].

In addition to the previously mentioned failure mechanisms, the physical variations in polymer structure under sliding are not clearly understood and should be further investigated as they strongly control friction and wear processes. The macromolecular systems are characterized by visco-elastic behaviour under stress, resulting in relatively small deformation forces and extremely high elongation without fracture. In contrast to other crystalline or amorphous materials, it is thermodynamically demonstrated that stresses acting on semi-crystalline networks cause a decrease in entropy rather than changes in internal energy. This implies that polymer chains, having different conformations of identical energy, exhibit decreasing number of possible conformations under deformation. The time and temperature dependency of those processes are widely known and investigated under uniaxial stress. However, the combination of shear stresses due to friction and frictional heating can also induce orientation near the polymer sliding surfaces, affecting wear resistance: they can yield higher transverse and rotational freedom of the molecular chains leading to re-orientation.

New potential to get additional insight in the tribochemical or tribophysical reactions on polymer surfaces under sliding is offered by a combination of surface analysis techniques[5]. A study of the variations in polymer structure on molecular scale will lead to additional insights in the tribophysical or -chemical phenomena at the interface and interference between mechanical, chemical and thermal effects. With respect to technical viability, characterization methods monitoring structural changes in a thin surface layer activated by wear should have (i) a sensitivity adequate to distinguish the surface structure from the bulk, (ii) a detection depth equivalent to the surface layer thickness to study the structure of interest, and (iii) a signal-to-noise ratio high enough to avoid possible interference originating from foreign matter adhering to the wear surface. Also the polymer wear debris undergoes structural changes through the combined effect of shear and temperature. While it is often assumed that melting of the polymer substance is the principle sliding mechanism, the separate effects of softening, crystallization and polymerization should be considered more in depth. Transitions between glassy and rubbery phase occur at specific temperatures and crystallization during sliding affects the melting behaviour. Bulk characterizing techniques and thermo-analytical methods should therefore be applied to the wear debris particles, relating the polymer sliding properties to the crystalline structure. Methods can be used for monitoring phase transitions and chemical reactions of polymer structures as a function of sliding parameters.

2. Experimental Techniques

2.1. *Raman Spectroscopy*

The polymer surfaces before and after sliding will be characterized by Raman spectroscopy[6]. Unlike infrared spectroscopy being an absorption process, Raman spectroscopy is a scattering process where molecular vibrations shift the wavelength of a laser beam striking a sample and result in sharp, well-resolved bands. The shifted light, called Raman scatter, is collected in a spectrum and can be used for qualitative and quantitative analysis of functional groups based on the intensity and wavelength position. The main strength of Raman spectroscopy is its short-time and flexible sampling by using appropriate probes coupled with optical fibers, both delivering the laser light and collecting the scattered light from the sample[7].

The phenomenon of inelastic light scattering or inelastic collisions between photons and molecules, known as Raman radiation, was first documented by the Indian scientists Raman and Krishnan[8]. When a surface is irradiated with monochromatic light, most of the scattered energy comprises radiation of the incident frequency (elastic or Rayleigh scattering). When energy is transferred from the incident photon to the molecule, light of lower energy (higher wavelength) will be scattered from the molecular centers, known as inelastic Stokes scattering. When energy is transferred from the molecule to the photon, light of higher energy (lower wavelength) will be scattered by the sample, known as inelastic anti-Stokes scattering. At room temperature, most polymer molecules are in the energetic ground state and they do not have sufficient thermal energy to overcome the energetic barrier to move into the first excited state. For this reason, Stokes scattering is more favourable than anti-Stokes scattering. Also, anti-Stokes scattering from the higher excited molecular modes is unlikely. The difference in frequency between the scattered light and the incident beam corresponds to the vibration and rotational energy states of the molecule. For large molecules, most Raman scattered light is due to changes in vibration energy modes, while rotation modes are lower in energy and occur around three orders of magnitude slower than vibration modes. Whether net energy absorption or emission is seen, depends on the population of the energy levels within a molecule. Quantitative analysis is based on the intensity of the Raman scattered light being proportional to the concentration. Qualitative analysis is based on the wavelength shifts being different for a wide range of molecular vibrations. The possibilities for molecular vibration and related Raman frequency are determined by the masses of the atoms (composition), their equilibrium spatial arrangement, their relative displacements during a vibration and the bonding forces. For vibration modes causing a frequency shift by Raman scattering, it should induce a change in the polarizability of the molecule, which depends on its orientation. This is the selection rule for Raman spectroscopy in contrast to infrared spectroscopy[9], where the permanent dipole moment of the molecule should change during a vibration cycle to be active in the infrared spectrum.

Any event that changes the equilibrium order, the composition, the orientation or the mutual forces between molecules (e.g. intermolecular interactions) can change the frequency and intensity of a spectral Raman band. Hence, vibration spectroscopy is strongly sensitive to the physical and chemical macromolecular structure. Different parts of the same molecule tend to be strongly coupled, so changing a single atom or orientation of a molecular segment can perturb many of its vibrations. This means that a subtle structural change can have large impact on the resulting spectrum, with the advantage that Raman spectra provide a fingerprint for a given molecule and its conformation. It is however not obvious to predict the numbers of bands to be expected in a spectrum for any molecule. The technique should rather be used as a relative evaluation tool in comparing different molecular structures before and after sliding. Laser Raman spectroscopy seems to be feasible as scattered photons originate from about 5 to 10 μm depth beneath the sample surface. This range typically includes the zone of the polymer that is affected by sliding and rising bulk and flash temperatures. Few applications of Raman spectroscopy in tribology were reported, mainly illustrating solvent-induced chemical degradation of e.g. PEEK under lubricated sliding[10], while variations under dry sliding will presently be further studied.

Measurements can be performed on a Brucker FT spectrometer Equinox 55S (Brucker Optik, Ettlingen, Germany), equipped with a

Raman module FRA 106 fitted to a nitrogen cooled (77 K) germanium high sensitivity detector D418-T. The applied laser wavelength during the experiments is the 1064 nm line from a Diode Laser Pumped Nd^{3+}: YAG laser. A stable laser light with relatively high wavelength is chosen as it reduces fluorescence and heating of the sample. The 1064 nm line corresponds to the frequency of light emitted by the atoms in the laser crystal. All spectra have a resolution of 3 cm^{-1} between wave numbers 0 and 3500 cm^{-1}, using a non-focused laser beam with a power of 70 mW (surface area approx. 3 mm^2). Each spectrum is collected as an average of 250 scans, so that the selected parameters result in an optimum noise to intensity ratio. The intensities are averaged from three scans. The dispersion of the spectrum over the total sliding surface is small. A quantitative interpretation of the spectra is made using the baseline theory for correction of individual spectra. Normalized or relative intensities should be used to compensate for any change in experimental conditions such as excitation intensity or sample positioning. There is ±2% variation in relative intensities.

2.2. *Thermal Analysis*

Simultaneous Differential Thermal Analysis (DTA) and Thermogravimetric Analysis (TGA) is done on a Stanton Redcroft 1500 Thermobalance for polymer bulk material and wear debris particles. The initial polymer sample weight is about 17 mg unworn polymer material and 3 mg polymer wear debris. Measurements are conducted in either air or nitrogen atmosphere during two subsequent heating cycles between room temperature and a maximum temperature, depending on the polymer in case (e.g. 300°C for polyamide or 700°C for polyimide). A constant heating rate of 10°C/min is usually applied.

In *Differential Thermal Analysis* (DTA), the polymer sample and an inert Al_2O_3 reference without no structural transitions in the temperature range of interest, are heated at the same rate, i.e. with a constant heat flux. The temperature difference between the polymer sample and the reference is measured and plotted as a function of the polymer sample temperature. If the polymer sample undergoes a thermal transition such as a glass transition and melting (endotherm) or crystallization (exotherm), a difference in temperature is observed: conventionally, the exotherm reaction is registered as an upwards peak and the endotherm reaction is a downwards peak in the thermograph. In case of *Differential Scanning Calorimetry* (DSC), a single heater element is used and the resulting sample temperature is a dependent parameter. Quantitative DTA analysis is more complicated than DSC (enthalpies are less accurate), but it is presently preferred for wear debris due to high degradation and mass loss under heating, damaging the DSC sample cells.

Using *Thermogravimetric Analysis* (TGA), decomposition during heating is studied from a sensitive balance that follows the weight changes as a function of the polymer sample temperature. The technique can characterize materials that exhibit weight loss or gain due to decomposition, oxidation or (de)hydration. The instrument is coupled to a mass spectrometer (MS) for analysis of the gaseous decomposition products.

2.3. *Other Techniques*

X-ray photoelectron spectroscopy (XPS) has detection depths of several nanometers beneath the sliding surface. It gives chemical information about the elements and/or molecules on the surface. In combination with image processing, the distribution of chemical elements over the surface can be detected in each pixel of the image. Local spectra can be made near depositions, grooves, transfer particles or degraded zones. Angular resolved spectroscopy is able to investigate the depth of the modified surface under sliding. Recent techniques such as secondary ion mass spectroscopy (SIMS) have ability for chemical surface mapping showing the distribution of molecules over a single monolayer, but are still in development in relation to tribological problems. It has potential to detect the direction of specific functional groups towards the surface or opposite conformation. The formation of radicals can be illustrated by means of election spin resonance spectroscopy (ESR). Infrared spectroscopy has generally stronger O-H response than N-H while the reverse is true for Raman spectroscopy. Surface-enhanced Raman scattering (SERS) offers signal enhancements of several orders of magnitude[11].

3. Polymer Surface Analysis

Worn polymer surfaces are analyzed with Raman spectroscopy and an overview of various surface modifications observed during sliding is presented in this paragraph. Polymers such as internally lubricated polyethylene terephtalate (PET/PTFE), polyamide (PA) or sintered polyimide (SP) and thermoplastic polyimide (TP) are included as examples. While it is not the intention to give a complete overview of sliding properties of those materials, rather the strength of spectroscopic methods in explaining various sliding mechanisms is illustrated. They can further be related to the transitions in friction and wear behaviour as a function of, e.g., normal loads, sliding velocities or sliding temperatures.

3.1. *Depositions of Internal Lubricant*

A SEM image of slid PET/PTFE surfaces (Fig. 1) shows the morphology of deposited internal lubricants on the polymer surface. It has separate flakes at low contact pressures that progressively build up and form a continuous film at high contact pressures.

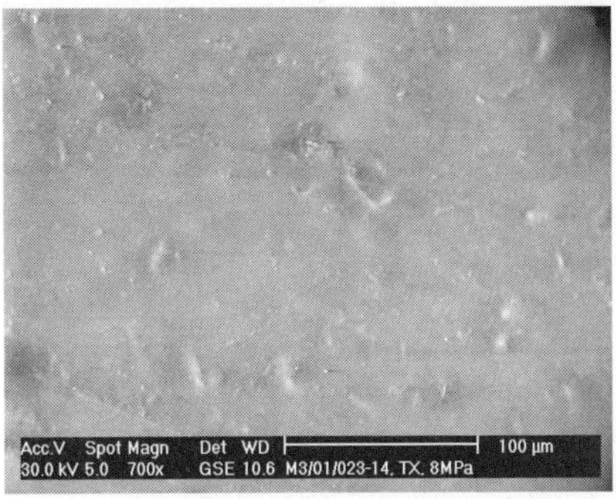

(a)

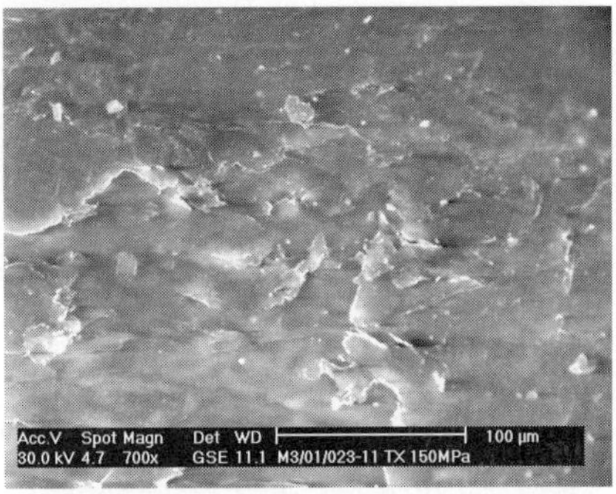

(b)

Fig. 1. SEM microscopy of PET/PTFE sliding surfaces with deposition of PTFE internal lubricant after sliding at (a) 8 MPa, 0.005 m/s, (b) 150 MPa, 0.005 m/s.

Raman spectra for unworn PET/PTFE and worn surfaces at different contact pressures between 8 and 150 MPa is shown in Fig. 2. Raman spectroscopy of PET was reviewed many times after unidirectional and biaxial stretching, assigning different Raman-active bands to molecular vibrations[12]: some bands are due to the backbone molecular bonds (e.g. benzene ring vibration) and others are attributed to the side groups (e.g. C=O stretching), most being stress-sensitive. Bands assigned to the aromatic structure at 702 cm^{-1}, 795 cm^{-1} or 1613 cm^{-1} can be used as internal reference bands[13]. Depositions of PTFE lubricant on the sliding surface are illustrated by a band at 732 cm^{-1} not appearing in spectra of pure PET. Compared to the neighbouring band at 702 cm^{-1}, representing the aromatic structure, the variation in relative intensity I_{732}/I_{702} is given in Fig. 3 and indicates an increasing intensity with normal load from 0.23 to 0.27, according to a sliding film of mixed PTFE/polyester debris developing on the polymer surface. The discontinuity at 16 MPa is in relation to conformational changes discussed in further on in this Chapter (Fig. 6). The Raman technique is also sensitive to the used counterface types: a higher intensity I_{732}/I_{701} is noted after sliding against stainless steel as it was observed that a more continuous and viscous polymer film has formed on the polymer surface, related to the lower surface energy of stainless steel (316L) compared to alloyed carbon steel (42 CrMo4).

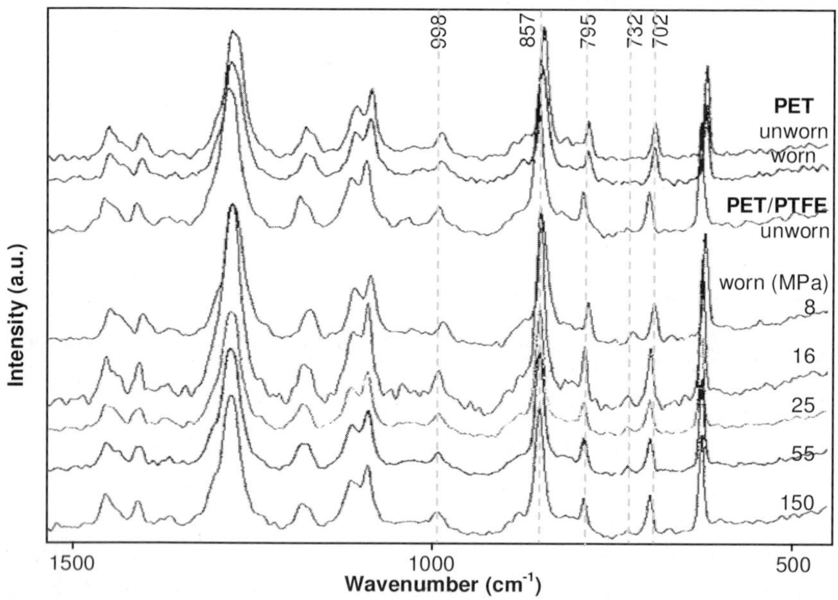

Fig. 2. Detail of Raman spectra of unworn and worn PET and PET/PTFE surfaces at different contact pressures.

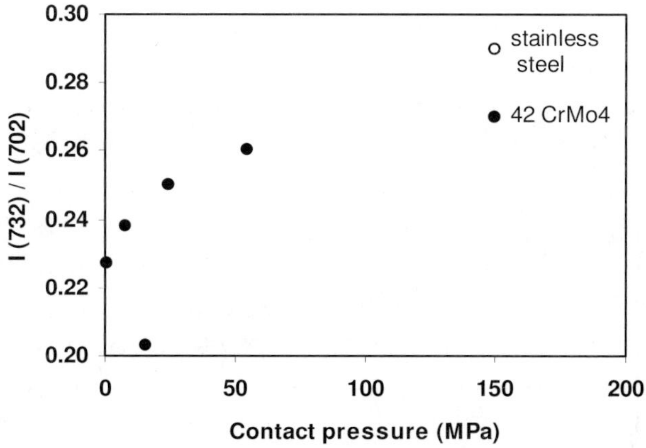

Fig. 3. Variation in PTFE-sensitive absorption band (732 cm^{-1}) after sliding of PET/PTFE at different contact pressures and steel counterface types.

The gradual build-up of a polymer sliding film contributes to extremely low coefficients of friction (0.10 at 8 MPa to 0.02 at 150 MPa) and controllable wear rates for PET/PTFE. The deposition layer is characteristic for the flat-on-flat contact geometry, where debris particles are entrapped in the sliding interface. Other geometries such as cylinder-on-plate cause easy evacuation of the debris from the sliding interface and depletion of PTFE lubricant was observed by Raman spectroscopy.

3.2. *Conformational Changes*

Orientation of thermoplastic polymer backbones occurs under sliding. Polyethylene terephtalate (PET) is characterized by a *trans* (stretched) or *gauche* (coiled) orientation of the ethylene glycol as shown in Fig. 4. Both conformations correspond to specific absorption bands at either 998 cm^{-1} (trans) or 857 cm^{-1} (gauche), for which intensities are calculated relatively to an aromatic band at 795 cm^{-1} (Fig. 2). Also the bands at 2908 cm^{-1} and 2960 cm^{-1} are assigned to trans and gauche isomers. Some relative band intensities are detailed in Fig. 5 after sliding as a function of contact pressure or temperature. In general, the intensity for trans bands increases and gauche bands accordingly decrease in intensity.

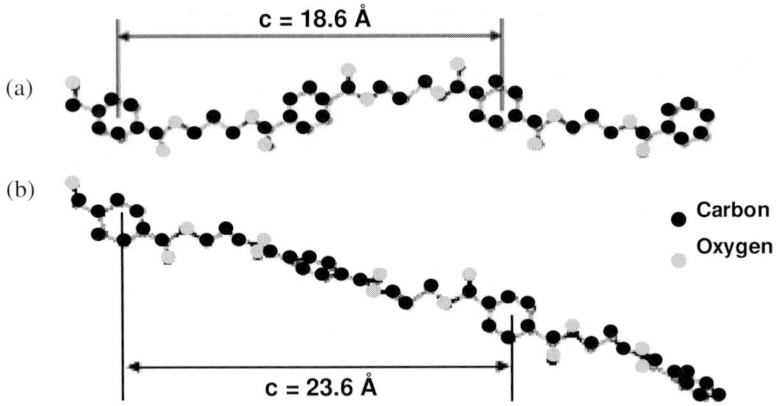

Fig. 4. Conformational arrangement of the polyethylene terephtalate (PET) molecules, (a) all-trans or 180° orientation, (b) gauche-trans or 60° orientation.

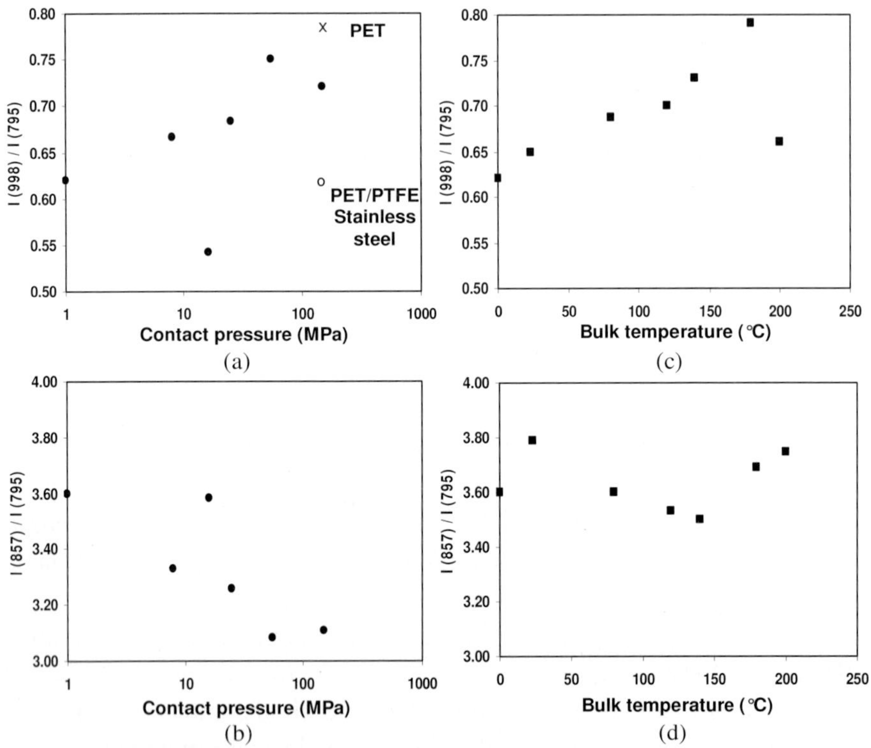

Fig. 5. Relative intensities of Raman bands related to trans (998 cm^{-1}) and gauche (857 cm^{-1}) conformations for PET/PTFE after contact pressure or temperature tests.

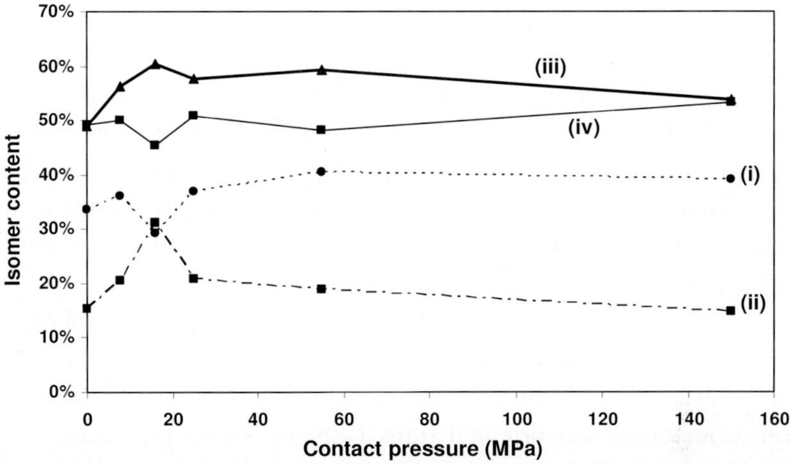

(a)

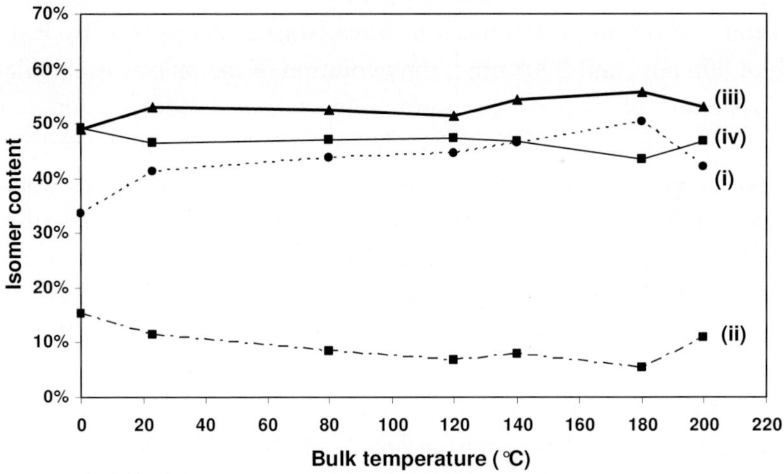

(b)

Fig. 6. Isomer distribution after sliding of PET/PTFE as a function of (a) contact pressure, (b) bulk temperature, with (i) crystalline trans, (ii) rigid amorphous trans, (iii) total trans, and (iv) gauche.

Both the composition of the polymer and counterface influence the orientation (see at 150 MPa): as the frictional forces are higher for sliding pure PET ($\mu = 0.25$) compared to PET/PTFE ($\mu = 0.05$), higher trans orientation is induced. Under the same conditions with stainless steel counter-faces, the friction is lower ($\mu = 0.03$) and therefore lower trans orientation is measured. The trans conformation after tests at high temperature increases from 40 to 48% and is larger than for tests at high contact pressure, increasing from 36 to 40%[14].

It is concluded that stressed PET contains three physical phases: a *crystalline*, an *amorphous* and a *rigid amorphous* phase, respectively containing trans isomers in a three-dimensional order, gauche isomers in random orientation and oriented trans isomers with a two-dimensional degree of order[15]. When macromolecular semi-crystalline networks are stressed, an increase in trans conformation is mainly favoured in the latter phase as it is an intermediate phase in the amorphous to crystalline transition[16]. From the differences in trans isomer represented by Raman bands at 998 cm^{-1} and 2908 cm^{-1}, the evolution of the phases is calculated as a function of contact pressure (Fig. 6a) or sliding temperature (Fig. 6b). It is generally concluded that an oriented or rigid amorphous phase offers better ductility to the polymer than a crystalline phase, leading to better tribological properties with stable and low friction or low wear rates and homogeneous transfer.

Highly aromatic polymers, such as polyimide, have more rigid polymer chains and orientation of the entire backbone structure therefore happens more difficultly. The molecular structure for either thermoplastic or thermosetting polyimide is shown in Fig. 7, with typically very high stiffness for the thermosetting polyimide (E = 3480 MPa) and some lower stiffness for the thermoplastic polyimide (E = 2760 MPa) due to inclusion of multiple ether links. The Raman spectra for both polyimide compositions are compared in Fig. 8. The characteristic absorption bands are located near 1788 cm^{-1} for the C=O side-group (imide I), near 1124 cm^{-1} for the C-N-C backbone groups (imide III) and near 1272 cm^{-1} for the C-O-C backbone groups[17,18].

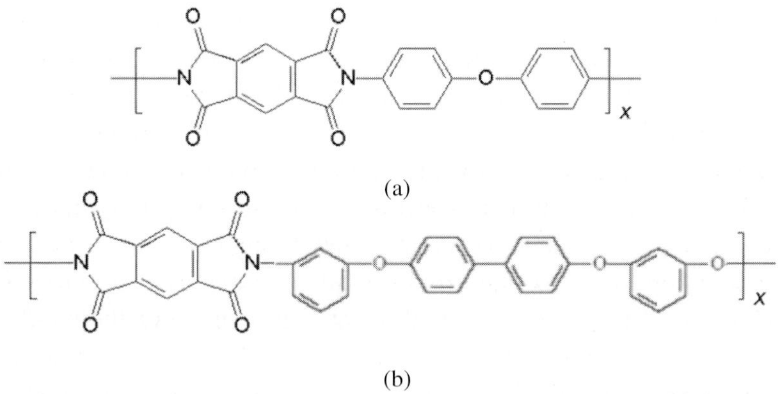

Fig. 7. Molecular structure of (a) thermosetting polyimide, (b) thermoplastic polyimide.

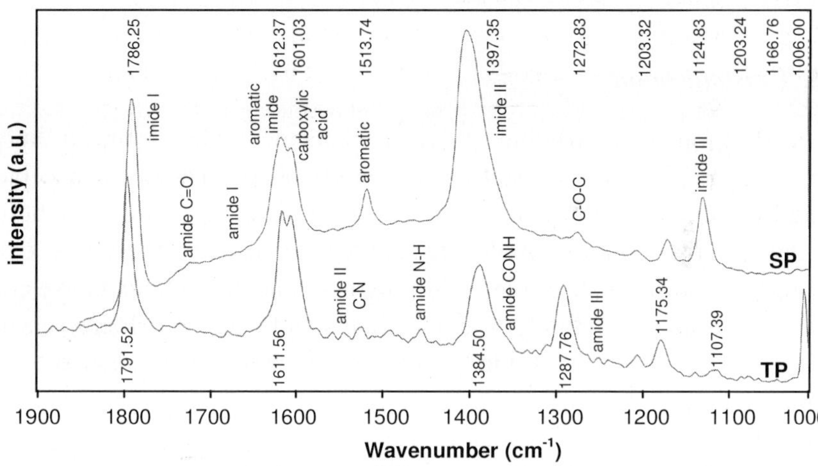

Fig. 8. Raman spectra of unworn thermosetting (SP) and thermoplastic (TP) polyimide.

Relative intensities for some Raman bands after sliding are shown in Fig. 9. The C=O group stretches through shear stresses and minor heating of the sliding surface, both enhancing polymer chain mobility. The thermally activated structural changes in polyetherimide by internal friction was investigated by Fukuhara[19] concluding that relaxation induced important C=O motions. The C-N-C transverse orientation progressively increases at higher sliding temperatures and higher normal loads. This structural evolution happens more gradually compared to the C=O group, because the C-N-C bonds are not directly influenced or constrained by additional chemical reactions. The C-O-C intensity above 180°C decreases, showing that higher molecular chain mobility above the transition temperature allows for tilting of the ether structure. From an in-depth analysis of Raman spectra for thermosetting polyimides[20], it is generally concluded that orientation after sliding at low loads is mainly concentrated on the C=O side-groups, while it is mainly concentrated on C-N-C and C-O-C backbone structures at high loads. For thermoplastic polyimides (Fig. 10), a strong variation in C-O-C stretch and slight variation in C=O stretch shows that molecular orientation during sliding at 50 N is mainly located in the molecular backbone rather than on side groups, in contrast to thermosetting polyimide. The conclusions confirm physical aspects of shear stresses acting on stiff or weak polymer chains.

3.3. *Crystallization*

Depending on their ordering, polymer molecules either form an amorphous (isotropic) or a crystalline (ordered) phase. The most accepted model for the crystalline phase is the folded chain structure: it is based on the fact that complex molecular aggregates, called spherulites, exist through the entire bulk of the solid polymer. The amorphous phase contains a mixture of randomly oriented chains and extended chains serving as links between various folded crystalline domains. If the distance between the folds is larger than 200 nm, the crystal is then designated as an "extended chain crystal". When no folds are present, it is called a "fully extended chain crystal". Semi-crystalline polymers undergo a partial transition from the amorphous into the crystalline phase in certain temperature range, eventually stimulated by mechanical stresses. Also during sliding, crystallization of the polymer surface happens and further controls sliding stability. Three crystalline structures of polyamide (PA) 6 are described in literature[21]: monoclinic α, pseudo-hexagonal β and mono-clinic γ.

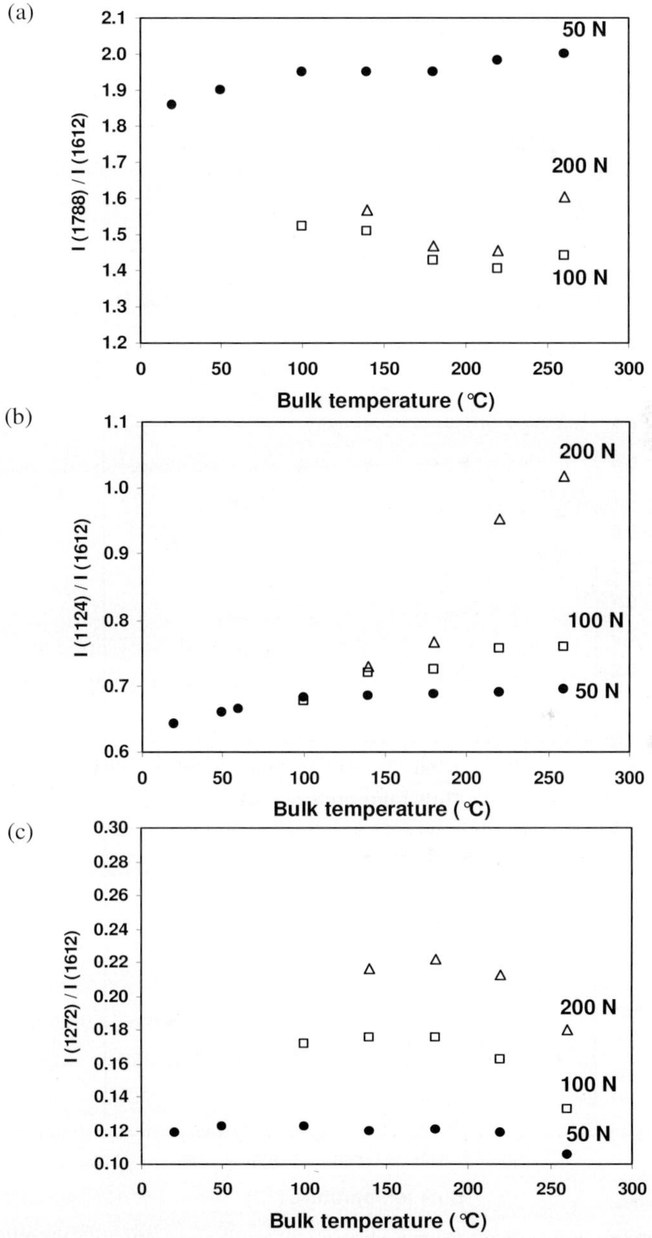

Fig. 9. Analysis of characteristic Raman absorption bands for thermosetting polyimide after sliding at different temperature and normal loads, (a) imide I (C=O), (b) imide III (C-N-C transverse), (c) C-O-C ether stretch band.

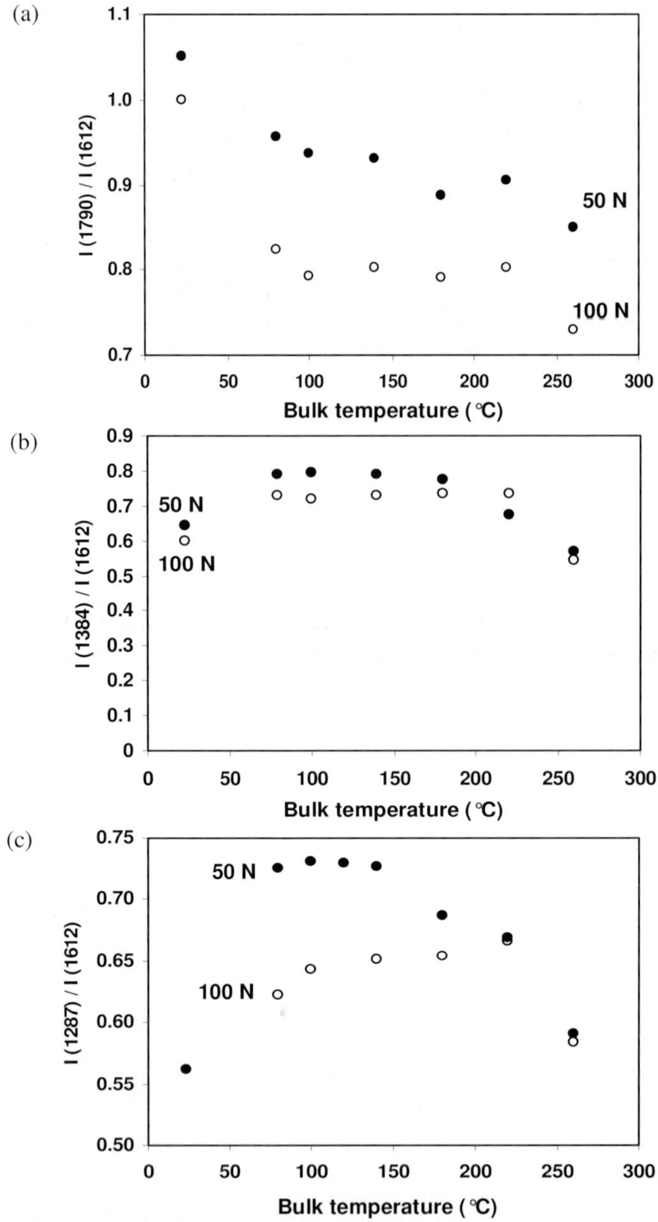

Fig. 10. Analysis of characteristic Raman absorption bands for thermoplastic polyimide after sliding at different temperature and normal loads, (a) imide I (C=O), (b) imide II (C-N-C axial), (c) C-O-C ether stretch band.

Based on reference Raman spectra for different structural composition (Fig. 11), the 1127 cm^{-1} band characterises a planar band representative for fully extended chains with the amide groups in the same plane as the methylene groups: it is only present in the α crystalline form. The band at 1079 cm^{-1} is characteristic for a twisted or non-planar configuration: it seems that the amide group in the γ structure is rotated 67° away from the main chain. Hydrogen bonds occur between parallel chains and the interchain distance for the γ form is higher than for the α form. Plasticity for PA 6 is basically governed by crystal slip parallel to the hydrogen bonded sheet structure[22], i.e. without rupture of the H bonds. The β structure is a mesomorphous phase obtained from quenching molten polymer.

The polymer surfaces and transfer films on 42CrMo4 steel after sliding of polyamide grades with different crystalline structure at 8 MPa, 0.005 m/s are shown in Fig. 12, indicating sliding instabilities for the α crystalline structure and smooth sliding for the γ crystalline structure. Consequently, the coefficients of friction for α-Pa are higher than for γ-Pa, indicating the effect of intrinsic material properties. However, overload for γ-PA is more frequently observed as it has a weaker structure[23].

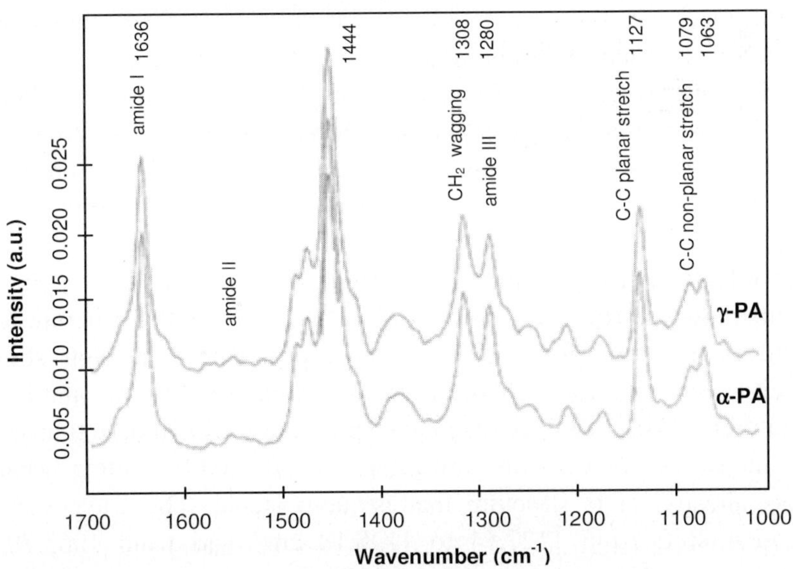

Fig. 11. Raman spectra of unworn polyamide with different crystalline structures.

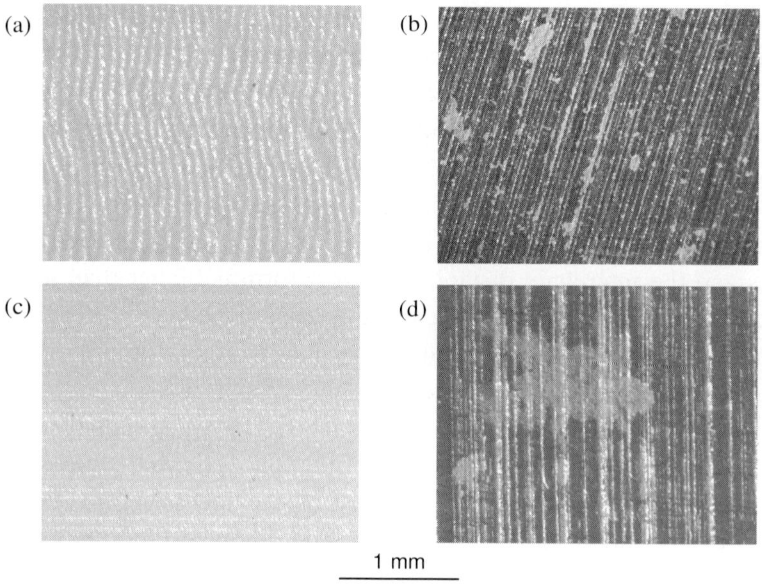

Fig. 12. Optical microscopy of polyamide surfaces and transfer film with different crystalline structure (a) α-PA surface, (b) α-PA transfer, (c) γ-PA surface, (d) γ-PA transfer.

The mechanical and thermal cycle during sliding affects the crystalline polymer structure. Strain induced γ to α transitions above 120°C has already been noticed during tensile testing[24], but it is not clear whether this process is mainly stress and/or thermal activated. The γ phase is reported to be mechanically reasonably stable and transformation into α is limited, not providing higher mechanical strength upon heating[25]. After sliding, however, weak γ into α transition can be demonstrated (Table 1). The γ related intensity I_{1079}/I_{1127} for γ-PA decreases from 0.50 to 0.46 and evolutes towards α-PA with low I_{1079}/I_{1127} intensity and α structure. The γ into α transition for PA-Mg is most pronounced at high sliding velocities allowing for high shear stresses, high temperatures and better orientation. Also 140°C temperatures promote the γ into α transition. The intensities for α-PA are lower and less affected by sliding because it has already an α structure that is more stable. The wavenumbers for γ-PA shift from 1127.54 to 1128.18 cm^{-1} and from 1063.70 to 1063.46 cm^{-1}. With the central band 1079 cm^{-1} representing γ-PA, both up- and down-ward shifts are asymmetric and directed away from 1079 cm^{-1} by weakening of its intensity. It is concluded that the γ structure slightly reverts into an α structure, most likely at 120 to 180°C reflected in an increase in coefficient of friction and a decrease in wear rates. Raman spectra at high temperatures indicated an increase in C=O stretching frequency due to crystallization and loosening intermolecular hydrogen-bonding.

Table 1. Relative intensities of some Raman bands for polyamide before and after sliding.

Tribological test conditions	Crystalline structures			
	I_{1079}/I_{1127}		I_{1063}/I_{1127}	
	α-PA	γ-PA	α-PA	γ-PA
Unworn	0.42	0.50	0.54	0.53
50 N, 0.005 m/s, 23°C	0.42	0.50	0.48	0.51
50 N, 0.3 m/s, 23°C	0.43	0.48	0.51	0.52
100 N, 0.3 m/s, 23°C	0.40	0.46	0.51	0.50
150 N, 0.3 m/s, 23°C	0.39	0.46	0.50	0.50
50 N, 0.3 m/s, 140°C	0.39	0.43	0.48	0.50

3.4. *Polymerization*

Depending on the polymer type, polymerization of the surface under sliding might happen through a radical mechanism (e.g. polyacetals) or polycondensation reactions (e.g. polyimides or polyethylene terephtalate). While crystallization often catastrophically deteriorates the sliding stability, polymerization reactions contribute to higher strength and ductility of the polymer, consequently improving the sliding properties.

A detail from Raman spectra of polyacetal surfaces worn at 25 and 150 MPa contact pressure, 0.005 m/s is shown in Fig. 13 and compared to the unworn polyacetal[26]: characteristic bands for aliphatic ethers are a symmetric C-O-C stretch at 930 cm^{-1} and O-CH_3, O-CH_2 deformations at 1480 cm^{-1} (assigned to end groups of the polymer chain). The relative intensity I_{1480}/I_{930} for ether bands changes from 0.45 in the unworn sample towards 0.41 for worn surfaces. The lower intensity for worn surfaces suggests a transition from $O-CH_3$ into C-O-C, or an increase in chain length due to further polymerization during sliding.

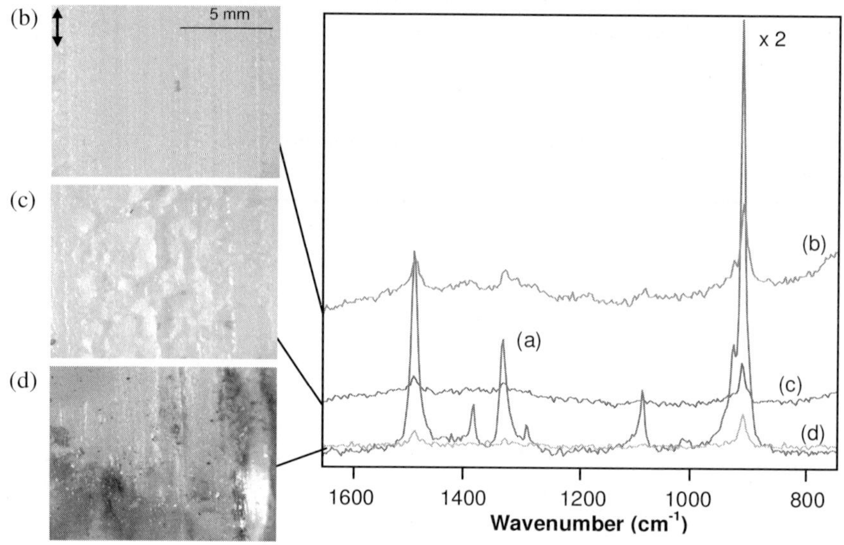

Fig. 13. Raman spectra of polyoxymethylene homopolymer (POM-H) surfaces, (a) unworn, (b) worn at 25 MPa, (c) worn at 55 MPa (d) worn at 150 MPa.

Those reactions are controlled by radical formation of polymer chains that fracture under mechanical stress. Bands only present in unworn polyacetal are: 1100 cm^{-1} (assigned to hydroxyl (OH) end groups of the polymer chains), 1300 to 1400 cm^{-1} (assigned to CH, CH_2) and 2800 cm^{-1} (assigned to the CHO group in aliphatic aldehydes originating from the formaldehyde monomers). The disappearance of the OH and CHO related bands suggests further polymerization during sliding and/or the thermal stabilization of the polymer chains by acetylation of the hydroxyl end groups. The observations of polymerization during sliding are confirmed by the calculated maximum polymer surface temperatures, ranging between 120 and 150°C under present sliding conditions.

For high-performance polymers such as polyimide, polymerization is observed as a conversion of polyamic acid precursor into polyimide above 180°C, affecting the sliding behaviour. Referring to the Raman spectrum in Fig. 8, the relation between the imidized structure and the hydrolyzed carboxylic acid structure is illustrated by comparing the 1612 and 1601 cm^{-1} band, respectively[27]. The relation between imide structure and hydrolyzed carboxylic acid structure is illustrated in Fig. 14 as a function of the applied sliding temperature. Enhancement of the imide structure through ring-closing of carboxylic acid is given by an upward shift of the 1612 cm^{-1} band[28] and establishes for thermosetting polyimide at 180 to 200°C. Dehydration or imidization happens most intensively at 180°C irrespective of the 50 to 150 N normal loads, represented by a maximum 1612 cm^{-1} band. The imide structure forming under sliding is characterized by reversion of the C-N-C imide chains from transverse into axial conformation[29-32]. It is known that polyimides show high friction (0.50 to 0.60) at low temperatures and low friction at high temperatures (0.20 to 0.30). The wear rates at low temperatures linearly increase with the applied sliding temperature while leveling at high sliding temperatures. As shown in Fig. 15, the polymerization reaction is visible on the surface by optical microscopy. It contributes to the formation of a transfer film, however, remaining fragmented own to the brittleness of polyimides. No transfer film was observed at low temperatures as lack of imidization results in low strength and coherence. For thermoplastic polyimides, imidization occurs at 120 to 180°C and decreases friction. The polymer surfaces become then irregularly shaped and opaque.

3.5. *Thermal Degradation*

While the sliding properties of thermosetting polyimides outlined in previous paragraph were controlled by polymerization, the behaviour of thermoplastics at high temperature is generally controlled by softening and melting. The aspects of thermoplastic polyimide sliding surfaces are illustrated in Fig. 16 for 120 and 260°C bulk temperatures, respectively corresponding to a maximum polymer surface temperature T* of 228°C (softening) and 320°C (melting). The softened surfaces are characterized by polymer flakes and formation of a patchy-like transfer film. Similar features were observed for polyacetals and polyethylene terephtalate. The molten surfaces become irregular while a thick transfer film develops on steel counterfaces. Roll-like wear debris particles are observed as an indication for melting.

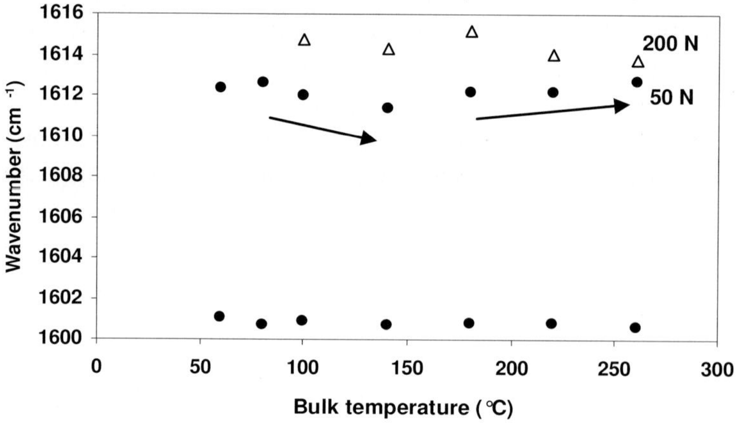

(a)

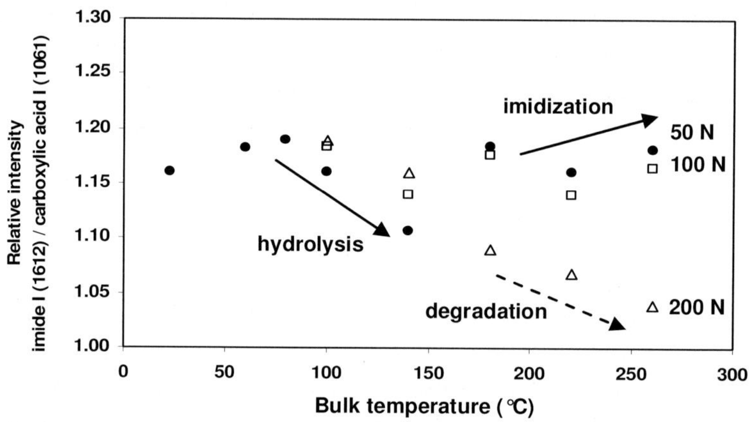

(b)

Fig. 14. Raman analysis of imide aromatic band (1612 cm^{-1}) and carboxylic acid band (1601 cm^{-1}) for thermosetting polyimide, (a) wavenumber position, (b) relative intensity.

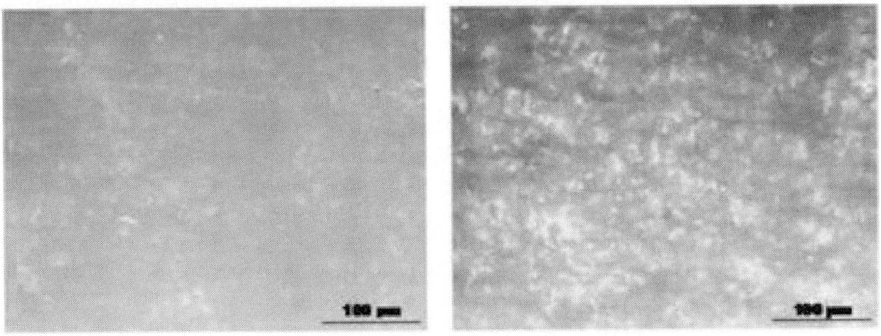

(a) (b)

Fig. 15. Optical microscopy for thermosetting polyimide surfaces after sliding at different temperatures, (a) 100 to 180°C no imidization, (b) 180 to 260°C with imidization.

(a) (b)

Fig. 16. Thermal effects on sliding surfaces of thermoplastic polyimide, (a) softening at surface temperatures T* = 228°C, (b) melting at surface temperatures T* = 320°C.

Thermal degradation is seen in the Raman spectra for thermoplastic polyimides after sliding at different temperatures (Fig. 17). Most significant differences in the spectra for sintered and thermoplastic polyimide wear surfaces are found at 1612 cm^{-1} bands, disappearing above 180°C. The positions and relative intensities of the 1612 cm^{-1} and 1601 cm^{-1} band (Fig. 18) indicate that the imide part for thermoplastic polyimide degraded severely and finally decomposes at 180 to 260°C through thermal degradation (while imidization happened for sintered polyimide at 180 to 260°C).

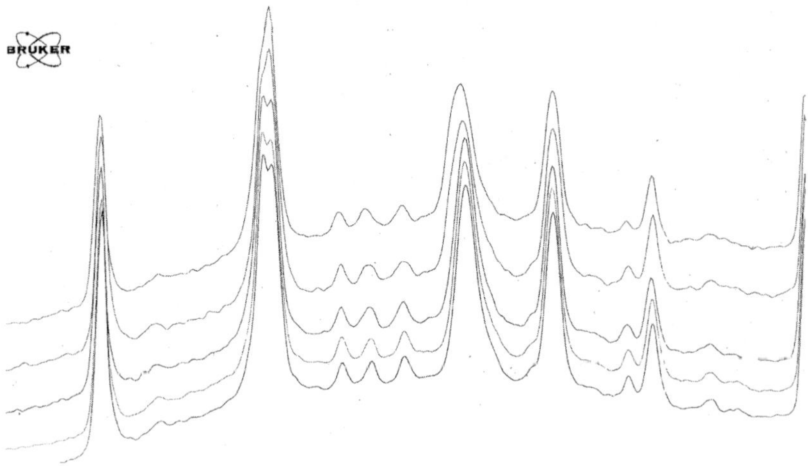

Fig. 17. Raman spectra of thermoplastic polyimide surfaces after sliding at different bulk temperatures and 50 N normal load, 0.3 m/s sliding velocity.

Thermal degradation at 180 to 260°C agrees to melting and unstable friction (Fig. 19). Thermal decomposition results in the reversion of polyimide into its monomers. The formation of CONH amide groups is confirmed by a shoulder around 1350 cm^{-1} at 220 and 260°C. The extrapolation of relative intensities I_{1612}/I_{1601} learns that imide completely disappears at 590°C, the decomposition temperature according to thermogravimetric analysis. Thermal degradation of aromatic structures into monomers weakens strength and results in high wear rates. The C=O stretching mode for thermoplastics varies stronger than for thermosetting polyimide[33,34]. The Raman shifts do not develop continuously, but they decrease mainly above 180°C in parallel to a transition in friction and wear regime. After sliding, the thermoplastic structure evolutes towards the thermosetting structure and it becomes equal for a sample hypothetically slid at 400 to 450°C. When thermoplastic polyimide is heated under a static conditions at 400°C in a hot air-oven, the position of the C=O band at 1788 cm^{-1} decreases relatively to the original material due to degradation. The intensities measured from combined thermo-mechanical actions are lower and indicate the influence of additional dynamic orientation effects.

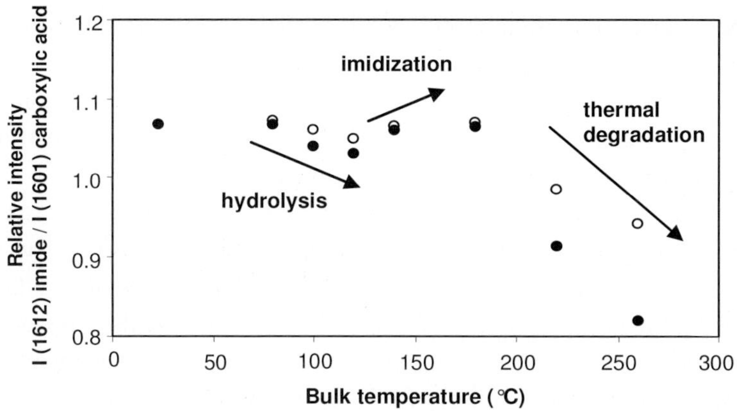

Fig. 18. Raman analysis of imide aromatic band (1612 cm^{-1}) and carboxylic acid band (1601 cm^{-1}) for thermoplastic polyimide.

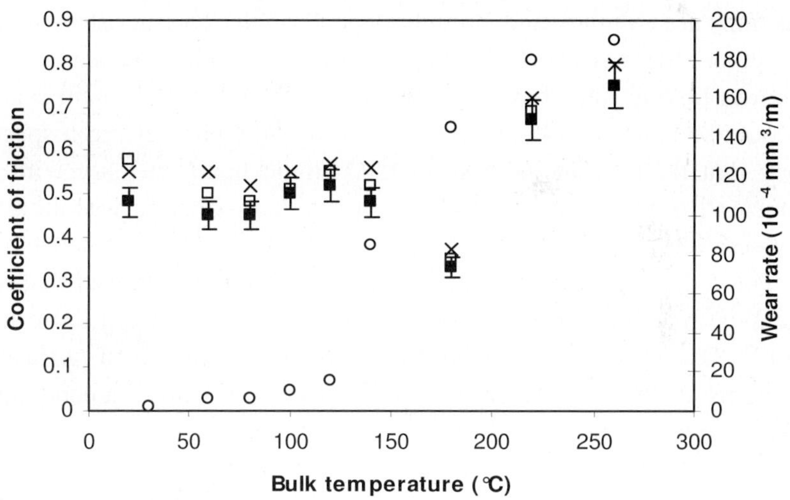

Fig. 19. Coefficients of friction (after x 30 m, □ 100 m and ■ 15000 m) and wear rates (○) for thermoplastic polyimide at different bulk temperatures and 50 N, 0.3 m/s.

3.6. *Physico-Chemical Degradation*

Physico-chemical changes involve reversible or irreversible degradation in molecular structure (depolymerization) due to the combined action of mechanical and chemical effects. Two important modifications that influence the sliding behaviour involve chain scission (e.g. for polyacetals) and hydrolysis (e.g. for polyimides). The latter reactions occur at temperatures far below the glass or melting temperature and constitute secondary β-transitions that are not clearly observed in tensile testing but that manifest clearly in shear (or friction) tests.

Referring to the spectra for polyacetal in Fig. 13, an increase in relative intensity I_{1480}/I_{930} from 0.45 to 0.51 is noted for surfaces worn at 150 MPa compared to unworn samples. The higher intensity of O-CH_3 and O-CH_2 groups relatively to the ether C-O-C bands suggests the degradation of ether links by chain scission and radical reactions with the environment into the formation of CH_3 end groups during sliding at high contact pressures. These reactions result in the formation of a brown sliding film on top of the polymer surface leading to low friction and, however, high wear due to degradation of the polymer structure. Radical reactions are complex and strongly depend on the polymer structure. For polyether-ether-ketone (PEEK) it was found that other mechanisms such as cross-linking must have accompanied or followed the chain scission. It could be concluded in combination with X-ray photoelectron spectroscopy that there is an increase in C-O single bond intensity with load compared to C=O intensity, which proves that oxidation was an important step during the wear test. Moreover, a study on the influence of the detection depth relatively to the observed phenomena showed that tribochemical reactions on the worn surfaces are different from those in the bulk[35]. Oxidation of the polymer is evident on the worn surface but reactions in the subsurface several micrometers below depend on the migration of free radicals. At lower pressures, the free radicals formed by scission should terminate quickly before transferring much. At higher loads, a higher mobility of the chain segments might make it easier for radicals to propagate and form branched, or even cross-linked structures in the sub-surface. The radical reactions have to end through recombination and/or disproportionation (transfer of substituents) at a certain surface depth that depends on a critical load allowing for mobility of the molecular segments. At even higher loads, however, scission of the polymer chains by mechanical stresses may become difficult because of disentanglements. Mechano-chemical transitions are thus driven by the activation of chain segments through frictional heating relatively to the glass transition. The above mentioned radicals may also interact with metal-oxides on the counterface to generate metal-organic compounds. Fluoride radicals or ions generated from the breakdown of polytetrafluoroethylene molecules strongly react with metal oxides or fresh metal to form metal fluorides that can be detected by XPS[5]. In the primary friction stage, the adherence of transfer substance to the counter-face strongly depends on the surface energy and mechanical embedment. This is a necessary condition for transfer film formation. If the polymer molecule chains can slide easily and the transfer film is chemically bonded to the counterface, then the transfer film would be difficult to destroy. The chemically bonded points between the polymer and the counterface increase gradually as rubbing goes on. Then, the adhesive strength of the transfer film to the counterface and the transfer film area both increase. This is a sufficient condition for transfer film formation.

For highly aromatic polymer chains with high stiffness, such as polyimide, chain mobility becomes less important while chemical degradation may occur through intramolecular ring-opening reactions[36]. Referring to Fig. 18, the downward shift in the 1612 cm^{-1} wavenumber at 100 to 140°C for a constant 1601 cm^{-1} position indicates hydrolysis. The lower frequencies represent deterioration of the imide structure, partially changing into the polyamic acid precursor. Hydrolysis at 100 to 140°C is also reflected in the relative intensity I_{1612}/I_{1601}. A maximum in hydrolysis intensity at 140°C is representative at different normal loads and corresponds to minimum wear rates with the acid groups acting as a lubricant. Minimum wear rates, however, do not correspond to minimum coefficients of friction because they the high polarity of COOH groups enlarges the adhesion component of friction. No transfer film is observed for the hydrolyzed polyimide structure, as low interaction between the wear debris particles results in limited strength. The molecular mobility has changed in the hydrolyzed structure, mainly affecting the C=O functional groups (Fig. 9a). Comparing friction at variable humidity, polyimides have high friction at high relative humidity and low friction at low relative humidity. It was similarly demonstrated that polyimides have high friction at low temperatures while it lowers above a critical temperature. Friction tendencies at high temperature are thus similar to low humidity. Chemical reactions as a function of temperature are thus in parallel revealed as a function of relative humidity.

4. Polymer Wear Debris Analysis

Thermo-analytical DTA/TGA analysis of wear debris particles for sintered and thermoplastic polyimide is made to illustrate changes in thermal stability, weight loss and position of endotherm reactions (Fig. 20). Measurements indicate that debris particles imidized in the interface during the frictional process at T* > 180°C. The structural variations for thermoplastic polyimide are more important than for sintered polyimides and correlate well to different transitions in friction and wear.

For sintered polyimide (Fig. 20a), no significant thermal degradation of wear debris is noted: the weight loss at each sliding temperature is restricted to 2% when heating to 450°C and it is 18% when heating to 590°C. The endotherm peak temperature or dehydration temperature T_{hydrat} (maximum dehydration intensity) shifts towards higher temperatures for wear debris relatively to the unworn SP-1. The dehydration reaction of wear products depends on sliding temperature and normal loads: the dehydration temperature generally increases at higher sliding temperatures while this trend is stronger at high normal loads; at low loads, a critical bulk temperature of 180°C must be exceeded to increase T_{hydrat}. The upward shift indicates chemical changes in wear debris structure after sliding, causing a delay in dehydration. When a modified polyimide network develops under sliding, higher temperatures or activation energy is needed for dehydration. This process remains reversible during subsequent heating-cooling-heating cycles. Imidized debris strongly interacts and allow for the formation of a transfer film.

For thermoplastic polyimides (Fig. 20b), structural changes mainly occur in the amorphous phase by crystallization and/or cross-linking. The shift in dehydration temperature at 180°C noted for sintered polyimides is less clear for thermoplastics through interference with complex phase changes. Thermal modifications have the upperhand for thermoplastics[37].

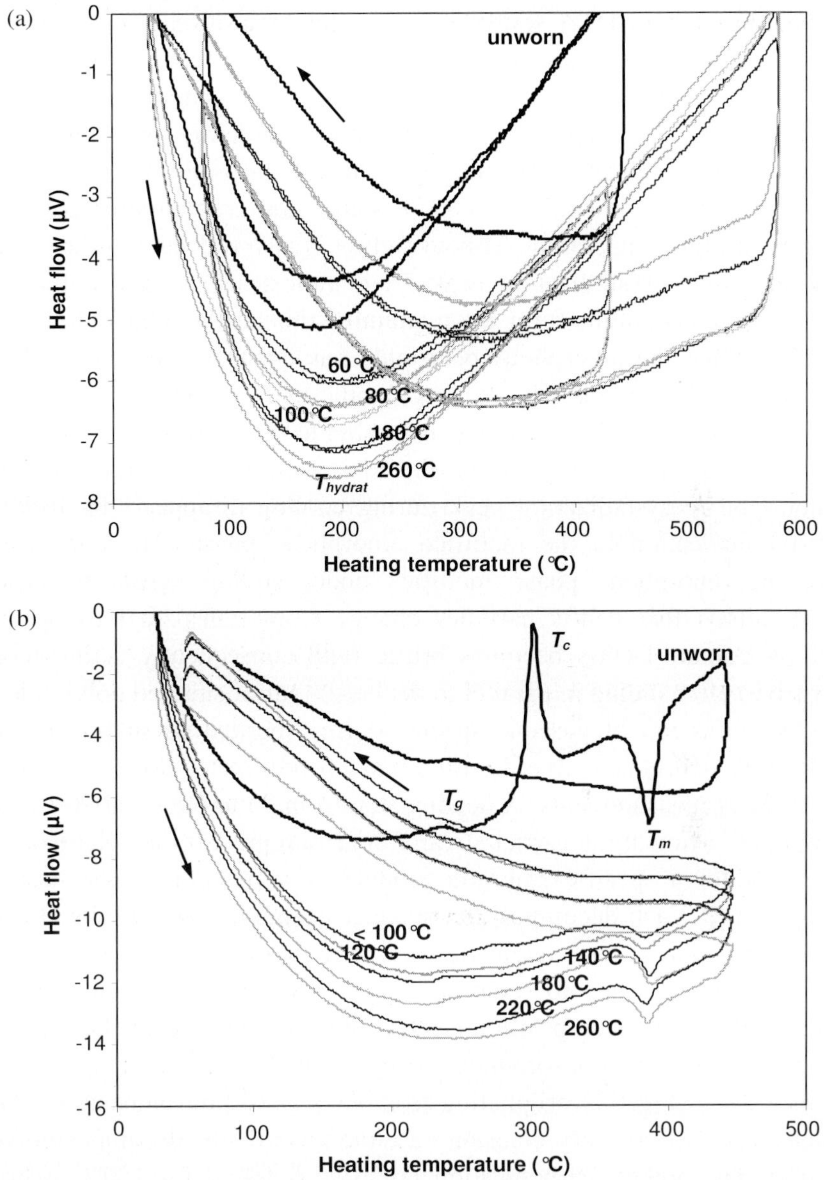

Fig. 20. Differential thermal analysis (DTA) of wear debris after sliding at different bulk temperatures and 50 N, 0.3 m/s, (a) thermosetting polyimide, (b) thermoplastic polyimide (note that the curves are shifted over a constant µV value for visualization).

A heating step in DTA shows that the glass transition temperature T_g increases or finally disappears when it becomes smoothened over a broad interval from 230 to 280°C. The temperature T_g is representative for linear molecular structures in an amorphous ordering and shifting or disappearance indicates that the amorphous zone is affected by cross-linking into the formation of a more stable and ordered structure after sliding at high temperature. This evolution is also reflected in the disappearance of a crystallization peak T_c: it indicates that most polyimide debris has a crystalline form when attaining this heating temperature and confirms that the amorphous polyimide has changed into a crystalline structure under sliding. The melting exotherm T_m decreases in intensity and its maximum value increases in parallel to the formation of a crystalline structure and strongly cross-linked structures with higher melting point. The recrystallization peak during cooling disappears by lacking crystal nucleation in the modified amorphous phase. The conclusion that the amorphous phase modifies under sliding agrees to visual observations that debris particles change from transparent to opaque yellow colour. Debris becomes brittle (and consequently acting more abrasive) after sliding in parallel to the behaviour of sintered polyimides.

The emission of gaseous species during thermal decomposition of polyimide debris is analyzed with a mass spectrometer (MS) coupled to DTA/TGA measurements. A heating step from 23 to 600°C at 20°C/min is applied, using argon carrier stream. Spectra represent the volatilization intensity for a specific emission product as a function of the heating temperature. Each decomposition product is characterized by its atomic mass unit (*a.m.u.*) and intensities are related to the ion current in the mass detector (scaled to sample weight), which is proportional to the concentration of decomposition product in the carrier stream. The totality of gaseous degradation products is represented by the *carbon* fraction (*a.m.u.* = 12, Fig. 21) originating from CO_2 or CO decomposition. The intensities have a nearly constant baseline towards the decomposition at 590°C. The highest carbonization intensity is found for 180°C debris, indicating that structures worn above 180°C become stable by imidization during sliding.

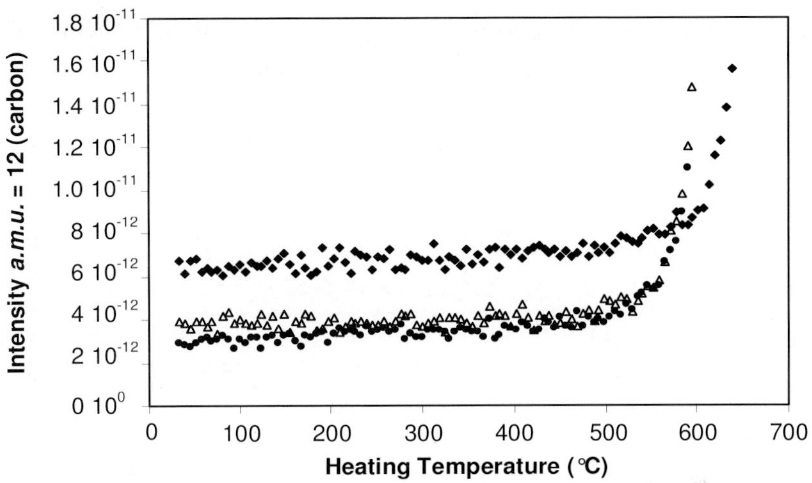

Fig. 21. Mass spectroscopy for degradation of polyimide wear debris related to carbon fraction (a.m.u. = 12) for • 100°C debris, ♦ 180°C debris, Δ 260°C debris.

Most important decomposition fraction for tribological performance is the production of *water* (*a.m.u.* = 18, Fig. 22), occurring at 180 to 200°C heating temperatures as a peak intensity for the debris samples after 100°C sliding or a maximum intensity for debris samples after 180 and 260°C sliding. Water volatilization at 200 to 600°C heating depends on the sliding temperatures, showing: (i) a constant or increasing water volatilization for 100°C debris indicating progressive imidization during heating, and (ii) a decreasing trend for 180 to 260°C debris indicating that debris is more inert during heating because of the imidization already happening during sliding. The peak in water emission for high temperature wear debris broadens through variations in structure (and molecular weight) after wear. Water volatilization agrees to the previously noted shift of dehydration temperature T_{hydrat} in DTA thermographs, rising for debris after 180 to 260°C sliding. The *nitrogen-monoxide NO* fraction (*a.m.u.* = 30, Fig. 23) has similar features to *nitrogen N* fraction (*a.m.u.* = 14). It is firstly stressed that the concentration of released nitrogen is a factor 10^{-2} to 10^{-3} smaller than water concentrations. The decomposition of polyimide wear debris into nitrogen monoxide is postponed and finally disappears with increasing sliding temperatures through formation of a strong imide structure after sliding. The nitrogen decomposition for polyimides occurs over a relatively long interval because it is a multiple-step process.

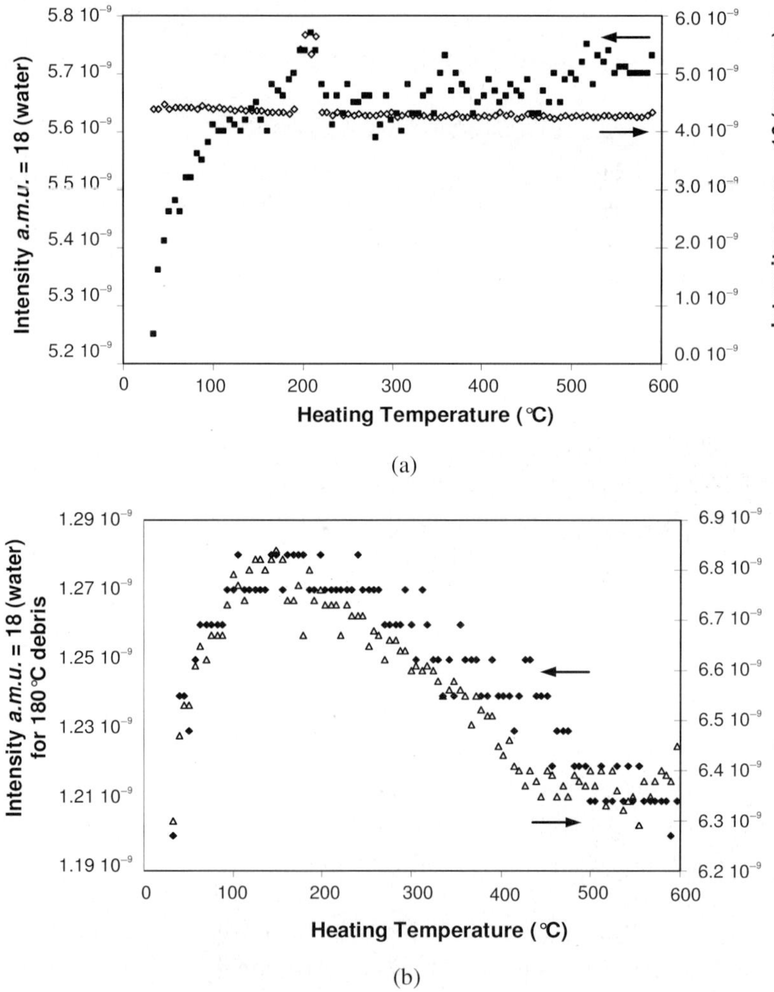

Fig. 22. Mass spectroscopy for dehydration of polyimide wear debris related to water fraction (a.m.u. = 18) and oxygen fraction (a.m.u. = 16) for (a) 100°C debris, (b) 180°C (♦) and 260°C (Δ) debris.

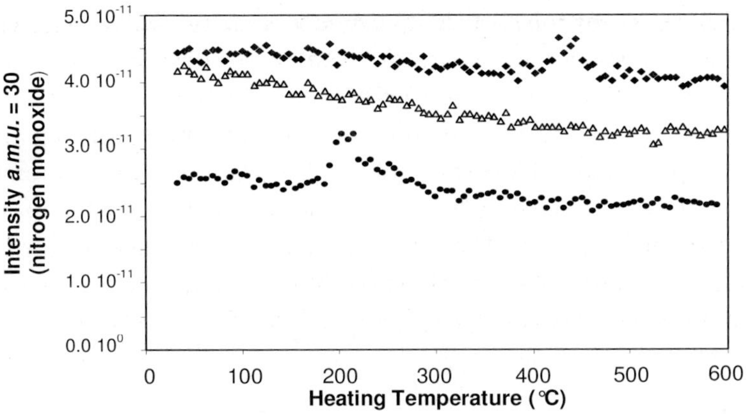

Fig. 23. Mass spectroscopy for degradation of polyimide wear debris related to nitrogen monoxide fraction (a.m.u. = 30) for • 100°C debris, ♦ 180°C debris, Δ 260°C debris.

5. Conclusions

An overview of tribophysical phenomena on polymer surfaces under wear and generated debris is presented, differentiating between mechanical interaction, thermal influences and chemical reactions in the interface. Such reactions were in general hypothetically assumed in literature, but they are more specifically demonstrated in this work.

Raman spectra of worn polymer surfaces show variations in polymer structure agreeing with transitions in friction and wear behaviour. Different sliding mechanisms are illustrated, such as polymerization or depolymerization reactions, radical reactions, degradation reactions, crystallization, orientation of side-groups or polymer backbone, orientation of a rigid amorphous phase, conformational changes and deposition of internal solid lubricant. Orientation for thermosetting polyimide is mainly concentrated on side groups at low load and on the molecular backbone at high load. Orientation for thermoplastic polyimide is located on the molecular backbone at low loads. Crystallization from γ into α structures is noted for polyamides. Orientation and crystallization for polyesters is sensitive to internal lubricants and applied counterface type: high friction causes higher molecular orientation. The orientation in a rigid amorphous phase is most favourable for smooth sliding.

Polymerization reactions such as imidization or polycondensation improve homogeneous transfer films, while crystallization often causes unstable sliding.

Thermal analysis of polymer wear debris indicates significant structural variations under sliding. For polyimides, DTA/TGA measurements agree with Raman analysis of the sliding surfaces and indicate hydrolysis and imidization reactions. The stepwise chemical degradation for thermosetting polyimides and a combination of thermal and chemical effects in the amorphous phase for thermoplastic polyimides are illustrated.

References

1. S. Bahadur, *Wear*, **245**, 92 (2000).
2. D. Tabor, in: *Surface physics of materials*, Ed. J.M. Blakely JM (Academic Press, New York, 1980).
3. L.S. Bark, *Wear*, **41**, 309 (1977).
4. J.M. Martin, Proceedings of International Tribology Conference, Kobe (2005).
5. J. Gao, S. Mao, J. Liu and D. Feng, *Wear*, **212**, 238 (1977).
6. P.J. Hendra C. Jones and G. Warnes (1991), Fourier transform Raman spectros-copy: instrumentation and chemical applications (Ellis Horwood, Chichester, 1991).
7. W.F. Maddams, *Spectrochim. Acta*, **50 A**, 1967 (1994).
8. C.V Raman and K.S. Krishnan, *Nature*, **121**, 501 (1928).
9. I.R. Lewis, *Handbook of Raman spectroscopy*, Ed. H.G.M. Edwards (Marcel Dekker, New York, 2001).
10. B.H. Stuart, *Tribol. Internat.*, **31**, 687 (1998).
11. J. Ge, G. Xue, F. Li, K. McCreight, S. Wang, F. Harris, S. Cheng, X. Zhuang, S. Hong and Y. Shen, *Macromol. Rapid Commun.*, **19**, 619 (1998).
12 V.B. Gupta, J. Radhakrishnan and S.K. Sett, *Polymer*, **34**, 3814 (1993).
13. G. Ellis, C. Marco, J. Del Pino, J. Lorente, M.A. Gomez and J.G. Fatou, *Vibratianal Spectrosc.*, **9**, 46 (1995).
14. P. Samyn, J. Van Craenenbroeck, J. Quintelier and G. Schoukens, *Surf. Interf. Anal.*, **38**, 868 (2006).
15. G. Schoukens and K. De Clerck, *Polymer*, **46**, 845 (2005).
16. J.C. Rodriguez-Cabello, J.C. Merino, L. Quintanilla and J.M. Pastor, *J. Appl. Polym. Sci.*, **62**, 1953 (1996).
17. T. Lippert, E. Ortelli, J.C. Panitz, F. Raimondi, J. Wambach and A. Wokaun, *J. Appl. Phys.*, **A 69**, 651 (1999).
18. J.M. Constantini, F. Couvreur, J.P. Salvetat and S. Bouffard, *Nucl. Instr. Meth. Phys.*, **B 194**, 132 (2002).
19. M. Fukuhara, *J. Appl. Pol. Sci.*, **90**, 759 (2003).

20. P. Samyn, dissertation, Ghent University (2007).
21. N. Vansanthan and D.R. Salem, *J. Polym. Sci.*, **B 39**, 536 (2001).
22. L. Lin and A.S. Argon, *Macromolecules*, **27**, 6903 (1994).
23. G. Rotter and H. Ishida, *J. Polym. Sci.*, **B 30**, 489 (1992).
24. L. Penel-Pierron, C. Depecker, R. Seguela and J.M. Lefebvre, *J. Polym. Sci.*, **B 39**, 484 (2001).
25. L. Penel-Pierron, R. Seguela, J.M. Lefebvre, V. Miri, C. Depecker, M. Jutigny and J. Pabiot, *J. Polym. Sci.*, **B 39**, 1224 (2001).
26. P. Samyn, I. Van Driessche and G. Schoukens, *J. Polym. Research*, in press (2007).
27. F. Raimondi, S. Abolhassani, R. Brutsch, F. Geiger, T. Lippert, J. Wambach, J. Wie and A. Wokaun, *J. Appl. Phys.*, **88**, 3659 (2000).
28. K.H. Yu, Y.H. Yoo, J.M. Rhee, M.H. Lee and S.C. Yu, *Bull. Korean Chem. Soc.*, **24**, 357 (2003).
29. A. Cossy-Favre, J. Diaz, Y. Liu, H.R. Brown, M.G. Samant, J. Stöhr, A.J. Hanna, S. Anders and T.P. Russell, *Macromolecules*, **31**, 4957 (1998).
30. J. Ge, C. Li, G. Xue, I.K. Mann, D. Zhang, S.Y. Wang, F.W. Harris, Z.D. Cheng, S.C. Hong, X. Zhuang and Y.R. Shen, *J. Amer. Chem. Soc.*, **123**, 5768 (2001).
31. S.H. Park and C.J. Durning, *J. Appl. Phys.*, **83**, 1270 (1998).
32. K. Sakamoto, R. Arafune, N. Ito, S. Ushioda, Y. Suzuki and S. Morokawa, *J. Appl. Phys.*, **80**, 431 (1996).
33. H. Ishida and M.T. Huang, *J. Pol. Sci. Pol. Phys.*, **32**, 2271 (1994).
34. H. Ishida and M.T. Huang, *Spectrochim. Acta*, **A 51**, 319 (1995).
35. T.Q. Li, M.Q. Zhang, L. Song and H.M. Zeng, *Polymer*, **40**, 4451 (1999).
36. M. Oh-e, D. Kim and Y.R. Shen, *J. Chem. Phys.*, **115**, 5582 (2001).
37. X.G. Li, M.R. Huang, H. Bai and Y.L. Yang, *Macromol. Mater. Eng.*, **286**, 421 (2001).

CHAPTER 3

SCALING EFFECTS IN TRIBOTESTING OF POLYMERS

Pieter Samyn[1], Gustaaf Schoukens[1] and Jan Quintelier[2]*

[1]*Department of Textiles, Ghent University*
Technologiepark 907, B-9052 Zwijnaarde, Belgium
[2]*Department of Mechanical Construction and Production, Ghent University*
Sint-Pietersnieuwstraat 41, B-9000 Gent, Belgium
**E-mail: Jan.Quintelier@UGent.be*

Correlations between meso-, small-, large- and full-scale tribotesting of engineering polymers are discussed in this chapter. A common parameter characterizing tribological data is the "pv-value" (product of contact pressure and sliding velocity), but its use is restricted to a single test scale. The influence of visco-elasticity and thermal effects should be taken into consideration. Three experimental extrapolation models are presented with up to two mechanical parameters. A macroscopic geometry model considers thermal effects, sample geometry and contact conditions. Coefficients of friction for bulk polymers can be extrapolated from small-scale to large-scale tests. Extrapolation of wear rates is impossible due to transitions between mild wear, softening and melting. Tribological data for solid-lubricated or internally oil-lubricated polymers are more difficult to extrapolate due to interactions of lubricant supply mechanisms with softening, melting and deformation. Different tribological behaviour obtained at different test scales is due to stress concentrations and debris mobility, which is confined within large contact areas and promotes film formation on the counterfaceor polymer s urface. Small-scale tests at high-temperature are irrepresentative for large-scale tests at high load due to variations in polymer structures: orientation concentrates in either a crystalline phase (high temperature) or a rigid amorphous phase (high load). Debris polymerization mainly improves the homogeneity of the transfer film at long resident time. Specifically designed large-scale tests agree with full-scale performance of bearing elements, as verified by finite element modeling.

1. Introduction

Tribological properties are not intrinsic material properties, in contrast to strength, elasticity modulus or other mechanical properties. Friction and wear performance is strongly influenced by the entire test system[1] (material pair, contact geometry, normal load, sliding velocity, mechanical stiffness, etc.). For analysis of polymer elements in relation to working conditions occurring in real constructions, test conditions should therefore simulate the dominant sliding mode as close as possible, including the structures of the material sliding couple, contact geometries, contact pressures, type of sliding motion, environmental conditions, mechanical stiffness, etc. As a result, many tribotest configurations are not standardized and customized test equipment is often built. The following considerations should be made to select a test program[2].

Material structure: The macro- and microstructure of the two contacting bodies (polymer wear sample and steel counterface) should agree with practical applications. For microstructures being small relatively to the contact surface, the material behaves homogeneously and test results do not strongly depend on the contact size. For large microstructures such as internally lubricated polymers and composites, the contact size influences test results importantly. Surface finishing by sand blasting, grinding or polishing should be adequately chosen with representative roughness. The surface conditions are often different at running-in compared to steady-state conditions due to wear and/or transfer[3].

Contact conditions: Conformal contacts (bodies with similar radius of curvature, e.g. flat-on-flat in pin-on-disc) and counter formal contacts (bodies with different radius of curvature, e.g. line and point contact in cylinder-on-plate or ball-on-disc) influence the contact pressure distribution over the contact area[4] and the mobility of wear debris in the sliding interface[5]. Debris particles can be immediately removed or interact to form a polymer transfer film. When wear debris remains circulating into the contact, it is called a closed tribosystem such as fretting contact[6]. When wear debris is immediately removed from the sliding contact, it is called an open tribosystem such as gear contact[7].

Sliding motion and sliding velocity: Unidirectional, reciprocating or multi-axial sliding motions and the distance of the sliding motion relative to the contact length (mutual overlap coefficient)[8] influence the wear debris mobility and dissipation of frictional heat. When contact is intermediately disrupted, chemical reactions with the atmosphere, cooling and/or relaxation of polymer samples influence friction and wear properties. The sliding velocity influences the visco-elastic behaviour of polymer molecules.

Energy dissipation: Friction and wear result from the dissipation of mechanical energy which is transformed into deformation (hysteresis) and frictional heating. Sliding temperatures predominantly influence the sliding performance of polymers due to transitions from the glassy into rubbery and from the rubbery into melting phase. The sliding properties for polymers are generally characterized by a *pv* sliding condition (contact pressure x sliding velocity)[9]. Additional heating or cooling of the contact surfaces is often required as dissipation of energy partially depends on the test-rig construction acting as a heat sink.

The most reliable testing of polymer components involves the real tribosystem, while it often cannot be tested due to high cost and practical obstructions (e.g. measuring accuracy, continuity of production process, no possibilities for parametric research). Design of specific equipment for component tests is only justified for mass production (e.g. engines). A scheme for scaling tribological research was given by Czichos[10], going from 'field tests' over 'large-scale simulation tests' on real components to 'laboratory tests' on artificial samples with simple geometry and numerical modeling. Mainly the importance of running-in conditions was stressed. An equivalent input of mechanical and thermal energy should then be provided on every test scale. This rule, however, seems too general for polymers as transitions in wear mechanisms can change conditions from being mild to severe. Decreasing testing time by artificially increasing the severity of one sliding parameter (e.g. contact pressure, sliding velocity or sliding temperature) is not an option, as the predominant polymer wear mechanism then transforms from mild (e.g. softening) into severe (e.g. melting) conditions. Also, the size of the contact area of test samples influences transfer film formation and is difficult to scale. The entrapment of wear debris particles in the sliding interface and film formation is strongly determined by the test configuration.

2. Tribotesting Facilities at Different Scales

Tribological data in existing literature pertaining to polymer/steel sliding are most commonly obtained from laboratory tests with standardized sample geometries, and provide fundamental information on friction and wear mechanisms. These methods, however, are only useful for a general classification of materials, and result in important errors and premature failure when extrapolated towards real working conditions[11]. In order to make a correct material selection, the consultancy of tribological literature or technical data based on small-scale tests is often not satisfying, since the layout of a tribological system and its test parameters have a strong influence on the material's sliding performance. Real criteria for material selection in a given application require the tribological characterization of polymer parts under conditions that are closely related to its practical functionality. Appropriate tribological data is often missing due to the lack of coherent research and experimental facilities for friction and wear tests on large-scale test specimens. A broad range of test-rigs as presented below should include different sliding applications.

2.1. *Meso-Scale Testing*

Meso-scale testing is in between traditional small-scale testing (typical normal loads ranging from 10 to 100 N) and nano-scale testing (typical normal loads below 1 N). Meso-scale testing is well-suited for investigating frictional stability of pure and internally-lubricated polymers, as thermal effects can be neglected through low contact pressures (in contrast to macro-scale testing) while the surface properties are taken into account through large contact areas (in contrast to nano-scale testing).

An example of a unidirectional *meso-scale flat-on-flat* tribotester is shown in Fig. 1. This tester has a circular steel counterface with diameter 60 mm that is used as a moving specimen and driven in a smooth and linear motion by a DC motor. A stationary rectangular polymer plate (30 mm parallel to sliding direction x 20 mm perpendicular to sliding direction) with thickness 1.5 mm is placed on top of the metal part. A normal load is applied on the polymer by dead weights. The polymer specimen is connected to the load cell with two thin metallic wires, monitoring the friction force. The normal loads vary between 1 and 5 N and the sliding velocity is low between 0.12 and 20 mm/s. Extremely low loads are applied in order to simulate nearly elastic flat-on-flat contact during one single sliding step. Stick-slip phenomena occurred at low sliding velocities and a quasi-harmonic oscillation happened at relatively high sliding velocities.

Meso-scale testing requires accurate motion at low speeds. The motion control system typically has a maximum stroke of 50 mm, maximum velocity of 50 mm/s, resolution less than 0.5 μm and repeatability less than 1 μm. A load cell (Futek L2357) with a range of 0 to 1 N (100 g), sensitivity of 1.5 mV/V, resolution of 0.001 N and non-linearity of 0.1%, is used. The friction force is recorded throughout the experiment by means of a data-acquisition card device (Multifunction I/O PCI-6035E from National Instruments Corporation). In order to investigate the transient friction behaviour, it is necessary to measure static and dynamic friction force. Assuming a hypothetical minimum coefficient of friction about 0.1 under dry sliding conditions, it results into a friction force of 0.5 N or less. This means that a precision load sensor with high accuracy is needed. The displacement of the bottom plate in the direction of movement of the drive unit is measured with a non-contact displacement sensor (Keyence LC-2440/LC-24000W) with a range of 5 mm and a resolution of 0.2 μm.

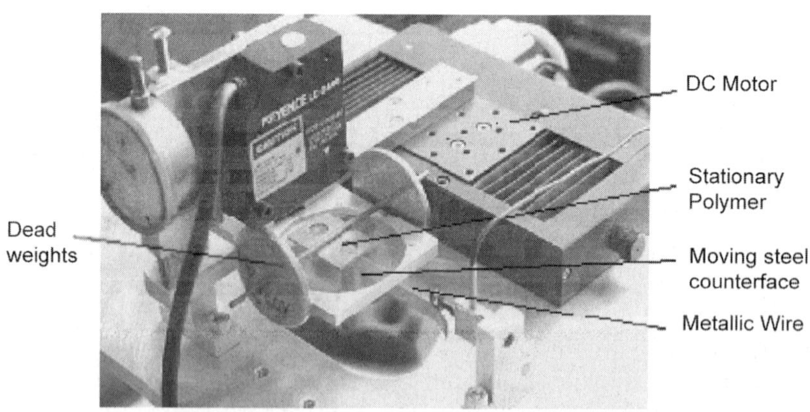

Fig. 1. Example of meso-scale test-rig.

2.2. Small-Scale Testing

In conventional tribotesting, small-scale tests are preferred due to cost- and time-effectiveness with easy handling of test specimens (apparent contact area 1 to 10^2 mm^2). An appropriate test set-up can be selected according to ASTM and some DIN standards for each wear mode. The National Physics Laboratory (UK) has identified more than 400 world-wide test methods with scientific or industrial interest[2]. Under adhesive conditions, pin-on-disc[12] (ASTM G99) or pin-on-plate[13] (ASTM G133) tests are mostly used in continuous or reciprocating motion. Block-on-ring[14] (ASTM G88) tests are applied for line contact as occurring in gears. Also small component tests are available simulating strong conformal contacts (Fig. 2).

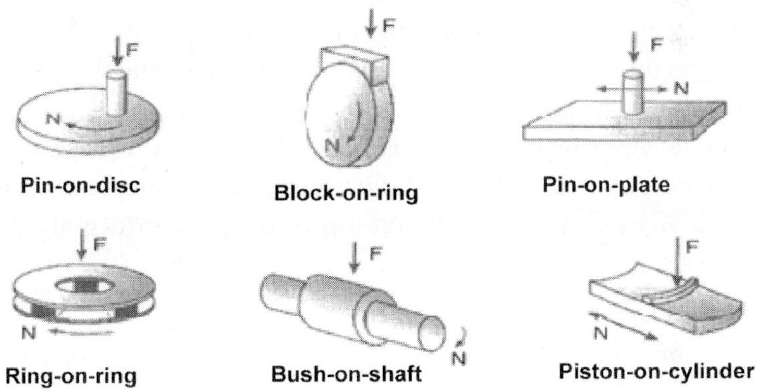

Fig. 2. Different conformations for small-scale testing.

A cylinder-on-plate tester used for present studies is schematically shown in Fig. 3. This tester is a universal PLINT TE 77 High Frequency tribotester for dry reciprocating sliding between a moving polymer cylinder and a fixed counterface plate. A line contact is chosen for easy alignment of the sliding pair and high contact pressures at running-in. The initial contact geometry is counter-formal, but the contact area increases with wear. This geometry favors polymer transfer during running-in. The evolution in contact geometry from line towards flat contact significantly influences friction and wear, stabilizing in parallel with the leveling of contact pressure[15]. Polymer cylinders (diameter

5 mm x length 15 mm) are mounted into a reciprocating moving arm, preventing it from rolling during the test so that pure sliding is guaranteed against a stationary steel specimen (58 mm x 38 mm x 4 mm). The oscillating motion of the arm is provided through an eccentric transmission, converting the continuous rotational motion of a DC electric motor into linear motion which provides the sliding stroke. The sliding velocity is varied between 0.3 to 1.2 m/s. Normal loads from 0 to 200 N are manually applied through a spring and lever system pulling down a transverse bridge structure. The stationary steel counterface is fixed onto a plate holder that is connected to the base plate by means of four leaf springs with high stiffness in vertical direction and appropriate flexibility in horizontal direction. A piezoelectric force transducer (range +/- 500 N, resolution 0.001 N) in contact with the stationary steel plate measures the horizontal friction force that the polymer specimen exerts on the counterface. A contactless inductive displacement transducer is positioned on top of the transverse bridge, measuring its displacement as a combined effect of wear and deformation of the polymer sample. Sliding temperatures are measured by a DIN 43710 K-type thermocouple on top of the counterface at the outer ends of the stroke.

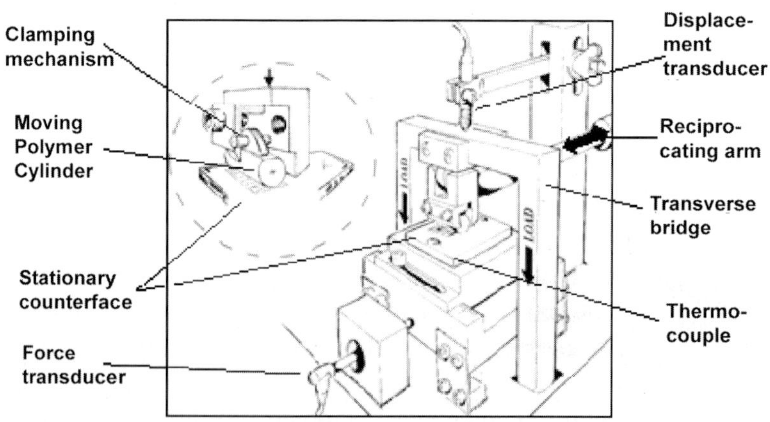

Fig. 3. Small-scale cylinder-on-plate testing configuration.

2.3. *Large-Scale Testing*

Large-scale wear samples (apparent contact area 10^4 to 10^5 mm^2) reduce the effect of stress concentrations near the sample edges due to a high bulk-to-edge material ratio, in addition to being sensitive to the effect of debris mobility into the sliding interface. Also in our experience, sliding instabilities, which are hidden at small-scales, show more frequently in large-scale tests[16]. As such, the 'real wear life' of polymer components can be determined more accurately through large-scale tests which are more representative to the actual working conditions. The disadvantages associated with these tests, however, is the requirement for test specimens with large dimensions and test equipment with high stiffness. The manipulation of these tests specimens will also be increasingly difficult as their weight increases.

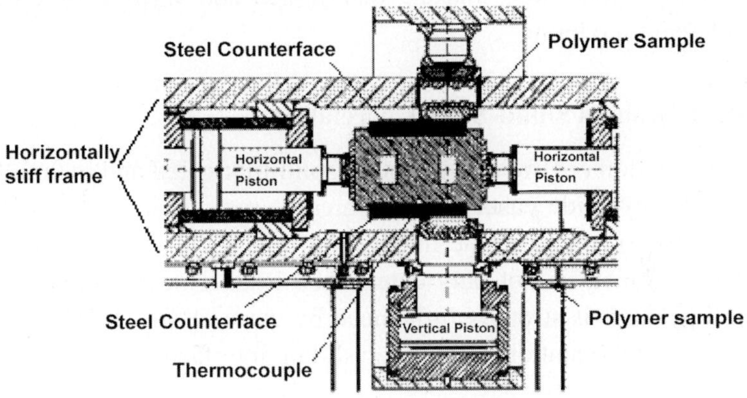

Fig. 4. Large-scale flat-on-flat testing configuration.

A tribotester for high-load testing of large specimens is schematically illustrated in Fig. 4. Two identical sliding couples (each consisting of a 150 x 150 x 20 mm^3 polymer specimen and a 410 x 200 x 20 mm^3 counterface steel plate) positioned at the top and bottom of a central sliding block, are tested simultaneously in a flat-on-flat linear reciprocating configuration. Normal loads up to 6000 kN are applied by the jack located in the vertical column. The central sliding block provides the reciprocating motion and is driven by two horizontal jacks on the left and on the right with maximum capacity of 2500 kN. Depending on the sliding direction, one of the jacks provides a counter pressure to increase the rigidity of the tester. The sliding velocity is controlled by a hydraulic circuit and is almost constant at 0.005 mm/s over the sliding stroke, except near the edges where the sliding direction reverses. These sliding conditions are typically found in bearing tests for off-shore construction[17].

2.4. *Field Testing*

Polymer components used in a bearing often perform as an integrated part of the entire bearing construction. Thus in addition to good tribological characteristics (i.e. dynamic properties), dimensional stability (i.e. static properties) should be ensured. The friction, wear and strength properties for an isolated part of the construction should be tested under representative conditions that are determined and verified by means of finite element modeling[18].

3. Meso-Scale to Small-Scale Correlation

Dynamic friction coefficients for pure, solid-lubricated and internally oil-lubricated polyamide casts are compared at meso-scale and small-scale as a function of the *pv*-value in Fig. 5 to Fig. 7. It is observed that the friction coefficient is not only determined by the mechanical *pv* sliding conditions, but it is strongly influenced by visco-elastic deformation and efficiency of lubricant disposal in the sliding interface.

3.1. *Influence of Deformation and Contact Situation*

There is good agreement in the relative classification of pure polyamide grades on both meso- and small-scale testing. As shown in Fig. 5, PA-Na (sodium-catalyzed) has higher friction than PA-Mg (magnesium-catalyzed) attributed to differences in mechanical material properties. While both materials have similar surface energies, the lower Young's modulus and tendency of plastification of PA-Mg relatively to PA-Na contribute to easy deformation. High tensile strength and strain at break for PA-Mg makes it more ductile than PA-Na with consequently lower friction.

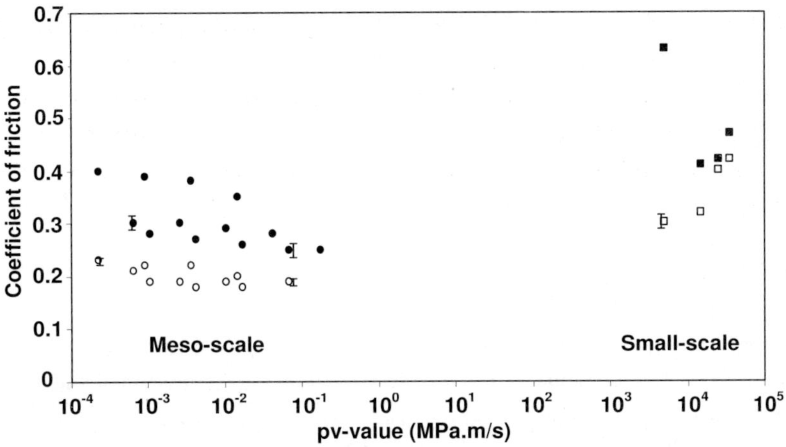

Fig. 5. Coefficients of friction of pure polyamides at meso-scale and small-scale tests (black dots = PA-Na, open dots = PA-Mg). Meso-scale tests are performed with flat-on-flat geometry at 1.15 N (0.0018 MPa), 3.30 N (0.0052 MPa), 5.16 N (0.0086 MPa) and 0.12, 0.5, 2.0, 8.0 m/s. Small-scale tests are perfomed with cylinder-on-plate geometry at 50 N, 100 N, 150 N, 200 N and 0.3 m/s.

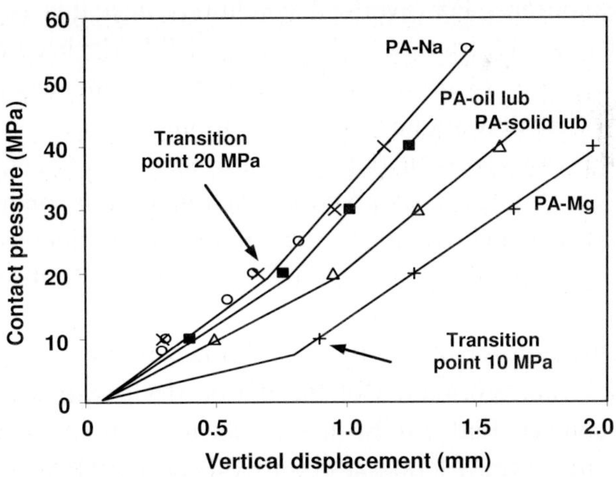

Fig. 6. Experimental deformation during static loading of different polyamide grades on a small-scale cylinder-on-plate contact configuration. Cylindrical polymer samples are statically loaded at 50 N, 100 N, 150 N in a Hertz contact against a steel plate (configuration, see Fig. 3).

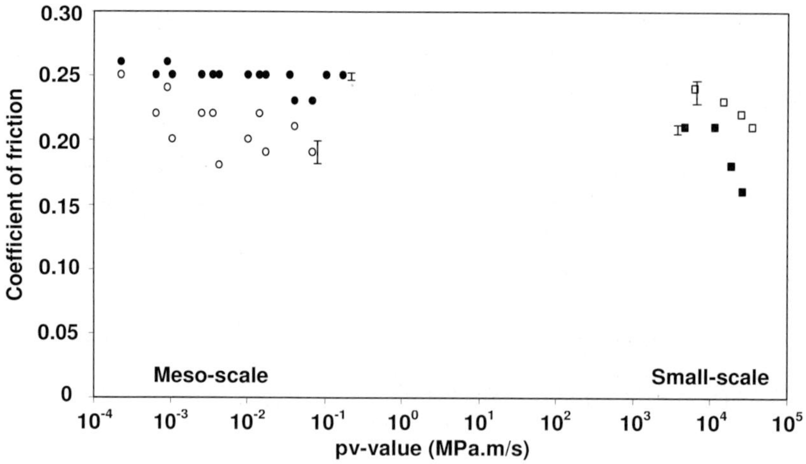

Fig. 7. Coefficients of friction of pure polyamides at meso-scale and small-scale tests (black dots = PA-solid lub, open dots = PA-oil lub). The test conditions are similar to those of Fig. 5.

The trend from meso-scale tests with decreasing friction at high normal load, is not uniquely transferred to small-scale tests due to different contact conditions. For overload conditions in a line contact, the dynamic coefficient of friction for PA-Na and PA-Mg becomes equal. The static coefficients of friction μ_{s1} disappear by changing PA-Na into PA-Mg on small-scale tests due to plastification, while this information was not included in meso-scale flat-on-flat testing.

It is not clear if polyamides give low friction at high contact pressures[19]. At low speeds and low loads, the coefficient of friction decreases gradually at higher loads. A maximum coefficient of friction is often found at certain load. According to Fig. 5, it is concluded that the coefficient of friction for pure polyamides either decreases or increases, depending on local deformation of the flat-on-flat (meso-scale) or line (small-scale) contact. Pure PA-Na has a transition on small-scale from decreasing to increasing friction at 100 to 150 N. Pure PA-Mg shows a transition on small-scale towards increasing friction at 50 to 100 N. Pure polyamides also show increasing friction for large-scale tests, indicating different sliding regimes than on meso-scale tests. While sliding over a single stroke, transfer or thermal effects do not explain different sliding behaviour. The counterface roughness is equal for every test scale and differences therefore mainly relate to mechanical material properties (strength, stiffness) or sliding parameters (*pv*).

Meso-scale flat-on-flat and small-scale cylinder-on-plate contact geometries cause different deformation of the contact surface, verified by static compression measurements (Fig. 6). For every testing scale, the relative classification with PA-Na highest stiffness and PA-Mg lowest stiffness is similar. Meso-scale tests run under elastic conditions, while deformation in a Hertz line contact is much larger: plastic deformation is observed from 45 MPa on for PA-Na (100 N) and from 30 MPa on for PA-Mg (50 N). Static deformation of internally lubricated polyamides is similar to PA-Na, as the latter serves as the bulk polymer. The limit between visco-elastic and plastic deformation observed in macroscopic deformation corresponds well to observed transitions in frictional behaviour as a function of normal loads. For a Hertz contact, the decrease in friction for PA-Na at 50 to 100 N is within the elastic region, while it increases at 100 to 200 N according to plastic deformation. Continuously increasing friction for PA-Mg at 50 to 200 N corresponds to plastic deformation over the entire load region. In the flat-on-flat test, the decrease in friction for PA-Na and PA-Mg on meso-scale at 1.15 to 5.16 N corresponds to elastic deformation under extremely low loads.

The investigation on macro-scale deformation was also performed at the micro-scale to explain the plastic-elastic transition by local deformations near rough asperities[20]. For contact pressures that are sufficiently low to neglect the interaction of the individual contact spots, the real contact area is proportional to the normal load. For high apparent contact pressures, the real contact area increases less then proportional to load. It seems from present work that the contact situation is also influenced by macro-parameters such as sample geometry, contact type and rigidity of specimen holder, which act to determine the stress conditions. Careful selection of tribotesters for the simulation of specific parameters should be made and extrapolation should take into account those specific transitions in contact situation.

3.2. *The Influence and Efficiency of Internal Lubrication*

Coefficients of friction for internally oil-lubricated polyamide and solid-lubricated polyamide are summarized in Fig. 7. Unlike in pure polyamides, the action of oil and solid lubricants in lowering friction is not unique and their efficiency strongly depends on the test scale. Solid lubricants result in lowest friction in the small-scale while they cause intermediate friction in the meso-scale test. The efficiency of solid lubricants was not demonstrated on meso-scale due to low contact pressures. Plastification of thermoplastic fillers (PTFE) at high loads is more favourable for low friction as observed on small-scale tests, with good loading capacity. Homogeneous oil lubricants resulted in lowest friction on meso-scale and intermediate friction on small-scale tests. The efficiency of the oil lubricant is largely limited by supply mechanisms at the sliding interface[21,22], which in turn is dependent on 'static' (migration, capillarity arising from surface roughness and Maragoni effects due to temperature gradients) and 'dynamic' (entrainment due to reciprocating sliding) effects. At low loads, the diffusion paths such as pores in the polymer structure are not blocked and allow for fluent lubrication. At high loads, deformation of the contact surface acts to obstruct internal diffusion of the oil towards the sliding interface, resulting in lumpy polyamide transfer. At extremely high loads, however, the inter-action between polymer bulk plastification and oil lubricant results in mixed transfer, favourably reducing friction at 30 to 40 MPa contact pressures.

In addition to homogeneous oil lubrication, oil injected into macroscopic surface depressions is more effective in stabilizing sliding, as captured oil remains in the sliding contact even if deformation through softening and/or melting occurs. Also, oil migration paths on the surface are larger than those through microscopic pores within the polymer bulk. Nevertheless, the coefficients of friction of filled polyamides are below those of pure polyamides on every test scale. The initial contact is mainly governed by adhesion. The presence of silicone oil or other thermoplastic lubricants such as PTFE decreases the surface energy and the polarity of the polymer surface (surface energy is 0.046 J/m^2 for pure PA-Na or PA-Mg and 0.037 J/m^2 for PA-oil lub) and also thermoplastic lubricants such as PTFE lower surface energy (e.g. 0.020 J/m^2 and less than 0.01 J/m^2

polar component). Lower adhesive forces resulting from oil lubrication is therefore responsible for the reduction in static and dynamic friction coefficient. As a result, stick-slip was not observed in tests with filled polyamides. The contact surface area of solid lubricated polyamides also influences the friction, as it determines the amount of lubricant in contact with the counterface and incorporated in the transfer film.

4. Small-Scale to Large-Scale Correlation

If a correlation between different test scales exists, extrapolation could provide more accurate design information. Due to the large number of parameters influencing the tribological behaviour, extrapolation is presently studied from experimental best-fits with parameters relating to mechanical loading, thermal heating, contact conditions and sample geometry. Three models are evaluated, based on (i) one single mechanical parameter (contact pressure), (ii) the contact pressure-sliding velocity (*pv*) model (temperature limit) and (iii) macroscopic geometry model. The last model is most extensive and considers thermal effects, sample geometry and contact conditions. In this evaluation, friction and wear data is captured during steady-state conditions for POM-H (polyacetal homopolymer) and PET/PTFE (polyethylene-terephtalate with PTFE lubricant) as examples.

4.1. *Single-Parameter Model: Effect of Contact Pressure*

The coefficients of friction for POM-H after small-scale and large-scale tests are plotted in Fig. 8 as a function of the steady-state contact pressure. Friction for polymers is not constant as a function of contact pressure in contrast to steel, due to visco-elastic deformation. Small-scale friction is higher than large-scale friction following different trends. Under thermally controlled conditions[23] (sliding speed v = 0.6 to 1.2 m/s), small-scale test results agree to the power-law equation $\mu = C.p^{-0.26}$, where μ = friction coefficient, C = constant and p = contact pressure. This trend was also found by other authors and it has been stated that a power-law parameter close to 1/3 indicates elastic deformation[24]. Small-scale friction at v = 0.3 m/s follows another relationship, because the sliding conditions are not yet thermally controlled, i.e. the maximum polymer temperatures T* are below the glass transition temperature T_g = 90°C. Large-scale test results at 16 to 150 MPa strongly depend on the contact pressure with a power-law equation $\mu = C.p^{-0.63}$. This normal load range is less frequently reported in literature, while power-law parameters close to 2/3 indicate larger influences of visco-elastic deformation[25].

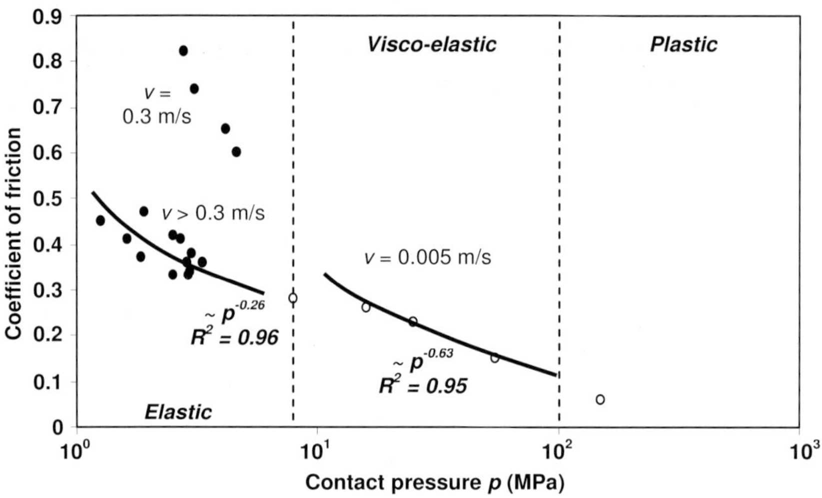

Fig. 8. Relation between coefficients of friction and steady-state contact pressure for POM-H on small-scale (•) and large-scale (○) tests. Small-scale tests (dry sliding) are performed with cylinder-on-plate geometry at 50 N, 100 N, 150 N, 200 N and 0.3, 0.6, 0.9, 1.2 m/s. Large-scale tests (dry sliding) are performed with flat-on-flat geometry at 190 kN (8 MPa), 380 kN (16 MPa), 560 kN (25 MPa), 1260 kN (55 MPa) and 3380 kN (150 MPa) and 0.005 m/s.

Extrapolation of the coefficients of friction over the entire contact pressure range for POM-H is impossible due to the transitions between elastic, visco-elastic and plastic deformation. These regimes were experimentally verified by static loading and recovery tests on line contact (small-scale) and constrained flat-on-flat (large-scale) samples similar to sliding tests. Upon removal of load, most deformations (> 95 %) are recoverable in small-scale tests and in large-scale tests with contact pressures up to 8 MPa. There remains significant permanent deformation after loading in large-scale tests at 16 to 150 MPa. At extremely high loads, the constrained polymer sample is under hydrostatic conditions[18]. This behaviour indicates that the limit between different deformation regimes is related to tribological performance: the coefficient of friction for large-scale tests at 8 MPa can be extrapolated from the small-scale sliding regime (both elastic regimes) while the coefficients of friction at 16 to 150 MPa are lower than expected from the extrapolated value. Gradual plastification of the sliding surface contributes to lower friction and stable sliding. The transition between visco-elastic and plastic deformation on micro-scale was explained previously. When two surfaces are brought into contact, real contact is only localized near some rough asperities. Mainly the size of existing contact spots grows at low contact pressure, while the number of contact spots increases at high contact pressures[26]. In the present research with extremely high contact pressures, the micro-asperity effects become less important and the real contact area approximates the apparent contact area.

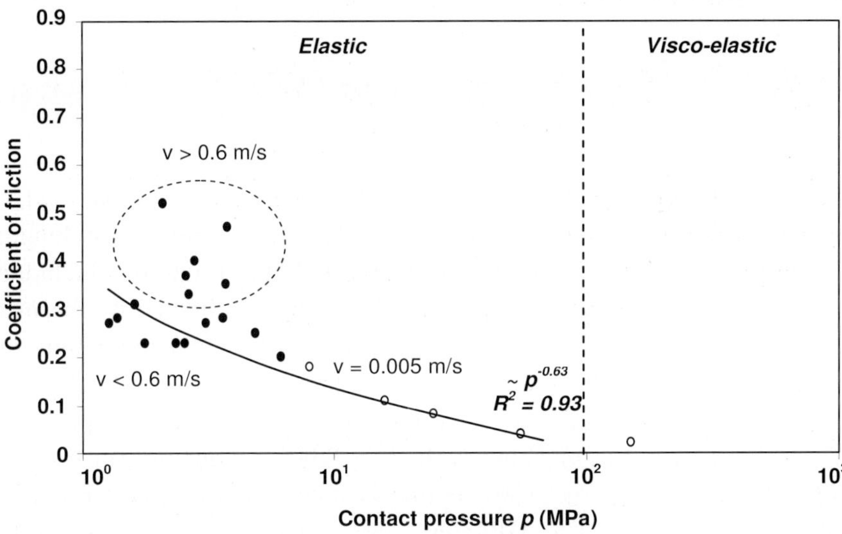

Fig. 9. Relation between coefficients of friction and steady-state contact pressure for PET/PTFE on small-scale (●) and large-scale (○) tests. The test conditions are similar to those of Fig. 8.

The coefficients of friction for PET/PTFE are given in Fig. 9, showing better extrapolation over the entire contact pressure range than for POM-H when thermally controlled conditions are considered. Elastic deformation with more than 95% recovery is experienced for small-scale and large-scale tests at 8 to 55 MPa. Visco-elasticity with permanent deformation only occurs at 150 MPa, due to high yield strength of PET/PTFE.

4.2. *Two-Parameter Model: pv-Limit*

The most common parameter to characterize polymer friction and wear is the *pv*-value (contact pressure x sliding velocity). For POM-H, a broad range of regime *pv*-values is covered: between 0.84 to 3.51 MPa.m/s for small-scale tests, and 0.04 to 0.75 Mpa.m/s for large-scale tests. Coefficients of friction and calculated maximum polymer surface temperatures, T*, are plotted in Fig. 10a. The wear rates are given in Fig. 10b. The temperature T* for reciprocating sliding is calculated using the following equation:

$$T^* = T_{env} + T_{bulk} + 4.10^{-4}\,\frac{\mu F_N \sqrt{v}}{b\sqrt{\ell}} \tag{1}$$

where μ is the coefficient of friction, v sliding velocity, F_N normal load, b contact width perpendicular to the sliding direction, ℓ semi contact length parallel to the sliding direction and thermal polymer properties included in the constant value. This temperature model was validated by thermal analysis of wear debris particles[27] confirming that crystallization or melting happened during sliding at temperatures T* = 120°C and 178°C.

For small-scale tests, the maximum polymer surface temperature T* fits linearly to the *pv*-value with $R^2 = 0.90$. A transition in friction (Fig. 10a) and wear (Fig. 10b) regime is noted for POM-H for small-scale tests at *pv* = 1.5 MPa.m/s, corresponding to a temperature T* = 90°C being the glass transition temperature T_g. This transition agrees with thermal softening and shear-lip formation due to overload on small-scale tests[28,29]. For large-scale tests, the maximum polymer surface temperatures T* varies with *pv*-values according to a linear relation, with $R^2 = 0.95$. The large-scale coefficients of friction decrease with higher *pv*-value and the wear rates increase at higher *pv*-values. However, friction and wear rates for large-scale tests cannot be directly related to the small-scale test conditions.

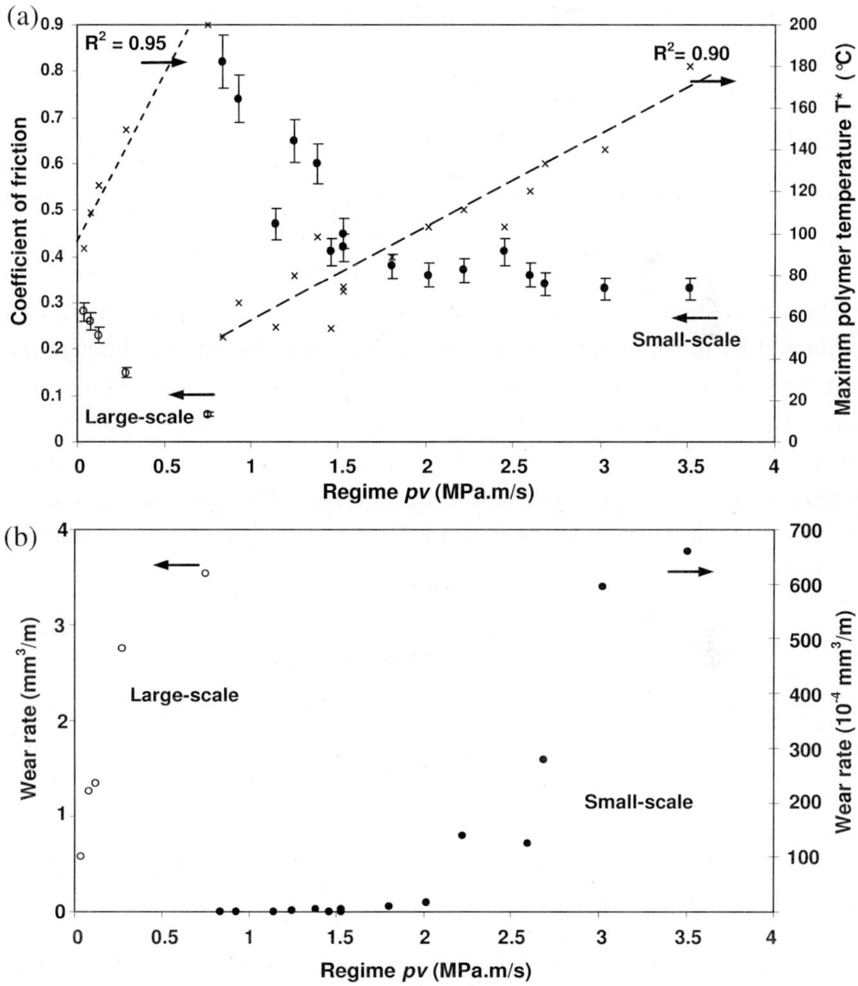

Fig. 10. Relation between small-scale (●) and large-scale (○) tests for POM-H as a function of regime *pv*-value, (a) coefficients of friction and temperature (x), (b) wear rates. The test conditions are similar to those of Fig. 8.

It is concluded that the *pv* temperature-limit model is useful to characterize transitions in friction, temperature and wear within a single testing scale, but extrapolation is impossible. Similar representations are obtained when tribological data is plotted as a function of the μpv-parameter. It is clear that the present model only considers the generation of frictional heat and does not account for the dissipation of energy within the sliding contact. The latter depends on test geometry and test conditions.

4.3. *Macroscopic Geometry Model*

The previous model only considers the generation of frictional heat (thermal energy input depending on the *pv*-value) and does not account for the dissipation of energy within the sliding contact. Scaling effects in friction, wear and maximum polymer surface temperature T* should be evaluated by introducing a scaling parameter that physically relates to the combination of mechanical sliding conditions, temperature generation and dissipation, contact situation and geometry. Macroscopic parameters are used as they can be determined a priori, rather than microscopic parameters such as surface roughness that constantly change during testing. Due to plastification under high loads, moreover, the concept of contact asperities disappears.

4.3.1. *Scaling Parameter*

A scaling factor is empirically defined to match the thermal sliding conditions on small-scale and large-scale tests, i.e. the maximum polymer surface temperatures T* should overlap for different testing scales. A macroscopic geometry parameter, G, and the Peclet number, Pe, previously used for numerical simulation of the sliding contact temperature[30], are therefore introduced here:

- The macroscopic geometry parameter, G, is dimensionless and is the ratio between contact area ($2\ell \times 2b$), and the product of sliding stroke, *s*, and contact contour ($2\ell + 2b$). It is given by: $G = 2\ \ell\ b/s\ (\ell + b)$. It is clear that large contact areas promote the accumulation of heat with a consequently higher temperature rise, while small contours and sliding strokes also promote higher interface temperatures. For large-scale samples, the contact geometry is constant while it changes from line contact into flat contact for small-scale tests.

- The dimensionless Peclet number, Pe, is given by: Pe = $v.\ell/2a$, and is constant for large-scale sliding (Pe = 1220) and variable for small-scale sliding ($580 < \text{Pe} < 8800$). Both sliding conditions range in the high-speed category with Pe larger than 10.

The maximum polymer surface temperature T* is plotted in Fig. 11 as a function of scaling factors $pv\sqrt{Pe}\,G$ (Fig. 11a) or $p\sqrt{vPe}\,G$ (Fig. 11b).

Best correlation between small-scale and large-scale temperatures T* occurs in Fig. 11b. Other combinations of scaling factors did not give any correlation with overlapping temperatures for small-scale and large-scale tests. The temperature T* fits a linear plot with $R^2 = 0.91$, over the 60 to 175°C. Extrapolation beyond the melting temperature range is impossible, however, as physical polymer properties such as thermal conductivity change. The present experimental model indicates that sliding temperatures depend on pv (dimensions W/m^2) related to heat generation, and G $(\ell/a)^{1/2}$ (dimensions $(m/s)^{-1/2}$) related to heat dissipation. The proportionality of temperature rise to $W/m^2.(m/s)^{-1/2}$ finds some evidence in recent work of Bhushan *et al.*, studying scaling effects of friction, wear and temperature between nano- and microscale[31]. Bhushan *et al.* considered the scale dependence of contact sizes, adhesion and deformation components of friction by using a fractal theory. In traditional statistical analysis, the real limit of detection is the distance at which two points on a contact surface have just attained the condition regarded as independent. A fractal approach considers a magnified image of a surface profile similar to the original profile. When an asperity comes into contact with another under sliding, the real area of contact starts growing while it is maximum when both asperities are above each other and it starts to get smaller when both asperities move away from each other. From numerical calculations of adhesion contact, with heat generation from a constant heat source, q (W/m^2), and heat flowing perpendicular to the sliding surface, the maximum temperature rise was found proportional to $(m/s)^{-1/2}$ as shown the following equation:

$$T \rho c\ v = 0.95\ q \left(\frac{2\, v\, \ell}{a} \right)^{1/2} \quad \text{or} \quad T \propto \frac{q}{v^{1/2}} \tag{2}$$

with ρc = volumetric specific heat, v = sliding velocity, ℓ = contact length and a = thermal diffusivity.

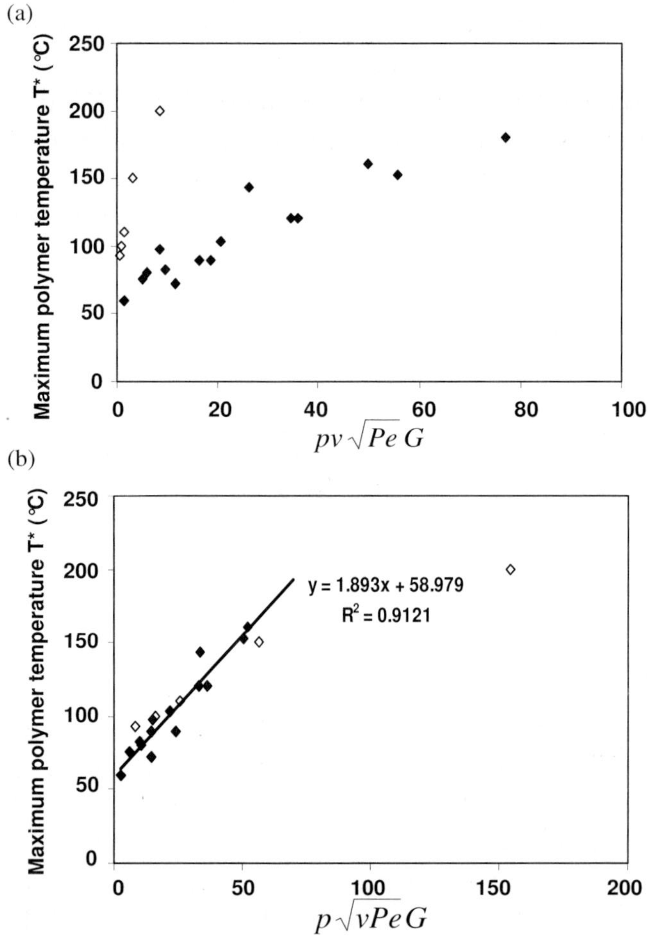

Fig. 11. Relation between scaling factors and maximum polymer surface temperature T* for small-scale (◆) and large-scale (◇) tests, (a) discrepancy between testing scales, (b) agreement between testing scales. The test conditions are similar to those of Fig. 8.

4.3.2. *Scaling of Friction and Wear for Bulk Polymers*

The small-scale and large-scale coefficients of friction for POM-H are plotted in Fig. 12a as a function of the empirical scaling factor. Although maximum polymer temperatures T* are identical, large-scale coefficients of friction are lower than small-scale values. One reason for this difference, as mentioned in previous discussions, is that small-scale tests are mainly under elastic conditions while large-scale tests behave under plastic conditions at 16 to 150 MPa. When a correction factor for plastic deformation is applied to the large-scale coefficients of friction, they agree to small-scale test results (Fig. 12b). The limit between elastic and plastic deformation occurs at a critical contact pressure p_o = 16 MPa according to recoverable or permanent deformation after static loading and creep tests[32]. Grouping thermal and contact conditions in experimental parameters allows agreement between coefficients of friction.

The small-scale and large-scale wear rates for POM-H are plotted in Fig. 12c as a function of the macroscopic scaling factor. Wear rates are related to coefficients of friction as it refers more directly to the frictional energy input. Under mild conditions, small-scale and large-scale wear rates agree, but softening on small-scale tests causes overload and shear lip formation resulting from stress concentrations near the sample edge. The contact stresses for large samples are more homogeneously distributed over the sliding surface while entrapment of wear debris particles stabilizes specific wear rates.

The wear experienced by POM-H at different testing scales is captured by optical microscopy. These images, shown in Fig. 13, indicate overload and shear-lip formation in small-scale tests, and in the case of large-scale tests, mechanical abrasion at T* < 90°C, softening at 90°C < T* < 175°C and melting at 175°C < T*.

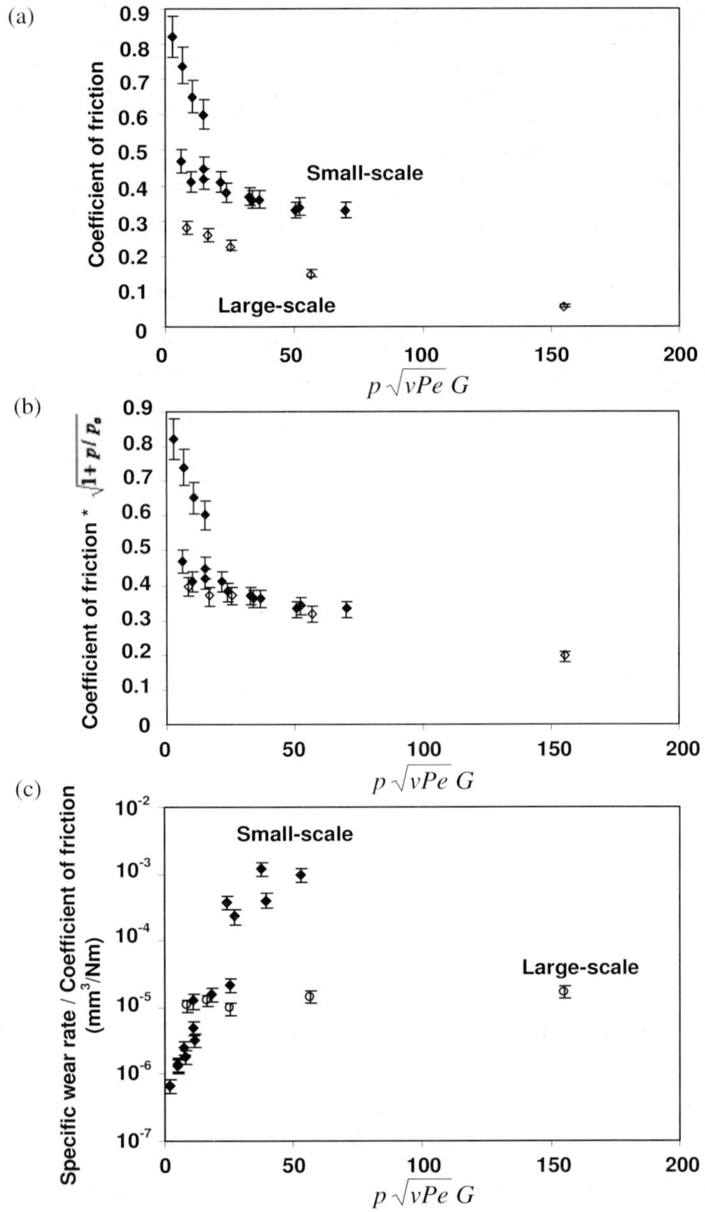

Fig. 12. Relation between small-scale (♦) and large-scale (○) friction and wear for POM-H as a function of the macroscopic scaling parameter, (a) coefficients of friction, (b) coefficients of friction corrected for contact deformation, (c) specific wear rates. The test conditions are similar to those of Fig. 8.

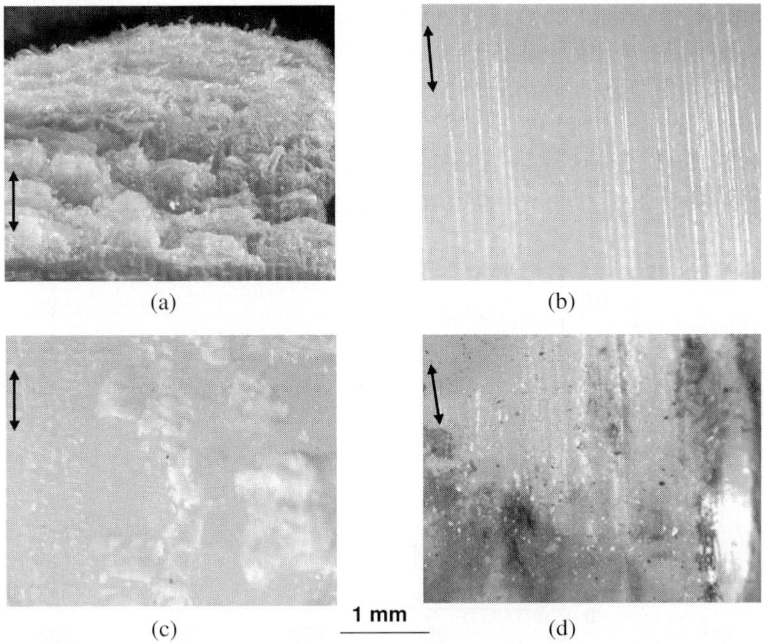

Fig. 13. Polymer surface aspect for POM-H after sliding at different testing scales, (a) small-scale overload, (b) large-scale mechanical interaction T*<90°C, (c) large-scale softening 90°C<T*<175°C, (d) large-scale melting 175°C<T*. The sliding direction is indicates by vertical arrows.

4.3.3. *Scaling of Friction and Wear for Solid Lubricated Polymers*

The small-scale and large-scale coefficients of friction for pure PET can-not be compared, because sliding becomes unstable due to softening or melting. The small-scale and large-scale coefficients of friction for internally lubricated PET/PTFE are plotted in Fig. 14. After correction for deformation, small-scale and large-scale values only agree for a limited number of sliding parameters: the coefficient of friction for large-scale tests at 8 MPa coincides with small-scale coefficients of friction, while friction becomes lower for large-scale tests at high load. For filled polymers, other effects than deformation and/or temperature favourably promote low friction during large-scale tests at high loads. Wear debris mobility becomes more important and is limited for large contact areas compared to small ones, enhancing the formation of a polymer/PTFE film on top of the polymer surface above 8 MPa (Fig. 15b). The formation of this polymer film is affected by the location of the polymer surface relative to the steel counterface. In the large-scale test (see Fig. 4), the bottom polymer samples displayed lower wear rates than those on top, due to debris accumulation on the polymer surface.

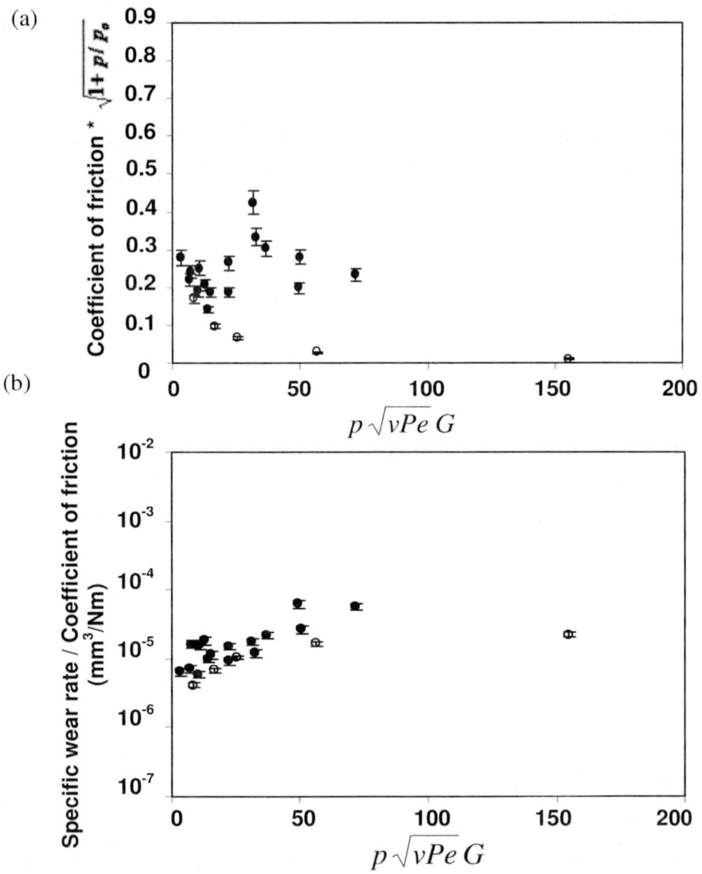

Fig. 14. Relation between small-scale (•) and large-scale (○) friction and wear for PET/PTFE as a function of the macroscopic scaling parameter, (a) coefficients of friction corrected for contact deformation, (b) specific wear rates. The test conditions are similar to those of Fig. 8.

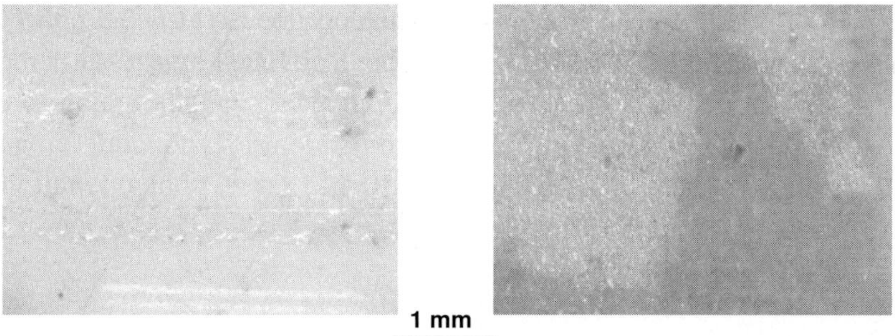

(a) (b)

Fig. 15. Polymer surface aspect for PET/PTFE after sliding at different test scales, (a) small-scale test at 100 N, 0.3 m/s without PTFE film formation, (b) large-scale test at 55 MPa, 0.005 m/s with PTFE film formation.

4.4. *Influence of Controlled Temperature on Small-Scale Testing*

The counterface bulk temperature for small-scale tests can be artificially controlled to be between 60 and 260°C for mild normal loads and sliding velocities (50 N, 0.3 m/s) in order to verify thermal transitions in friction and wear. It can be used to determine if high temperatures on small-scale tests can compensate for high loads on large-scale tests, causing low coefficients of friction of 0.10 to 0.04. Friction and wear rate results for PET/ PTFE at high temperature are presented in Fig. 16 along with the calculated temperature T*. Low coefficients of friction between 0.10 and 0.15 are obtained at temperatures T* between 85 and 90°C, while softening leads to unstable sliding at 90°C < T* < 250°C. An increase in sliding temperature above the softening point or melting point is not able to simulate low friction on a small-scale tester as it was obtained during large-scale tests: high temperatures and high loads cause different sliding regimes, confirming the usefulness of large-scale testing.

The polymer surfaces of PET/PTFE after testing were analyzed with Raman spectroscopy, indicating different polymer structures after either small-scale or large-scale tests. Besides the mechanical effects explained before, variations in polymer structure also contribute to differences in friction and wear with testing scale and explain why high-temperature tests are not equivalent to high-load tests.

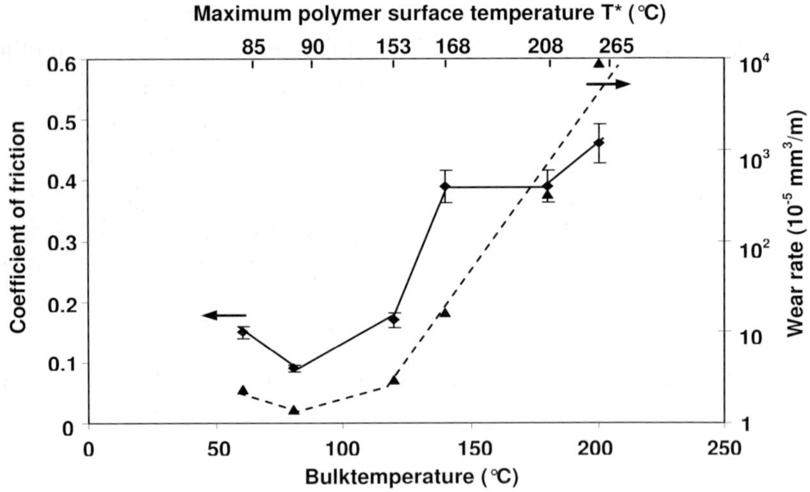

Fig. 16. Influence of controlled bulk temperatures applied to small-scale tests with PET/PTFE cylinders of diameter 5 mm and length 15 mm on friction (♦) and wear rates (▲).

For details on the spectroscopic evaluation method, we refer to more extensive works[27]. Some important conclusions are summarized below:

- After small-scale tests at variable normal loads and sliding velocities or at high temperatures, conformational changes (trans/gauche orientation) of the molecular chains under shear is mainly concentrated in the formation of a crystalline phase. It illustrates that higher mechanical energy input is concentrated on smaller areas resulting in a rather brittle surface structure. After large-scale tests, the orientation is mainly concentrated in a rigid amorphous phase that gives certain ductility to the polymer surface.
- The deposition of internal solid lubricants on the polymer surface is characterized by a separate absorption band in the Raman spectra, respectively decreasing or increasing with severity of one sliding parameter on small-scale or large-scale tests. It is verified that the depositions and thickness of a lubricating film gradually increase with time and contact pressure during large-scale tests, while debris is removed out of the sliding interface during small-scale tests.

The wear debris particles after testing were analyzed with thermo-analytical analysis (DTA/TGA). For details on the evaluation method, we refer to more specific work[27]. It is verified that the resident time of the wear debris particles within the sliding interface is larger during large-scale sliding, leading to important chemical reactions between separate debris particles and formation of a polymer transfer film on the counterface. The specific effect of polymerization in the sliding interface for POM-H or PET/PTFE particles is to enhance smooth and thin transfer film formation. In contrast, crystallization of the wear debris particles occurs during small-scale testing and results in fragmented transfer causing unstable sliding.

5. Large-Scale to Full-Scale Correlation

Bearings in off-shore construction often function under extremely high loads and low sliding velocities. Hydrodynamic lubrication is impossible under those conditions, as external oil is squeezed out of the contact interface causing failure due to seizure. Soft lubricating coatings such as molybdenium-disulphide provide low friction, but failure of the film leads to delamination and cold-welding after a single sliding step.

Polymers can sometimes be used as an elastic bearing layer under extremely high loads exceeding their yield strength. Stable friction for pure polymers was demonstrated in previous sections, showing that the overload concept for polymers makes them function as a 'liquid'. Ensuring dimensional stability, ultra-high molecular weight polyethylene elements (UHMWPE) reinforced with a carbon fiber/epoxy ring were designed by Solico BV (The Netherlands)[33]. The concept has been tested at the large-scale at Ghent University (Belgium) and was implemented by the Nederlandse Rijkswaterstaat for use in the ball-joints of a movable storm surge barrier, the Maeslantkering near Rotterdam. For a maximum transmitted bearing load of 350.10^6 N, the sliding surfaces of the ball-joint were originally covered with a thin sliding film containing solid lubricants. The film provides low friction, but failed due to details such as shape tolerances that occurred in practice which is difficult to simulate in laboratory scale tests. The application of an elastic polymer bearing layer changes the functional characteristics of the entire construction.

From a global analysis of the ball-joint by finite element modeling, test conditions for large-scale testing can be accurately determined. The resultant bearing force on an elastic bearing interlayer changes with lower elasticity, higher friction and higher thickness. Through the combined rolling/slip behaviour of the convex surface in contact with concave supports, the bearing force attains extreme positions more rapidly and loads on the construction increase. The concave ball-joint surfaces were modified and contain 500 separate polymer pads. Using discrete sliding pads, the load becomes more homogeneous over the sliding surface and local variations in stiffness or dimensional tolerances are redistributed. The maximum contact pressure of 150 MPa for a single bearing element depends on the distribution of the pads and local deformation.

A single hybrid UHMWPE bearing element with a nominal diameter 250 mm is illustrated in Fig. 17. From a local analysis of the bearing pads, strength and deformation is verified by means of finite element modeling. Much attention is given to the design and strength of the reinforcing carbon fiber/epoxy ring, for which small-scale selection tests were developed. The bearing elements are incorporated in machined holes on the steel concaves and have a free surface height of 8 mm in contact with the ball counterface with a Zn-phosphate coating protecting against corrosion. A polymer lip on top of the surface prevents direct contact between the composite ring and the sliding counterface, which would lead to catastrophic wear. Under static loading at 150 MPa, regular cold-flow of the polyethylene lip was experienced on full-scale compressive tests

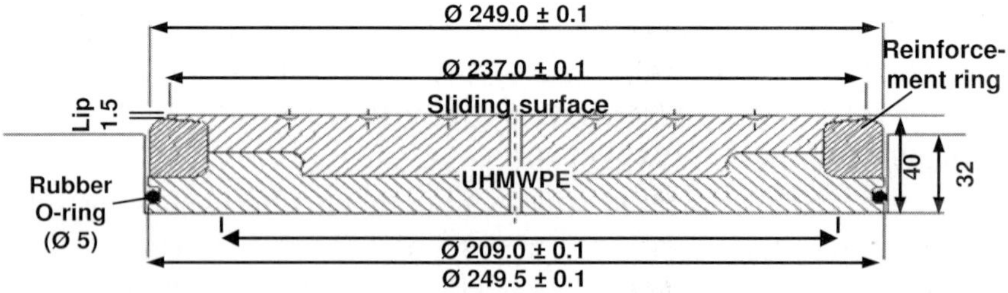

Fig. 17. Design of hybrid UHMWPE pads for sliding at high loads (150 MPa), preventing creep (Solico BV (Oosterhout, NL), Nederlandse Rijkswaterstaat, NL).

and numerically calculated by finite element analysis. The dimples on the sliding surface may contain lubricant, but they are not effective at high loads during which low dry sliding friction was demonstrated.

Large-scale sliding tests on the hybrid UHMWPE pads at 15 MPa to 150 MPa contact pressures were done to determine static coefficients of friction (μ_{s1} at first stroke and $\mu_{s,max}$ at following strokes) and dynamic coefficients of friction (μ_d) over ten reciprocating sliding strokes, corresponding to a 300 m wear path determined from the ball-joint performance. Besides coated counterfaces, bare steel was used as polymer/steel contact may occur after coating wear. Test results are given in Table 1.

Table 1. Coefficients of friction for a hybrid UHMWPE pad sliding against steel and Zn-coated counterfaces at 15 to 150 MPa over ten sliding strokes (300 m).

Contact pressure (MPa)	Steel counterface				Zn-phosphate coating			
	μ_{s1}	$\mu_{s,max}$	$\mu_{d,min}$	$\mu_{d,max}$	μ_{s1}	$\mu_{s,max}$	$\mu_{d,min}$	$\mu_{d,max}$
15	0.12	0.12	0.10	0.11	0.10	0.12	0.08	0.09
30	0.10	0.11	0.09	0.10	0.10	0.09	0.07	0.08
60	0.08	0.08	0.06	0.07	0.05	0.07	0.04	0.04
90	0.06	0.06	0.05	0.05	0.04	0.05	0.03(5)	0.03(5)
120	0.05	0.05	0.04	0.04	0.03(5)	0.04	0.03	0.03
150	0.04	0.04	0.03(5)	0.03(5)	0.03	0.03	0.02(5)	0.02(5)

For sliding conditions of 30 MPa, the largest deviation occurs in static friction at start μ_{s1} (± 8%). Other friction values show max ± 6% variation, being somewhat lower than small-scale tests. A design coefficient of friction for the entire ball-joint can be determined, using a modeling factor $\chi_m = 1.25$ on the experimental friction values:

- The global design criterion for friction assumes that not every pad on the bearing surface is simultaneously subjected to maximum static friction, but roll and slip motion of the ball-joint causes 80% of the pads to slide dynamically and 20% of the pads to break from the ball counterface and overcome static friction. The global design coefficient of friction can then be estimated from following equation:

$$\mu_{G;des} = \chi_m \left(0.80\, \mu_{d,max} + 0.20\, \mu_{s1} \right) = 0.22 \tag{3}$$

- The local design criterion for friction considers the maximum coefficients of friction on one single hybrid UHMWPE pad on the bearing surface during the first sliding stroke following Eq. (4) and during subsequent sliding strokes following Eq. (5):

$$\mu_{L;s1,des} = \chi_m \; \mu_{s1} \quad = 0.26 \tag{4}$$

$$\mu_{L;sn,des} = \chi_m \; \mu_{s,max} = 0.21 \tag{5}$$

The design criterion indicate a global coefficient of friction below the design limit of $\mu_{G;des} < 0.25$. Present solution incorporating polymer elements fits the strength requirements of the steel structure in between the holes. Shear stresses on the UHMWPE sliding surface cause a complex stress distribution in the carbon fiber/epoxy reinforcing ring. The tensile hoop stress (1094 MPa) and radial compressive stresses meet strength requirements. The radial-axial shear stresses (43 MPa) are most critical and should be controlled by modification of the production process for reinforcing composite rings. On-the-field test operations of the storm surge barrier (Fig. 18) showed favourable functioning of the Zn-phosphate coating and deformation of the hybrid UHMWPE pads agrees exactly to the phenomena observed on full-scale laboratory tests.

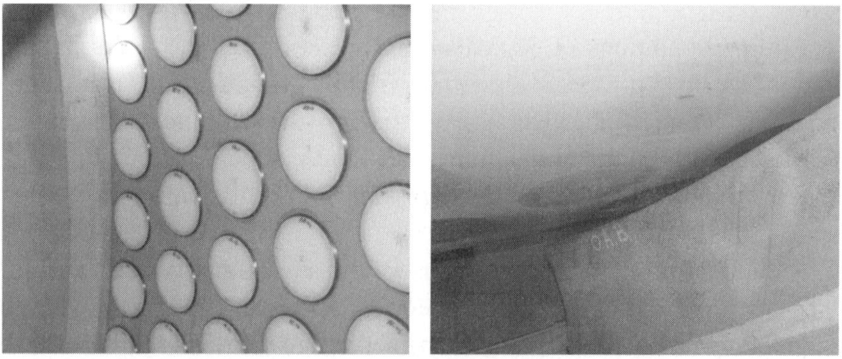

(a) (b)

Fig. 18. On-the-field evaluation of a ball-joint structure with discrete polymer bearing pads after a test operation sliding over 300 m, (a) convex bearing surface with polymer pads, (b) concave steel bearing surface with Zn-phosphate coating.

6. Summary and Conclusions

Meso-scale to small-scale correlations indicate higher friction for pure PA-Na than for pure PA-Mg on every test scale, due to intrinsic material properties. A transition from decreasing to increasing friction at high loads coincides with a transition from visco-elastic into plastic deformation, determined from in-situ stress-strain characteristics for small-scale tests, while no transition was found for meso-scale tests due to elastic deformation. For filled polyamide grades, oil lubricants provide lowest friction on meso-scale testing while solid lubricants are most efficient on small-scale testing. Oil lubricant is favourably squeezed out of the polymer by small deformation, but diffusion paths are blocked at high loads. At low loads, a polymer matrix with homogeneous oil dispersion is preferred, while at high loads oil lubricating holes on the sliding surface are preferred. Solid lubricants are not efficient on meso-scale compared to small-scale due to lack of plastification and debris entrapment.

Small-scale to large-scale correlations for friction of pure polymers can be found empirically after correction for thermal effects (heat generation and dissipation) and contact deformation. A scaling factor containing the Peclet number and a geometrical parameter was introduced. For filled polymers, film formation on large-scale polymer surfaces causes very low friction due to limited wear debris mobility. The wear rates on small-scale and large-scale samples cannot be extrapolated due to stress concentrations causing overload. Besides mechanical effects, also the worn polymer structure is different: it is either crystalline after small-scale testing or has an oriented amorphous phase after large-scale testing.

Large-scale to full-scale correlations can be made for constrained polymer bearing elements, when the contact situation and testing conditions are determined and verified by finite element modeling. While polymers provide low friction when loaded above their yield strength, dimensional stability should be ensured by a reinforcing composite ring.

In general, it seems that small-scale tests predict the sliding behavior of polymers with good reliability only under limited conditions. Large-scale tests are more useful for testing specific designs and filled polymer composites, where test conditions can be applied in agreement with real sliding applications.

Acknowledgments

The authors acknowledge Nederlandse Rijkswaterstaat and Solico BV for design and full-scale testing of the ball-joint in the Maeslantkering.

References

1. H. Endo and E. Marui, *Wear*, **258**, 1525 (2005).
2. M.J. Neale and M. Gee, *Guide to Wear Problems and Testing for Industry* (Professional Engineering Publishing Limited, St. Edmundsbury Press Ltd, Suffolk, 2000)
3. V.K. Jain and S. Bahadur, *Wear*, **46**, 177 (1978).
4. J.H. Horng, M.L. Len and J.S. Lee, *Wear*, **253**, 899 (2002).
5. H.M. Hawthorne, *Wear*, **149**, 169 (1991).
6. M.G. Jacko, P.H.S. Tsang and S.K. Rhee, *Wear*, **133**, 23 (1989).
7. J.K. Lancaster, *J Lubrication Technology*, **2**, 187 (1997).
8. S. Abarou, D. Play and F.E. Kennedy, *ASLE Transactions*, **30**, 269 (1986).
9. Y. Yamaguchi and K. Kasshiwagi, *Polym. Eng. Sci.*, **22**, 248 (1982).
10. H. Czichos, Polymer wear and its control, in: *ACS Symposium Series*, Ed. Lee LH (ACS Press, Washington, 1985).
11. M. Vermeulen, *Wear*, **132**, 287 (1989).
12. H. Unal, U. Sen and A. Mimaroglu, *Tribol. Internat.*, **37**, 727 (2000).
13. S.E. Franklin and A. de Kraker, *Wear*, **255**, 766 (2003).
14. L.C. Seabra and A.M. Baptista, *Wear*, **253**, 394 (2002).
15. P. Samyn, J. Quintelier and G. Schoukens, *Experimental Mechanics*, in press (2007)
16. P. Samyn and T. Tuzolana, *Polymer Test.*, **26**, 660 (2007).
17. P. De Baets, *Wear*, **180**, 61 (1995).
18. P. Samyn, W. Van Paepegem, J.S. Leendertz, E. Suister, L. Van Schepdael, J. Degrieck and P. De Baets, *Engineering. Structures*, in press (2007).
19. M. Watanabe, M. Karasawa and K. Matsubara, *Wear*, **12**, 185 (1968).
20. N. Adams, *J. Appl. Polym. Sci.*, **7**, 2075 (1963).
21. T. Yabe, T. Takajo, S. Kato and F. Ueki, *Tribol. Trans.*, **43**, 453 (2000).
22. M. Marchetti, H.M. Meurisse, P. Vergne, J. Sicre and M. Durand (2003), *Tribol. Letters*, **10**, 163 (2003).
23. C.M. McEttles, *ASLE Trans.*, **30**, 149 (1987).
24. J.F. Archard, *Proc. Roy. Soc. London*, **A 243**, 190 (1957).
25. M.W. Pascoe and D. Tabor, *Proc. Roy.. Soc. London*, **A 235**, 210 (1955).
26. J.A. Greenwood and J.B.P. Williamson, *Proc. Inst. Mech. Eng.*, **A 295**, 300 (1966)
27. P. Samyn, PhD dissertation, Ghent University (2007).
28. C. Subramanian, *Wear*, **126**, 57 (1988).
29. S. Odi-Owei and D.J. Schipper, *Wear*, **148**, 363 (1991).

30. K. Friedrich, J. Flöck, and K. Varadi, *Wear*, **225-229**, 368 (1999).
31. B. Bhushan and M. Nosonovsky, *Nanotechnology*, **15**, 749 (2004).
32. P. Samyn, P. De Baets, G. Schoukens and J. Quintelier, *Tribol. Internat.*, **40**, 819 (2007).
33. J.S. Leendertz, L. Van Schepdael, W. Van Paepegem, P. Samyn, P. De Baets and J. Degrieck, *Stahlbau*, **75**, 45 (2006).

CHAPTER 4

SCRATCH EXPERIMENTS AND FINITE ELEMENT SIMULATION: FRICTION AND NONLINEARITY EFFECTS

Hervé Pelletier, Vincent Le Houerou, Christian Gauthier and Robert Schirrer

Institut Charles Sadron, UPR CNRS 22
23, rue de Loess, BP 8404, F-67034 Strasbourg Cedex, France
E-mail: pelletier@ics.u-strasbg.fr

The scratch behaviour of an amorphous polymer was investigated to determine how the characteristics of the material affect the contact geometry during scratch experiments with spherical indenters. A thermoplastic polymer (Polymethylmethmcrylate) was studied using both an experimental device allowing *in situ* observation of the contact area during scratching with spherical tips and finite element modelling (FEM). The rheological properties of the PMMA surface were approximated by a simplified bilinear law, while the friction at the interface between the indenter and the material was modelled with Coulomb's friction coefficient varying between 0 and 1, for each computed ratio a/R from 0.3 to 0.6 (where a is the contact radius and R the radius of the tip). FEM results for elastoplastic contacts indicate that the contact geometry is directly related to the plastic strain field in the deformation beneath the indenter during scratching. We show that the dimensions of the plastically deformed volume and the gradient of plastic strain both depend on the ratio a/R and also on the friction coefficient. An equivalent average plastic strain is calculated by FEM over a representative plastically deformed volume. The average plastic strain increases with a/R, as predicted by Tabor's empirical rule, and with the local friction coefficient for a given ratio a/R. Clear correlations are demonstrated between the average plastic strain and the geometrical parameters classically used to describe the shape of the contact area.

1. Introduction

Scratch and mar resistance is an important characteristic for polymeric surfaces in numerous applications like ophthalmology or the automobile industry. Marring is associated with a high density of small, shallow scratches distributed over the surface, although scratching can also involve a single, deeper plastic groove. In industrial applications, where gloss and transparency are important parameters, marring and scratching affect more or less such attributes of appearance, according to the ability of the customer to perceive these flaws. The relations between the material and surface properties, scratch resistance and surface damage of amorphous polymers are still poorly understood despite the development of numerous experimental test methods.[1-5] Up until now, surface mechanical testing, including indentation and hardness tests, scratch tests and mar resistance tests[6-8] have been used with limited success and primarily for qualitative comparisons and quality control.

Recently, progress has been made in understanding the surface deformation occurring during scratch experiments with the aid of a specific experimental device.[9-13] This new apparatus was built to investigate the scratch properties of polymers over temperatures ranging from -70°C to +120°C and scratching speeds of 1 to 10^4 μm/s. Unlike conventional scratch testers,[4,5,8] this device allows *in situ* observation of the contact area and the groove left on the surface. The main advantage of such equipment is the possibility of (i) calculating the mean contact pressure generated by the moving tip and (ii) recording the profile of the residual groove from the beginning of its life time.

During scratch tests, the polymeric surface and the material near the contact surface are subjected to a complex stress field arising from the sliding of the tip under controlled normal force and velocity. Scratch testing appears to be a local deformation process leading to compressive stresses in front of and tensile stresses behind the rigid indenter with shearing at the contact surface. In the case of purely elastic sliding contact, the complex stress state and notably the distribution of normal and shear stresses at the surface have been analytically described for bulk materials[14] and bilayer systems[15] as a function of the elastic properties (elastic modulus) and friction coefficient. In the case of scratching, several authors[11,16-19] have proposed three dimensional analytical models for the ploughing, adhesive and apparent frictions of various hard pyramidal or conical shaped indenters ploughing along a metallic surface, assumed to be a rigid plastic material. However, owing to the large elastic recovery observed during the scratching of amorphous polymers, an analysis would appear to be far more difficult for the latter than for metals.

This chapter presents a numerical simulation analyzing some scratch transitions observed during the scratching of a polymeric surface with a spherical indenter, as a function of the ratio a/R (where a is the contact radius and R the radius of the indenter) and the local friction coefficient varying between 0 and 1.

2. Experimental Observations

The experimental device for the scratch test, called the 'microvisioscratch', has been described previously.[9,10] It consists of a commercial servomechanism bearing a small, temperature controlled transparent box which contains the sample and the scratching tip. Control of the moving tip and recording of the normal F_n and tangential F_t loads, scratching speed V_{tip} and temperature T are computer driven. A built-in microscope allows *in situ* observation and measurement of the groove left on the surface.[9,10] Scratching over a wide range of speeds (1 to 10^4 μm/s) and within a temperature range covering the polymer relaxation peaks (-70 to +120 °C) are the main innovative features of the system. The normal load F_n applied to the moving tip can be selected from 0.05 to 35 N by adjusting the compression of a supple spring. In the present experiments, performed at 30°C, the speed of the tip was kept constant at 0.03 mm/s. A cone-shaped diamond indenter with a spherical extremity having an apex angle of 60° and a tip radius R of 116 μm was used to scratch a commercial grade of cast poly (methylmethacrylate) (PMMA). The tip geometry and the sliding velocity have been selected to get at room temperature elastic-plastic contact during scratching. The Young's modulus of PMMA is about 3.3 GPa at room temperature.

The PMMA surfaces were prepared and cleaned using different procedures to obtain different values of the local friction coefficient, which was determined from scratch experiments according to the reverse analytical method of Lafaye *et al.*[11,16] The apparent friction, defined by the ratio of the tangential force F_t on the applied normal force F_n (Fig. 1-a) is equal to the local friction coefficient only for elastic sliding, corresponding to a symmetric contact. As soon as the contact becomes dissymmetric, the apparent friction is higher than the local friction, and a ploughing friction has to be considered in order to take into account for the plastic deformation. With increasing the applied normal load at room temperature, the apparent friction increases progressively although the local friction is shown to be constant.[11,16]

In scratch tests using a spherical indenter, for polymeric surfaces the geometry of the contact between the moving tip and the deformed surface is much more complex to predict than for other classes of materials such as, for example, ceramics, which mostly display elastic contacts, or metals which rapidly exhibit fully plastic contacts. In the case of polymers, the shape of the contact area (S_n) changes with the temperature, tip speed and applied normal load for a given tip radius. As described schematically in Fig. 1-a, the contact area has (i) a front side with the formation of a frontal pile-up pad and (ii) a rear side due to the high level of elastic recovery. A lateral pile-up pad can also be observed along the residual groove. With increasing normal load F_n, the corresponding ratio a/R increases and the contact surface varies from a perfectly elastic contact to a fully plastic contact at constant sliding velocity V_{tip} and room temperature. As a function of the nature of the contact, the size and shape of the front and rear areas evolve. These changes in contact geometry are known to depend mainly on the plastic strain imposed around the contact area, just beneath the indenter. Using the 'microvisioscratch', the variation of the projected contact area S_n and the residual groove can be analyzed *in situ* as a function of the scratching conditions, the surface behaviour which depends mainly on the local friction coefficient and the bulk properties of the test material, as shown in Fig. 1-b. Some characteristic geometrical parameters such as the plastic contact radius, a_f, the rear contact radius, a_{rear}, the thickness of the frontal push pad e_p and the rear angle ω (Fig. 1-a) have been introduced here to describe the evolution of the contact area.

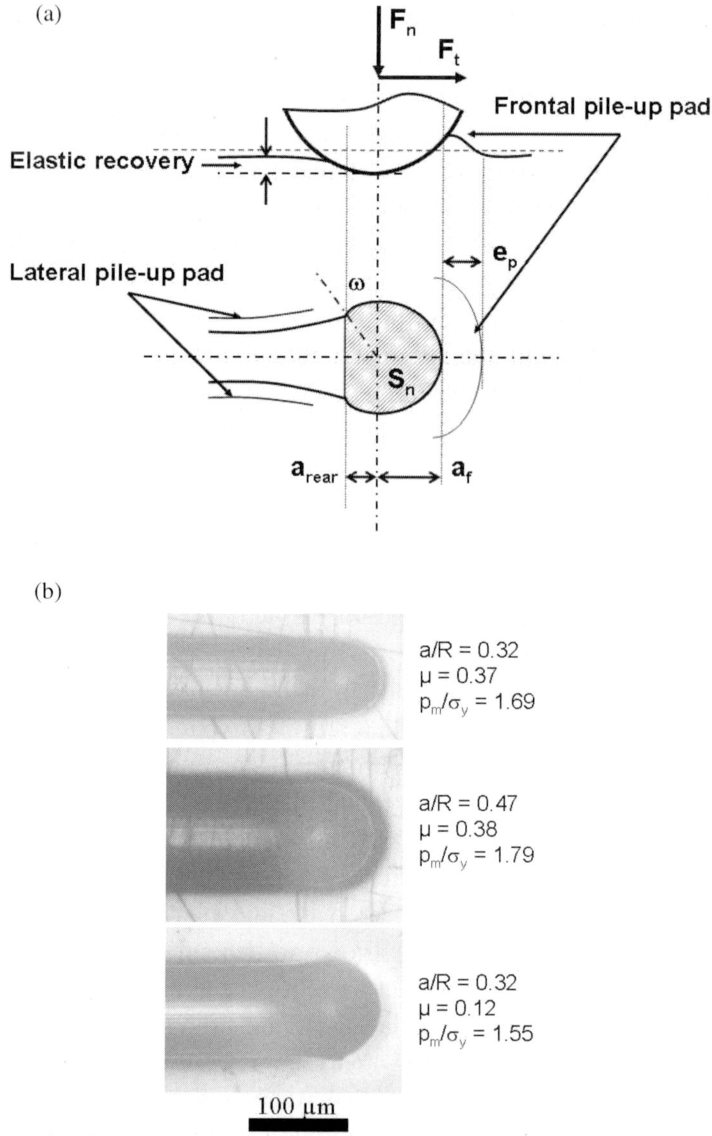

Fig. 1. (a) Schematic description of the contact geometry between a rigid spherical indenter and a polymeric surface in a scratch experiment and (b) evolution of the true contact areas and the grooves left immediately after passage of the tip as a function of the ratio a/R and the local friction coefficient for PMMA.

In the case of PMMA, for a local friction μ = 0.15 and up to a ratio a/R of 0.15-0.2, the contact surface is symmetric with $\omega = \pi/2$, corresponding to a perfectly elastic contact with no residual groove (Fig. 2-a). When the ratio a/R lies in the range 0.2–0.4, the rear angle decreases progressively with increasing plastic deformation imposed during the contact and the elastic recovery at the rear diminishes. At a/R = 0.3, plastic deformation clearly occurs under the tip and a permanent groove remains on the surface. At a higher ratio of a/R, greater than 0.4, the contact surface becomes fully dissymmetric with ω less than 0.4, corresponding to a fully plastic contact for an elastoplastic material. The evolution of the size and shape of the frontal push pad with the ratio a/R presents similarities to the evolution of the rear angle. While the contact remains elastic and elastoplastic, no frontal push pad can be observed in Fig. 2-b, which depicts the evolution of the normalized push pad e_p/a_f as a function of the ratio a/R. However, as the contact becomes mainly plastic (a/R > 0.4), a frontal push pad appears. As seen in Fig. 2-b, for a/R = 0.47, the frontal push pad is less pronounced than the lateral pile-up pad along the permanent groove.

In Fig. 1-b, the effect of the friction coefficient on the residual groove and the contact area can be described qualitatively. At a given ratio a/R = 0.32, reduction of the friction coefficient by a factor of 3 modifies the contact geometry and decreases the normalized contact pressure p_m/σ_y, where p_m and σ_y are respectively the average contact pressure (defined by the ratio between the applied normal load and the true contact area), and the yield stress of the tested material determined in compression uniaxial test. The contact appears to be less plastic, with a high value of the rear contact angle and no frontal push pad. However, the contact between a moving tip and a polymeric surface is generally very complex due to the number of dependent parameters (friction, strain hardening, tip geometry, sliding velocity etc.) and therefore difficult to describe analytically. FEM is needed to study separately the influence of each experimental parameter or material property.

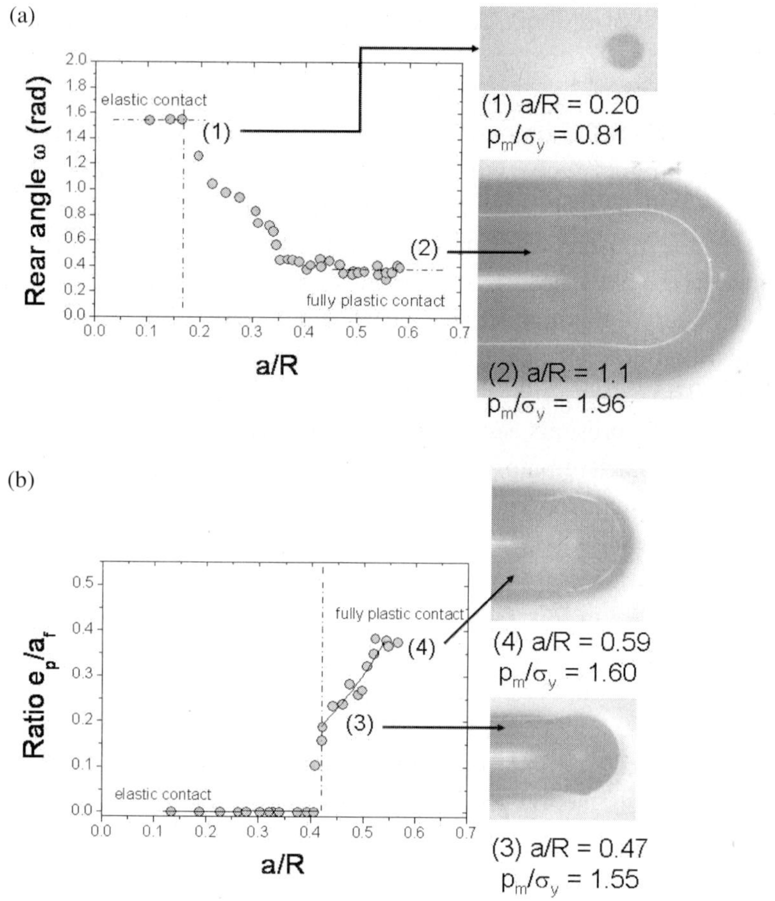

Fig. 2. (a) Rear angle ω and (b) frontal pile-up pad defined by the ratio e_p/a_f as a function of a/R for PMMA with a local friction coefficient $\mu = 0.15 \pm 0.02$. Optical *in-situ* photographs obtained for increasing values of a/R show the evolution of the contact geometry.

3. Finite Element Modelling

The scratch tests using a spherical indenter having different radii R and different penetration depths were modelled with a three dimensional finite element code. All calculations were carried out with the implicit FEM package MSC Marc® and schematic illustrations of the FEM model for a ratio a/R of 0.3 are presented in Fig. 3. The domain was modelled

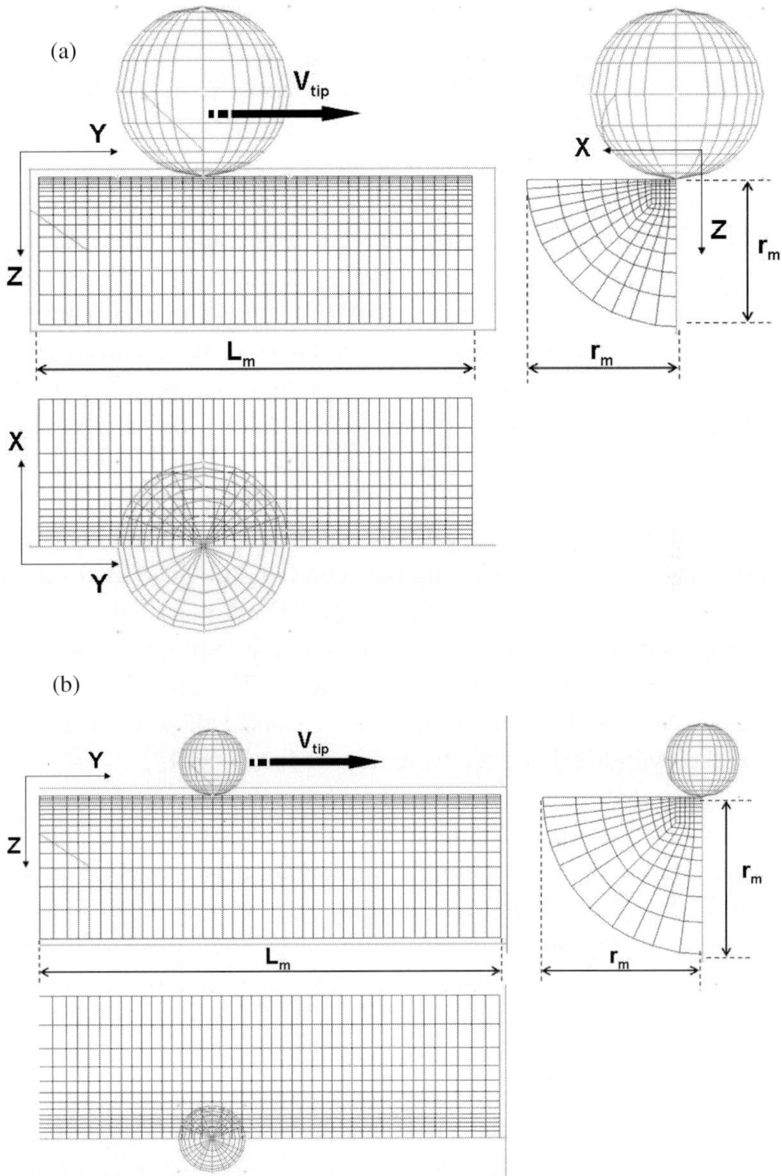

Fig. 3. Views of the spherical indenter and the mesh for simulation of scratching at $a/R = 0.3$ (a) with a tip radius $R = 5.4$ μm and a penetration depth $h = 0.25$ μm (see Table 1) and (b) with a tip radius $R = 2.2$ μm and a penetration depth $h = 0.1$ μm (see Table 2).

as a half of a cylinder and to limit the number of elements, we reduced it to half by introducing a symmetry plane (x = 0). In contrast to previous studies using the FE approach to scratch tests,[19-21] remeshing boxes during simulation runs were not employed in the present work.

Thus, we developed a specific finite element mesh defined by 3 zones, with different sizes of bilinear isoparametric 8-noded brick elements, allowing linear interpolation (full integration procedure). The simulations required 5240 elements and 6232 nodes for $a/R = 0.3$, and 10,873 elements and 12,768 nodes for $a/R = 0.45$ or 0.6. During the indentation and scratching phases, the thickness of the smallest element in the contact area was at least 0.5 times the imposed penetration depth. The lengths (along the y-axis) and widths (along the x-axis) of the elements in the contact area are given in Table 1. The domain size was chosen to be sufficiently large so that the boundary effects did not affect the results. The tip radius R and penetration depth h were selected to obtain different ratios a/R varying between 0.3 and 0.6. The distance of the indentation from the edge of the sample (along the y-axis) was more than 6 times the contact radius of the indentation, while the thickness of the sample (along the z-axis) was at least 26 times the depth of penetration. As boundary conditions, the x, y and z displacements of the nodes on the cylindrical surface were required to be null.

Assuming that the experimental results are obtained at a given temperature and constant sliding speed, the analysis may be performed for simple elastoplastic behaviour. The situation is modelled as being quasi-static and as such time-independent, with no influence of the strain rate. In order to reproduce experimental tests, the kinematics may be divided into two distinct phases: (i) a first step, corresponding to the indentation process (along the z-axis) at a given penetration depth and (ii) a second step, corresponding to the scratching process (along the y-axis) at constant relative velocity and fixed indentation depth. The length of the scratch L_R is chosen so that the normal and tangential loads applied to the indenter reach a steady state. L_R is fixed and corresponds to a ratio L_R/a_0 of more than 5, depending on the simulated ratio a/R (see Table 1). The theoretical radius a_0 is determined by the geometrical relationship $a_0 = \sqrt{2Rh - h^2}$, assuming no elastic deflection and no piling up phenomenon around the contact area. In all simulations performed in this study, the spherical indenter was assumed to be rigid and the rheology of the material elastoplastic. The stress–strain curve was specified according to a bilinear von Mises model with isotropic hardening. Since PMMA has a very low hardening tendency at room temperature and low strain rate, a bilinear model with a constant tangent modulus provided a suitable fit for the material parameters. For an elastic, linear hardening plastic material, the stress–strain relation may be represented by:

$$\sigma = \begin{cases} E\varepsilon & (\sigma \le \sigma_y) \\ \sigma_y + E_T(\varepsilon - \varepsilon_y) & (\sigma > \sigma_y) \end{cases} \tag{1}$$

where E is the elastic modulus, σ_y the yield stress, ε_y the yield strain satisfying $\varepsilon_y = \sigma_y / E$ and E_T the tangent modulus, corresponding to the constant work hardening slope. The elastic modulus, the yield stress and the tangent modulus are fixed at E = 3.5 GPa, σ_y = 100 MPa and E_T = 350 MPa. This simplified model is close to the true rheology of PMMA.

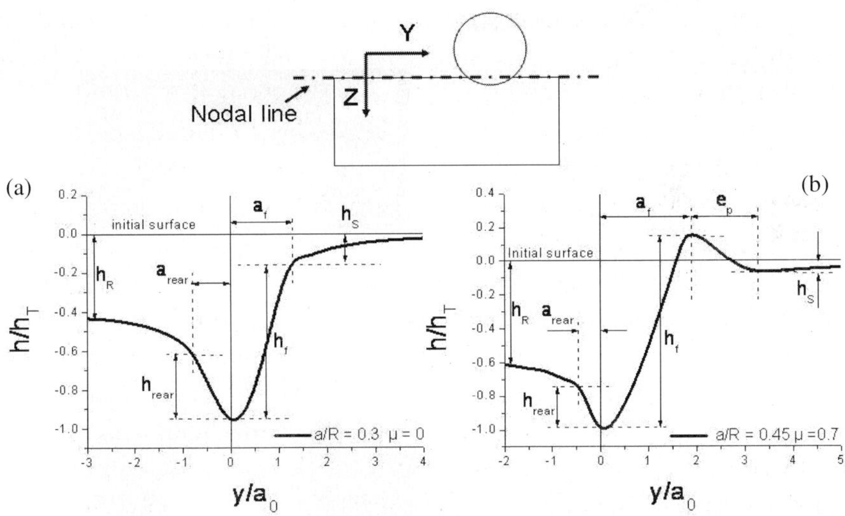

Fig. 4. Frontal path plot determined by FEM describing the node displacement along the z-axis and definition of the geometrical parameters used to characterize the contact geometry, for two scratching conditions (a) $a/R = 0.3$ and $\mu = 0$ (sinking phenomenon) and (b) $a/R = 0.45$ and $\mu = 0.7$ (piling up phenomenon).

The contact between spherical radii and the surface of the sample is enforced using a penalty function method and a geometrical description of slave and master surfaces. In contrast to previous work,[19] the interface between the rigid indenter and the deformable surface is not assumed to be frictionless and frictional effects are included in the analysis by means of an isotropic Coulomb model. Thus, the maximum local shear stress acting at the interface τ_c is given by:

$$\tau_c = \mu\, p \tag{2}$$

where p denotes the local normal pressure and μ is the coefficient of the local friction between the surfaces in contact. The contact model was implemented within a finite-sliding formulation, where any arbitrary sliding and rotation between the surfaces could occur. At each ratio of a/R, simulations were performed for different values of the friction coefficient in the range 0 to 1.

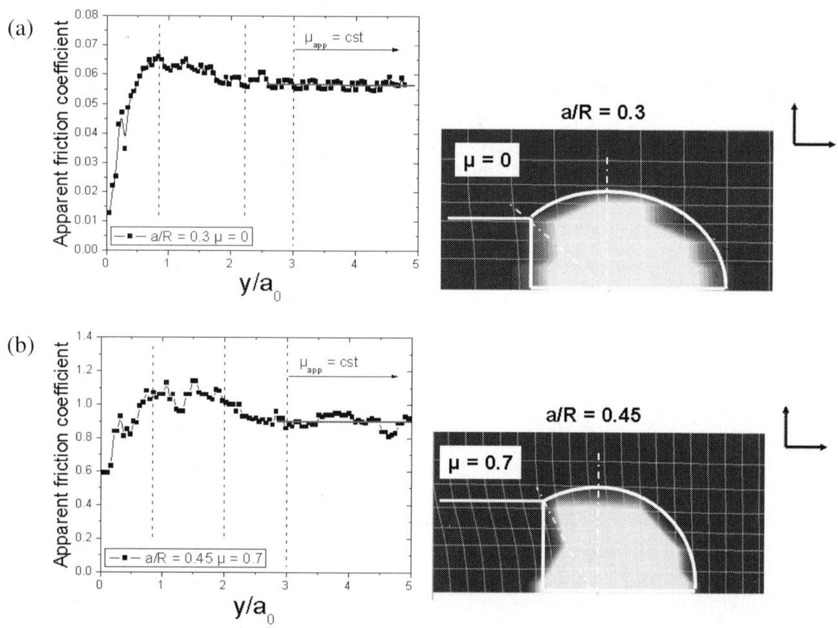

Fig. 5. Apparent friction coefficient determined by FEM for two scratching conditions (a) $a/R = 0.3$ and $\mu = 0$ and (b) $a/R = 0.45$ and $\mu = 0.7$.

Using a three dimensional finite element code for the tip, similar geometrical parameters may be calculated from specific path plots along nodal lines. The plastic contact radius a_f, the rear contact a_{rear} and the thickness of the frontal push pad e_p can be directly determined as a function of the scratching conditions $(a/R,\ \mu)$ from the nodal displacement in the x, y and z directions, as seen in Fig. 4. The rear contact angle is then calculated using the following geometrical relationship:

$$\tan\omega = \frac{a_{rear}}{\sqrt{a_f^2 - a_{rear}^2}} \tag{3}$$

As described in previous studies,[10,13] the evolutions of the rear angle ω and the ratio a_{rear}/a_f are similar as a function of the ratio. The apparent friction coefficient μ_{app}, defined as the ratio of the tangential load F_t to the normal load F_n, can be derived in each simulation run, as shown in Fig. 5. μ_{app} increases rapidly to reach a maximum at a specific scratch length y/a_0 of about 1.5 and then decreases slowly to become constant for y/a_0 > 3. In the following sections, the values of the apparent friction coefficient determined by FEM correspond to an average value in the stabilized part of the curve μ_{app} versus y/a_0. The top view of the contact area can also be analyzed as a function of a/R and μ. As compared to experimental results, the main advantage of FE simulations is the possibility of obtaining an accurate description of the elastic and plastic strain tensors beneath the moving tip and in the residual groove.

Table 1. Principal geometrical characteristics of the finite element model.

a/R	Element size in contact	Geometrical parameters (μm)			Domain size		
		R	h	a_0	L_m/h	r_m/h	r_m/a_0
0.3	0.65×0.45	5.4	0.25	1.62	16	37	5
0.45	0.26×0.32	3.3	0.36	1.50	18	26	6
0.6	0.26×0.32	1.25	0.25	0.75	36	37	12

Note: R, h and a_0 are respectively the tip radius, the penetration depth and the theoretical contact radius assuming non elastic deflection and no piling up phenomenon around the contact area (see Section 3).

Table 2. Principal geometrical characteristics of the second FE model developed for a ratio $a/R = 0.3$ (see Fig. 3-b).

a/R	Element size in contact	Geometrical parameters (μm)			Domain size		
		R	h	a_0	L_m/h	r_m/h	r_m/a_0
0.3	0.65×0.45	2.2	0.1	0.65	42	92	14

Note: R, h and a_0 are respectively the tip radius, the penetration depth and the theoretical contact radius assuming non elastic deflection and no piling up phenomenon around the contact area (see Section 3).

4. Finite Element Results

4.1. *Plastic Strain Field*

During scratch tests, elastic and plastic energy is consumed mainly in two distinct zones: (i) a very thin layer at the interface between the rigid indenter and the surface, which is subjected to extremely high shear strain due to the high strain rate imposed by the relative movement and adhesive slipping arising from local friction, and (ii) a larger and deeper volume, where elastoplastic yielding occurs. By analogy with indentation, the second volume is assumed to be roughly spherical and its size is comparable to that of the groove left on the surface after scratching. In the case of the elastic sliding of an indenter with a spherical tip, the distribution of the stress component and hence of the corresponding strain on the surface of a semi-infinite solid is known and the influence of the local friction coefficient has been clearly demonstrated.[14] Conversely, in the case of elastoplastic contact with friction, such analytical equations are not yet available. Figures 6 and 7 show the influence of the local friction coefficient on the plastic strain gradient beneath the indenter and in the residual groove in a scratch experiment, for $a/R = 0.3$ and 0.45 respectively. At low values of a/R and μ, as observed in Fig. 6-a for $\mu = 0.1$, the plastic strain does not reach the surface and is maximal at a specific depth of about $z = a_f/2$, just beneath the indenter, in good agreement with the data of Hamilton and Goodman for elastic sliding contact.[14] This point corresponds to the region of maximal shear stress, where plastic deformation may occur. An elastically deformed volume is observed between the rigid indenter and the plastic gradient and therefore at the surface beneath the indenter and in the residual groove, the contact appears to be mainly elastic. At higher a/R and low friction coefficient (Fig. 7-a) the plastic strain reaches the surface, but the maximal value of the equivalent plastic strain is always located at a depth of $z = a_f/2$. The shape and gradient of the plastic strain are similar to those observed at a/R = 0.3 (Fig. 6-a). The plastically deformed volume extends gradually in each case to a maximal depth of $1.5\,a_f$, the only difference being in the range of plastic strain which differs by a factor of about 2 between the two cases.

However, for a high value of the friction coefficient and a/R = 0.3 (Figs. 6-b and c) or 0.45 (Figs. 7-b and c), an important difference can be seen in the shape and gradient of the plastic strain. The increase in the friction coefficient generates an increase in the maximal equivalent plastic strain, but also a displacement of its location with respect to the initial surface and to the indenter. At the two ratios a/R in Figs. 6 and 7, the maximal value of the equivalent plastic strain increases by a factor of 3 for μ = 0.4 and by a factor of 7 for μ = 0.7, as compared to its value for μ = 0.1. Moreover, the maximal plastic strain is located beneath the rear part of the indenter directly under the surface while the plastically deformed volume extends to a penetration depth of less than the true contact radius a_f. In the residual groove, the plastic strain gradient is mainly confined to the sub-surface region for $z \leq a_f/3$. Our results, especially for frictionless contact, are in good agreement with the findings of several authors[19-23] using three dimensional finite element approaches and modelling by stress and strain simulation of the mechanical behaviour of bulk materials[19-21] or hard coated surfaces[22-23] in scratch tests. For bulk materials assumed to be elastic and perfectly plastic, Felder and Bucaille[21] derived topographical equivalent plastic strain maps for conical indenters and frictionless contact, showing the influence of the semi-angle θ of a conical indenter. The maximal plastic strain attained near the indenter tip varied from 1.5 to 8.4 as the semi-apical angle θ decreased from 70.3° to 30°. In scratch tests on polymers with a spherical tip, for a constant friction coefficient (μ = 0.3) and different ratios a/R varying between 0.15 and 0.64, Bucaille *et al.*[20] demonstrated that the plastic strains involved in the deformation during scratching increase as the penetration depth (and hence the ratio a/R) increases. The maximal value of the equivalent plastic strain was about 0.35 at $a/R = 0.17$ and rose to 1.75 at $a/R = 0.64$. The differences between the plastic strain gradient described in the present study and those reported in references[19-21] are mainly due to the stress-strain curve used to model the isotropic mechanical behaviour of the simulated material. Bucaille *et al.*[20] introduced a flow stress described by a G'sell-Jonas law[24] with more pronounced strain hardening at large plastic strain. Moreover, these authors showed the influence of the strain hardening on the plastic strain field and found notably an important decrease in the maximal value of the equivalent plastic strain with increasing ability of the material to strain harden. Finally, a similar evolution of the plastic strain gradient with friction, leading to a reduction in the depth of the strained volume and an increase in the average plastic strain in the material, was reported by Felder and Bucaille[21] for scratching of steel with a conical indenter.

A description of the plastic strain field during scratching with a spherical tip is thus highly complex, depending not only on the ratio a/R as initially proposed by Tabor[25] for elastic, perfectly plastic materials but also on the local friction coefficient. For given bulk properties, the equivalent plastic strain maps show the influence of surface properties such as friction on the mechanical response of surfaces during a scratch test. As observed experimentally (Fig. 1-b), a decrease in the friction coefficient for a given ratio a/R leads to a more elastic contact.

4.2. *Mean Contact Strain*

The behaviour of the material in a scratch test, from quasi elastic to ductile ploughing and then severe damage,[1-7] depends on several factors governing the level of deformation of the sample surface. As seen in Figs. 6 and 7, the deformation under the indenter tip is not homogeneous. Close to the indenter, the plastic strain is maximal below the tip and decreases with increasing distance from the indenter tip. Moreover, the plastic strain beneath the indenter is rapidly affected by the local friction and the scratching contact strain must be defined as a function of two

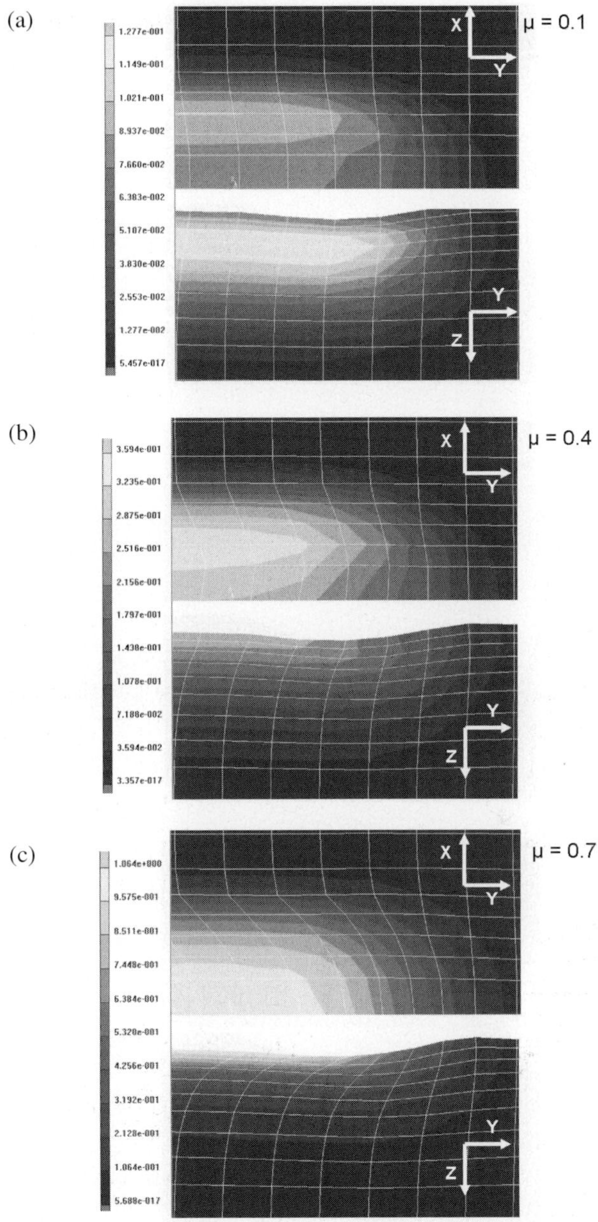

Fig. 6. Equivalent plastic strain maps as a function of the friction coefficient for (a) μ = 0.1, (b) μ = 0.4 and (c) μ = 0.7, for a ratio a/R = 0.3. For each friction coefficient, the maps are shown in the planes $z = 0$ (top view) and $x = 0$ (side view).

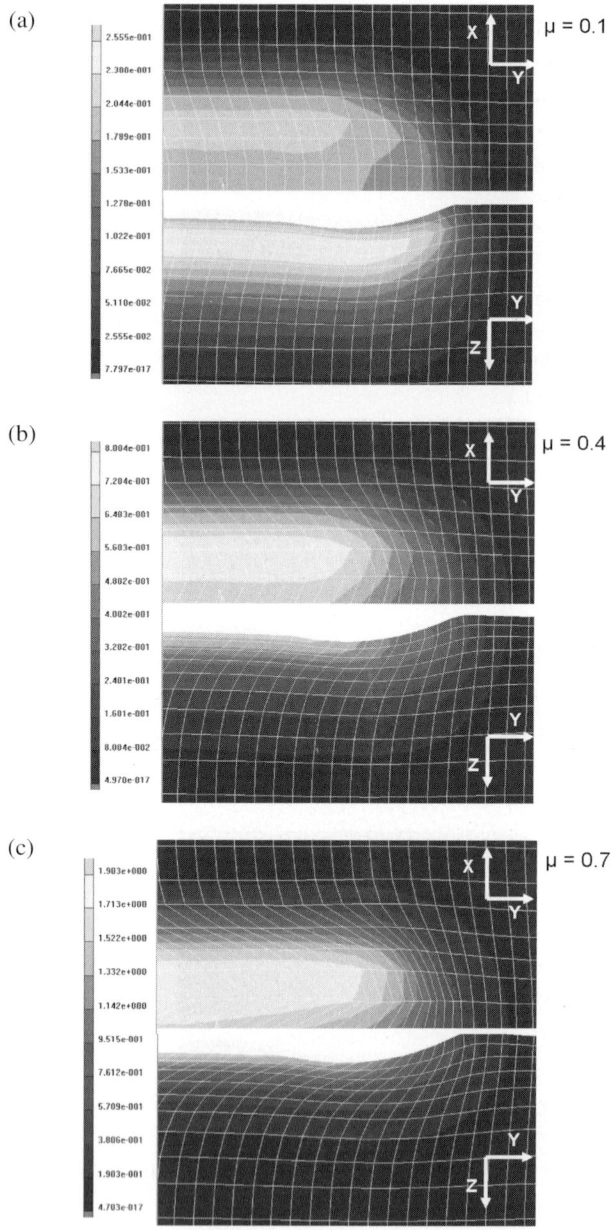

Fig. 7. Equivalent plastic strain maps as a function of the friction coefficient for (a) μ = 0.1, (b) μ = 0.4 and (c) μ = 0.7, for a ratio a/R = 0.45. For each friction coefficient, the maps are shown in the planes $z = 0$ (top view) and $x = 0$ (side view).

variables: (i) the geometrical strain which is related to the ratio a/R and (ii) the local friction coefficient μ.[13] A representative strain is often introduced to characterize the level of strain imposed in indentation[26] and by extension in scratch experiments.[19] In the case of normal indentation with a conical indenter, Tabor[25] has shown that the constant ratio of hardness (mean contact pressure) to yield stress for a perfectly plastic solid may be applied to a solid hardening under strain. The initial yield stress, corresponding to the onset of plasticity, is replaced by the flow stress measured under compression at a specific representative geometrical strain, proportional to $\cot\mathrm{an}\,\theta$ where θ is the semi-apical angle of the indenter:

$$\bar{\varepsilon} = 0.2 \cot\mathrm{an}\,\theta \tag{4}$$

A similar expression for the representative geometrical strain has been derived for a spherical indenter:

$$\bar{\varepsilon} = 0.2 \frac{a}{R} \tag{5}$$

where a and R are the surface contact radius and the radius of the tip, respectively. In the case of a spherical indenter, this equation can only be used for an indentation depth or contact radius small by comparison with the tip radius ($a \ll R$). Furthermore, Briscoe *et al.*[7] have demonstrated that these expressions are also valid for scratching, but the proportionality constant may be different for materials like polymers which have large elastic strains as compared to metals. In a first approximation, a recent analysis of scratching with spherical tips[13] has shown that the level of deformation is governed by the relation proposed by Tabor and the average deformation can be expressed for all penetration depths and for a given spherical indenter by:

$$\bar{\varepsilon} = 0.2 \frac{a}{\sqrt{R^2 - a^2}} \tag{6}$$

As reported by Jayaraman *et al.*[27] for indentation and recently by Bucaille *et al.*[19] for scratch tests with conical indenters, an average equivalent plastic strain beneath the indenter can be computed from FE simulations. The average equivalent plastic strain ε_{av}^{p} in the plastically deformed volume V_p is defined as:

$$\varepsilon_{av}^{p} = \frac{1}{V_p} \sum_{1}^{m} \varepsilon_i^p \, \Delta V_i \tag{7}$$

where ε_i^p and ΔV_i are the equivalent plastic strain and the volume of the centroid of element i, respectively. m is the total number of elements having a non zero centroidal equivalent plastic strain (above 0.002) in the plastically deformed volume V_p. Thus,

$$V_p = \sum_{1}^{m} \Delta V_i \tag{8}$$

For elastic, perfectly plastic materials defined by the rheological parameter $X = E/\sigma_y \cot an\theta$ and frictionless contact, Bucaille *et al.*[19] arbitrarily introduced a critical value of the equivalent plastic strain, denoted ε_C^p, to define the elements which have to be taken into account for calculation of the average equivalent plastic strain during scratching. This critical value ε_C^p was found to depend on the rheological factor X and the corresponding value of ε_{av}^p on the chosen equivalent plastic strain ε_C^p. To improve this procedure in the present study, we selected only the elements within a fixed volume, called the representative volume, to compute ε_{av}^p by FEM. Since only elements near the indenter have a significant equivalent plastic strain, the geometrical shape of this volume was a right angled parallelepiped centred on the location of the indenter at the end of the scratching phase. The dimensions of the representative volume were defined from studies of the volume affected during indentation.[28-30] Thus, the length L, width l and height H of the parallelepiped (Fig. 8-a) were chosen to be proportional to the estimated contact radius a_0, with $L = 6a_0$ and $l = H = 3a_0$.

Figure 8-b shows the variation of the average equivalent plastic strain computed by FEM using equations (7) and (8) as a function of the true friction coefficient for the different simulated ratios a/R. An increase in local friction coefficient leads to an increase in the average equivalent plastic strain for all imposed ratios a/R computed by FEM up to a horizontal asymptotic value ε_{av}^p = 0.14. This asymptote is reached at different values of the friction coefficient, with μ = 0.6 for a/R = 0.45 and μ = 0.4 for a/R = 0.6. The slope of the linear part of

(a)

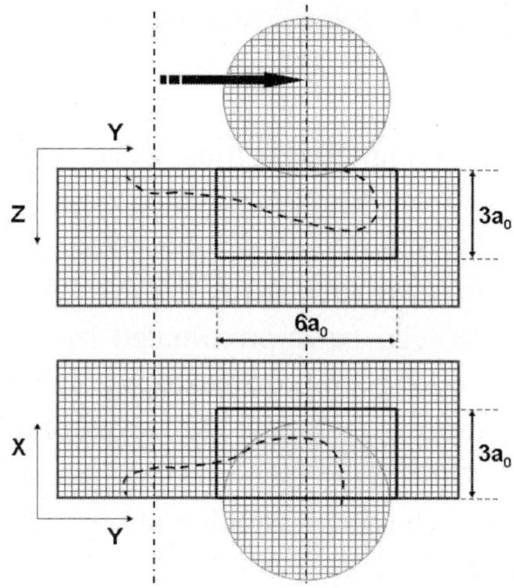

(b)

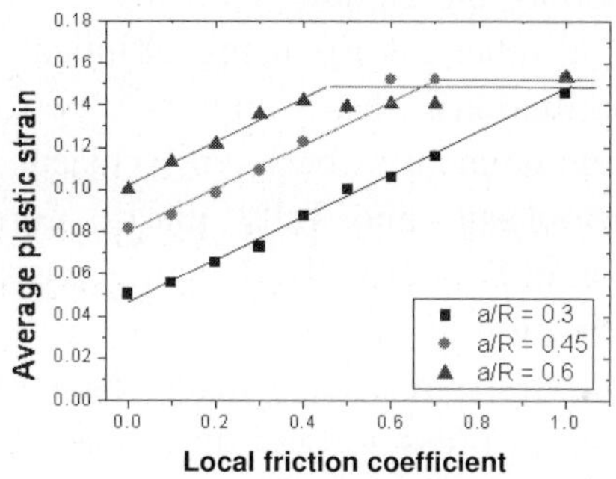

(c)

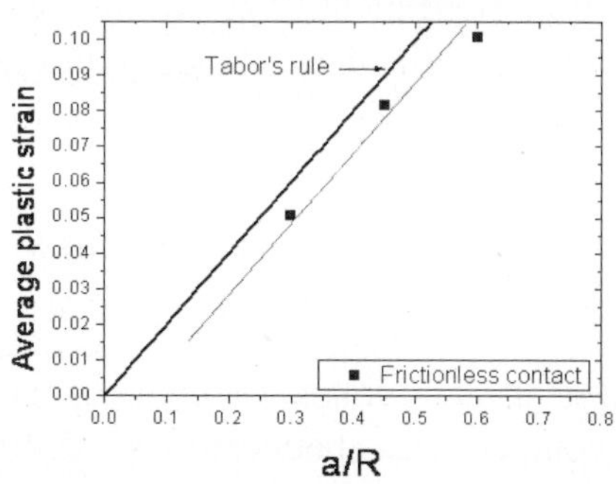

Fig. 8. (a) Definition of the representative volume beneath the indenter used to evaluate the average equivalent plastic strain in a scratch test. (b) Evolution of the average plastic strain as a function of the local friction coefficient and (c) average plastic strain as a function of the ratio *a/R* for frictionless contact.

the curves ε_{av}^{p} versus μ in Fig. 8-b is constant for all tested ratios a/R with a mean value of about 0.1. The evolution of the average plastic strain for frictionless contact ($\mu = 0$) is shown in Fig. 8-c, where the symbols represent values computed in this work while the bold line corresponds to the level of deformation as initially proposed by Tabor[25] and defined by Eq. (5). At each ratio a/R, the value of ε_{av}^{p} is very close to the representative strain estimated using Eq. (5) and the evolution of ε_{av}^{p} as a function of a/R can be approximated by a linear fit with a slope of about 0.17. The relationship proposed by Tabor for indentation of metals can thus be used to estimate the level of plastic strain, as suggested by Gauthier *et al.*[13] This relation may be extended to scratch tests on polymeric materials having a small strain hardening ability and a low friction coefficient ($\mu \leq 0.1$).

These original results may be correlated with the recent findings of Gauthier *et al.*[13] Modelling the contact behaviour of a polymer surface in scratch tests with a spherical tip using elliptical contact pressure and shear stress distributions, these authors[13] performed numerical simulations to locate the boundaries between (i) elastic and elastoplastic contact and (ii) elastoplastic and fully plastic contact. The curve describing the evolution of the normalized contact pressure (p_m/σ_y) as a function of the local friction at the second boundary, revealed that the apparition of the fully plastic regime beneath the moving tip depends both on the applied contact pressure (and by extension on the ratio a/R) and on the local friction coefficient. The mean contact strain during scratching must therefore be defined as a function of these two variables. To a first approximation, the mean contact strain may be considered to be proportional to μ for a given ratio a/R.

4.3. *Contact Geometry*

As depicted on the optical micrographs of Figs. 1-b and 2, the size and shape of the true contact area between the sliding indenter and the polymer surface are modified as the contact radius increases with the applied normal load. These *in-situ* observations illustrate clearly: (i) the evolution of the symmetry of the contact as estimated from the rear angle

ω or the ratio of the rear length a_{rear} to the frontal length a_f and (ii) the formation of a plastic pad pushed in front of the indenter, defined as the ratio of the thickness of the pad e_p to the radius of the contact, assimilated in a first approximation to the frontal length a_f. Figures 9-a and b show the evolution of the normalized push pad and the rear angle respectively with the local friction coefficient, as computed by FEM for the different ratios a/R. As a function of the scratching conditions, one observes a more or less strong influence of a/R and μ on the elastic recovery of the contact (i.e. the rear angle) and on the elastic deflection and material accumulation in front of the indenter. At a given ratio a/R, the simulations show that there is no push pad in front of the indenter until reaching a specific friction coefficient, denoted μ_{cr}. At higher local friction, the frontal push pad appears quickly and its thickness increases rapidly, especially for a/R = 0.6. The critical value of the friction coefficient at which the frontal push pad appears decreases with increasing ratio a/R, with μ_{cr} = 0.6 at a/R = 0.3 and μ_{cr} = 0.3 at a/R = 0.6. An increase in the local friction seems to produce a more important accumulation of material than an increase in a/R. This conclusion is consistent with the data of the literature[13,20,21] and Bucaille *et al.*[20] observed similar tendencies for the growth of the frontal push pad with the ratio a/R.

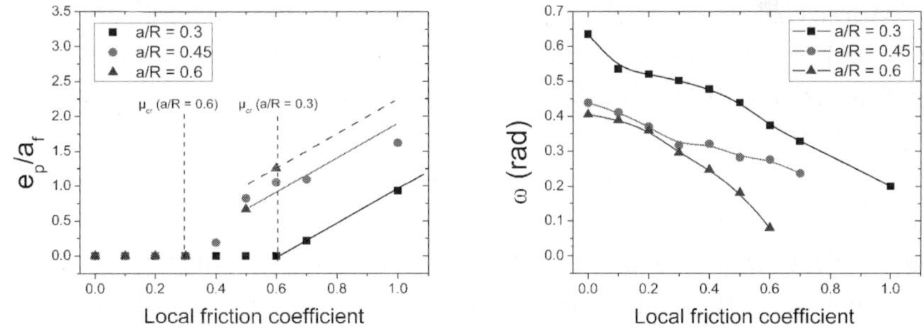

(a) (b)

Fig. 9. (a) Frontal push pad defined by the ratio e_p/a_f (where e_p is the thickness of the push pad and a_f the plastic contact radius) and (b) rear angle ω of the contact as a function of the local friction coefficient μ.

Similarly (Fig. 9-b), the rear angle ω computed by simulation decreases progressively with the friction coefficient for a given ratio a/R, while for a given friction coefficient the rear angle ω decreases with a/R. The calculated values of the rear angle are in good agreement with those found experimentally (Fig. 2-a).

5. Discussion

5.1. *Relationship between Contact Geometry and Mean Contact Strain*

The above figures indicate that the local friction coefficient plays a similar role to the geometrical ratio a/R in the evolution of the plastic strain during a scratch test with a rigid spherical indenter. The FE results reveal that with increasing local friction there is:

- an increase in the level of plastic strain imposed beneath the moving tip and in the residual groove;
- an important modification of the shape and size of the plastically deformed volume and hence of the corresponding plastic strain gradient;
- a displacement of the location of the maximum plastic strain to the sub-surface region just beneath the indenter

Moreover, geometrical parameters such as the normalized frontal push pad e_p/a_f or the rear contact angle ω evolve similarly as a function of the local friction coefficient (for a given ratio a/R) as observed experimentally as a function of a/R for a given friction coefficient (Fig. 2). As suggested by Gauthier *et al.*,[13] for a given material such geometrical parameters are directly related to the mean contact strain imposed during scratching and are therefore representative of the nature of the contact between the rigid indenter and the deformed polymeric surface. The latter authors also demonstrated that the size of the frontal push pad and the contact yielding, which defines the boundary between elastoplastic and fully plastic contact, depend mainly on the local friction coefficient for a given normalized contact pressure.

In Fig. 10, the normalized push pad and the rear contact angle, shown in Figs. 9-a and b respectively as a function of the friction coefficient, are plotted as a function of the average plastic strain determined using equation (7) for different scratching conditions (a/R, μ). Figs. 10-a and b demonstrate a strong relationship between the geometrical parameters e_p/a_f and ω and the average plastic strain. These results prove that the average plastic strain can be estimated for different scratching conditions ($a/R, \mu$), as seen in Fig. 8-b. However, the contact geometry defined by the parameters e_p/a_f and ω is related to a unique value of the average plastic strain. In the case of PMMA, where the mechanical behaviour may be approximated with a simple bilinear stress-strain curve, no frontal push pad can be observed before a constant average plastic strain of about 0.12 (Fig. 10-a), while the rear angle decreases in a first approximation linearly with the average plastic strain (Fig. 10-b).

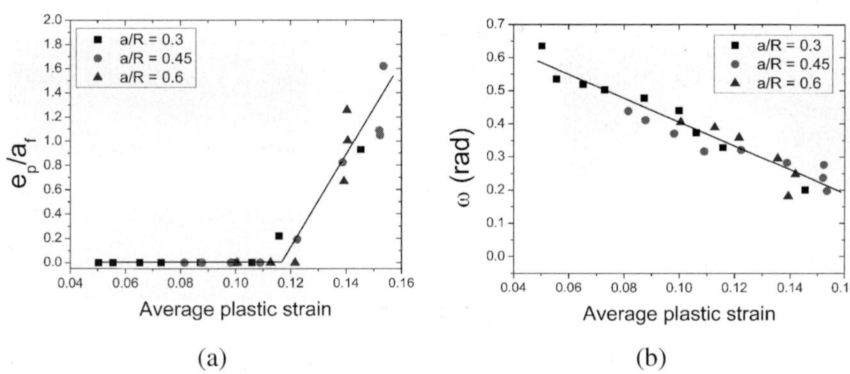

Fig. 10. (a) Frontal push pad defined by the ratio e_p/a_f (where e_p is the thickness of the push pad and a_f the plastic contact radius) and (b) rear angle ω of the contact as a function of the average plastic strain determined using Eq. (7).

All the qualitative results presented here were obtained by FEM. Since some of these results, especially those describing the evolution of the contact geometry, are in good agreement with experimental observations, it is not easy to imagine that FE results are more or less sensitive to the model used to generate them. In the following sections, we look at the influence of the mesh size and of the definition of the representative volume introduced to calculate the average plastic strain. The last section deals with the effect of the strain hardening ability of the material on the average plastic strain.

5.2. *Influence of the Mesh Size*

To demonstrate the influence of the mesh size, complementary numerical simulations were performed using the same domain division as for $a/R = 0.3$ with a tip radius R of 5.4 μm (Table 1). Adjusting the penetration depth and the tip radius as described in Table 2, we were able to simulate scratch tests at an imposed ratio a/R of 0.3, but with a smaller contact radius ($a_0 = 0.65$) and then with division of the number of elements in the contact area by a factor of 3 (Fig. 3-b). The contact status for frictionless contact and $a/R = 0.3$ for the two FE models developed is shown in Fig. 11.

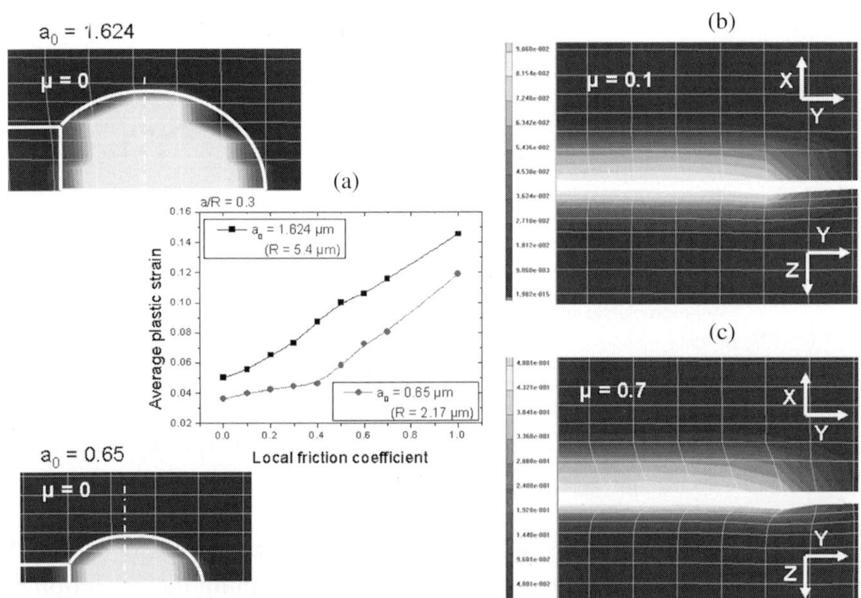

Fig. 11. (a) Evolution of the average plastic strain computed with the different models for $a/R = 0.3$ (Tables 1 and 2) as a function of the friction coefficient and corresponding equivalent plastic strain maps for (b) $\mu = 0.1$ and (c) $\mu = 0.7$. For each friction coefficient, the maps are shown in the planes $z = 0$ (top view) and $x = 0$ (side view).

Using Eq. (7) for a representative volume defined by a representative length $r = 3a_0$, we estimated the average plastic strain as a function of the friction coefficient in this new FE model. With at least 10 fewer elements in the contact zone, one observes a diminution of the estimated plastic strain for all simulated friction coefficients, as compared to the values found using the previous model with $a/R = 0.3$. The shift between the two curves is quite constant as a function of μ and in both cases the average plastic strain increases with the local friction coefficient.

An analysis of the corresponding equivalent plastic strain maps in Figs. 11-b and c shows that the plastic strain gradient is located in only 3 or 4 elements in the z direction. In this indentation direction, the number of plastically affected elements decreases with increasing friction coefficient. The plastic strain maps obtained for the model with a contact radius $a_0 = 0.65$ μm have to be compared to the maps obtained for the model with a contact radius $a_0 = 1.624$ μm (Figs. 6-a and c). At high values of μ, the plastic gradient is very high with a maximum equivalent plastic strain of about 0.48 and confined to elements in the sub-surface region (first layer of elements), as seen in Fig. 11-c.

5.3. *Definition of the Representative Volume*

Determination of the average plastic strain from numerical simulations is in fact not as easy as it might appear in the previous Section 4.2. The main problems concern the definition of the representative volume and the choice of the plastically deformed volume within this representative volume. As seen on the plastic strain maps of Figs. 6 and 7, only elements near the indenter at the front of the contact or in the rear part have a significant equivalent plastic strain. Unlike in the work of Bucaille *et al.*[19] who analyzed scratch tests with a conical indenter, values of the equivalent plastic strain do not fall to zero in the groove after several time increments. As explained by the authors,[19] this loss of plastic strain is mainly due to the re-meshing procedure. Moreover, the numbers of elements in the contact area and in the plastically deformed volume vary with the scratching conditions and the simulated materials.

Some differences are often observed between the size of the elements in the groove and that of the elements in contact with the indenter. To estimate the average plastic strain, Bucaille *et al.*[19] introduced a critical value of the equivalent plastic strain, denoted ε_C^p. These authors have clearly shown that the computed value of the average plastic strain depends on the initial chosen value of ε_C^p, which has to be adjusted as a function of the ratio E/σ_y. In the present study, we chose a critical value of the equivalent plastic strain of about 0.002, in order to limit the number of elements with a very low level of plastic strain. This critical value corresponds more or less to the error due to numerical integration during convergence for elasto-plastic contact at large strain.

The last key parameter is the size of the plastically deformed volume during scratching. In Johnson's expanding cavity model,[29] for the indentation of elastic, perfectly plastic materials with a spherical indenter, a hemispherical hydrostatic core of radius a, corresponding approximately to the contact radius, is surrounded by an incompressible plastic zone of outer radius r_c, defined by the following equation:

$$\left(\frac{r_c}{a}\right)^3 = 0.25\frac{E}{\sigma_y}\frac{a}{R} \qquad (9)$$

According to the above equation with E/σ_y = 35 (see Section 3), the ratio r_c/a varies between 1.4 and 1.8 as a function of the ratio a/R. To study the influence of the size of the representative volume, the average plastic strain was calculated for different volumes having the form of a right angled parallelepiped and a characteristic length successively decreasing from $r = 3a_0$ to $r = 1a_0$. Figure 12 presents the evolution of the average plastic strain as a function of the friction coefficient for a/R = 0.3 (with a_0 = 0.65 μm, see Table 2) and for different sizes of the representative volume. For the volume with a characteristic length $r = 1a_0$, the average plastic strain increases rapidly with μ to reach a maximum value of 0.35 for μ = 1. In addition, for a given friction coefficient, the average plastic strain is higher for the volume with $r = 1a_0$ than for those with $r = 2a_0$ and $r = 3a_0$, especially when the friction coefficient is high. This result does seem to be obvious because for a representative volume defined by $r = 1a_0$, only elements with the highest level of equivalent plastic strain on the plastic strain map are considered. In Fig. 12, we have included a side view of the elements selected for calculation of the average plastic strain for two values of the friction coefficient, $\mu = 0.1$ and $\mu = 0.7$.

With a further increase in the size of the representative volume, we do not observe any variation of the average plastic strain. The difference between the two curves for the volumes with characteristic lengths of $r = 2a_0$ and $r = 3a_0$ is less than 5%, even for a high friction coefficient ($\mu > 0.5$). These results are in good agreement with the expanding cavity model, which assumes a radius of the plastic zone greater than 1.35 for $a/R = 0.3$ according to Eq. (9).

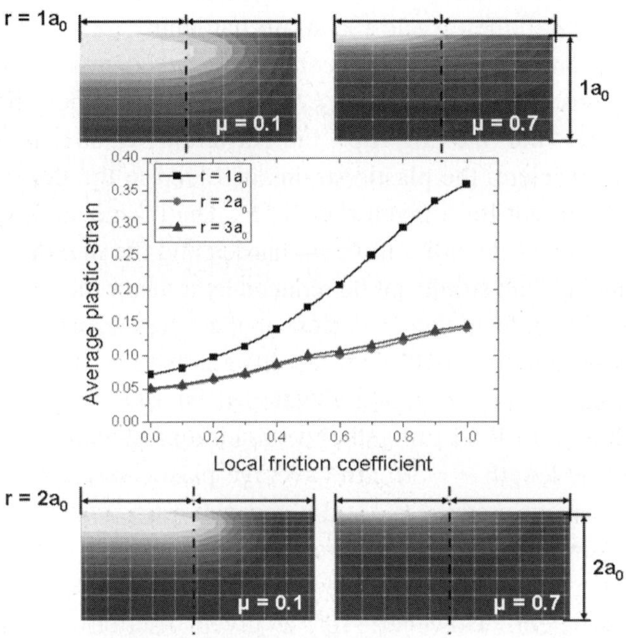

Fig. 12. Evolution of the average plastic strain determined using Eq. (7) as a function of the local friction coefficient for different sizes of the imposed representative volume.

5.4. *Influence of the Strain Hardening*

Plastic deformation during scratching with a spherical indenter depends mainly on the geometry of the tip (i.e. the ratio a/R) but also on the surface behaviour, especially the local friction coefficient. Although the results deduced from numerical simulations are more or less sensitive to the description of the domain in the finite element model, the observed trends correlate with *in situ* observations. Thus, as reported by several authors,[13,19-21] the mean contact strain during scratching is also known to depend on the rheology of the test material. In the case of an elastic, perfectly plastic material, Bucaille *et al.*[19] have shown by FEM that the average volume deformation for a conical indenter is a function of both the conical semi-apical angle and the ratio E/σ_y. Recently, Bucaille *et al.*[20] described the influence of the strain hardening ability of the material in scratch experiments with a spherical indenter for ratios a/R in the range 0.2 – 0.6 and $\mu = 0.3$. Introducing a flow stress defined by a G'sell-Jonas law,[24] they found the strain hardening to be a fundamental parameter among the scratching conditions, in contrast to the sliding contact response where the ratio E/σ_y controls the behaviour of the material under the moving tip. For different values of the strain hardening coefficient, the plastic strains involved in the deformation are completely different for a given ratio a/R. The plastic strain is lower in a material with a high ability to strain harden and the maximum value of the equivalent plastic strain can be reduced by a factor of 3.

Figure 13-a depicts the evolution of the average plastic strain as a function of the friction coefficient for different strain hardening abilities of the substrate. Using the model developed for $a/R = 0.3$ with a small contact radius $a_0 = 0.65$ μm and choosing a representative volume with a characteristic length $r = 3a_0$, the average plastic strain was calculated for various tangent moduli E_T and a constant ratio E/σ_y. As observed previously, for all input stress – strain curves with ratios E_T/E varying between 0.025 and 0.1, the average plastic strain increases with the local friction coefficient. However, for a given friction coefficient and especially for high values of μ, the average plastic strain increases as the work hardening slope E_T decreases. As a function of the local friction coefficient, there is a roughly constant factor in the range 2.5 to 3 between the average plastic strains calculated for the two materials with $E_T/E = 0.025$ and $E_T/E = 0.1$. Materials which may be considered in a first approximation to be elastic and perfectly plastic (i.e. E_T/E = 0.025) display a very high level of plastic deformation, as observed in

Fig. 13-b where the equivalent plastic strain maps (side view) are shown for $\mu = 0.4$. The maximum value of the equivalent plastic strain in the sub-surface region is close to 0.8. Conversely, the material with $E_T/E =$ 0.1 (see Fig. 13-c) exhibits a much lower level of equivalent plastic strain under the same scratching conditions, with a maximum value of about 0.3. The size of the plastically deformed volume diminishes slightly and it migrates towards the surface region as the tangent modulus decreases.

These findings are in good agreement with previous studies reported in the literature[13,19-21] and confirm that the mean contact strain during scratching with a spherical indenter is a complex function of the ratio a/R and the local friction coefficient, but also of the strain hardening and more generally the rheological properties of the material.

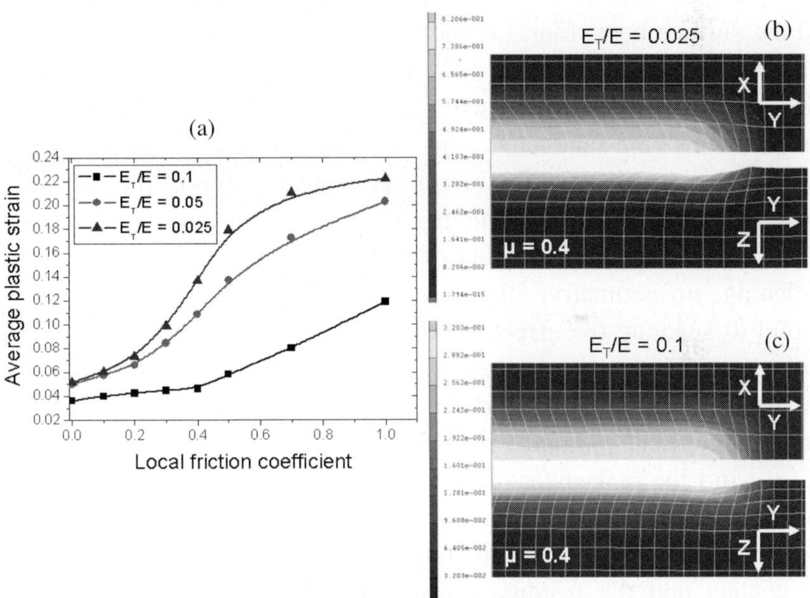

Fig. 13. (a) Average plastic strain computed by FEM using Eq. (7) with a representative volume $r = 3a_0$ as a function of the friction coefficient for different work hardening abilities. Equivalent plastic strain maps for (b) $E_T/E = 0.025$ and (c) $E_T/E = 0.1$ for a given friction coefficient $\mu = 0.4$.

6. Conclusion

A detailed study of the effect of the local friction coefficient on scratch testing was carried out. The results of numerical simulations reveal the influence of the local friction coefficient on the plastic deformation in scratch experiments with a spherical indenter. The local friction between the moving spherical indenter and an elasto-plastic surface not only increases the maximal value of the equivalent plastic strain, but also modifies the shape of the plastically deformed volume and the corresponding plastic strain gradients in the contact zone and the residual groove. For frictionless contact and materials with no strain hardening, the rule initially proposed by Tabor can be used as a first approximation to evaluate the mean contact strain imposed during the test. However, for contacts with friction the mean contact strain appears to be described by a complex function of the imposed geometrical strain, i.e. the ratio a/R, and the surface behaviour, i.e. the friction coefficient. The mean contact strain can be estimated from FE simulations by calculating the volume average plastic deformation, using a model with at least 15 elements in contact with the indenter and a representative volume with a characteristic length of at least 2 times the mean radius of the contact.

The contact between a moving indenter and a polymeric surface is very complex due to the number of dependent parameters (friction, strain hardening, tip geometry, sliding velocity,...) and FE modelling is very helpful to separate the effects of the different parameters on the scratch response. If the results obtained by FEM are generally subject to caution due to their sensitivity to the description of the domain, the evolutions of some geometrical parameters such as the normalized frontal push pad e_p/a_f and the rear angle of contact ω as a function of the ratio a/R and the friction coefficient are in good agreement with those observed experimentally in a specific scratch tester allowing *in situ* observation of the contact and the residual groove. The FE results indicate that the geometrical parameters e_p/a_f and ω are directly and linearly related to the level of plastic deformation in the contact, or more exactly to the mean contact strain. For a given rheology, the geometrical parameters seem to be unique for a specific mean contact strain, whatever the scratching conditions $(a/R, \mu)$. This suggests that it will soon be possible to estimate the mean contact strain in a scratch test using a reverse method based on the analysis of such geometrical parameters. Nevertheless, as described in the last section of this chapter, the level of elastoplastic deformation depends mainly on the rheology of the material, especially the strain hardening coefficient and the ratio E/σ_y. Hence, new scratch experiments on various amorphous polymeric surfaces and the corresponding numerical simulations are now in progress to study the influence of rheological parameters and the strain hardening rule.

References

1. B. J. Briscoe, P. D. Evans, E. Pelillo, S. K. Sinha, *Wear*, **200**, 137 (1996).
2. J. A. Williams, *Tribol. Int.*, **29**, 675 (1996).
3. L. Lin, G.S. Blackman, R.R. Matheson, *Progress in Organic Coatings,* **40**, 85 (2000).
4. C. Xiang, H.-J. Sue, J. Chu, B. Coleman, *J. Polym. Sci. Polym. Phys.*, **39**, 47 (2001).
5. M. Wong, G. T. Lim, A. Moyse, J. N. Reddy, H.-J. Sue, *Wear*, **256**, 1214 (2004).
6. B. J. Briscoe, P. D. Evans, S. K. Biswas, S. K. Sinha,, *Tribol. Int.*, **29**, 93 (1996).
7. B. J. Briscoe, E. Pelillo, S. K. Sinha, *Polym. Eng. Sci.*, **36**, 2996 (1996).
8. V. Jardret and P. Morel, *Progress in Organic Coatings*, **48**, 322 (2003).
9. C. Gauthier and R. Schirrer, *J. Mater. Sci.*, **35**, 2121 (2000).
10. C. Gauthier, S. Lafaye, R. Schirrer, *Tribol. Int.*, **34**, 469 (2001).
11. S. Lafaye, C. Gauthier, R. Schirrer, *Tribol. Int.*, **38**, 113 (2005).
12. I. Demirci, C. Gauthier, R. Schirrer, *Thin Solid Films*, **479**, 207 (2005).
13. C. Gauthier, A.-L. Durier, C. Fond, R. Schirrer, *Tribol. Int.,* 39, 88 (2006).
14. G. M. Hamilton and L. E. Goodman, *Journal of Applied Mechanics*, **33**, 371 (1966).
15. R. B. King and T. C. O'sullivan, *Int. J. Solids Structures*, **23**, 581 (1987).
16. S. Lafaye, C. Gauthier, R. Schirrrer, *J. Mater. Sci.*, **41**, 6441 (2006).
17. F. P. Bowden and D. Tabor, Friction and lubrification of solids, Oxford University Press, London, 1951.
18. D. F. Moore, The friction and lubrification of elastomers, Pergamon Press, London, 1972.
19. J. L Bucaille., E. Felder, G. Hochstetter, *Wear*, **249**, 422 (2001).
20. J. L. Bucaille, C. Gauthier, E. Felder, R. Schirrer, *Wear*, **260**, 803 (2006).
21. E. Felder and J. L. Bucaille, *Tribol. Int.*, **39**, 70 (2006).
22. K. Holmberg, A. Laukkanen, H. Ronkainen, K. Wallis, S. Varjus, *Wear*, **254**, 278 (2003).
23. J. Li and W. Beres, *Wear*, **260**, 1232 (2006).
24. C. G'Sell and J. J. Jonas, *J. Mater. Sci.*, **14**, 583 (1979).

25. D. Tabor, The hardness of solids, Proceedings of the Institute of Physics F, Physics in Technology, **1**, 145 (1970).
26. J. L. Bucaille, E. Felder, G. Hochstetter, *J. Mater. Sci.*, **37**, 3999 (2002).
27. S. Jarayman, G. T. Hahn, W. C. Oliver, C. A. Rubin, P. C. Bastias, *Int. J. Solids Structures*, **35**, 365 (1998).
28. D. M. Marsh, *Proc. R. Soc. Lond.*, **A 279**, 420 (1964).
29. K. L. Johnson, *J. Mech. Phys. Solids.*, **18**, 115 (1970).
30. X. L. Gao, X. N. Jing, G. Subbash, *Int. J. Solids Structures*, **43**, 2193 (2006).

CHAPTER 5

NANOINDENTATION AND INDENTATION CREEP OF POLYMERIC MATERIALS

Kaiyang Zeng

Department of Mechanical Engineering, National University of Singapore
9 Engineering Drive 1, Singapore-117576
E-mail: mpezk@nus.edu.sg

Indentation, in general, is a well-known experimental technique to determine hardness of materials. Nanoindentation is a later development of the technique which can be used to determine elastic modulus and hardness at submicron to nanometer levels. However, application of the nanoindentation to polymeric materials is still a challenging issue as for polymeric materials, the rate- and time-dependent properties of the materials are not included in the conventional nanoindentation analysis, and this usually results in over estimation of the elastic modulus and hardness from the nanoindentation tests for the polymeric materials. In addition, the indentation unloading curve of polymers appears to be dependent on both the holding time at the maximum load and the unloading rate; this further leads to certain uncertainty and errors during the nanoindentation experiments and analysis.

This paper will review the current understanding of the experimental and analysis of nanoindentation of polymeric materials. First, traditional analysis of nanoindentation load-displacement data will be presented, followed by discussions of the uncertainties and problems when applying the traditional analysis to the nanoindentation of polymeric materials. The paper will then focus on some of the central issues with applying nanoindentation techniques to the polymeric materials including: (i) how to modify the commonly-used nanoindentation unloading analysis; (ii) how to characterize the time-dependent properties through indentation creep experiments; (iii) how to analysis the rate-dependent properties during nanoindentation experiments; and (iv) the use of the atomic-force microscopy or scanning probe microscope to the polymeric materials as well as the comparison with the conventional nanoindentation experiments where appropriate. When appropriate, the discussions are aided with the results from several research projects in author's group.

1. Introduction

Polymers and polymer-based composites are the most common engineering materials used as functional and structural applications as well as in our daily life, they are also potentially ideal nanoscale building blocks because of their length scale, well-defined architecture, controlled chemicals and synthesis, ease of processing and wide range of functionality that can be incorporated. However, polymeric materials are susceptible to damage by indentation, scratching and abrasive wear as well as other modes of mechanical contact. Such processes impair the appearance and also reduce the mechanical properties of the materials by introducing surface or bulk flaws. In general, mechanical properties of bulk polymeric materials can be determined through conventional mechanical testing methods, such as tensile or bending tests. However, these bulk properties are generally not the same as the properties at surface or sub-surface region, since the polymer structures at surface or sub-surface region are very different from that of the bulk. It is therefore necessary to study the surface or sub-surface mechanical properties of polymeric materials in more detail.

Nanoindentation, also known as ultra-low load indentation, or depth-sensing indentation, or instrumented indentation, or controlled indentation, is a technique developed over the last twenty years for probing the mechanical properties of materials at very small scale or at surface or sub-surface regions. It is perhaps the simplest and most direct way to probe the surface or sub-surface mechanical properties of a material, and can be used to measure the elastic and plastic properties of the materials from a very small volume of testing materials.[1-16] Nanoindentation is now used routinely to characterize the mechanical properties of thin films, thin surface layers, secondary phases (in composite materials), or any new materials which the large size specimens are not available or difficult to be obtained. Similar to microindentation tests, nanoindentation technique uses a sharp pyramid indenter tip; a load is applied to the tip so that the tip is pressed into the surface of the testing materials. In addition, nanoindentation technique uses high-resolution-controlled system to measure the applied load *(P)* to an indenter tip, the penetration depth *(h)* into the sample from the surface by the indenter tip under the applied load, or strain rate (defined by the ratio of the penetration rate and penetration depth as *dh/dt/h)* and time *(t)*. Nanoindentation technique provides the capabilities to record the applied load *(P)*, the penetration depth *(h)*, and indentation time *(t)* continuously during loading and unloading cycle of the indentation. The recorded load and penetration depth forms so-called load-penetration curve (also called load-displacement curve, or *P-h* curve) which can then be used to derive elastic modulus *(E)* and hardness *(H)* of the tested materials in general.

The nanoindentation technique can also provide an understanding of the elastic and plastic deformation behavior of materials at the surface or sub-surface region in a sub-micron scale. The elastic-plastic deformation behavior of the materials during contact can be correlated to many other mechanical properties of the materials, leading to many applications of the nanoindentation technique. This technique has been used in the mechanical characterization of thin films and thin surface layers, such as those modified by ion implantation and other energetic beam techniques, or those structures or systems used in microelectronic devices, MEMS devices or structures, display, and biology devices, which thin films or multilayer structures are widely used. There are now several commercially available nanoindentation systems in the market. In additional, some of the nanoindentation systems are based on scanning probe microscopy (SPM) technique, in which the applied force as small as a nano-Newton and displacement of about an Angstrom can be accurately measured. One of the great advantages of the nanoindentation technique is that many mechanical properties can be solely determined by just analyzing the indentation load-penetration curve or data alone, thereby avoiding the need to image the indentation impression and therefore facilitating the property measurement at the sub-micron scale. Another advantage of the technique is that nanoindentation measurements can be made without having to remove the film or surface layer from its substrate. This simplifies the sample preparation procedures and makes the properties measurements possible in the systems which would otherwise be difficult to test, as is the case for most ion beam modified materials. The third advantage is that a large variety of mechanical properties can be obtained through small amount of samples and few experimental runs. These mechanical properties include elastic modulus, hardness, yield strength, strain hardening, residual stress, interface toughness and many others. This technique is also especially suitable for characterizing mechanical properties of new materials, as for any newly developed materials, their availabilities are usually limited in the development stage.

There are many excellent review articles, journal special issues or books on the topic of nanoindentation experiments and analysis.[6-16] This chapter will therefore only focus on the topic of nanoindentation of polymeric materials, as nanoindentation of polymers, both experiments and analysis, are very different from the cases of nanoindentation of metals or ceramics. Therefore, in this paper, special techniques developed for nanoindentation of polymers will be discussed and summarized. This paper consists of following sections: after this introduction section, a brief discussion of basic nanoindentation analysis will be summarized in general; followed by the discussions on the problems when using nanoindentation techniques to polymeric materials; some general approaches used for nanoindentation of polymers will then be discussed; thereafter, two specific problems: rate-dependent and time-dependent properties will be discussed, in particular, indentation creep will be discussed in detail; some theoretical analysis of nanoindentation of polymers and nanoindentation based on scanning probe microscopy will be also discussed briefly, and at last, some general summary, conclusions and implications will be given.

2. Brief Summary of Basic Nanoindentation Analysis

Nanoindentation systems are now being built by several companies and many research groups in the world have also built custom-made systems for studying indentation characteristics at the sub-micron scales. Typically, very sharp three-side pyramid diamond indenters (named as Berkovich indenter[17]) are used in these instruments to allow the indentation to be made in a very-localized region. The Berkovich indenter tip is used in most of nanoindentation systems because the Berkovich tip is usually much sharper than other type of indenter tips (Vickers, cone or Knoop indenters). Figure 1 shows some typical nanoindentation load-displacement curves (P-h curves) taken from several bulk polymeric materials such as polycarbonate (PC), polymethylmethacrylate (PMMA), Polyester (PET); and Epoxy (Figs. 1(a) to 1(d)), which are indented using the Berkovich indenter tip. In Fig. 1, the characteristics of nanoindentation of those polymeric materials are also compared with nanoindentations of both soft and hard metals (Al and W), as well as ceramics (Sapphire and Si) (Figs. 1(e)-1(h)). The general features of the Fig. 1 show that, regardless of the material being tested, as the load, P, increases, displacement (penetration depth by indenter tip into the sample surface), h, increases due to both elastic and plastic deformation of the testing material under a the indenter tip. If the indenter is held under constant load, depending on the materials, deformation due to viscoelasticity in the material could also occur. Finally, elastic deformation is recovered as the load is removed.

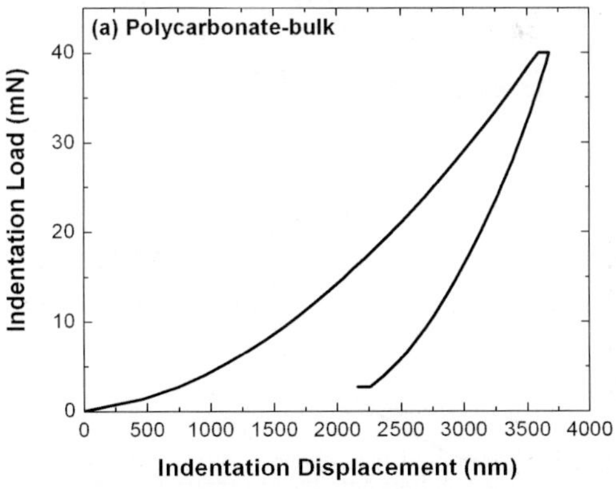

Fig. 1. Nanoindentation load-displacement curves of several polymeric materials: (a) PC; (b) PMMA; (c) PET; (d) epoxy, as well as the comparison with that of the metals and ceramics: (e) Al as a soft metal; (f) W as a hard metal; (g) Sapphire as a ceramic material; and (h) Si as a ceramic material.

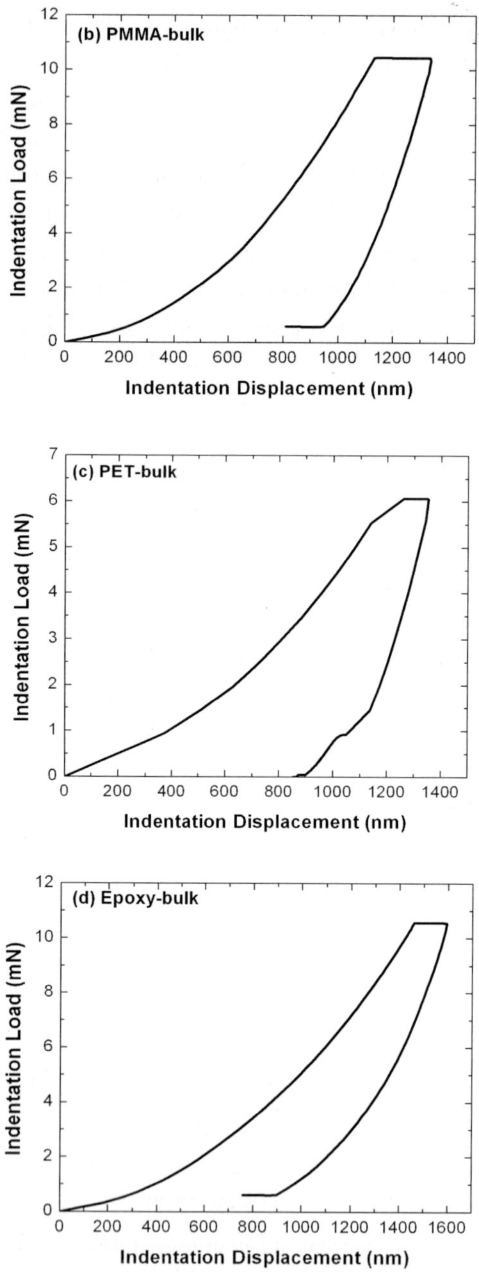

Fig. 1. (*Continued*)

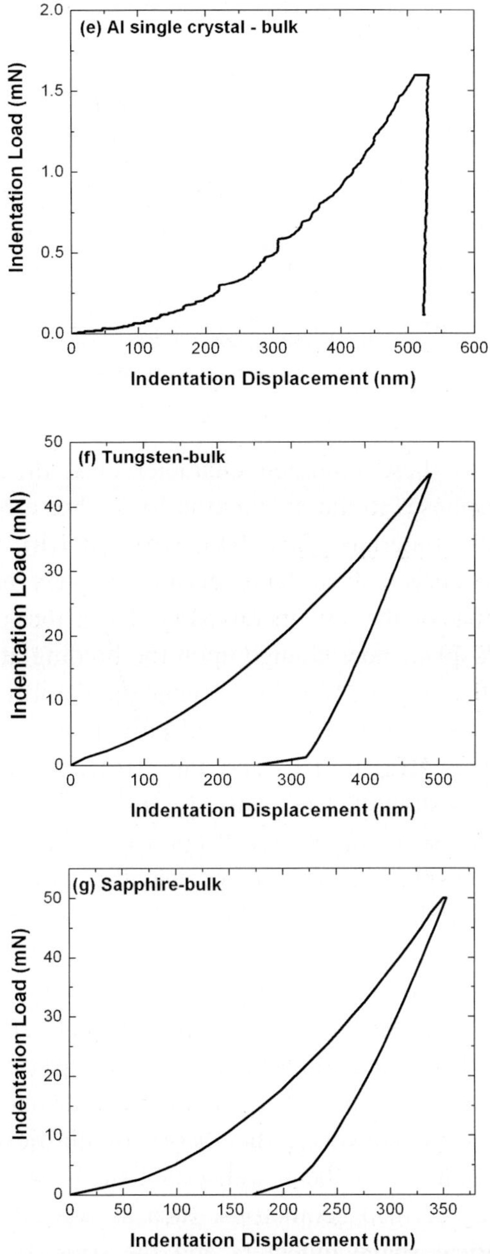

Fig. 1. (*Continued*)

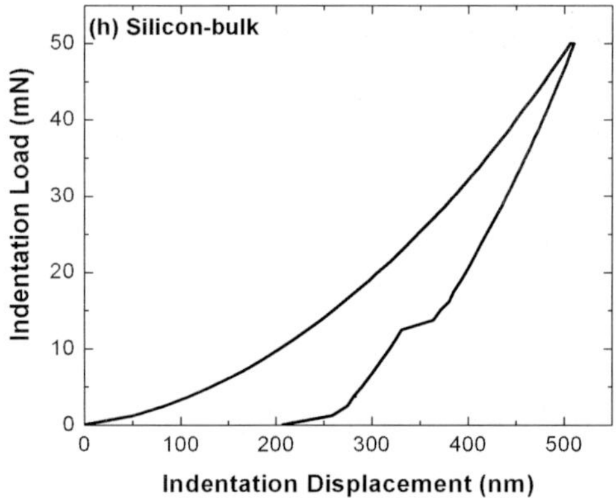

Fig. 1. (*Continued*)

In additional to these common characteristics, different materials show different responses to the indentation loads. For example, for hard metals and ceramic materials, Figs. 1(f), 1(g) and 1(h), the indentation load-displacement curves show large elastic recovery upon unloading comparing with that of the soft metals (Fig. 1(e)); these materials also show very little displacement change upon the holding at the maximum load, indicating the neglectable time-dependent deformation for these materials, whereas for the soft metals and polymeric materials, such as those shown in Figs. 1(a) to 1(d) and Fig. 1(e), the indentation load-displacement curves show that not only the materials are softer, but the elastic recovery is also relatively small upon unloading (in comparison with the total displacement), there usually have a significant displacement change during the holding at the maximum load, showing creep-like behavior, or time-dependent deformation in these materials.

Indentation by sharp pyramid tips creates a very complex three-dimensional stress state in the material. An adequate analytical model, which includes both plastic and elastic deformation processes, has yet to be developed, especially for sharp pyramid indenter tips such as Berkovich or Vickers. However, the theory of elastic contact of the axisymmetric indentation has been well-established, especially for those indenters with well-defined geometries, such as conical, spherical, and flat-ended cylindrical shape indenters and the stress-fields created by these indenter tips can be simplified as two-dimensional cases. Thus, elastic contact model has been adopted in the interpretation of nanoindentation load-displacement curves.[18-19] In the following sections, we will briefly discuss the two methods that were used to analyze nanoindentation load-displacement curves, i.e., Doerner and Nix method,[20] and Oliver and Pharr's method.[21]

2.1. *Doerner and Nix's Analysis Method*

Doerner and Nix were the first to develop a comprehensive method for interpreting nanoindentation data by providing a means of determining the contact area during nanoindentations.[20] They have observed that, for many materials, especially for those metallic materials, the data fall very linearly on a straight line during the initial stages of the unloading, at least within 1/3 of the maximum load. They thus adapted the Sneddon's elastic analysis[18] based on the following assumptions:[22-23]

- The indenter can be modeled as a rigid flat punch, i.e., the contact area remains unchanged during the initial stages of unloading; and
- During the unloading, all of the material in contact with the indenter is plastically deformed and that outside the contact only elastic deformation occur at the surface.

Doerner and Nix[20] therefore applied their model by fitting a straight line to the top 1/3 of the unloading curve and extrapolating this fitted line to the depth axis. The slope of the fitted line gives directly the contact stiffness, the fitted line can be written as:

$$P_{fit} = Sh + h_0 \tag{1}$$

where S is the slope of the upper 1/3 unloading curve, it is also called unloading stiffness.

The intercept of the fitted line at the depth axis, h_i, can be found by setting $P_{fit}=0$, therefore:

$$h_i = \frac{-h_0}{S} \tag{2}$$

The examples of the linear fitting and the intercept are shown in Figs. 2(a) and 2(b) for nanoindentation of bulk PMMA and epoxy.

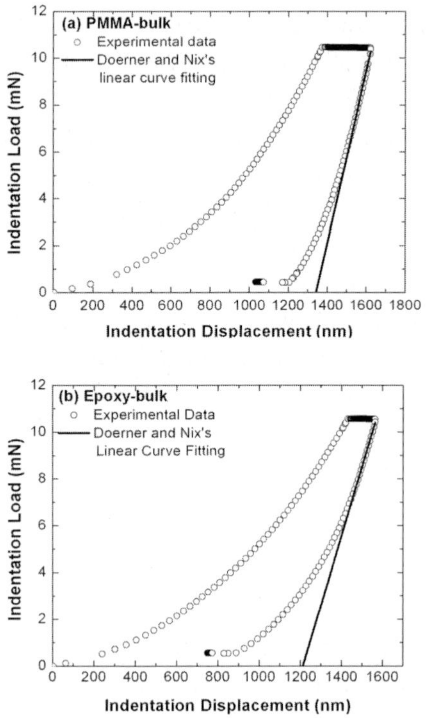

Fig. 2. Experimentally measured nanoindentation load-displacement curves (data points) and the linear fitting of the top 1/3 of unloading curve (solid lines) as proposed by Doerner and Nix's method[20] for (a) PMMA and (b) Epoxy.

Doerner and Nix's method[20] indicated that the intercept at the depth axis is the plastic contact depth, and this plastic contact depth can be then used to determine the contact area, so that the hardness of the testing materials can be determined as:

$$H = \frac{P_{\max}}{A} \tag{3}$$

where H is hardness, P_{max} is the maximum indentation load and A is the corresponding projected area of the indentation impression at the maximum indentation load. The elastic modulus is determined as:

$$S = \frac{dP}{dh} = 1.167 E_r \sqrt{A} \tag{4}$$

In Eq. (4), S is the unloading stiffness and it is determined through the linear curve fitting of the upper *1/3* of unloading curve. The constant *1.167* was proposed by King[24] based on finite element simulation of the elastic-half space indented by a flat-ended punch with triangular shape. This constant was also confirmed by Larsson *et al.* for Berkovich tip using a three-dimensional finite element analysis.[25] E_r is called reduced modulus and it is determined as:

$$\frac{1}{E_r} = \left(\frac{1-\nu_i^2}{E_i} \right) + \left(\frac{1-\nu_m^2}{E_m} \right) \tag{5}$$

where E and ν are elastic modulus and Poisson's ratio, and the subscripts i and m refer to the indenter and sample, respectively, for the commonly used diamond Berkovich indenter tip in the nanoindentation experiments, $E_i = 1141\ GPa$ and $\nu_i = 0.07$.[21] In Eqs. (4) and (5), the contact area A is determined through the geometry of the indenter tip based on the true indentation depth. Ignoring the pile-up or sinking-in of the materials around the indentation impression, also assuming the indenter tip is perfectly sharp, the indentation area can be written as:

$$A = 24.56 h_i^2 \tag{6}$$

Based on Doerner and Nix's analysis,[20] if the shape of the indenter is accurately known, the contact area A can be calculated directly from the plastic depth as shown in Eq. (6). Hence the elastic modulus and hardness of the materials can be determined using Eqs. (3)-(5). This method was widely used in the earlier stage of the development of nanoindentation techniques in 1980s, and provided relative accurate results for most of metallic materials.

Doerner and Nix[20] assumed that the unloading curve can be modeled as a linear curve, at least for the initial part of the unloading data, this assumption does not have significant deviation if the deformation of the materials is dominated by plastic deformation such as in the case of nanoindentation of metallic materials, especially for those soft metals, such as Al or Cu. However, it is observed by many researchers that the unloading data never fall exactly along a straight line all the way to zero load, even at the beginning of the unloading curve, non-linear behavior could be dominated for many of the materials such as ceramics or other metallic materials, therefore the linear unloading is, for the most, a simplification, and this simplification could lead to a large error particularly for nanoindentation of hard materials such as ceramics or hard metals. The slope of the unloading data, i.e., the contact stiffness, and consequently, the contact area, A, decrease continuously during unloading due to elastic recovery outside and within the indentation contact, this change occurs immediately and continuously as the indenter is withdrawn from the maximum load. Therefore, the flat-ended punch approximation is not an entirely adequate description of real material behavior during the nanoindentation. These problems have led to further development and modification of the analysis method. The most well-known and accepted method used today is the one developed by Oliver and Pharr in early 1990s.[21]

2.2. *Oliver and Pharr's Analysis Method*

Oliver and Pharr[21] were the first to account for above-mentioned problems and they have carefully examined the nanoindentations of six different materials including Aluminum, Tungsten, soda-lime glass, Fused silica, Quartz and Sapphire, representing soft to hard materials. An important observation of their work is:

- The unloading curve of the nanoindentation can be accurately described by a power-law relation:

$$P = \alpha\left(h - h_f\right)^m \tag{7}$$

  where h_f, the "final depth", is the depth at which the indenter is last in contact with the material, and m and α are constants, P is the load during unloading, therefore, the quantity $(h-h_f)$ is the elastic deformation.

In viewing of this analysis, Oliver and Pharr proposed a new insight of the indentation process,[21] their analysis on the unloading curve of the nanoindentation load-displacement curves of two polymers (PMMA and epoxy) are shown in Figs. 3(a) and 3 (b). The unloading curve is fitted by the power-law relation as expressed by Eq. (7). Comparing to Fig. 2, the power-law equation, Eq. (7), can fit the unloading curve well for most parts of the unloading data, except at the very last portion of the unloading curve. In this method, the initial unloading slope is found by analytically differentiating Eq. (7), and evaluating the derivative at the maximum load and displacement, i.e.,

$$S = \frac{dP}{dh} = \alpha m \left(h_{\max} - h_f\right)^{m-1} \tag{8}$$

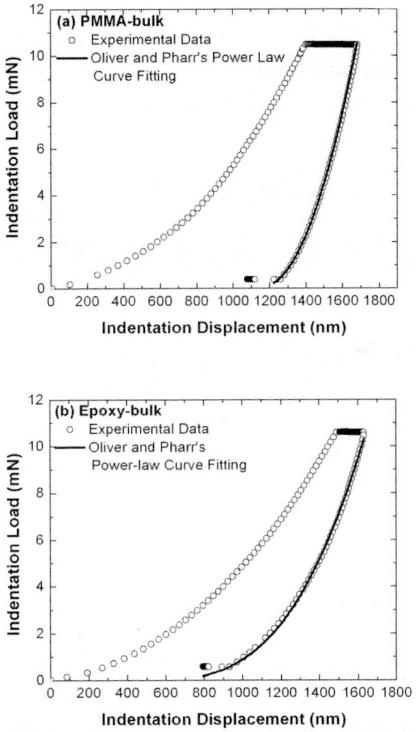

Fig. 3. Experimentally measured nanoindentation load-displacement curve (data point) and the power law fitting of the unloading curve (solid lines) as proposed by Oliver and Pharr's method[21] for (a) PMMA and (b) Epoxy.

The contact depth can be determined as:

$$h_c = h_{max} - \varepsilon \frac{P_{max}}{S} \tag{9}$$

where ε is a geometric constant, and the value of which is given as:[21]

for conical indenter:	$\varepsilon = 0.72$
for paraboloid of revolution:	$\varepsilon = 0.75$
for flat-ended punch:	$\varepsilon = 1$

Oliver and Pharr's method[21] is consistent with the Doerner and Nix's approach[20] when the flat-ended punch geometry is assumed. However, for the conical and paraboloid indenters, the contact depths are greater than those for the flat-ended punch, and this must be taken into account for the analysis in using these indenter geometries if accurate measurements are to be obtained.

Once the contact depth is determined, the contact area A is then determined as:

$$A = F(h_c) \tag{10}$$

Table 1. Summary of the results from Oliver and Pharr's experiments and analysis (data from Ref. 21).

Material	Modulus determined from Oliver and Pharr's method (GPa)	Literature modulus (GPa)	Poisson's ratio
Aluminum	68.0 ± 0.93	70.4	0.347
Quartz	124.0 ± 0.54	95.0	0.077
Soda-lime glass	69.9 ± 0.22	70.0	0.23
Sapphire	441 ± 4.70	403.0	0.234
Fused Silica	69.3 ± 0.39	72.0	0.17
Tungsten	410 ± 4.70	09.8	0.280

A careful calibration of the indenter tip gives the indenter area function is a polynomial equation as follows:

$$A = 24.56h_c^2 + C_1h_c + C_2h_c^{1/2} + C_3h_c^{1/4} + \ldots + C_8h^{1/128} \tag{11}$$

where C_1 through C_8 are fitting constants. The leading term describes a perfect Berkovich indenter geometry; and the others describe deviations from the perfect Berkovich geometry due to the tip blunting. These constants can usually be determined through a calibration test on Fused Silica (SiO_2) by assume the elastic modulus of 72 GPa and it is now a standard procedure in most of control and analysis software for the nanoindentation instruments. Oliver and Pharr have thoroughly analyzed the nanoindentation results of six materials and their results can be summarized in Table 1.[21] This method has been extensively used and verified for large variety of materials. However, the Oliver and Pharr's method was derived based on the experiments on six materials, ranging from metals to ceramics; none of the materials are polymeric materials.

3. Problems of Nanoindentation of Polymeric Materials

Nanoindentation experiments and analysis are widely used in characterizing the mechanical properties of advanced materials, in principle; it should be also applicable to the polymeric materials, other soft materials, and polymer-based composites provided that an appropriate analysis is conducted on the nanoindentation data.[26-36] However, a numbers of problems exist when applying the nanoindentation techniques to polymeric materials using a conventional nanoindenter or a scanning probe microscopy (SPM) system. These problems have been summarized by VanLandingham *et al.* in their review papers.[33-34] In general, indentation measured elastic modulus, E, is significantly higher than the values measured using tensile testing or dynamic mechanical analysis (DMA). In addition, due to so-called indentation size effects, the elastic modulus at the surface is always higher than the values obtained from the bulk materials; this has led to large errors and uncertainties in the nanoindentation measurement of polymer properties.

Another problem is that, beneath the sharp indentation tips, as well as the effects of localized stress fields created by the indentations, polymeric materials exhibit creep-like behavior even at the room temperature, i.e., the indenter tips continuous penetrated into the materials during holding at a constant load, or even during unloading, which may be manifest as displacement at fixed load, or as a forward-going "nose" during unloading. The second case results in negative unloading stiffness, therefore the elastic modulus cannot be calculated in the conventional way as that using Oliver and Pharr's method.[21] In the first case, the degree of creep-like behavior depends on the material, loading level, rate and time, these factors lead to the properties measured by nanoindentations, such as elastic modulus and hardness, to be time-dependent, or rate-dependent, and are not consistent values.

The third problem is that, the point taken as representing initial contact *(h=0)* generally corresponds to a significant applied load, such that the penetration of the indenter tip into a polymeric surface could be significant relative to the maximum indentation displacement, h_{max}. This could lead to a significant error in estimating the contact area, hence to the values of hardness and elastic modulus, this is especially true if one indenting very soft polymeric materials, such as polymer films, or gels, or at a very small indentation depth, such that by using SPM systems. Usually, using SPM systems may improve the accuracy of the determination of the initial contact point.

Those problems associated with nanoindentation of polymeric materials, in general, do not exist when indenting ceramic or metallurgical materials, except some very soft metals, such as tin or lead, and some solder alloys (lead-tin, or other lead-free solder materials). For nanoindentation of these soft metals, similar behaviors, as those in the nanoindentation of polymeric materials, are observed, i.e., significant creep-like behavior during nanoindentation or time/rate-dependent indentations. The approaches to the nanoindentation of these soft metals are somewhat similar to those used for nanoindentation of polymers. Therefore, the analysis to be described in this chapter, in principle, should also be applicable for analyzing the nanoindentation of soft metals or alloys, especially, for various solder alloys.

4. Approaches to the Nanoindentation of Polymers or Other Soft Materials

In general, nanoindentation can be used to study the time-dependent properties of the materials. The analysis methods for such purposes are proposed by numbers of researchers.[35-46] There are several approaches that are usually used to evaluate the time-dependent nature, or rate-dependent nature of the deformation processes that occur in the polymeric materials or other soft materials during nanoindentations.

4.1. *Load Relaxation Experiments*

The first procedure is load relaxation or stress relaxation experiments using indentation experiments and was summarized by LaFontaine *et al.*[42] This experimental procedure is analogous to the experiment in bulk specimens using conventional mechanical testing procedure. In the uniaxial deformation test, the crosshead motion of the testing machine is suddenly stopped and the system is then held at a constant strain – the stress (or load) decreases with time according to the combined modulus of both the specimen and the testing machine. This method can also be applied to other materials such as thin films.[47-48]

In an *indentation load relaxation* test, or by short ILR test, the indenter is first pushed into the test sample at a fixed displacement rate until a predetermined load or displacement is achieved. The position of the indenter is then fixed. The material below the indenter is elastically supported and will continue to deform in a non-elastic manner, thereby tending to push the indenter further into the material. Load relaxation is achieved by the conversion of elastic strain in the machine and of the sample to inelastic strain in the sample. During the test, the load and position of the indenter and the specimen are continuously monitored. Normally, the indenter motion is held constant and the changes in the load are monitored as a function of time. It is possible to obtain the plastic indentation rate from the indentation load and total depth information during a relaxation test run. The data are typically plotted as a double logarithmic plot of the indentation pressure versus the normalized plastic indentation rate (plastic indentation rate/plastic indentation depth at the start of the relaxation test). The indentation data from experiments were modeled by a flow equation which has the form of:[42]

$$\ln\left(\frac{\Re^*}{\Re}\right) = \left(\frac{\dot{x}^*/h}{\dot{x}/h}\right)^{\lambda} \tag{12}$$

where

$$\Re(x,t) = P(t)/A(x,t) \tag{13}$$

is identical to the hardness obtained in a continuous indentation test, *P(t)* is the indentation load as a function of time, *A(x,t)* is the projected area of indentation, *x* represents the plastic indentation depth, and can be obtained by:

$$x = h - P/s(P) \tag{14}$$

in Eq. (14), *h* and *P* are the total indentation depth and applied load, respectively, *s(P)* is a spring constant, representing the machine and the elastic part of the specimen in series. In Eq. (12), $\Re^*$ can be interpreted as state variable hardness parameter [31], $\Re$ represents indentation stress, $\dot{x}^*/h$ is a rate parameter, and $\dot{x}/h$ is the plastic indentation rate and λ is a shape parameter. This model has been used to fit the indentation load-relaxation data measured on (111) single crystal of aluminium and 0.75 μm Al-2% Si film with a shape parameter λ=0.15, and showing good agreement with the experimental data.[42] However, this test cannot be done with the most popular nanoindentation instruments since most of the nanoindentation systems are load-control instruments, it is possible when using a nanoindenter with a proper displacement actuator (displacement control indentation).

4.2. *Time-Dependent Properties*

A second technique for analyzing the time-dependent properties during indentation was introduced by Mayo and Nix.[37-39] Two experimental procedures were proposed by them. In the first one, known as the *Constant Rate of Loading Test,* or in short CRL test, individual indentations are performed at a pre-specified loading rate that is varied from indentation to indentation by a factor of two or so. The data are analyzed by defining two quantities: (a) an indentation stress:

$$\sigma = \frac{P}{A} \tag{15}$$

where P is the load on the indenter and A is the projected area under the indenter at the current depth; and (b) indentation strain rate as:

$$\dot{\varepsilon} = \left(\frac{1}{h}\right)\left(\frac{dh}{dt}\right) \tag{16}$$

In Eq. (16), h represents the plastic depth; the plastic depth at the point of maximum load is obtained by extrapolating the linear portion of the load versus depth unloading curve to zero load, or using Oliver and Pharr's analysis.[21] The values of the stress and strain rate from tests performed at different loading rates were compared at a common indentation depth. The strain rate sensitivity, m, is then determined as:

$$m = d[\log \sigma] / d[\log \dot{\varepsilon}] \tag{17}$$

The second experimental procedure proposed by Mayo and Nix[37-39] is termed as *Loading Rate Change Test,* or in short as LRC test. In this test, the loading rate is held constant until a specified depth, but is suddenly changed to a new value and the subsequent changes in stress and plastic strain rate are monitored. Through a determination of the steady state stress exponents, it was possible to distinguish between pure metal and alloy type creep behavior. Mayo and Nix used these procedures to determine the steady state stress exponents in Pb, Sn and its alloy.[39] However, these procedures have two principal difficulties for some materials such as TiO_2. First due to the high hardness of some materials, measurements at low loading rates were complicated by thermal drift problems and therefore low strain rate tests could not be performed. Second, the statistical variations in hardness from point-to-point swamped the systematic increases in stress associated with changes in loading rate. In order to overcome these difficulties, Mayo *et al.* has proposed another procedure.[39] In this procedure, as soon as the indenter contacts the sample surface, the indenter loading rate is instantaneously increased to a high value (e.g. 1.85 mN/s) and this value is maintained until the indenter reaches a depth of 1 μm. The load is held constant at this point, typically for about 50 – 60 s. The indentation stress is calculated by dividing the applied load by the projected area, same as described before. The initial fast descent rate of the indenter produces a substantial amount of creep during the constant hold period. A range of descent rates are realized ranging from high values at the beginning of the hold period to smaller values as the material stops deforming. These correspond to a range of strain rates and by dividing the hold period into 10-s segments, it was possible to obtain at least five stress-strain rate pairs. Again, the slope of a double logarithmic plot of stress versus strain rate yields the strain rate sensitivity. These testing procedures were later classified and summarized by Raman and Barriche.[40]

In performing constant load holding experiments, especially for these materials sensitive to the time of loading (stressing) such as polymeric materials, there are some problems associated with this experiment. One of the problems is that different holding time leads to the different maximum indentation depth and hence the true contact depth determined by using either Doerner and Nix's method,[20] or Oliver and Pharr's method,[21] also different unloading stiffness resulting from the different holding time, therefore, elastic modulus and hardness decrease with increasing the holding time. This leads to the elastic modulus and hardness to depend on the experimental conditions, such as holding time and maximum force. It is therefore needed to analyze the constant load-hold segment more closely. This will be discussed in more details in the later sections of this chapter.

4.3. *Nanoindentation with Continuous Stiffness Measurement*

Above-mentioned approaches are based on single-cycle of indentations, i.e., loading-holding-unloading cycle. Another approach that can be used for nanoindentation of polymeric materials is to use the so-called Continuous Stiffness Measurement (CSM) technique. CSM technique is accomplished by applying a small oscillation signal to the force signal. The amplitude of the force oscillation is kept sufficiently small so that the deformation process is not significantly affected by its additional signal. The corresponding displacement oscillation is monitored at the excitation frequency using a two-channel, phase-sensitive detector (lock-in amplifier) which can accurately measure displacement as small as 0.001 nm using frequency specific amplification. The phase difference between the displacement signal and the force signal can also be measured. Solving for the in-phase and out-of-phase portions of the response results in the explicit determination of the contact stiffness, *S*, as a continuous function of depth.[49]

To calculate the contact stiffness, the dynamic response of the indentation system has to be determined, and this can be done by using a dynamic model including both springs and dashpot.[21] An analysis of this model reveals that the contact stiffness, *S*, can be calculated from the amplitude of the displacement signal:

$$\left|\frac{P_{os}}{h(\varpi)}\right| = \sqrt{\left[\left(S^{-1} + C_f\right)^{-1} + K_s - m\varpi^2\right]^2 + \varpi^2 D^2} \tag{18}$$

or from the phase difference between the force and displacement signal:

$$\tan(\phi) = \frac{\varpi D}{\left(S^{-1} + C_f\right)^{-1} + K_s - m\varpi^2} \tag{19}$$

where:

C_f= the compliance of the load frame
K_s= the stiffness of the column support springs
D= the damping coefficient
P_{os}= the magnitude of the force oscillation
$h(\omega)$= the magnitude of the resulting displacement oscillation
ω = frequency of the oscillation
ϕ = the phase angle between the force and displacement signals
m = mass of the indenter.

By solving Eqs. (18) and (19), the contact stiffness and damping can be expressed as:

$$S = \left[\frac{1}{\frac{P_{os}}{h(\varpi)}\cos\phi - \left(K_s - m\varpi^2\right)} - K_f^{-1}\right]^{-1} \tag{20}$$

and

$$\varpi D = \frac{P_{os}}{h(\varpi)} \sin\phi \tag{21}$$

The constants that must be known for the analysis are C_f, K_s and D. The constants K_s and D are experimentally determined by examining the motion of the indenter according to the dynamic model[21] when the indenter is not touching the surface ($S=0$). This was usually done as standard calibration procedure during the initial set-up of the nanoindentation instruments.

The CSM technique gives direct measurement of contact stiffness, elastic modulus and hardness as function of indentation depth during loading, therefore, unloading data is not really needed. The relation between contact stiffness and indentation depth can be derived from Eq. (4), assuming the contact area is perfect, then:

$$A = 24.56 h_c^2 \tag{22}$$

Substituting Eq. (22) into Eq. (4), we have

$$S = \frac{dP}{dh} = 1.167 \times \sqrt{24.56} E_r h_c \tag{23}$$

Equation (23) indicates that for a homogeneous material with a constant elastic modulus value of E_m, the reduced modulus E_r should also be a constant (Eq. (5)), therefore, the relation between S and h_c should be linear, with the slope corresponding to the reduced elastic modulus. If the elastic modulus changes with the indentation depth, as the case of thin film or non-homogeneous materials (the changes of elastic modulus in the case of thin film is actually due to the effects of substrate during indentation), then the linear relation does not exist anymore. Therefore, the CSM technique can be used to study the mechanical properties of graded materials and multilayered structures by monitoring the changes in contact stiffness, elastic modulus and hardness as a function of indentation contact depth. This principle has been applied to analyze the indentation results for a number of materials by Li and Bhushan.[49-51] For example, for indenting homogeneous materials, the resulting S-h_c data should be a linear line, and the slope of the line is corresponding to the elastic modulus of the testing materials; whereas for indenting nonhomogeneous materials, the S-h_c relation is no longer linear and will be either a convex curve (for increasing elastic modulus with depth) or a concave curve (for decreasing elastic modulus with depth), depending on the materials tested.[49]

The CSM technique can also be used to characterize the viscoelastic properties of polymeric materials. The amplitude and phase relationship between the excitation and the response are a composite of the response of the indenter and materials being indented. The dynamic response during CSM indentation can be separated into two parts.[52] The portion of the displacement signal which is "in phase" with the excitation is the elastic response of the contact. It is related to the stiffness *(S)* of the contact and to the elastic modulus, or storage modulus *(G')*, of the material. The "out of phase" portion characterizes the energy being absorbed by the materials, i.e., the damping *(ωD)* of the contact, and thus the loss modulus *(G'')* of the materials. Both the "in phase" and "out of phase" portions of the response scale according to the square root of the contact area *(A)*. Using the *Principle of Correspondence* of the linear viscoelastic theory, Loubet *et al.*[52] has proposed following two relations for determination of the storage modulus *(G')* and the loss modulus *(G'')*[52] from nanoindentation testing:

$$G' = \frac{\sqrt{\pi}}{6} \frac{S}{\sqrt{A}} \tag{24a}$$

and

$$G'' = \frac{\sqrt{\pi}}{6} \frac{\omega D}{\sqrt{A}} \tag{24b}$$

4.4. *Fitting of Unloading Curve of Nanoindentation Polymers*

A key issue of analyzing nanoindentation P-h relation is the unloading curve analysis. For nanoindentation of polymeric materials, the unloading curve might not be entire elastic, time-dependent deformation could be an important factor affecting the unloading behavior. VanLandingham *et al.* have examined several different fitting routines to fit the nanoindentation of unloading curve obtained from nanoindentation of a benzocyclobutene (BCB) polymer,[33] and found out that power law fit (Oliver and Pharr's method) using the commercial nanoindentation system software did not converge for any of the data set they obtained, but the values for S, h_c and E were output based on the values of the fitting parameters for the final iteration. It was also observed that for non-linear power law fits, the power law exponents were all greater than 2, ranging from 2.2 to 2.7, similar to the results presented by Briscoe and Sebastian.[53] Therefore, VanLandingham[33] proposed to use a smooth spline fitting routine to fit the unloading curves, and have shown that this method has yielded excellent fits to the unloading data. Following this approach, the nanoindentation unloading data of bulk PMMA and epoxy are fitted by the smooth spline fitting routine and the results are shown in Fig. 4. Besides producing a good fit to the unloading data, this routine also produces a direct calculation of the first derivative of the fitting curve corresponding to each data point, as shown in Fig. 5 for the same materials as a function of indentation displacement. The values of S and E obtained from the smooth spline fitting are less dependent on indentation load and human judgements, such as deciding on the numbers of points to include in a linear fit to the high load portion of the unloading curve, or the final indentation depth, h_f, used in the power-law fitting.

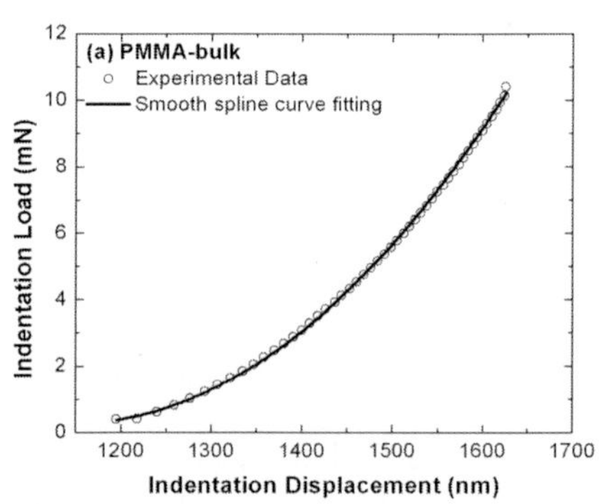

Fig. 4. Examples of nanoindentation unloading curves for (a) PMMA and (b) epoxy along with the smooth spline fitting and power law fitting.

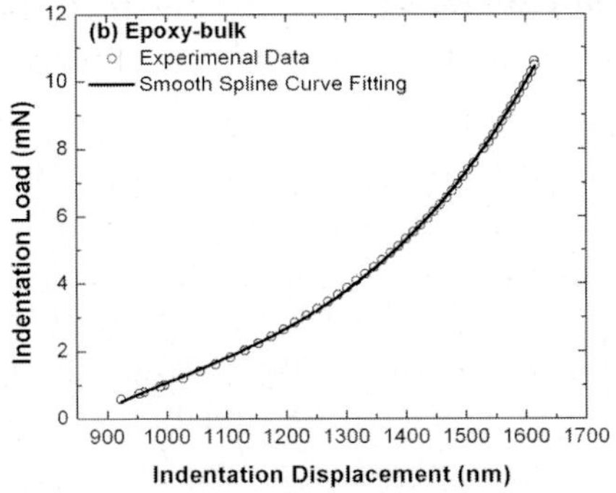

Fig. 4. (*Continued*)

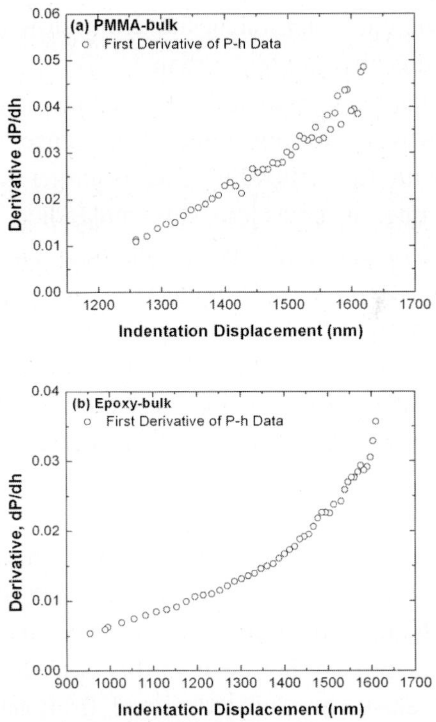

Fig. 5. Plots of the first derivative of the smooth spline curves in Fig. 4 as a function of load for (a) PMMA and (b) epoxy.

4.5. *Rate-Dependent Properties*

Rate dependent properties refer to the changes of indentation loading/unloading characteristics with the changes of the indentation loading or unloading rate. This is because the effects of loading/unloading rates have changed the characteristics of the nanoindentation of polymeric materials. Figures 6(a) and 6(b) shows the effects of loading times on the characteristics of the nanoindentation loading curve for epoxy and PMMA and Fig. 6c shows the effects of unloading time to the characteristics of the unloading curves for epoxy. Similar results were also observed by Yang *et al.*,[45] and it has been shown that the powers of both the loading and unloading curves were rate dependent. It is also observed that as the loading rate increases, the power of the loading curve approaches to 2, for slow loading, the power of the loading curve is always greater than 2.

For unloading curve, it is also observed that a fast unloading rate results in less recovery during unloading, hence the slope of the unloading curve with fast unloading rate is larger than that from the slower unloading rate, whereas the slow unloading rate results more recovery during unloading and smaller slope of the unloading curves (Fig. 6(c)). The changes of unloading curve with the unloading rate will affect the determination of the elastic modulus from the unloading curve. Table 2 summarized the elastic modulus and hardness values of epoxy samples indented with different loading/holding/unloading times by using the conventional Oliver and Pharr's method, and Table 3 summarized the same values for polycarbonate (PC) indented with different maximum loads and loading/holding/unloading times. It can be seen from Tables 2 and 3 that the values of elastic modulus and hardness are not consistent in these conditions. They are dependent on the loading/holding/unloading time used in the experiments; however, the trends of the dependence are not clear in this case. It should also notice that the values of elastic modulus obtained from nanoindentation are generally higher than the actual values for both PC and epoxy.

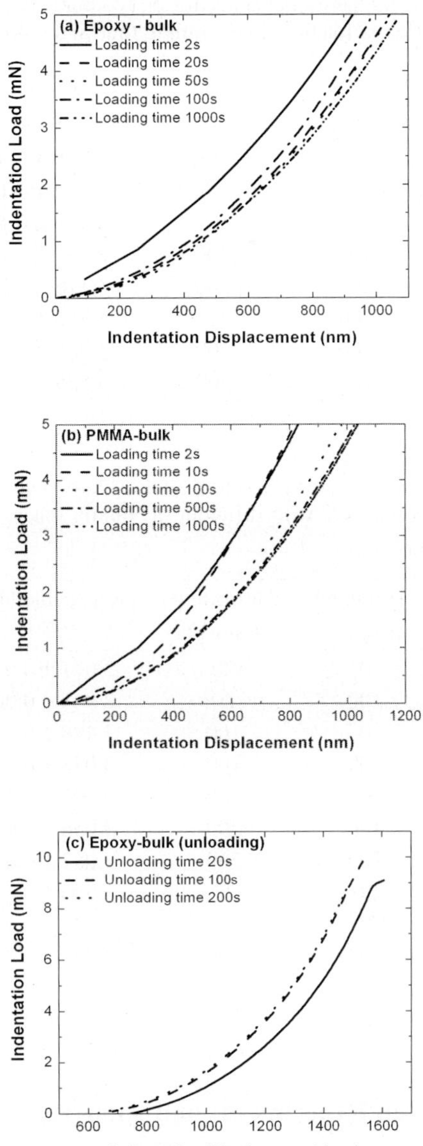

Fig. 6. Influence of the different loading rates on loading curve of polymeric materials during indentation with a sharp Berkovich indenter tip: (a) Epoxy; (b) PMMA; and (c) Influence of the unloading rate on unloading behavior of epoxy during indentation with a sharp Berkovich indenter tip.

Table 2. Elastic modulus and hardness values of Epoxy sample obtained using Oliver and Pharr's methods with different loading time, holding time and unloading time.

Max. load (mN)	Loading time (s)	Holding time (s)	Unloading time (s)	Elastic modulus (GPa)	Hardness (GPa)
10	20	0	20	3.466 ± 0.079	0.249 ± 0.015
10	20	100	20	3.272 ± 0.144	0.214 ± 0.016
10	20	0	100	3.925 ± 0.155	0.265 ± 0.009
10	20	100	100	3.367 ± 0.112	0.246 ± 0.014
10	20	0	200	4.170 ± 0.222	0.278 ± 0.023
10	20	100	200	3.356 ± 0.128	0.248 ± 0.017
10	100	100	100	3.285 ± 0.143	0.242 ± 0.016
10	100	100	200	3.555 ± 0.380	0.284 ± 0.057

Table 3. Elastic modulus and hardness values of Polycarbonate (PC) sample obtained using Oliver and Pharr's methods with different maximum loads, loading time, holding time and unloading time.

Max. load (mN)	Loading time (s)	Holding time (s)	Unloading time (s)	Elastic modulus (GPa)	Hardness (GPa)
10	20	0	20	3.071 ± 0.017	0.180 ± 0.001
10	20	100	20	2.988 ± 0.061	0.168 ± 0.001
10	20	0	100	3.488 ± 0.069	0.184 ± 0.001
10	20	100	100	3.078 ± 0.236	0.172 ± 0.009
10	20	0	200	3.605 ± 0.063	0.178 ± 0.006
10	20	100	200	3.056 ± 0.039	0.167 ± 0.003
10	100	100	200	3.082 ± 0.020	0.165 ± 0.001
100	20	20	20	2.940 ± 0.016	0.162 ± 0.001
100	20	20	100	2.824 ± 0.133	0.144 ± 0.019
100	20	0	100	2.840 ± 0.040	0.163 ± 0.013
100	100	100	100	3.071 ± 0.004	0.140 ± 0.001
100	100	100	500	2.816 ± 0.136	0.139 ± 0.001

5. Indentation Creep Experiments and Analysis

There are a number of works in using indentation technique to characterize the creep behaviors of the materials including polymers.[43-46,54-57] In this paper, indentation creep refers to the changes of indentation displacement during holding segment at a constant indentation load. The indentation displacement during the holding segment can be analog to the creep deformation, as the indentation displacement continues increasing with the time although the load is kept as a constant. However, there are also several important differences from a conventional tensile creep test. The conventional tensile creep test was usually carried under constant stress and therefore the entire specimen is subjected to the uni-axial tensile stress, as the creeping time increases, if the cross-sectional area of the specimen decreases, then the load has to be reduced in order to keep the constant stress. Therefore, in a conventional tensile creep testing, a specimen subjected to a constant load creep will fail much faster than the specimens subjected to a constant stress creep test (Fig. 7). However, indentation creep is under localized multi-axial stress field. Under the constant load condition, the indentation creep is actually under a decreased stress condition if using spherical or sharp tips since the contact area increases with the time, when using flat-ended punch tip, the indentation creep is under a constant stress condition as the contact area remains a constant. In a conventional creep testing, there are usually three stages: a primary creep stage, which is characterized with increasing strain and with decreasing strain rate; a secondary creep or steady state creep stage, in this stage, the strain rate is a constant; and finally, a tertiary stage, which the final failure will occur. However, comparing with the indentation creep, only primary and secondary creep stages can be observed during an indentation creep, and the final tertiary creep cannot be seen under the indentation conditions, this is because that indentation creep usually do not lead to the failure of the entire materials, it can only create localized creep deformations. With these differences in mind, we can now analyze the indentation creep experiments. It is worth noticing that the experimental and analysis methods describe below can also be used to study the creep behavior of other non-polymer materials, such as solder alloys, or other soft materials.

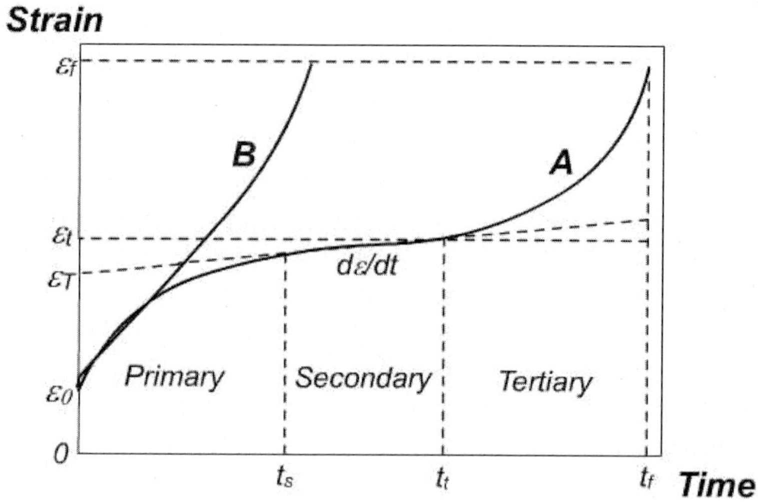

Fig. 7. Schematic of creep deformation curves under constant stress and constant load using conventional uni-axial tensile creep test.

5.1. *Basic Indentation Creep Analysis*

Figure 8 shows the changes of indentation displacement with time during holding at the maximum indentation load for eight (8) different polymeric materials. It is clear that the indentation creep, in terms of indentation displacement as function of the holding time, has two stages, i.e., a primary stage and a steady state stage. In the primary stage, the indentation displacement increases with the time but with a decreased rate, the displacement rate decreases with time and approaches to a constant value, in which is the creep rate of the steady state stage.

This phenomenon was found in many different polymeric materials, which were indented with a flat-ended punch tip, spherical tips and a sharp Berkovich tip. The similar behavior was also found when indenting some soft-alloys, such as solder materials. The changes of indentation displacement with time can be fitted with a non-linear curve fitting method. There are at least two different models that can be used to describe the non-linear creep behavior shown in Fig. 8. Interesting enough, both models can lead to the same functional form to describe the non-linear curve formed during the indentation creep experiments.

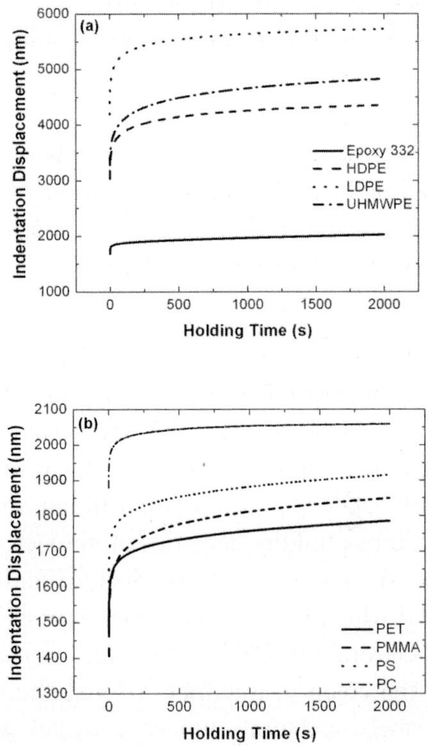

Fig. 8. Indentation displacement as function of time during the holding at the maximum indentation load for various polymeric materials: (a) Epoxy 332, HDPE, LDPE and UHMWPE; and (b) PET, PMMA, PS and PC. The maximum indentation load is 10 mN with 2000 s holding time, the loading rate is 2 s to maximum load for all of the materials.

The first model was developed based on a generalized Kelvin model. In line with this, Yang *et al.*[45] has used this model to derive a general form of creeping function and it was used to fit the nanoindentation creep of a number of polymeric materials indented with both flat-ended punch and sharp Berkovich indenter tips. The function can be written as following:

$$h = h_0 + \sum_{1}^{n} h_i \left[1 - \exp\left(-t/t_i\right)\right] + mt \tag{25}$$

In this equation, h_0 is defined as the indentation deformation when the indentation creep started. h_i, t_i and m are fitting constant and can be determined by the non-linear curve fitting procedure. t is the indentation creep time and h is the total deformation during indentation, including the deformation in both loading and holding segments. When using generalized Kelvin model, $m=1/\eta$ is related to the viscosity coefficient of the single dashpot element, in which η is the steady-state viscosity of the material. This equation is analog to the situation of the tensile creep when using the generalized Kelvin model, in tensile creep under a constant stress, the generalized Kelvin model gives:

$$\varepsilon = \varepsilon_0 + \sum_1^n \varepsilon_i \left[1 - \exp\left(-t / t_i\right)\right] + t / \eta \tag{26}$$

Equation (25) has been used by Yang *et al.* [45] to fit many of the experimental results, in addition, Eq. (25) can also be used further to derive creep compliance and retardation spectrum for the testing materials. Furthermore, Zeng[16] has demonstrated that the molecular weights of PMMAs, cross-linking densities of epoxies, as well as testing temperature, all have significant effects on the indentation creep data.

The second model which leads to the same expression as Eq. (26) can be named as Kraus-Garofalo model.[58-59] Garofalo model[58] was originally derived to describe the creep of a number of metallic materials, such as austenitic stainless steels, cadmium, tungsten, nickel, zinc, iron, copper-aluminum alloys, and nickel-chromium alloy etc. The original Garofalo's equation only includes one exponential term, as:

$$\varepsilon = \varepsilon_0 + \varepsilon_p \left[1 - \exp\left(-t / t_p\right)\right] + \dot{\varepsilon}_s t \tag{27}$$

where ε_p is the transient strain during the primary stage, t_p is the time constant for the primary stage, and $\dot{\varepsilon}_s$ is the secondary-creep rate. Kraus[59] modified this equation and wrote as:

$$\varepsilon = \varepsilon_0 + \varepsilon_1 \left[1 - \exp\left(-t / t_1\right)\right] + \varepsilon_2 \left[1 - \exp\left(-t / t_2\right)\right] + \dot{\varepsilon}_s t \tag{28}$$

Comparing Eqs. (26) and (28), we found that, although the two equations are derived based on two different models, and also the creeping mechanisms in metals and polymers are very different, the functional form of the creeping equation is the same. Regardless the physical mechanisms behind the creep deformations in polymer and metals, the constant $1/\eta$ in Eq. (26) can be defined as the rate of the secondary creep as in Eq. (28), and it is related to the viscosity of the materials. In addition, in the Garofalo's model, Eq. (27), the time constant, t_p, is the ratio of the transient creep rate to the transient creep strain, and ε_p is the limiting transient strain. However, after the modification by Kraus, when using two exponential terms, Eq. (28), the physical meanings of the constant ε_1, ε_2, t_1 and t_2 are not clear.

When fitting the indentation creep deformation curves using Eq. (25), it is found that one exponential term cannot fit the curve, especially at the beginning and at the end of the creep curve, whereas the two exponential terms function can fit the curve much better, both at the beginning and at the end of the creep deformation, this was shown in Fig. 9 for three different polymers (PMMA, epoxy and UHMWPE (Ultra-high Molecular Weight Polyethylene)). It also shows that if one only uses the two exponential terms but without the linear term (*mt* in Eq. (25)), the fitting is not good at the end of the creep deformation. Therefore, the suitable functional form for indentation creep should have the following form:

$$h = h_0 + h_1\left[1 - \exp\left(-t/t_1\right)\right] + h_2\left[1 - \exp\left(-t/t_2\right)\right] + mt \qquad (29)$$

Where h_0, h_1, h_2, t_1 and t_2 are fitting parameters and each should have its own physical meanings. It is therefore necessary to explore the physical meanings of each fitting parameters in Eq. (29). Based on the comparison of the one and two exponential term equations, we would expect that one of the exponential terms should be associated with the initial creep deformation, whereas the linear term is associated with the later part of the creep curve, or steady-state creep deformation. To understand the meanings of each parameter, we first perform parametric studies of the fitting parameters in Eq. (29).

5.2. *Meaning of the Fitting Parameters*

First of all, as the creeping time, *t*, increases, Eq. (29) approaches a straight line, which can be written as:

$$h = A + mt \qquad (30)$$

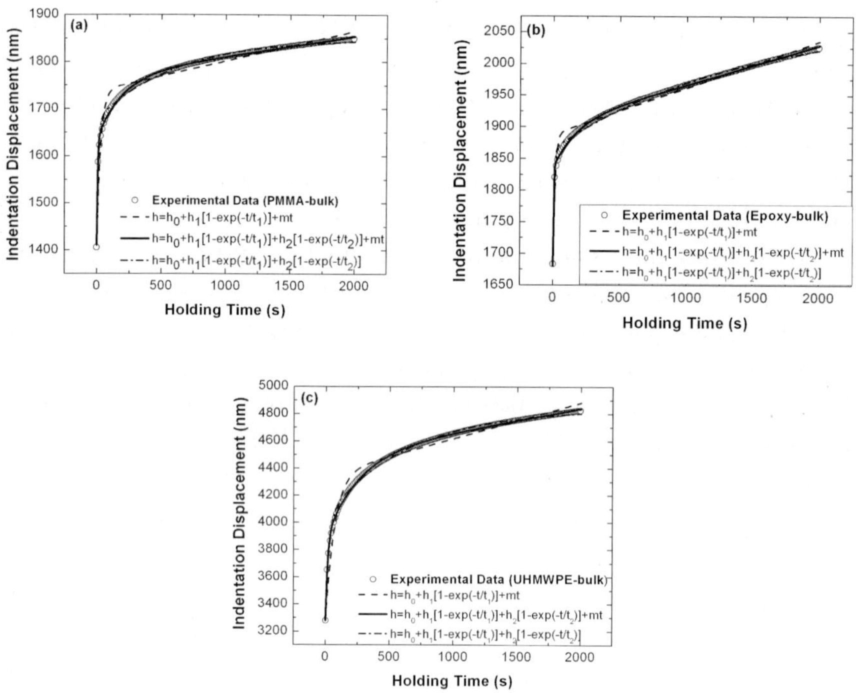

Fig. 9. Indentation creep data and curve fitted with different forms of Eq. (25) for polymeric materials: (a) PMMA, (b) Epoxy and (c) UHMWPE. Each set of data is fitted with three functions based on Eq. (25): (1) one exponential term with one linear term, Eq. (F9.1); (2) two exponential terms with one linear term, Eq. (F9.2); and (3) two exponential terms without the linear term, Eq. (F9.3) as following:

$$h = h_0 + h_1[1 - \exp(-t/t_1)] + mt \tag{F9.1}$$

$$h = h_0 + h_1\left[1 - \exp\left(-t/t_1\right)\right] + h_2\left[1 - \exp\left(-t/t_2\right)\right] + mt \tag{F9.2}$$

$$h = h_0 + h_1\left[1 - \exp\left(-t/t_1\right)\right] + h_2\left[1 - \exp\left(-t/t_2\right)\right] \tag{F9.3}$$

where $A=h_0+h_1+h_2$ *(if $t \to \infty$)*. This equation describes the secondary stage creep, that is, when the creeping time is sufficient long, the creep deformation is completely determined by the characteristics of the secondary creep. In other words, for polymeric materials, the secondary creep is controlled by the viscosity of the materials. This analysis further suggests that the exponential terms are associated with the transient creep in the primary stage.

In this analysis, h_0 is defined as the indentation displacement at the beginning of the holding segment; and it is independent of the creep segment during indentation. Therefore, the quantity of (h_1+h_2) is the limit of the creep deformation at the primary stage. The total creep deformation during indentation is therefore the quantity of (h_1+h_2+mt) for both primary and secondary stages.

Secondly, we perform the parametric analysis on the two time constants, t_1 and t_2, by definition, $t_1<t_2$. During the parametric study, one of the parameters is systematically changed whereas the other one is fixed. The results are summarized in Fig. 10 for the effects of t_1 and Fig. 11 for the effects of t_2. From Fig. 10, it is clear that the smaller time constant, t_1, has a larger influence on the initial stages of the primary creep, in particular, it determines the gradient of the initial creep curve, or the initial creep rate. A smaller value of t_1 will give rise to a steeper gradient on the initial creep curve, in other words, a higher initial creeping rate with a smaller value of the time constant, t_1. In addition, a smaller value of t_1 (such as $t_1=1.5s$) causes a sharp transition in the creep curve, whereas a larger value of t_1 (such as $t_1=15s$) leads to a more smooth transition in the creep curve.

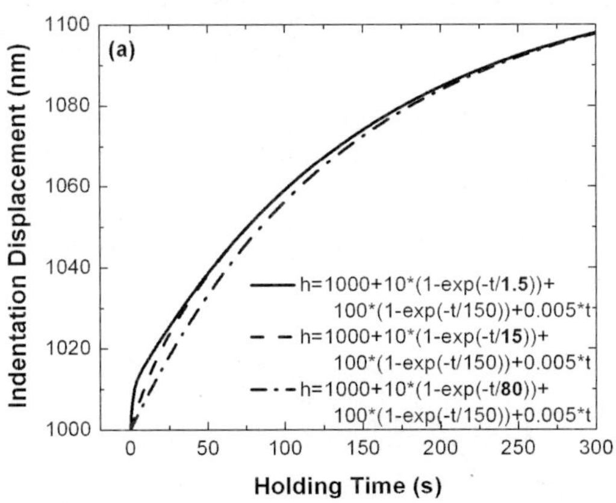

Fig. 10. Effects of variation of parameter t_1 in Eq. (29) (where other parameters are fixed): (a) Overall creep curves and (b) initial creep curves.

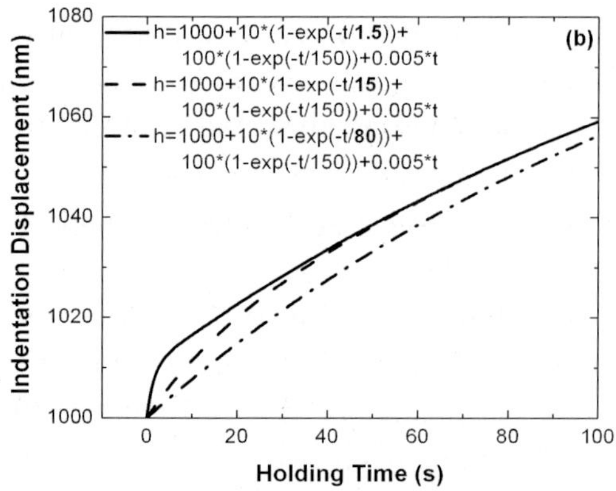

Fig. 10. (*Continued*)

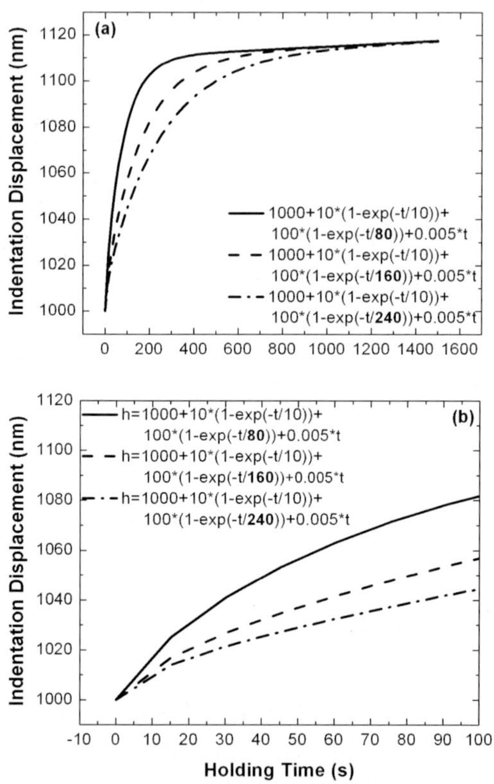

Fig. 11. Effects of variation of parameter t_2 in Eq. (29) (where other parameters are fixed): (a) Overall creep curves and (b) initial creep curves.

Likewise, similar analysis is applied to the time constant t_2, as shown in Fig. 11, it is found that t_2 has a distinct effect on the later part of the creep in the primary stage; it has small effects on the initial part of the creep. A larger value of t_2 causes a smooth transition to the secondary stage, where as a smaller value leads to more sharp transition to the secondary stage. In other words, large value of t_2 means primary creep stage is longer. It is noticed that both parameters have no effects on the creep deformation at the secondary stage. After a certain time, all of the curves with different values of t_1 or t_2 are merged into one linear curve if the linear term has the same coefficient, *m*, in Eq. (29). Basically, when the holding time, *t*, is larger than the time parameter, t_2, the primary stage creep ends and the secondary creep stage starts. Based on this analysis, we can summarize the meaning of the fitting parameters in Eq. (29) as following:

(1) h_0 is the indentation displacement at the beginning of the creep, and it is independent of creep deformation during holding segment; it can be determined by indentation P-h relations based on the geometry of the tips used in the experiments;
(2) The sum of the two parameters: (h_1+h_2) describes the limit of the primary creep at a long enough time;
(3) t_1 characterizes the initial creep rate;
(4) t_2 characterizes the period of the primary creep, in other words, it describes the transition from the primary creep to secondary creep; and
(5) *m* is the steady-state creep rate, and it is related to the viscosity of the materials.

The deformation rate during indentation creep can be found by differentiating Eq. (29), i.e.:

$$\frac{dh}{dt} = \frac{h_1}{t_1}\left[\exp(-t/t_1)\right] + \frac{h_2}{t_2}\left[\exp(-t/t_2)\right] + m \tag{31}$$

Therefore, the initial creep rate at the time $t=0(+)$ is:

$$\left.\frac{dh}{dt}\right|_{t=0+} = \frac{h_1}{t_1} + \frac{h_2}{t_2} + m \tag{32}$$

At later stage of the creep, i.e., when $t \to \infty$, the creep rate is therefore:

$$\left.\frac{dh}{dt}\right|_{t \to \infty} = m \tag{33}$$

In general, $t_1 < t_2$, therefore, the initial creep rate is dominated by the parameter of t_1, The creep rate at the intermediate state is determined by the parameters t_1 and t_2 and at the later stage of the creep is determined by m alone. For the longer creeping time, m is the only parameter governing the indentation creep.

With the understanding of the meaning of the fitting parameters, in additional to the measurement of the elastic modulus and hardness, we can now use nanoindentation creep experiments to characterize the initial creep rate, primary creep deformation and time, as well as the rate of the steady-state creep, and the factors affecting these properties. This will be discussed in the next section.

5.3. *Factors Affecting the Indentation Creep*

In this section, we will discuss several factors that will affect the indentation creep analysis; these factors include (i) different applied stress states created by different indenter tips, (ii) maximum indentation loads, and (iii) temperatures.

5.3.1. *Indenter Tips*

The term "applied stresses" is used here to refer the effects of the different indenter tips, which will generate different stress states. For example, flat-ended punch tip mainly creates constant elastic contact stresses, the displacement is also mainly elastic, and the plastic deformation under indenter tip is very small, whereas the sharp Berkovich indenter tip creates plastic dominated deformation and therefore the stresses are also higher, but it is not constant during the indentation creep experiments. While using spherical indentation tips, the stress states are in between flat-ended punch and sharp Berkovich tips. In order to study the effects of these factors, indentation creep experiments with different indenter tips have been performed and the results will be discussed here. Figure 12 shows the nanoindentation creep results and curve fitting of Eq. (29) for PMMA samples indented with a flat-ended punch tip (10 μm in diameter), spherical tips with 1 μm and 100 μm radius, respectively, and a standard Berkovich tip. The fitting parameters are also summarized in the Fig. 12. Comparing the creep behaviors from all of the four indenter tips, we can find that: (i) the creep using the sharp Berkovich tip has highest value of the steady state creep rate (*m=0.076*); this shows that the secondary stage creep is more dominated when using sharp indenter tip; (ii) the creep deformation during holding segment are

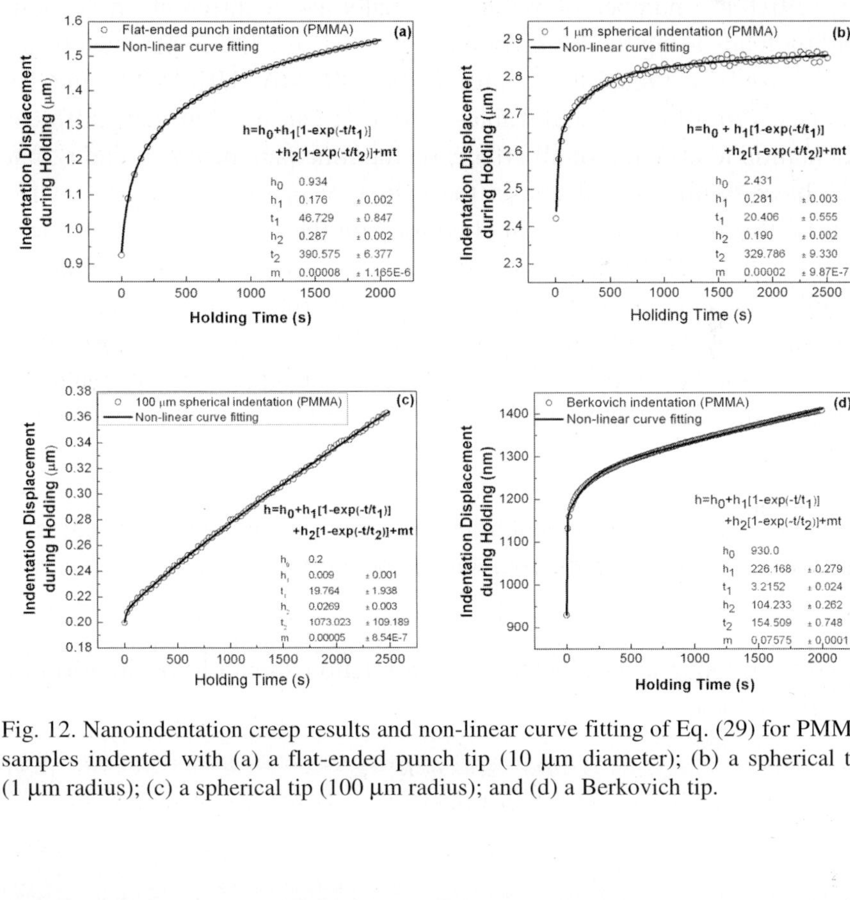

Fig. 12. Nanoindentation creep results and non-linear curve fitting of Eq. (29) for PMMA samples indented with (a) a flat-ended punch tip (10 μm diameter); (b) a spherical tip (1 μm radius); (c) a spherical tip (100 μm radius); and (d) a Berkovich tip.

smallest when using 100 μm spherical tip, where the creep using flat-ended punch tip has higher deformation during the creep, this is probably due to the reason that the creep under the flat-ended punch is under constant stress condition, whereas the other tips are under decreased stress condition, therefore, the creep deformation are smaller; (iii) the larger radius spherical tip leads to more significant secondary stage creep and also longer period of the primary stage creep (associated with the parameter of t_2), however, the initial creep rate (associated with t_1) is less sensitive to the radius of spherical tip used; and (iv) sharp Berkovich tip also leads to higher initial creep rate than those using other indenter tips.

5.3.2. *Maximum Indentation Loads*

Figure 13 shows the nanoindentation creep results and curve fitting of Eq. (29) for a number of polymeric materials at different maximum holding loads using the Berkovich tip, whereas the fitting parameters are summarized in Tables 4 to 7 for epoxy, PMMA, UHMWPE and PS samples. From Fig. 13 and Tables 4 to 7, it can be seen that, different indentation loads have small effects on the time parameters, t_1 and t_2, but they have significant effects on the values of h_0, h_1 and h_2. Except for UHMWPE, the effect of the applied load on the values of m is also less significant; however, it has significant effects on the values of m for UHMWPE. This further suggested that the time constants are load-independent, therefore, they might be considered as materials properties which might be used to characterize the indentation creep behavior. Whereas the other three parameters (h_0, h_1 and h_2) are dependent on the indentation loads, therefore, they should follow the relationship between load and indentation penetration, especially, h_0 can be determined by indentation load-displacement relations. For example, for conical and Berkovich indenter tips, the relation can be written as:

$$P_{\max} = Ch_0^2 \tag{34}$$

whereas for spherical indenter tips, the relation should have the form of Hertzian contact as:

$$P_{\max} = \frac{4}{3} E_r R^{1/2} h_0^{3/2} \tag{35}$$

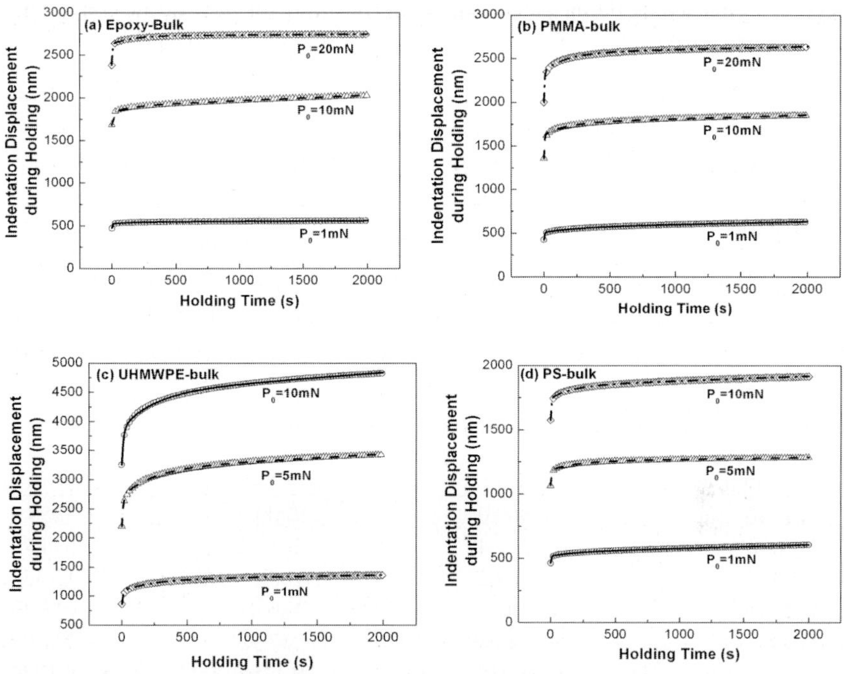

Fig. 13. Nanoindentation creep results and non-linear curve fitting of Eq. (29) for numbers of polymeric materials at different maximum holding loads: (a) Epoxy (indentation loads: 1, 10 and 20 mN); (b) PMMA (indentation loads: 1, 10, and 20 mN); (c) UHMWPE (indentation loads: 1, 5, and 10 mN); and (d) PS (indentation loads: 1, 5 and 10 mN).

Table 4. Summary of fitting parameters using Eq. (29) for nanoindentation creep of several epoxy sample with different maximum holding loads as shown in Fig. 13(a).

Fitting parameters	$P_0 = 1$ mN	$P_0 = 10$ mN	$P_0 = 20$ mN
h_0 (nm)	472.0	1680.0	2380.0
h_1 (nm)	51.29	155.76	247.26
t_1 (s)	1.98	3.00	2.42
h_2 (nm)	21.70	68.71	99.04
t_2 (s)	142.77	169.70	161.91
m (nm/s)	0.0073	0.0063	0.011

Table 5. Summary of fitting parameters using Eq. (29) for nanoindentation creep of several PMMA sample with different maximum holding loads as shown in Fig. 13(b).

Fitting parameters	$P_0 = 1$ mN	$P_0 = 10$ mN	$P_0 = 20$ mN
H_0 (nm)	420.0	1360.0	2000.0
H_1 (nm)	89.01	267.07	360.65
T_1 (s)	5.08	4.61	5.78
H_2 (nm)	59.18	139.09	207.48
T_2 (s)	286.17	176.39	185.32
m (nm/s)	0.03	0.044	0.035

Table 6. Summary of fitting parameters using Eq. (29) for nanoindentation creep of several UHMWPE sample with different maximum holding loads as shown in Fig. 13(c).

Fitting parameters	$P_0 = 1$ mN	$P_0 = 5$ mN	$P_0 = 10$ mN
h_0 (nm)	860.0	2200.0	3250.0
h_1 (nm)	223.49	501.92	624.72
t_1 (s)	10.62	10.82	13.45
h_2 (nm)	206.67	480.89	617.46
t_2 (s)	240.89	237.92	251.08
m (nm/s)	0.037	0.1260	0.1730

Table 7. Summary of fitting parameters using Eq. (29) for nanoindentation creep of several PS sample with different maximum holding loads as shown in Fig. 13(d).

Fitting parameters	$P_0 = 1$ mN	$P_0 = 5$ mN	$P_0 = 10$ mN
h_0 (nm)	465.0	1065.0	1575.0
h_1 (nm)	53.01	119.06	172.18
t_1 (s)	4.74	4.32	4.29
h_2 (nm)	30.03	67.31	96.23
t_2 (s)	188.96	182.10	181.57
m (nm/s)	0.029	0.017	0.037

Equations (34) and (35) suggest that the term h_0 can be determined analytically based on the maximum indentation load, elastic modulus of the materials, and indenter tip used. In practice, this value is determined experimentally.

5.3.3. *Temperatures and Holding Time*

Nanoindentation creep is usually performed at room temperature, since it is difficult to change temperatures during nanoindentation experiments. However, a meaningful creep experiments should be carried out at elevated temperatures. This can be done using microindentation techniques at different temperatures. We have therefore performed microindentation creep experiments using a Berkovich tip mounted on a high-precision universal testing machine with a temperature chamber up to 250°C. It was found that Eq. (29) can still be applied to the microindentation creep experiments performed at various elevated temperatures. This technique and analysis can actually apply to both polymeric materials and metallic materials such as solder alloys. Figure 14 shows the microindentation creep experiments (at the maximum holding load of 10 N) and non-linear curve fitting for epoxy samples at 25°C and 150°C testing temperatures and with two different holding times (5 min and 60 min), and the fitting parameters are also summarized in the figure. It can be seen that, at a short holding time (5 min), the temperature has some effects, but less significant, on all of the parameters, however, as the holding time increases (60 min), the effects on various parameters become significant, especially on the time parameters, t_1 and t_2, one can see that both t_1 and t_2 has increased significantly compared with that of the short holding time. It is also interesting to notice that for short holding time (5 min) and higher loads (10 N), the time constant t_1 is similar to that of nanoindentation creep experiments, however, the time constant t_2 has reduced significantly, this suggests that, as the load increases, the primary creep period is reduced and the secondary creep period increases, however, for epoxy samples, the loading effects on the values of m are not clear, and need more studies.

We can now summarize the effects of the indentation loads, indentation stress states (indenter tips), holding time, and temperatures to the fitting parameters of Eq. (29) as follows:

(1) The fitting parameters are dependent on the indenter tips used, in other words, they are dependent on the indentation stress states, this

most likely suggested that the creep under different indenter tips might have different mechanisms (Fig. 12).

(2) The indentation load significantly affects the creep deformation (h_1+h_2), however, it has less effect on the time constant t_1 and t_2. In other words, its effects on initial creep rate and the length of the primary stage creep is less significant (Fig. 13 and Tables 4 to 7).

(3) The effects of indentation load to the steady state creep rate is not obvious unless the loads are very high (such as in the range of microindentation) or the materials are very soft (such as UHMWPE) (Fig. 13 and Tables 4 to 7).

(4) The testing temperature affects all of the parameters, but it is more significant to the two time constants, especially when the testing time is long (such as 60 min). At short testing time (such as 5 min), the effect is not as significant as that in the long creeping time. In addition, increasing testing temperatures will also increase the steady state creep rate (Fig. 14).

5.4. *Modeling of Indentation Creep*

The indentation creep experiments and analysis described so far are based on experimental observations and the analysis is analog with the tensile creep analysis expressed on stress and strain. Zhang *et al.*[60-63] have performed a series simulation and theoretical studies using finite element analysis and Laplace transformation. The main findings of these analyses are summarized here.

In using generalized Kelvin model in modeling the indentation creep of polymers, one should use at least two Kelvin elements in the model, so that the model can fit the experimental data well, this is in agreement with the above discussion on the fitting of Eq. (29) to the indentation creep data. It also shows that Eq. (29) can be used to fit the indentation creep of thin polymeric films.[60-61]

Indentation creep using sharp indenter tip is a very complicated process. To simplify the problem, one can use so-called "effective indenter shape" concept developed by Pharr and Bolshakov[64] as well as Sakai,[65] so that the sharp indentation solution can be equivalent to either

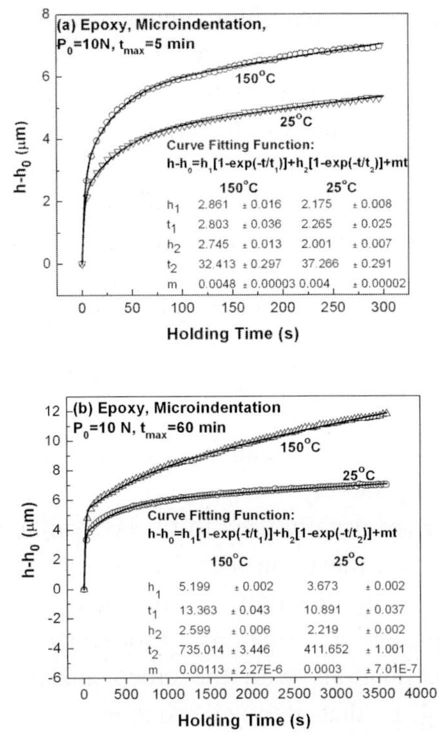

Fig. 14. Microindentation creep results and curve fitting of Eq. (29) for epoxy samples at different temperatures with the maximum indentation holding load of 10N: (a) holding time 5 min, at 25°C and 150°C, respectively; and (b) holding time 60 min, at 25°C and 150°C, respectively.

spherical indenter tip or conical indenter tip. Using this concept, the functional form of indentation creep is still in the form of Eq. (29), but the fitting parameters are different, especially, for the parameters of h_0, h_1 and h_2.[62-63]

5.5. *Other Creep Functions and Comparison*

There are several other indentation creep functions in different forms. For example, Oyen[66] has given following equation for the displacement-time *(h-t)* relation for the indentation creep using spherical indenter tip as:

$$h^{3/2}(t) = B_0 - B_1 \exp(-t/\tau_1) - B_2 \exp(-t/\tau_2) \tag{36}$$

where B_0, B_1, B_2, τ_1 and τ_2 are fitting constants and determined from the non-linear curve fitting algorithm. Another example is that developed by Lu *et al.*,[67] they have developed two equations for indentation creep analysis. For conical indenter tips, the creep function is:

$$h^2(t) = \frac{1}{4}\pi(1-\nu)v_0 \tan\alpha\left[\left(J_0 + \sum_{i=1}^{N} J_i\right) - \sum_{i=1}^{N} J_i\tau_i\left(1-\exp(-t/\tau_i)\right)\right] \tag{37}$$

whereas for spherical indenter, the creep function has the form of:

$$h^{3/2}(t) = \frac{3(1-\nu)v_0}{8\sqrt{R}}\left[\left(J_0 + \sum_{i=1}^{N} J_i\right) - \sum_{i=1}^{N} J_i\tau_i\left(1-\exp(-t/\tau_i)\right)\right] \tag{38}$$

where v_0 is the loading rate, J_k and τ_k are fitting creep coefficients and retardation times, respectively.

These equations are usually used to fit the loading and holding curve together. Therefore one must consider the relationship between load and displacement when using different indenter tips, i.e., h^2 relationship for conical indenter tips, and $h^{3/2}$ relationship for spherical indenter tips. The similarity of these relations with the creep analysis described in the Section 5.1 to 5.4 is that they all contain a term in the form of $A_i\sum\left[1-\exp(-t/\tau_i)\right]$. The difference in the creep analysis described in Section 5.1 to 5.4 are only for holding segment and the loading curve was not included in the analysis, as well as the using of h, h^2 or $h^{3/2}$ in the creep analysis.

In order to find out the difference between the various models using h, h^2 or $h^{3/2}$ in the creep analysis, the creep curves presented in Figs. 12(b) to 12(d) are re-analyzed. For the creeps with spherical indenter tips (1 μm and 100 μm radius, respectively), the creep curves are re-fitted with the function of $h^{3/2}$ versus t, and for the creep curve with Berkovich indenter tip, the creep curve is re-fitted with the function of h^2 versus t. The re-plotted creep curves, in terms of $h^{3/2}$ *vs* t, or h^2 *vs* t, together with the fitting parameters are shown in Fig. 15. Comparing with the fitting parameters presented in Fig. 12, it can be seen that for using Berkovich indenter tip, whereas the function uses h or h^2, the effects on the time constants t_1 and t_2 are very small, all other parameters changes with the functions used. For spherical indenters, again, the effects on the time constant t_1 and t_2 are small for the creep data using 1 μm tip, but the effects on time constants is relatively large for the data using 100 μm tip. The effects on other parameters are also large, similar to the case of using Berkovich indenter. This suggests that when using the relatively sharp indenter tips, if one only wants to analyze the indentation creep data during the indentation holding segment, whether one uses h or h^k (k=3/2, or 2), the time constants are essentially the same, the depth constants are scaled with the power used. For the uses of the large radius spherical tip, the errors on all of the parameters are large. These changes are probably related to the sizes of the contact area during the indentation creep experiments. This might, in turn, suggest that the indentation creep experiments should be performed with the relative sharp indenter tips.

For indentation creep using flat-ended punch tip, Cheng *et al.*[43] has also proposed a function in which is essentially the same with Eq. (29). They have also related the fitting parameters with the elastic properties and the viscosity of the materials. In using the flat-ended punch, the relation between indentation depth and load is linear, therefore the stress for creep is a constant and the analytical solution is readily available.

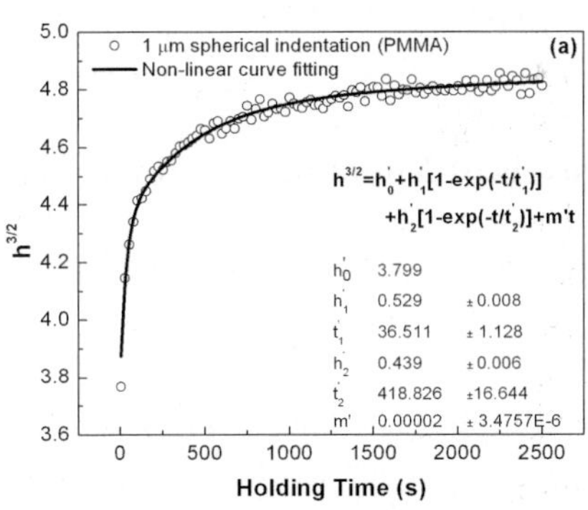

Fig. 15. (a) Indentation creep data plotted in the form of $h^{3/2}$ vs t and the non-linear curve fitting results for using spherical indenter tip (1μm radius); (b) Indentation creep data plotted in the form of $h^{3/2}$ vs t and the non-linear curve fitting results for using spherical indenter tip (100 μm radius); and (c) Indentation creep data plotted in the form of h^2 vs t and the non-linear curve fitting results for using Berkovich indenter tip.

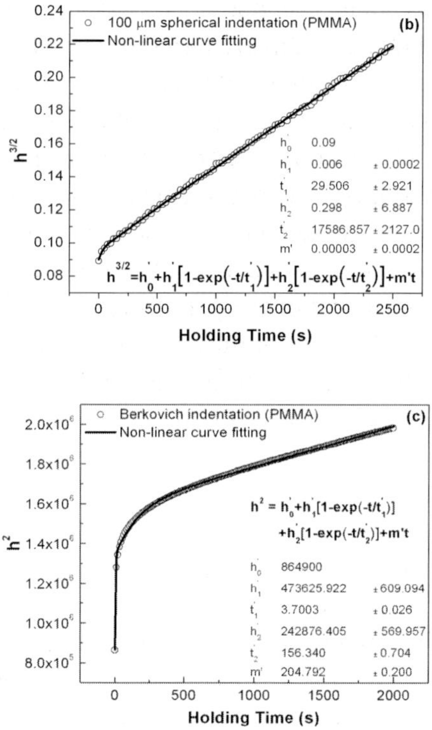

Fig. 15. (*Continued*)

6. Theoretical Analysis of Nanoindentation of Polymers

It has been a long history for some theoretical analyses related to the indentation or contact mechanics of polymeric materials.[68-75] In additional to those methods that we have discussed in various part of this chapter, these theoretical analyses are mainly used to derive the viscosity of the polymeric materials. Earlier works by Lee and Radok,[68] Yang[69] or Ting[70-71] provided some fundamentals on the contact mechanics of polymeric materials, in which were used in many of the later analysis works. Finite element analysis was also used to perform some of the numerical analyses related to the contact mechanics problems for polymeric materials. For example, Anand and Ames[74] has recently developed a new continuum model for the viscoelatic-plastic deformation of amorphous polymers and applied this model to the microindentation of PMMAs. In additional, Chang *et al.* have extended their scaling analysis to the nanoindentation of unloading data of polymeric materials.[75] A number of other works[36,44,46,60-63] have developed the analysis for either the loading curve or the displacement-time relationship during the holding of the indentation experiments. By combining these analytical methods, one should be able now to perform a complete analysis of nanoindentation loading/holding/unloading data obtained from either nanoindentation or microindentation of polymeric materials.

7. Nanoindentation Using Scanning Probe Microscope

The scanning probe microscope (SPM) is useful for evaluating the mechanical properties of polymeric materials on the scale even smaller than what can be done by conventional nanoindenters. The initial contact load by a SPM is much smaller than that used by a normal nanoindenter. In SPM, a probe consisting of a sharp tip (nominal tip radius in the order of 10 nm) located near the end of a cantilever beam is scanned across the sample surface using piezoelectric scanners. The SPM can also be operated in force mode to perform indentation tests. A force curve is produced, which is a plot of tip deflection as a function of the vertical motion of the SPM scanner. This curve can be analyzed to provide information about the local mechanical response.[33-34] By changing the spring constant of the probe, forces applied during indentation can be varied from a few nano-Newtons to tens or perhaps hundreds of micro-Newtons. However, accurate measurement of the probe spring constant can be difficult to achieve. This difficulty combined with a lack of information regarding the tip shape of the SPM probes makes the indentation by SPM a relative measurement.

In the analysis of the nanoindentation by SPM, Hertzian contact mechanics is usually used for the loading and unloading curves. Some examples of nanoindentation of BCB (benzocyclobutene) polymer using SPM systems have been summarized by VanLandingham *et al.*[33-34] Another work is that by Tranchida *et al.*[76] on using SPM to indent a variety of polymers ranging from amorphous polymers (PC and PMMA) to semicrystalline polymers (such as polypropylene) samples. During the nanoindentation by SPM, the mechanical properties of all the polymers studied are governed by an elastic response just like rubber, therefore, the indentation is largely dominated by elastic deformations. The elastic modulus of the amorphous polymers is found to be of similar value to that obtained from other testing methods.

8. Summary, Conclusions and Implications

This chapter summarized recent developments on the experimental methods and analyses of nanoindentation of polymeric materials. The main focus is on the effects of viscoelasticity to the indentation results. Unloading analysis and nanoindentation creep experiments are discussed in detail to illustrate the effects of viscoelasticity, as well as how to obtain various mechanical properties through nanoindentation experiments, as well as the discussions on what new information can be obtained through the indentation creep experiments. The results presented in this paper suggest that, with a proper analysis, nanoindentation can be applied to characterize a number of mechanical properties of the polymeric materials.

Indentation creep is a new experimental technique to characterize the time-dependent properties of polymeric materials, even for other soft materials. Indentation creep experiments have advantage of short testing time, small amount of samples needed, and large amount of information can be obtained. It also shows that indentation creep experiments can be performed with different indenter tips and testing temperatures, and eventually, the indentation creep can be modeled by the same functional form.

Acknowledgments

The author would like to acknowledge the support from National University of Singapore during the course of this work. I would also express my sincere appreciation and gratitude to many of my colleagues and students who have made great contributions to this paper. Special thanks to Ms. L. Shen, Institute of Materials Research and Engineering

(Singapore) for the help of many of the indentation experiments; to Prof. Y.-W. Zhang at Department of Materials Science and Engineering, National University of Singapore (Singapore) on many of the discussions related to the issues on the modeling of indentation creep; and to Prof. T. X. Liu, Laboratory of Advanced Materials, Fudan University (China) on the discussions related to the issues of polymeric materials. A numbers of post-graduate and undergraduate students are contributed significantly to this work, special thanks to Ms. S. H. Lim, Mr. C. Y. Zhang, Mr. S. Yang, Mr. S. T. Kian, Mr. K. S. Tan, Mr. Y. H. Chan, Mr. H. X. Chua, Mr. A. Y. R. Tan, Mr. Q. H. Tan, and Mr. C. Y. Chia.

References

1. J. B. Pethica, R. Hutchings and W. C. Oliver, *Philo. Mag. A*, 48, 593, (1983).
2. W. C. Oliver, R. Hutchings and J. B. Pethica, in *Microindentation Techniques in Materials Science and Engineering,* P. J. Blau and B. R. Lawn, Eds., Am. Soc. Testing & Mater., ASTM STP 889, (1986), p.90.
3. W. C. Oliver, *Mat. Res. Soc., Bull.*, 11, 15, (1986).
4. G. M. Pharr, *Mater. Sci. Eng. A*, 253, 151, (1998).
5. W. D. Nix, *Mater. Sci. Eng., A*, 234-236, 37, (1997).
6. Special issue: Nanoindentation: From Angstroms to Microns, *J. Mater. Res.*, 14, 6, (1999).
7. Special Issue: Fundamentals and Applications of Instrumented Indentation in Multidisciplinary Research, *J. Mater. Res.*, 19, Issue 1, (2004).
8. Special Issue: International Indentation Workshop, *Philo. Mag. A*, 74, Issue 5, (1996).
9. Special Issue: *Second International Indentation Workshop*, Phil. Mag., A, 82, Issue 10, (2002).
10. A. C. Fischer-Cripps, *Nanoindentation,* (Springer, New York, 2002).
11. B. Bhushan, *Handbook of Micro/Nano Tribology,* (CRC Press, Boca Raton, FL, 1995).
12. Mat. Res. Soc. Symp. Proc., *Fundamentals of Nanoindentation and Nanotribology,* N. R. Moody, W. W. Gerberich, N. Burnham, and S .P. Baker, Eds., Mat. Res. Soc., (1998), Vol. 522.
13. Mat. Res. Soc. Symp. Proc., *Fundamentals of Nanoindentation and Nanotribology II,* S. P. Baker, R. F. Cook, S. G. Corcoran, and N. R. Moody (Eds), Mat. Res. Soc., (2001), Vol. 649.

14. D. T. Smith, Eds., *Conference Proceedings: International Workshop on Instrumented Indentation*, San Diego, CA, April, 22-23, 1995, NIST Special Publication 896, NIST, (1996).
15. F. J. Baltá Calleja and S. Fakirov, *Microhardness of Polymers,* (Cambridge Univ. Press, Cambridge, 2000)
16. K. Y. Zeng, in *Handbook of Theoretical and Compoutational Nanotechnology,* M. Rieth and W. Schommers Eds., American Scientific Publisher, Vol. 4, (2006), p. 387.
17. E. S. Berkovich, *Ind. Diamond Rev.*, 11, 129, (1951).
18. I. N. Sneddon, *Int. J. Eng. Sci.*, 3, 47, (1965).
19. K. L. Johnson, *J. Mech. Phys. Solids*, 18, 115, (1970).
20. M. F. Doerner and W. D. Nix, *J. Mater. Res.*, 1, 601, (1986).
21. W. C. Oliver and G. M. Pharr, *J. Mater. Res.*, 7, 1564, (1992).
22. S. P. Baker, in Thin Films: Stresses and Mechanical Properties IV, P. H. Townsed, T. P. Weihs, J. E. Sanches, Jr., and P. Børgesen, Eds., Mat. Res. Soc. Proc., Vol. 308, *Mat. Res. Soc.*, (Pittsburgh, PA, 1993), p. 209.
23. S. P. Baker and T. P. Weihs, in Thin Films: Stresses and Mechanical Properties IV, P. H. Townsed, T. P. Weihs, J. E. Sanches, Jr. and P. Børgesen, Eds., *Mat. Res. Soc. Proc.*, Vol. 308, Mat. Res. Soc., (Pittsburgh, PA, 1993), p. 217.
24. R. B. King, *Int. J. Solids Structures*, 23, 1657, (1987).
25. P.-L. Larsson, A. E. Giannakopoulos, E. Söderlund, D. J. Rowcliffe and R. Vestergaard, *Int. J. Solids Structures*, 33, 221, (1996).
26. S. M. Chia, K. W. Leong, J. Li, X. Xu, K. Y. Zeng, P. N. Er, S. Gao, and H. Yu, *Tissue Engg.*, 6, 481, (2000).
27. K. Zeng and C.-H. Chiu, *Acta Mater.*, 49, 3539, (2001).
28. L. Shen, I. Y. Phang, L. Chen, T. X. Liu and K. Y. Zeng, *Polymer*, 45, 3341, (2004).
29. M. H. Nai, C. T. Lim, K. Y. Zeng, and V. B. C. Tan, J. Meta. Nanocry. Mater., 23, 363, (2005).
30. K. Y. Zeng, Z.-K. Chen, L. Shen and B. Liu, *Thin Solid Films*, 477, 111, (2005).
31. L. Shen, I. Y. Phang, T. X. Liu, and K. Y. Zeng, *Polymer*, 45, 8221, (2004).
32. J. C. Huang, P. C. Lim, L. Shen, P. K. Pallathadka, K. Y. Zeng and C. B. He, *Acta Mater.*, 53, 2395, (2005).
33. M. R. VanLandingham, J. S. Villarrubia, W. F. Guthrie and G. F. Meyers, *Macromol. Symp.*, 167, 15, (2001).
34. M. R. VanLandingham, J. S. Villarrubia, and G. F. Meyers, *ACS Polymer Preprints*, 41, 1412, (2000).
35. B. J. Briscoe, L. Fiori, and E. Pelillo, *J. Phys. D: Appl. Phys.*, 31, 2395, (1998).
36. P.-L. Larsson, and S. Carlson, Polymer Testing, 17, 49, (1998).
37. M. J. Mayo, R. W. Siegel, Y. Z. Liao and W. D. Nix, J. Mat. Res., 7, 793, (1992).
38. M. J. Mayo and W. D. Nix, Acta Metall. Mater., 36, 2183, (1988).
39. M. J. Mayo, R. W. Siegel, A. Narayanasamy and W. D. Nix., J. Mater. Res., 5, 1073, (1990).
40. V. Raman and R. Berriche, J. Mater. Res., 7, 627, (1992).

41. S. Shimizu, T. Yanagimoto and M. Sakai, J. Mater. Res., 14, 4075, (1999).
42. W. R. LaFontaine, B. Yost, R. D. Black and C. Y. Li, J. Mater. Res., 5, 2100, (1990).
43. L. Cheng, X. Xia, W. Yu, L. E. Scriven and W. W. Gerberich, J. Poly. Sci. B: Polym. Phys., 38, 10, (2000).
44. M. L. Oyen and R. F. Cook, J. Mater. Res., 18, 139, (2003).
45. S. Yang, Y.-W. Zhang and K. Y. Zeng, J. Appl. Phys., 95, 3655, (2004).
46. M. Sakai, S. Shimizu, N. Miyajima, Y. Tanabe, and E. Yasuda, Carbon, 39, 605, (2001).
47. S.-P. Hannula, D. Stone and C.Y. Li, in *Electronic Packaging Materials Science,* E. A. Giess, K.-N. Tu and D. R. Uhlmann, Eds., Mat.Res.Soc.Proc., Vol.40, Mat. Res. Soc., (Pittsburgh, PA, 1985), p.217.
48. T. W. Wu, M. Moshref and P. S. Alexopoulos, Thin Solid Films, 187, 295, (1990).
49. X. Li and B. Bhushan, Mater. Charact., 48, 11, (2002).
50. X. Li and B. Bhushan, Script Mater., 42, 929, (2000).
51. X. Li and B. Bhushan, Thin Solid Films, 377-378, 401, (2000).
52. J.-L. Loubet, B. N. Lucas and W. C. Oliver, in Conference Proceedings: International Workshop on Instrumented Indentation, San Deigo, CA, April 22-23, 1995, NIST Special Publication 896, D. T. Smith, Ed., (NIST, USA, 1996), p.31.
53. B. J. Briscoe and K. S. Sebastian, Proc. Roy. Soc. Lond., A 452, 439 (1996).
54. T. H. Hyde, K. A. Yelnia and A. A. Becker, Int. J. Mech. Sci., 35, 451, (1993).
55. J. C. M. Li, Mat. Sci. Eng. A, 322, 23, (2002).
56. J. Kolařík and A. Pegoretti, Polymer Testing, 23, 113, (2004).
57. C. A. Tweedie and K. J. Van Vliet, J. Mater. Res., 21, 1576, (2006).
58. F. Garofalo, Fundmentals of Creep and Creep-Rupture in Metals, MacMillan, New York, (1965).
59. H. Kraus, *Creep Analysis,* John Wiley & Sons, (1982).
60. C. Y. Zhang, Y.-W. Zhang, and K. Y. Zeng, J. Mater. Res., 19, 3052, (2004).
61. C. Y. Zhang, Y.-W. Zhang, K. Y. Zeng, and L. Shen, J. Mater. Res., 20, 1597, (2005).
62. C. Y. Zhang, Y.-W. Zhang, K. Y. Zeng, L. Shen and Y. Y. Wang, J. Mater. Res., 21, 2991, (2006).
63. C. Y. Zhang, Y. W. Zhang, K. Y. Zeng and L. Shen, Philos. Mag., 86, 4487 (2006).
64. G. M. Pharr and A. Bolshakov, J. Mater. Res., 17, 2660, (2002).
65. M. Sakai, J. Mater. Res., 18, 1631, (2003).
66. M. L. Oyen, Acta Materialia, 55, 3633, (2007).
67. H. Lu, B. Wang, J. Ma, G. Huang, and H. Viswanathan, Mech. Time-Depend. Mater., 7, 189, (2003).
68. E. H. Lee and J. R. M. Radok, Trans. ASME, E: J. App. Mech., 27, 438, (1960).
69. W. H. Yang, Trans. ASME E: J. App. Mech., 33, 395, (1966).
70. T. C. T. Ting, Trans. ASME E: J. App. Mech., 33, 845, (1966).
71. T. C. T. Ting, Trans. ASME, E: J. App. Mech., 35, 248, (1968).
72. R. J. Roberts, R. C. Rowe, and P. York, J. Mater. Sci., 29, 2289, (1994).

73. B. Beake, J. Phys. D: App. Phys., 39, 4478, (2006).
74. L. Anand and N. M.Ames, Int. J. Plasticity, 22, 1123, (2006).
75. Y. T. Cheng, C.-M. Cheng and W. Ni, Mater. Sci. Eng., A, 423, 2, (2006).
76. D. Tranchida, S. Piccarolo, and M. Soliman, Macromolecules, 39, 4547, (2006).

CHAPTER 6

EFFECTS OF PHYSIOLOGICAL FACTORS ON WEAR OF UHMWPE FOR JOINT PROSTHESIS

Yoshinori Sawae

Faculty of Engineering, Kyushu University
744 Motooka, Nishi-ku, Fukuoka 819-0395, Japan
E-mail: sawa@mech.kyushu-u.ac.jp

Wear of Ultra-high molecular weight polyethylene (UHMWPE) has been the most important tribological issue of artificial joint prostheses since fine wear particles released from UHMWPE components exert serious adverse reactions on living tissues. Those tissue reactions finally induce the aseptic loosening of prosthetic joint components. Therefore, a large number of studies have been conducted to understand the *in vivo* wear characteristics of UHMWPE and to extend the longevity of artificial joints by improving wear resistance of polyethylene components. However, the polyethylene wear mechanism in human body is very unique and complex because it is inevitably affected by the complicated joint kinematics and various constituents contained in the joint environment. This chapter focuses on physiological factors involved in the *in vivo* wear mechanism of UHMWPE. We first review representative studies about effects of kinematical and environmental factors on the *in vivo* wear mechanism of UHMWPE components of joint prostheses. Subsequently, some experimental results are presented to discuss about the considerable effects of multidirectional nature of the sliding motion in prosthetic joints and biological macromolecules contained in the periprosthetic fluid.

1. Introduction

Various material combinations are now utilized in bearing surfaces of joint prostheses. However, ultra-high molecular weight polyethylene

(UHMWPE) is still the most successful bearing material for artificial joints. UHMWPE has been used in a wide range of joint prostheses since Prof. Charnley first introduced it in an acetabular component of his hip joint prosthesis in 1962.[1] He selected UHMWPE as an alternative for the failed previous polymeric material, polytetrafluoroethylene (PTFE), because UHMWPE had superior wear resistance. Although UHMWPE seemed to have good biocompatibility as well, it had become evident that UHMWPE components released very fine wear particles during articulation and these fine wear particles could exert serious adverse reactions to living tissues. They activate macrophages and cause a necrosis and resorption of surrounding bone tissues. These tissue changes finally induce the aseptic loosening of prosthetic joint components.[2,3,4] Therefore, the wear characteristic of the UHMWPE used in bearing surfaces is now recognized as a controlling factor deciding durability and the long term survivorship of artificial joint prostheses.

Many attempts have been made to extend joint longevity of hip and knee prostheses through improving the wear resistance of UHMWPE which is used as an acetabular component of hip prostheses or a tibial insert and a patella component of knee prostheses.[5-10] The wear characteristic of newly developed materials has been examined in laboratory wear tests before their clinical application. The preclinical wear estimation of prosthetic joint materials usually has several steps, and simple screening test devices, such as a pin-on-disk or a pin-on-plate wear tester, are preferably used in the very early phase of the material development due to the convenience in the specimen preparation and the low cost of experiments. However, actual *in vivo* wear behaviour of artificial joint materials can be hardly predicted by hasty screening wear tests with too simple test configurations. For example, Charnley already reported in early 80's that the polyethylene wear rate obtained with a pin-on-disk device and a non-physiological lubricant was unacceptably smaller than the clinical value, by two orders of magnitude.[11] Subsequently, the discrepancies between the laboratory wear test result and the physiological wear behaviour of UHMWPE was emphasized as a serious technical problem by several researchers[12] in mid 90's, because such an unsuccessful wear estimation could mislead the material development scheme. Actually, inappropriate laboratory evaluations could not prevent serious failures of clinical products in some cases.[13,14,15] For example, carbon fiber-reinforced polyethylene was introduced in the clinical application of joint prostheses since it had shown the increased creep resistance and also improved wear behaviour in early laboratory wear tests, in which distilled water was used as lubricant.[16] However, it did not show the enhanced wear resistance in the later reciprocating pin-on-plate wear test with serum lubricant[16] and induced several early clinical failures.[13] Consequently, carbon fiber-reinforced polyethylene was withdrawn from the clinical application about seven years after its introduction.[17] More recently, UHMWPE with higher crystallinity and thicker lamella was developed and introduced in hip and knee prosthesis as an alternative to the conventional UHMWPE.[17] The modified UHMWPE had improved mechanical properties and its improved wear resistance was also expected from results of pin-on-disk[14] and ring-on-disk[18] tests. However, this material revealed the higher clinical wear rate compared with the conventional UHMWPE and induced early osteolysis and join failures.[14,15]

The problem of the screening wear tests mentioned above would be brought about by the difficulty in the reproduction of the *in vivo* polyethylene wear mechanism in *in vitro* wear experiments. The polyethylene components of joint prostheses have experienced quite unique wear mechanism in human body, since the wear of prosthetic joint components is inevitably affected by both the complicated chemical composition of the physiological environment and characteristic multiaxial joint kinematics. Joint prostheses are lubricated with the periprosthetic fluid, a kind of body fluid secreted into joint capsule after the joint implantation surgery. The periprosthetic fluid contains a very large number of constituents including several kinds of electrolytes, radicals and biological macromolecules.[19-22] These constituents should be entrained into a contact zone during joint movement and possibly affect the friction and wear behaviour of materials through some chemical and mechanical interactions with the bearing surfaces. Moreover, the physiological kinematics of hip and knee joints contains 3 dimensional motions. Therefore, polyethylene surfaces are exposed to the multidirectional sliding motion in prosthetic hip and knee joints during activities of daily life.[23-28] These environmental and kinematical factors make the *in vivo* wear mechanism of UHMWPE complex and unclear. Consequently, the physiological wear behaviour of UHMWPE could not be reproduced in the simple pin-on-disk wear test with non physiological lubricant, such as distilled water or saline.

The *in vivo* wear mechanism of UHMWPE components has attracted strong interest of researchers. Many studies, including *ex vivo* analysis of retrieved polyethylene components and *in vitro* polyethylene wear experiments, have been conducted to examine details of the *in vivo* polyethylene wear process and to elucidate the role of each physiological factors playing in the polyethylene wear. As a result, plenty of important knowledge has been accumulated during past few decades and a number of predominant factors which has considerable influence on the *in vivo* UHMWPE wear have been identified. These findings are quite important for the material development to improve the wear characteristics of prosthetic joint components. Furthermore, they can be utilised to establish an ideal wear test procedure for prosthetic joint materials, since the appropriate reproduction of the *in vivo* wear mechanism is essential to estimate the *in vivo* wear characteristics of materials and derive clinically reliable wear data from *in vitro* laboratory wear tests.

In this chapter, we would focus on two predominant physiological factors involved in the *in vivo* wear mechanism of UHMWPE, the kinematical factor based on the kinematics of joint movements and the environmental factor related to the constituent of synovial environment. Representative studies about effects of kinematical and environmental factors on the UHMWPE wear would be first reviewed. Subsequently, some experimental results obtained in our laboratory are shown to discuss about the considerable effects of multidirectional nature of sliding motion in the prosthetic joint and biological macromolecules contained in the periprosthetic fluid.

2. Physiological Factor Relating to Joint Kinematics

Synovial joints in a human lower limb make intricate 3 dimensional movements during a normal gait. In the case of a natural hip joint, the joint movement consists of three independent rotations about the three axes, i.e. flexion-extension, abduction-adduction and internal-external rotation. As a result, surfaces of artificial joint components experience the multidirectional or cross-shear sliding motion during a gait cycle and it has been evident that the shape of the sliding pathways of points on the articulating surface and associated transition of the sliding direction has a great impact on the wear of UHMWPE components. The primary concept of the relationship between the polyethylene wear and the sliding kinematics is based on the friction-induced molecular orientation of semi-crystalline polymers which has been indicated first by Pooley and Tabor.[29] If UHMWPE surface was exposed to unidirectional linear sliding motion, randomly oriented polyethylene molecules would be stretched by the tangential force and aligned parallel to the sliding direction. The resultant molecular orientation in the direction of sliding induces the anisotropic strain hardening of the polyethylene surface and subsequently increases the wear resistance to the unidirectional sliding. However, the molecular orientation causes the strain softening in the perpendicular direction as well and oriented molecules can be easily removed from the articulating surface by applying shear stress transversely. Therefore, the polyethylene wear become very sensitive to the shape of the sliding pathways which defines the transition of the direction and magnitude of the shear stress vector imposed on polyethylene surface and affects the surface molecular orientation and subsequent wear behaviour.

2.1. *Shape of Sliding Pathways in Joint Prostheses During Walking*

The actual shape of sliding pathways exerted in natural joints has attracted great interests in connection with the wear problem of UHMWPE joint components. Several research groups have succeeded in calculating the shape of the sliding pathways in hip joint surfaces from the kinematic data obtained by a gait analysis. Ramamurti *et al.*[23] estimated the sliding loci of 20 points on the femoral head of the hip joint prosthesis by using kinematic data from the gait analysis of Johnston and Smidt.[30] They found that most of the sliding loci had distorted elliptical shape which would exert the multidirectional shear stress on the surface of UHMWPE acetabular cup. However, both shape and length of the sliding loci varied with the location on the femoral head. Similar results have been obtained by Wang *et al.*[24] and Barbour *et al.*[25] by using kinematic data of Ramakrishnan[31] and Paul[32], respectively. Based on their simulation results, Wang *et al.*[24] categorised the shape of sliding pathways into three types; irregular ellipses, tear-drops and figure eights. Wang *et al.*[24] also graphically represented the transition pattern of shear stress vector imposed on the acetabular cup surface, which could indicate the transition of both magnitude and direction of the stress vector, by using the joint loading pattern reported by Paul.[32] On the other hand, Bennett *et al.*[26,33] conducted the similar computer simulation with joint kinematic data of 19 total hip arthroplasty (THA) patients and quantified the individual variation in the multidirectional sliding motion of hip joints. Their results indicated that the shape of the sliding loci in the hip joints of THA patients varied significantly in individuals. The average sliding distance during one gate cycle ranged from 10.0 to 18.1 mm while the average aspect ratio of sliding pathways ranged from 2.5 to 9.2. Saikko *et al.*[27] also calculated the shape of sliding pathways in hip joints from the kinematic data of Johnston and Smidt[30] by using unique computational method based on Euler angles. Especially, they verified their simulation results experimentally in their joint simulator using specially designed acetabular cup with embedded sharp pins which clearly depicted the sliding pathways on the femoral head. For knee joint prostheses, Hamilton *et al.*[28] calculated sliding loci of contact points between the femoral component and the polyethylene tibial insert from the fluoroscopic kinematic data of a total knee arthroplasty (TKA) patient. In this case, the *in vivo* kinematics of artificial joint components could be determined directly from X-ray fluoroscopic image with the image matching technique[34,35] and used to determined the joint sliding pathways.

2.2. *Effect of Multidirectional Sliding Motion on Wear of UHMWPE*

The relationship between the multidirectional nature of joint articulation and wear of UHMWPE acetabular component was pointed out by Wang *et al.*[12,24,36] and several research groups.[37-39] Since then, many studies have been conducted during the last decade to understand the role of the multidirectional sliding motion in the physiological polyethylene wear mechanism. Most of them also aimed to establish the reliable *in vitro* wear test method. Wang *et al.*[12,36] focused on the apparent difference in both the wear rate and wear mechanism of UHMWPE between a linear reciprocating wear test and a hip joint simulator test with physiological joint motion. Their test results revealed the adverse effect of the multidirectional sliding motion associated with distorted elliptical sliding loci on polyethylene wear. They explained about the motion-dependent wear behaviour of UHMWPE by the strain-induced molecular orientation at the articulating surface and resultant strain softening of the polyethylene.[24,40] Bragdon *et al.*[37] examined the effect of the sliding motion on the result of their hip joint simulator test. Their experimental result was clearly indicating the crucial importance of the multidirectional sliding motion with the physiological shape of sliding pathways in the UHMWPE wear estimation. With the multidirectional sliding motion exerted by the physiological joint movement, they obtained wear rate of polyethylene acetabular cup of 24.8 mg/million cycles; it was very close to the clinically obtained polyethylene wear rate. However, the polyethylene wear rate became negligible with the unidirectional sliding motion. Barbour *et al.*[25] also demonstrated the significant effect of the sliding pathway shape on the results of joint simulator tests. Statistically significant increase in the wear rate of polyethylene cup was observed in their hip joint simulator test by changing the shape of sliding pathways from liner to elliptical. Bennett *et al.*[33] evaluate the *in vivo* wear rate of the polyethylene acetabular cup from a radiograph of THA patients and examined the correlation between the characteristics of their hip joint sliding pathways and the *in vivo* polyethylene wear rate. As a result, positive correlations between the *in vivo* wear rate of UHMWPE acetabular cup and the inverse of aspect ratio of the sliding pathways was found. It meant that the wear rate of UHMWPE increased with decreasing aspect ratio of elliptical sliding pathways. The same research group[41,42] has recently observed the orientation of polymer chains at the articulating surface of retrieved polyethylene cups by FT-IR and confirmed the strong correlation to the aspect ratio of sliding loci. They also reported the effect of multidirectional sliding motion on material properties of UHMWPE, such as crystallinity, degree of oxidation and mechanical properties. The material property changes detected by their experiments might be contributing to the increased wear under the multidirectional sliding.

Based on the knowledge of the motion-dependent wear process of UHMWPE, modified screening wear test devices have been proposed for the wear estimation of prosthetic joint materials.[38,39,43,44] Saikko *et al.*[38] developed an unique multidirectional sliding pin-on-disk wear tester called circularly translating pin-on-disk (CTPOD) device. It has a similar specimen configuration to a conventional pin-on-disk device, however, the disk specimen is moving circularly, not rotating, in this test device. Therefore, the polyethylene pin specimen translates on the disk surface along a circular path and the direction of shear stress vector changes continuously relative to the polyethylene surface. They applied elliptical sliding pathways to polyethylene specimens as well in this test device with a little modification and examined the relationship between the aspect ratio of elliptical sliding loci and wear rate of UHMWPE.[45] Their results indicated that the wear rate of UHMWPE decreased with increasing the aspect ratio of sliding loci and morphological characteristics of polyethylene worn surface were also changed with the aspect ratio significantly. The wear rate and wear mechanism of polyethylene became similar to the clinical results if the aspect ratio was less than 5.5. They also examined the effect of contact pressure on the polyethylene wear rate and indicated that the contact pressure should be less than 2 MPa to reproduce the clinical wear rate and wear mechanism of UHMWPE with this test device.[46] Bragdon *et al.*[44] proposed to use a rectangular sliding pathway for the wear estimation of UHMWPE and successfully obtained the wear rate and worn surface morphologies consistent with the retrieved polyethylene acetabular component. A similar screening test device was also used by Turell *et al.*[47,48] and Mazzucco *et al.*[49] to examine the effect of the aspect ratio of rectangular paths[47], counterface roughness[48] and contact conditions.[49]

2.3. *Wear Characteristics of UHMWPE Evaluated in Multidirectional Sliding Wear Test*

2.3.1. *Comparison of UHMWPE Wear Behaviour between Unidirectional Sliding and Multidirectional Sliding*

The author of this chapter has also conducted some experimental studies about the relationship between the multidirectional sliding motion and the wear behaviour of UHMWPE.[50-54] Two types of multidirectional sliding wear tester has been constructed in our lab and used in the wear estimation of UHMWPE. First one was developed from a conventional pin-on-disk wear tester by replacing a specimen movement system. A schematic drawing of this wear tester is shown in Fig. 1(a). The basic principle of the multidirectional sliding mechanism is identical to that of the CTPOD device developed by Saikko.[38] A cylindrical pin specimen is fixed to the loading arm and pressed against a plate specimen, which is installed in the liquid bath, by dead weights. The liquid bath was mounted on a table placed upon a vertical crank and directly driven by a ring-cone motor. Two rods are attached to the liquid bath and the tips of the rod are connected to two linear guides to prevent the rotation of the plate specimen. Subsequently, the pin specimen articulates upon the plate surface along a circular path with the diameter of 40 mm. A sliding motion arising between the pin and plate specimen in this apparatus is shown schematically in Fig. 1(b). In this case, the velocity vector is varied from v to v' while the centre of the plate specimen moves from o to o'. As a result, the direction of the sliding velocity vector changes continuously relative to the UHMWPE pin surface whereas the magnitude of the vector remains constant at any point on the pin surface.

In Fig. 2, the wear behaviour of UHMWPE obtained from the above-mentioned multidirectional sliding wear tester is compared with the polyethylene wear in the conventional pin-on-rotating disk test device. Test configurations of both unidirectional and multidirectional sliding tests were common except the sliding kinematics. UHMWPE specimens were cut from a ram extruded GUR415 bar stock and were not sterilized. Polished 316 stainless steel plate with the surface roughness of 0.02 μm R_{rms} was used as a sliding counterface and fetal bovine serum diluted to

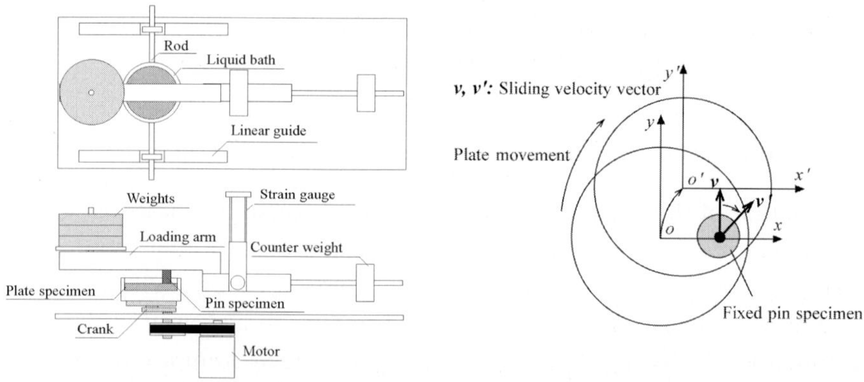

Fig. 1. (a) A schematic drawing of pin-on-plate wear tester developed from conventional pin-on-rotating disk apparatus and (b) the circular motion of plate specimen and resultant transition of sliding velocity vector.

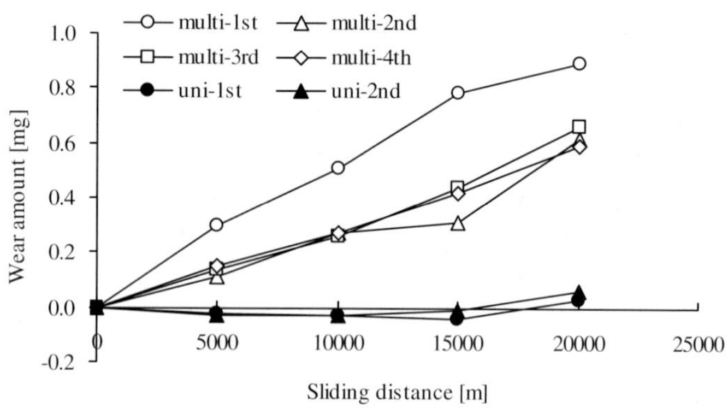

Fig. 2. Comparison of wear behaviour of UHMWPE between the multidirectional sliding pin-on-plate test and unidirectional sliding pin-on-rotating disk wear test.

30 vol% with distilled water and supplemented with 0.3 wt% sodium azide was used as test lubricant. Contact pressure and average sliding speed were set to 8 MPa and 20 mm/s, respectively. All experiments were conducted at room temperature. Weight of UHMWPE pin specimen was measured every 5 km sliding to an accuracy of 0.01 mg and the measured weight loss of specimen was corrected with the weight gain of the soak control pin to cancel the effect of fluid uptake from lubricant.

The wear test was iterated twice in the unidirectional sliding pin-on-disk tester and four times in the multidirectional sliding pin-on-plate tester. Results are clearly indicating the crucial role of the multidirectional sliding motion in the realistic wear behaviour of UHMWPE. The wear amount of UHMWPE increased linearly during the wear test and average wear rate became 2.3 x 10^{-7} mm^3/Nm in the multidirectional sliding tests. On the other hand, the wear amount of UHMWPE was negligible in the unidirectional sliding pin-on-disk test even after 20 km sliding.

The apparent difference in the morphological characteristic of polyethylene worn surface was also observed between two sliding motions. The polyethylene surface became glossy and microscopically smooth after the multidirectional sliding test (Fig. 3(a)) and those characteristics were consistent with the worn surface of retrieved polyethylene cup.[55,56] On the other hand, the large area of polyethylene worn surface was covered with fine ripples aligning perpendicular to the sliding direction after the unidirectional sliding test (Fig. 3(b)).

The difference in the surface morphology became more significant in the molecular scale observation with the atomic force microscopy (AFM, Fig. 4). The primal characteristic of the worn surface associated with the unidirectional sliding motion was fine parallel ripples aligned in transverse direction with an interval of 2 to 3 μm in the 20 μm x 20 μm image (Fig. 4(a)). However, the distinct molecular-scale structure could be identified in 4 μm x 4 μm image; molecular bundles with the thickness less than 100 nm aligned in the sliding direction (Fig. 4(b)).

It would be indicating the friction-induced molecular orientation at the sliding surface. The polyethylene wear was therefore suppressed by the strain hardening of the polyethylene surface in the unidirectional sliding wear test. The similar molecular orientation in the polyethylene surface and resultant reduction in the wear rate had been observed by Wang *et al.*[24] and other researchers.[37]

On the other hand, such an oriented molecular structure could not be found on the polyethylene worn surface after the multidirectional sliding test. Instead, very fine fibrils with a length of 2 to 3 μm were formed on

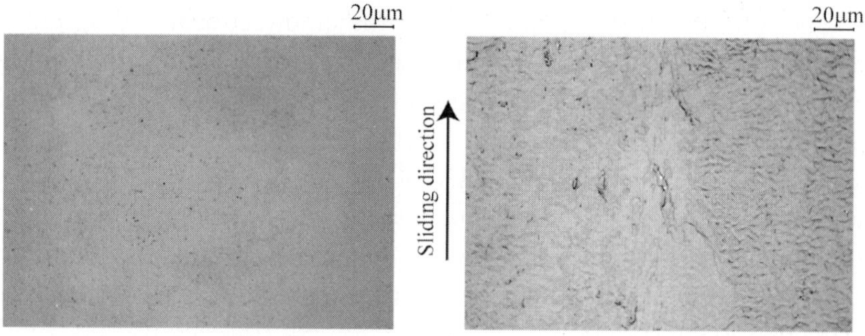

(a) Multidirectional sliding (b) Unidirectional sliding

Fig. 3. Comparison of microscopic image of polyethylene worn surface taken by a confocal laser scanning microscope.

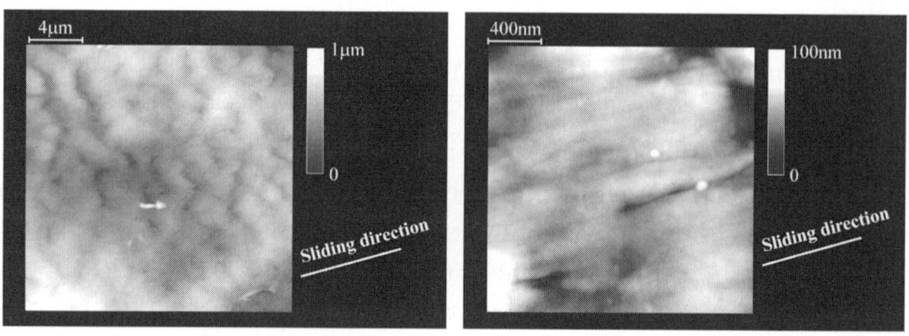

(a) Unidirectional sliding
20μm x 20μm scan

(b) Unidirectional sliding
2μm x 2μm scan

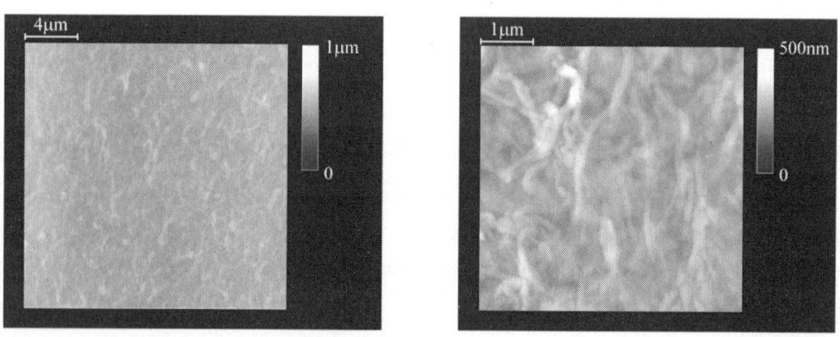

(c) Multidirectional sliding
20μm x 20μm scan

(d) Multidirectional sliding
2μm x 2μm scan

Fig. 4. Comparison of molecular scale structure of polyethylene worn surface observed by atomic force microscope.

the UHMWPE surface after the multidirectional articulation (Fig. 4(c), (d)). The similar microstructure had been recognised in the worn area of retrieved polyethylene cups.[12,55,56] McKellop *et al.*[55] reported that fibrils of elongated polymer were found within the glossy worn area of retrieved cups and the average length of fibrils was 1.3 μm. Wang *et al.*[12] and Oonishi *et al.*[56] also reported the identical morphological characteristics of the weight bearing area of the retrieved polyethylene cup surface. Therefore, the fibrillated microstructure in the UHMWPE worn surface would be suggesting the successful reproduction of the physiological wear mechanism by the multidirectional sliding motion.

Hall *et al.*[57] reported that the mean value of the clinical UHMWPE wear rates estimated from retrieved Charnley acetabular cups was 2.1×10^{-6} mm^3/Nm. The average UHMWPE wear rate obtained in our multidirectional sliding wear test was 2.3×10^{-7} mm^3/Nm and it was significantly higher than that measured in the unidirectional sliding test. However, it was still one order of magnitude less than the clinical value. Experimental results of Saikko[46] indicated that the UHMWPE wear rate in CTPOD device, which has an identical articulating mechanism to our multidirectional sliding wear tester, was consistent with the clinical value with the contact pressure of 2 MPa, and the polyethylene wear rate decreased with increasing contact pressure. In this study, wear tests were conducted with an average contact pressure of 8 MPa and such an elevated contact pressure might be responsible for the relatively lower wear rate compared with the clinical value.

2.3.2. *Wear Test of Cross-Linked UHMWPE with Multidirectional Sliding Device*

The dose dependent wear characteristic of gamma-irradiated cross-linked UHMWPE was evaluated by using another multidirectional sliding wear tester. Necessity of using a multidirectional sliding apparatus in the wear characterization of cross-linked UHMWPE has already indicated by Wang *et al.*[12] In their linear reciprocating wear tests, gamma-irradiated UHMWPE showed inferior wear behaviour compared with unirradiated UHMWPE. However, the ranking of wear resistance inverted in the physiological joint simulator test; irradiated UHMWPE exhibited superior wear resistance compared with the unirradiated material under the multidirectional sliding condition. Of course, the latter result was more reliable as discussed previously. The schematic drawing of the newly designed multidirectional sliding pin-on-plate wear tester is shown in Fig. 5. The wear tester also has pin-on-plate configuration, however, a PC-controlled two dimensional servo actuator was incorporated as a drive unit to realize the multidirectional sliding motion along the various shaped pathways. During the wear test, two cylindrical UHMWPE pin specimens are loaded against stationary plate specimens by air cylinders which were controlled by solenoid valves to maintain constant contact pressure. Pin specimens were then moved by the servo actuator and articulated on the plate surfaces along defined two dimensional sliding loci. Each plate specimen was installed in the lubricant bath and immersed in test lubricant. Temperature of the lubricant was adjusted to 37°C in a temperature-controlled water bath and circulated to the lubricant bath.

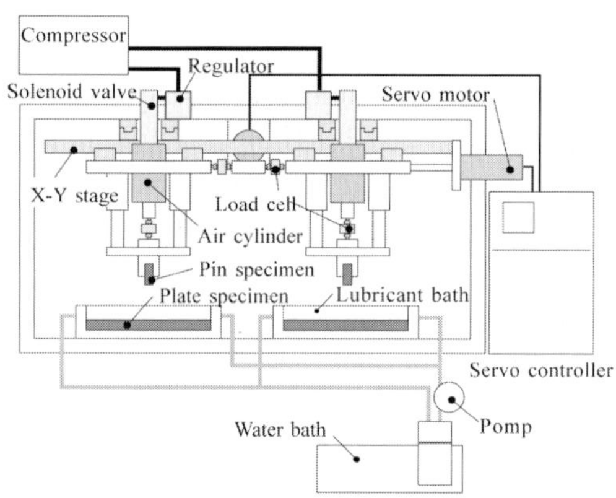

Fig. 5. Schematic drawing of the multidirectional sliding pin-on-plate wear tester coupled with a PC-controlled two dimensional servo actuator.

Prior to the wear evaluation of cross-linked UHMWPE, a series of wear tests with non-cross-linked UHMWPE and four different sliding loci was conducted to confirm the effect of the shape of sliding pathways. The sliding pathway shape examined in this study were linear reciprocating, square, circle, and distorted ellipse which consisted of two symmetric arcs with curvature radius of 50 mm. All sliding loci have a sliding length of 140 mm and their shape is depicted in Fig. 6. Same as the previous experiments, pin specimens were cut from the non-cross-linked GUR415 block. Polished 316 stainless steel plates and 30 vol% diluted bovine serum were used as sliding counterface and test lubricant, respectively. The applied contact pressure was 10 MPa, and average sliding speed was set at 45 mm/s.

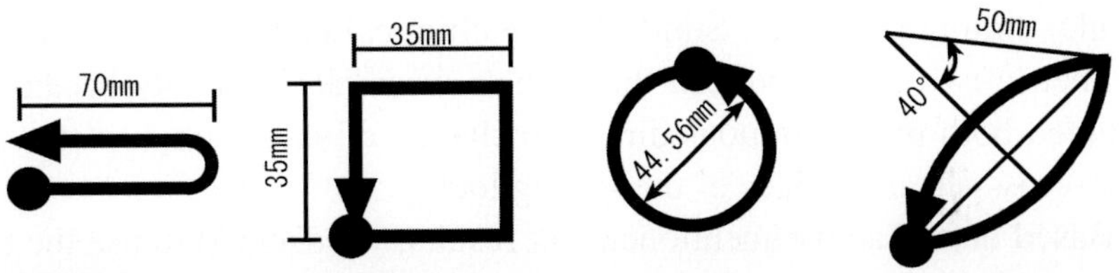

(a) Linear (b) Square (c) Circle (d) Ellipse

Fig. 6. Illustration of the sliding pathways used to examine the effect of its shape on the wear rate of non-cross-linked UHMWPE in the multidirectional pin-on-plate wear test.

The resultant wear rates of UHMWPE were compared in Table 1. The wear rates in Table 1 are still considerably smaller than the clinical value.[57] As mentioned previously, the relatively high contact pressure of 10 MPa in our wear test would be one of the possible factors lowering the value of specific wear rate. However, the relationship between the shape of sliding pathways and wear behaviour of UHMWPE obtained in this study are consistent with previous studies. Multidirectional sliding pathways produced higher and more realistic polyethylene wear rate compared with the unidirectional linear reciprocation. Furthermore, the wear rate of UHMWPE highly depended on the shape of the multidirectional sliding pathways.

Table 1. Comparison of average specific wear rate of non-cross-linked UHMWPE (n=2) among four different shape of sliding pathways.

	Linear	Square	Circle	Ellipse
Wear rate $\times 10^{-7}$ [mm^3/Nm]	0.13	0.45	0.78	1.12

Although, square and circular pathways had identical aspect ratio, the polyethylene wear rate became higher with circular pathways compared with the square pathway. Wang[40] suggested that the wear amount of UHMWPE was determined by the frictional work in the direction perpendicular to the molecular orientation, not the geometric characteristics like the aspect ratio of pathways. The continuous transition in the direction of the shear stress vector and associated frictional work might be responsible for the greater wear with the circular sliding pathways. Similarly, the elliptical pathways consisting of symmetric two arcs produced the highest wear rate probably because it involved both the continuous direction change of shear stress vector and sharp cross-shear at edges of the sliding loci.

Based on the above-mentioned test results, we decided to use the two symmetric arcs loci in the wear evaluation of cross-linked UHMWPE. In this series of experiment, the wear of gamma irradiated cross-linked polyethylene with the irradiation dose level ranged from 5 Mrad to 100 Mrad was evaluated and the relationship between the dose level of gamma irradiation and wear characteristic of cross-linked UHMWPE was examined. Polyethylene pin specimens were prepared from GUR1050 compound. Medical grade CoCr alloy and alumina ceramics were used as plate specimens to examine the effect of sliding counterface material on the wear of cross-linked UHMWPE. The effect of the surface roughness of the metal couterface was also examined by preparing some CoCr plate specimens with different surface roughness values. The length of the sliding pathway and the curvature radius of arcs were reduced to 125.7 mm and 45 mm, respectively since the plate specimen size was reduced. The average contact pressure and sliding speed were set to 10 MPa and 45 mm/s, respectively and diluted bovine serum (30 vol% fetal bovine serum + 70 vol% distilled water + 0.3 wt% sodium azide) was used as test lubricant.

Effects of gamma dose on material properties of UHMWPE are shown in Fig. 7. Cross-links between polyethylene molecular chains were induced by gamma irradiation, and cross-linked UHMWPE became harder and more brittle exponentially with increasing dose level. Wear rate of cross-linked UHMWPE was plotted against dose level of gamma

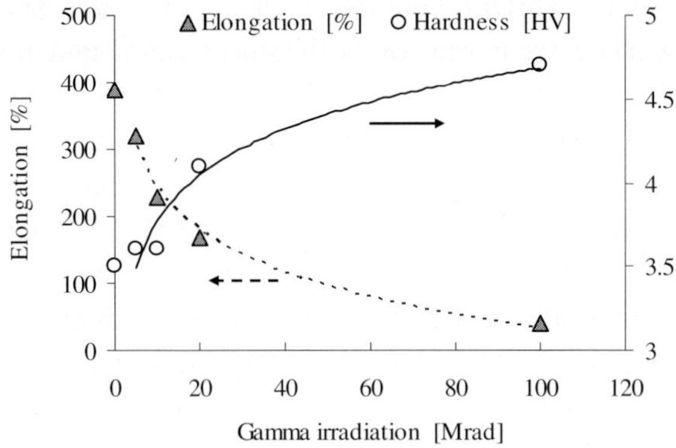

Fig. 7. Relationship between gamma-irradiation dose level and mechanical properties, elongation at break [%] and Vickers hardness [HV], of cross-linked UHMWPE.

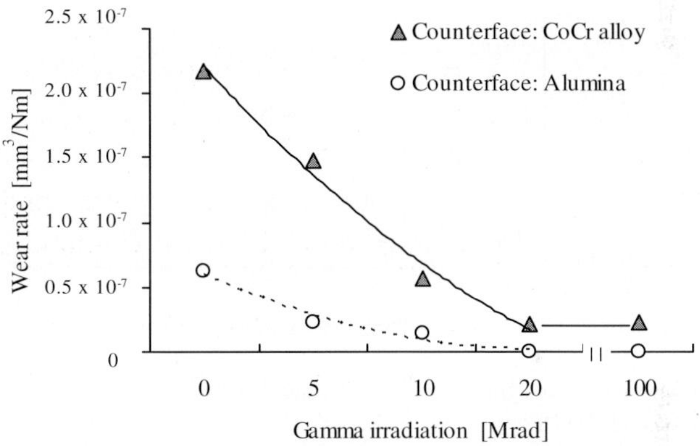

Fig. 8. Relationship between gamma-irradiation dose level and specific wear rate of cross-linked UHMWPE sliding against medical grade CoCr alloy or alumina ceramics.

irradiation in Fig. 8. The wear rate of cross-linked polyethylene could be reduced by increasing the irradiation dose up to 20 Mrad. However, the difference could not be identified between 20 Mrad dose and 100 Mrad dose in our wear test. The counter face material also had apparent effects on wear of cross-linked polyethylene. Wear amount of cross-linked UHMWPE could be reduced by changing counter face material from CoCr alloy to alumina ceramics and became negligible with the 20 Mrad dose.

The effects of the surface roughness of CoCr plate specimen were shown in Fig. 9. The wear rate of both unirradiated and highly cross-linked UHMWPE was varied with R_a value of the CoCr plate surface. However, the relationship between the polyethylene wear rate and the counterface roughness was inconsistent with the previous studies. It is generally accepted that the wear rate of UHMWPE increases with surface roughness of the sliding counterface under lubricated condition.[58,59] Even in the similar multidirectional sliding wear test with serum based lubricant, Saikko *et al.*[60] has reported that the wear rate of conventional and cross-linked UHMWPE increased with R_a value of the sliding counterface.

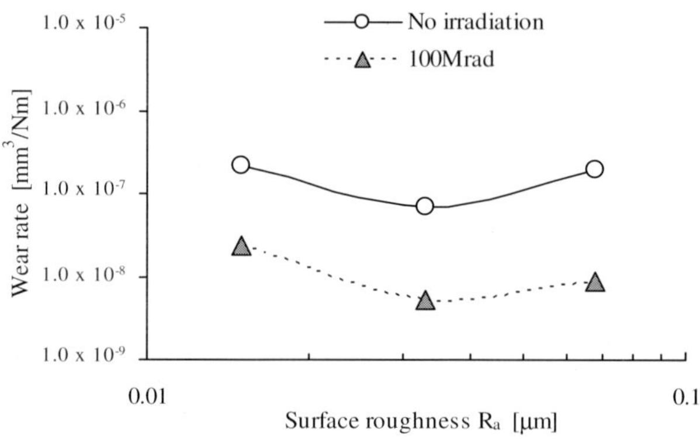

Fig. 9. Effect of surface roughness of CoCr plate on the wear rate of cross-linked and non-cross-linked UHMWPE.

On the other hand, both unirradiated and highly cross-linked UHMWPE indicated a minimum wear rate with the moderate surface roughness of the plate specimen and the polyethylene wear rates increased significantly if the R_a of plate surface was reduced to 0.015μm in this study. Similar trend can be observed usually in wear of UHMWPE under dry condition.[58] In this study, we used flat surfaced pin specimens with a diameter of 6 mm and applied average contact pressure of 10 MPa. Therefore, sufficient amount of lubricant might not be entrained into the contact region due to the relatively large contact area and high contact pressure. Consequently, the effect of direct asperity contacts might become significant like a dry sliding condition.

3. Physiological Factor Relating to Joint Environment

Articulating surfaces of diarthrodial joints are surrounded by joint capsule. The inner surface of the joint capsule is lined with synovial membrane which secretes synovial fluid and fills internal space of the joint capsule with this biological fluid. Synovial fluid plays important roles to maintain the joint homeostasis. It would have two main functions; lubrication of joint surfaces to keep friction and wear low and nutrition of cells living within cartilage and ligaments to sustain the viability of joint tissues.[19] The composition of synovial fluid is very complicated involving many kinds of electrolytes, radicals, nutrients and biological macromolecules. Especially, macromolecule constituents of synovial fluid consist of those derived from brood, i.e. serum proteins and lipids, hyaluronic acid (HA) which is secreted from synovial membrane and catabolic products from joint tissues, such as chondroitin sulphate.[19]

The joint capsule and synovial membrane are usually removed during total joint arthroplasty. However, these tissues are reconstructed around the articulating part of joint prosthesis after the surgery and the regenerative synovial membrane will secret periprosthetic fluid.[20] DesJardins *et al.*[22] analyzed the composition of the periprosthetic fluid obtained from patients undergoing revision surgery of total knee arthroplasty. Their results indicated that the periprosthetic fluid has similar serum proteins and lipid composition to the synovial fluid obtained from the osteoarthritic (OA) knee joints. Joint prostheses are immersed and articulating within this complex body fluid during normal activities of daily life. Therefore, their friction and wear behaviours would be inevitably influenced by the constituents of the surrounding fluid because they should be entrained into a contact zone during joint movement and possibly have some chemical and mechanical interactions with the bearing surface materials. The appropriate evaluation of the environmental effects from the surrounding joint fluid would be inevitable to understand the *in vivo* wear mechanism.

3.1. *Degradation of UHMWPE in Human Body*

Oxidative degradation of UHMWPE has been recognized as one of the adverse reactions deteriorating mechanical properties of UHMWPE after the implantation and increasing *in vivo* polyethylene wear.[61-63] The primary cause of the polyethylene degradation is supposed to be free radicals generated during gamma-irradiation which has been used for the sterilization of medical implants and more recently for cross-linking polyethylene molecular chains.

Some *ex vivo* studies indicated the considerable effects of the physiological environment on the oxidative degradation of UHMWPE and subsequent wear behaviour of the bearing surface materials. Eyerer and co-workers[64,65] analysed the chemical composition of retrieved polyethylene acetabular cups by FT-IR in detail and indicated the accelerated oxidative degradation of the polyethylene articulating surface in physiological environment. They suggested that the peroxide constituents of synovial fluid could attack the UHMWPE and reduce the average molecular weight at the surface. The chain scission of polyethylene molecules resulted in the increase in the density and consequently caused mechanical property changes in UHMWPE. Costa *et al.*[66] also examined the oxidative degradation of retrieved polyethylene cups by FT-IR microscopy and reported the accelerated oxidation of polyethylene worn surfaces. They called this material degradation in the physiological environment as biodegradation and postulated the possible mechanism of the accelerated oxidation.[66,67] They suggested that the peroxide radicals could be generated *in vivo* both biologically by the contact between inflamed cells and polyethylene and mechanically by the bond scissions in polyethylene molecules caused by the mechanical abrasion. These radicals would subsequently contribute to the oxidative degradation of UHMWPE surface. On the other hand, Ambrosio *et al.*[68] used XPS to analyse the oxidation of retrieved UHMWPE surfaces and reported similar findings. In this case, they described these environmental adverse effects as “biological corrosion”. The IR adsorption possibly related to oxidation products was also recognized in the UHMWPE worn surface after our laboratory wear test only when the diluted bovine serum was used as lubricant.[69]

However, controversial results were reported by James *et al.*[70] They immersed retrieved UHMWPE specimens in hexane to extract fatty acids adsorbed from surrounding periprosthetic fluid. Then, the FT-IR spectra from the immersed specimen were compared to the spectra from original specimen. As a result, they concluded that some peaks in FT-IR spectra, which had been supposed to be the evidence of the increased oxidation, were indicating the adsorption of esterified fatty acids. Shen *et al.*[71] also examined the effect of the constituent uptake from the surrounding fluid on the results of FT-IR measurement and they indicated that contaminants increased the apparent level of oxidation and induce the overestimation of oxidative degradation. Therefore, there are still some debates about the relationship between the oxidative degradation of UHMWPE and physiological environment in the joint space.

3.2. *Effect of Macromolecules on Friction and Wear*

As mentioned previously, the physiological environment of the periprosthetic fluid contains several kinds of macromolecules including proteins, lipids and HA. Macromolecule contents of synovial fluid and periprosthetic fluid reported in previous studies[21,22,72] are listed in Table 2. Effects of these biological macromolecules on the UHMWPE wear have been examined mainly by *in vitro* experimental methods. Wang *et al.*[43] focused on protein molecules contained in the physiological solution and examined effects of the protein concentration of lubricant on the wear behaviour of UHMWPE acetabular cup in their hip joint simulator test. Their simulator test results clearly suggested the important roles of protein molecules in the UHMWPE wear; the wear rate of the serum lubricated UHMWPE cup clearly depended on the protein concentration of the lubricant, and the wear of UHMWPE became negligible if they used pure water as a lubricant. Similar experiments were conducted by Saikko[73] using CTPOD device and consistent results were obtained. Wang *et al.*[74] also indicated the effect of protein type and their relative concentration in the lubricant. The wear rate of UHMWPE cup was reduced by increasing the albumin/globulin ratio in their simulator test. The significant effects of protein concentration and protein type on the friction and wear of UHMWPE were also reported from several research groups.[75-77] Heuberger *et al.*[76] suggested that the protein molecules in lubricants could be denatured during a sliding test and unfolded proteins preferably adsorbed onto hydrophobic polyethylene surface. The adsorbed proteins might increase adhesive friction which enhanced microscopic plastic deformation of polyethylene surface and finally brought some changes in the polyethylene wear behaviour. The trace of adsorbed protein molecules has been identified on the surface of retrieved UHMWPE components by XPS[68] and FT-IR.[70,78] These findings support the considerable interaction between protein molecules contained in the periprosthetic fluid and the polyethylene articulating surface.

Table 2. Macromolecule content of joint fluids reported in previous studies.OA synovial fluid[21,22], Periprosthetic fluid[21,22] and pig synovial fluid.[72]

	Concentration [g/l]			
	Total protein	Albumin	Phospholipid	HA
OA synovial fluid[21]	27 ± 10	NA	0.52 ± 0.18	1.3 ± 0.5
Periprosthetic fluid[21]	34 ± 13	NA	0.52 ± 0.19	0.9 ± 0.4
OA synovial fluid[22]	24.75 ± 7.13	20.00 ± 4.98	0.41 ± 0.14	NA
Periprosthetic fluid[22]	28.00 ± 5.03	21.75 ± 3.20	0.50 ± 0.06	NA
Pig synovial fluid[72]	34	19	0.19	1.2

On the other hand, there are not many previous reports on the effects of HA and lipid molecules on the wear of UHMWPE. Wang *et al.*[74] examined the effect of HA on the wear rate of UHMWPE in their hip joint simulator test. They prepared a HA-contained test lubricant by dissolving HA to alpha-calf serum (ACS) and subsequently diluting it with deionised water to yield a protein concentration of 35 g/l and a HA concentration of 0.34 g/l. However, the significant difference in the wear rate of UHMWPE acetabular cups was not observed between test lubricants with and without HA in their simulator test. Later on, DesJardins and his co-authors[22,79] examined the effect of HA on wear and fatigue behaviour of UHMWPE components of artificial knee joints. Results of their knee joint simulator test indicated that 1.5g/l HA supplementation to 50 % bovine serum solution with a protein concentration of 35 g/l induced pitting damage on CoCr femoral surface and increased wear of UHMWPE tibial insert 6.88 times.[22] Meanwhile, the addition of HA to the serum solution delayed crack initiation and growth.[79] Saikko and Ahlroos[80] dispersed phopholipid in water and used as lubricant in their wear test of UHMWPE-on-CoCrMo alloy sliding pair. Their experimental results suggested the remarkable boundary lubrication effect of phospholipid. The boundary lubrication effect of phospholipid for the artificial joint was also reported by Bell *et al.*,[81] although the phospholipid concentration of their lubricants was higher than the physiological lipid concentration of synovial fluid. Greenbaum *et al.*[82] examined the interaction between lipid constituents of synovia and UHMWPE experimentally. They doped cholestene and squalene to UHMWPE as representatives of lipid constituents of synovia and examined their effects on the mechanical properties and wear behaviour of UHMWPE. Their test results indicated that the lipid molecules could diffuse into the bulk of material and had plasticising effect on UHMWPE. Their wear test results could not show the apparent effect on the polyethylene wear behaviour. However, results of several *ex vivo* analysis of retrieved components supported the *in vivo* lipid diffusion into UHMWPE[70,78] and tribological effects of diffused lipid molecules should be postulated carefully.

3.3. *Effect of Protein and Lipid on Wear of UHMWPE*

Among biological macromolecule constituents of the periprosthetic fluid, we focused on serum proteins and phospholipids. To explore the role of these macromolecules in the UHMWPE wear mechanism, we prepared several lubricants with different protein and phospholipid concentrations for the multidirectional sliding wear test and examined the effect of the lubricant composition on the UHMWPE wear.[51,53,83]

The schematic illustration of the wear tester used in this study is shown in Fig. 10. The multidirectional sliding pin-on-plate wear tester developed from conventional pin-on-rotating disk apparatus (Fig. 1) was extended to be able to test three specimens simultaneously. The basic mechanism is identical to the previous one; fixed pin specimens are articulating upon the plate surface along a circular path. In this series of experiments, the diameter of the circular sliding pathways and mean contact pressure were reduced to 30 mm and 7 MPa, respectively, and the average sliding speed was set to 20 mm/s. Cylindrical polyethylene pin specimens with a diameter of 6 mm were machined from GUR415 bar stock. They were not sterilized prior to the wear tests. Medical grade cast Co-Cr-Mo alloy (ASTM-F75) was used as plate specimens. The plate surface was polished by diamond slurry and the resultant average surface roughness was 0.009 μm (R_a). At intervals of 5 km sliding, the pin specimens were weighed to an accuracy of 0.01 mg and the volumetric specific wear rate was determined from their weight loss. A soak control pin was prepared to estimate the weight gain due to a fluid uptake from the lubricant.

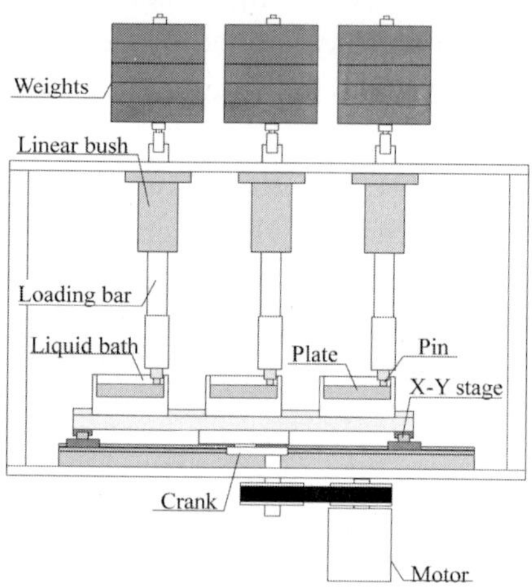

Fig. 10. Schematic illustration of three-station multidirectional sliding pin-on-plate wear tester.

Several phosphate buffered saline (PBS) solutions containing protein and lipid were prepared and used as lubricants in the wear test. In this study, bovine serum γ-globulin and synthetic phospholipid (L-α-dipalmitoyl phosphatidylcholine (DPPC)) were used as a representative protein and phospholipid constituents of the periprosthetic fluid. DPPC molecules were dispersed in PBS as fine colloidal particles and γ-globulin was subsequently dissolved in the DPPC dispersion. 0.3 wt% sodium azide was also added to prevent bacterial growth. PBS that contained no macromolecule and diluted bovine serum (30 vol% fetal bovine serum + 70 vol% distilled water) were also used as control lubricants. The diluted bovine serum used in this study has the protein concentration of 1.1 wt% including the albumin content of 0.7 wt% and the globulin content of 0.4 wt%.

Results are plotted against the protein concentration of test lubricants in Fig. 11. The UHMWPE wear rate clearly depended on the protein and lipid concentrations of test lubricants. The UHMWPE wear was almost negligible in the macromolecule-free PBS; the specific wear rate was down to 5×10^{-9} mm^3/Nm. The wear rate was still very small if the protein concentration was below 0.2 wt%. However, it increased with increasing the protein concentration of the lubricant and the lubricant containing 2.0 wt% protein and 0.02 wt% phospholipid produced the highest wear rate of 3.74×10^{-7} mm^3/Nm that was comparable to the specific wear rate of 3.47×10^{-7} mm^3/Nm yielded by the diluted bovine serum.

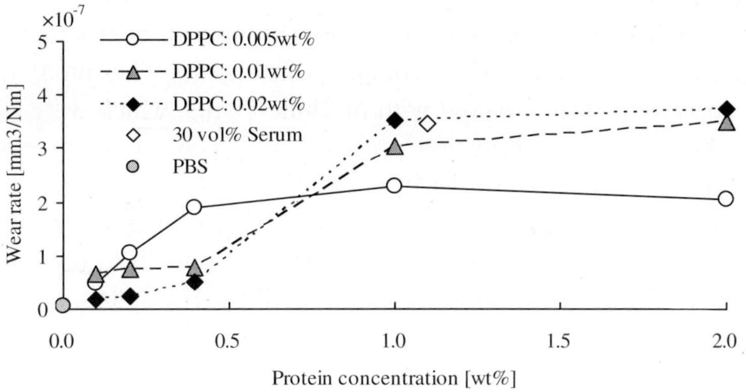

Fig. 11. Comparison of volumetric wear rate of UHMWPE plotted against protein concentration of test lubricant.

Figure 12 shows the relationship between the lipid concentration of the test lubricant and the specific wear rate of UHMWPE with different protein concentrations. The results were apparently divided into two groups. If the protein concentration was relatively low, i.e. 0.1, 0.2 and 0.4 wt%, the UHMWPE wear rate decreased with increasing lipid concentration. However, the wear of UHMWPE was clearly enhanced by increasing lipid concentration in the lubricants with high protein concentrations, 1.0 and 2.0 wt%.

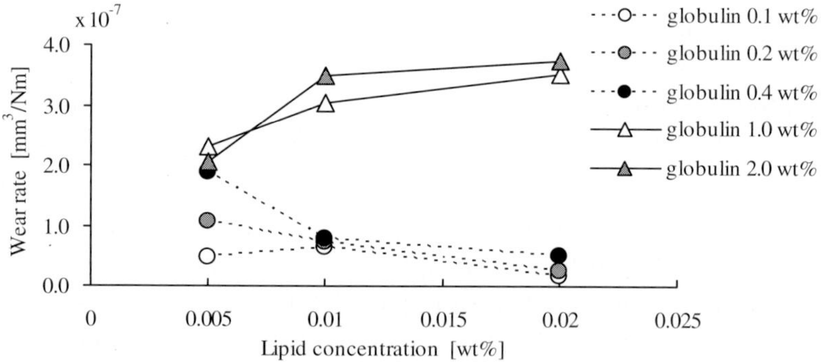

Fig. 12. Comparison of volumetric wear rate plotted against phospholipid concentration of test lubricant.

Morphological characteristics of the polyethylene worn surface were examined by optical microscope and AFM to discuss the wear mechanism of UHMWPE in the test lubricant. Optical micrographs from polyethylene worn surfaces are compared in Fig. 13. The initial surface of pin specimens were covered with machine-marks, which were formed during the specimen fabrication process, and many of them remained on the polyethylene worn surface in the macromolecular-free PBS (Fig. 13(a)). However, visible machine-marks were completely worn out and the polyethylene worn surface became very smooth in the lubricant containing 0.2 wt% globulin and 0.02 wt% DPPC (Fig. 13(b)).

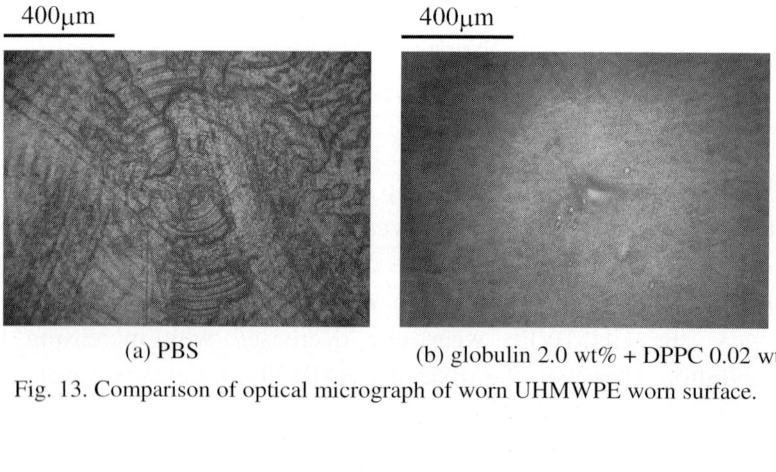

(a) PBS (b) globulin 2.0 wt% + DPPC 0.02 wt%

Fig. 13. Comparison of optical micrograph of worn UHMWPE worn surface.

Apparent differences in the worn surface morphology could be also recognized in the molecular scale by using AFM. The polyethylene worn surfaces were covered with small asperities and fine fibrillated structures in the lubricant containing 0.2 wt% globulin and 0.02 wt% DPPC (Fig. 14(b)). However, such microscopic structures could not be found on the worn surface in the macromolecular-free PBS (Fig. 14(a)).

Our test results clearly indicated the important effect of protein and lipid molecules on UHMWPE wear. The specific wear rate of UHMWPE primarily depended on the protein concentration of the test lubricant. Moreover, the noticeable influence of protein molecules on the microscopic deformation of polyethylene surface was indicated by AFM observation. Heuberger *et al.*[76] suggested that the protein molecules denatured easily and preferably adsorbed onto hydrophobic polyethylene surface. Therefore, the adsorbed proteins might increase adhesive friction which enhanced microscopic plastic deformation of polyethylene surface and finally brought some changes in the polyethylene wear behaviour.

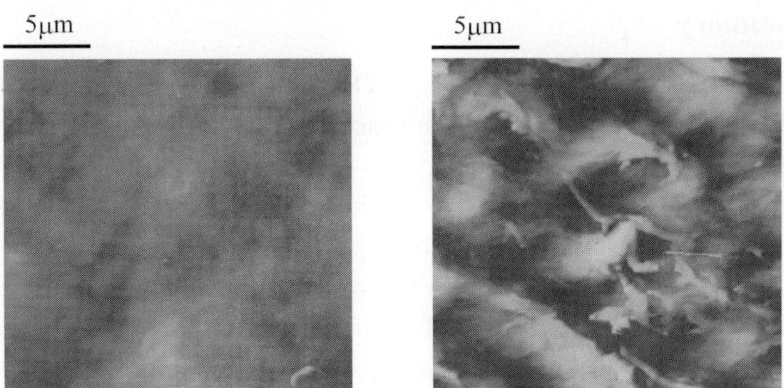

(a) PBS (b) globulin 2.0 wt% + DPPC 0.02 wt%

Fig. 14. Comparison of AFM image of worn UHMWPE worn surface.

The relationship between the phospholipid concentration and the polyethylene wear rate depended on the coexisting protein concentration. In the low protein solutions, phospholipids played a role of boundary lubricant and reduced the UHMWPE wear. The boundary lubrication effect of phospholipid was also reported by Bell *et al.*[81] and Saikko and Ahlroos[80]. However, the wear rate of UHMWPE increased by increasing the lipid concentration of the high protein lubricants. As mentioned above, protein molecules preferably adsorbed on the polyethylene surface. In addition proteins can easily bind to lipid molecules to form lipoproteins. Therefore, the characteristics of macromolecule adsorbed layer may be altered by increased protein concentration and lost its boundary lubrication ability. Furthermore, several *ex vivo* analyses of retrieved polyethylene acetabular cups revealed that the lipid constituents of synovial fluid adsorbed on the polyethylene component and diffused into the bulk material *in vivo*. Greenbaum *et al.*[82] demonstrated that the lipid molecules diffused into bulk of UHMWPE has significant plasticizing effect on mechanical properties. Therefore, the polyethylene surface might be plasticized by the phospholipid molecules added to the lubricant during the wear test and that would be responsible for the formation of characteristic molecular scale surface morphology and increased polyethylene wear rate.

4. Summary

Wear behaviour of implanted UHMWPE components has been affected by several physiological factors associating to joint kinematics and *in vivo* environment. Especially, the shape of multidirectional sliding pathways exerted on the polyethylene surface and biological macromolecules contained in the periprosthetic fluid are playing predominant roles in the UHMWPE wear mechanism. The appropriate understanding of the relationship between these physiological factors and *in vivo* UHMWPE wear is essential for the reliable *in vitro* wear estimation of new materials, such as cross-linked UHMWPE.

Acknowledgments

This study was financially supported by the Grant-in-Aid for Scientific Research of Japan Society for the Promotion of Science. All pin and plate specimens were supplied by Japan Medical Materials Corporation.

References

1. J. Charnley, *J Bone Joint Surg Br*, 54B (1972).
2. H. G. Willert and M. Semlitsch, *J Biomed Res,* 11(2) (1977).
3. P. A. Revel, B. Weightman, M. A. R. Freeman, B. Roberts, *Arch. Orth. Traum. Surg.*, 91 (1978).
4. E. Ingham and J. Fisher, *Proc Instn Mech Engrs Part H,* 214(1) (2000).
5. J. S. Bradley and E. J. Evans, *Engineering in Medicine*, 6(4) (1977).
6. R. H. Schmidt, *Orthopedics*, 17(9) (1994).
7. H. McKellop, F. W. Shen, B. Lu, P. Campbell and R. Salovey, *J Othodaedic Res,* 17(2) (2003).
8. O. K. Muratoglu, C. R. Bragdon, D. O. O'Connor, M. Jasty, W. H. Harris, *J Arthroplasty*, 16(2) (2001).
9. H. Oonishi, Y. Kadoya and S. Masuda, *J Biomed Mater Res,* 58(2) (2001).
10. T. Moro, Y. Takatori, K. Ishihara, T. Konno, Y. Takigawa, T. Matsushita, U. I. Chung, K. Nakamura and H. Kawaguchi, *Nature Materials.* 3(11) (2004).
11. J Charnley, in *Proceedings of One Day Course on Laboratory and Clinical Evaluation of Joint Replacement*, (University of Leeds, 1982).
12. A. Wang, C. Stark and J. H. Dumbleton, *Proc Instn Mech Engrs Part H*, 210 (1996).
13. T. M. Wright, D. J. Astion, M. Bansal, C. M. Rimnac, T. Green, J. N. Insall and R. P. Robinson, *J Bone Joint Surg Am.*, 70 (1988).
14. M. J. Chmell, R. Poss, W. H. Thomas and C. B. Sledge, *J Arthroplasty*, 11 (1996).
15. B. J. Livingston, M. J. Chmell, M. Spector and R. Poss, *J Bone Joint Surg Am.*, 79-A (1994).
16. H. McKellop, I. Clarke, K. Markolf and H. Amstutz, *J. Biomed Mater Res*, 15 (1981).
17. S. Li and A. H. Burstein, *J Bone Joint Surg Am.*, 76 (1994).
18. J. Huber, W. Plits, A. Walter and H. J. Refior, *Orthopade*, 26 (1997).
19. D. A. Swann, in *The Joints and Synovial Fluid*, Ed. L. Sokoloff (Academic Press, New York, 1978), p. 407.
20. J. Delecrin, M. Oka, S. Takahashi, T. Yamamuro and T. Nakamura, *Clinical Orthop and Related Res*, 307 (1994).
21. D. Mazzucco, R. Scott and M. Spector, *Biomaterials,* 25 (2004)
22. J. DesJardins, A. Aurona, S. L. Tanner, T. B. Pace, K. B. Acampora and M. LaBerge, *Proc Instn Mech Engrs Part H*, 220 (2006).
23. B. S. Ramamurti, C. R. Bragdon, D. O. O'Connor, J. D. Lowenstein, M. Jasty, D. M. Estok and W. H. Harris, *J Arthroplasty*, 11 (1996).
24. A. Wang, D. C. Sun, S. S. Yau, B. Edwards, M. Sokol, A. Essner, V. K. Polineni, C. Stark and J. H. Dumbleton, *Wear*, 203-204 (1997).
25. P. S. M. Barbour, M. H. Stone and J. Fisher, *Proc Instn Mech Engrs Part H*, 213 (1999).
26. D. B. Bennett, J. F. Orr and R. Baker, *J Arthroplasty*, 15 (2000).

27. V. Saikko and O. Calonius, *J Biomech*, 35 (2002).
28. M. A. Hamilton, M. C. Sucec, B. J. Fregly, S. A. Banks and W. G. Sawyer, *J Tribology*, 127 (2005).
29. C. Pooley and D. Tabor, *Proc R Soc Lond*, A.329 (1972).
30. R. C. Johnston and G. L. Smidt, *J Bone Joint Surg Am*, 51 (1969).
31. H. K. Ramakrishnan and M. P. Kadaba, *J Biomech*, 24(10) (1991).
32. J. P. Paul, *Proc Instn Mech Engrs*, 181(3J) (1967).
33. D. Bennett, J. F. Orr, D. E. Beverland and R. Baker, *Proc Instn Mech Engrs Part H*, 216 (2002).
34. S. A. Banks, G. D. Markovich, and W. A. Hodge, *Am. J. Knee Surgery*, 10 (1997).
35. S. A. Banks, G. D. Markovich, and W. A. Hodge, *J Arthroplasty*, 12 (1997).
36. A. Wang, V. K. Polineni, A. Essner, M. Sokol, D. C. Sun, C. Stark and J. H. Dumbleton, *J Testing and Evaluation*, 25(2) (1997).
37. C. R. Bragdon, D. O. O'Connor, J. D. Lowenstein, M. Jasty andW. D. Syniuta, *Proc Instn Mech Engrs Part H*, 210 (1996).
38. V. Saikko, *J Biomed Mater Res*, 41 (1998).
39. H. Marrs, D. C. Barton, R. A. Jones, I. M. Ward, J. Fisher and C. Doyle, *J Mater Science: Material in Medicine*, 10 (1999).
40. A. Wang, *Wear*, 248 (2001).
41. S. M. Davey, J. F. Orr, F. J. Buchanan, J. R. Nixon and D. Bennett, *Strain*, 40 (2004).
42. S. M. Davey, J. F. Orr, F. J. Buchanan, J. R. Nixon and D. Bennett, *Biomaterials*, 26 (2005).
43. A. Wang, A. Essner, V. K. Polineni, C. Stark and H. Dumbleton, *Tribology International*, 31 (1998).
44. C. R. Bragdon, D. O. O'Connor, J. D. Lowenstein, M. Jasty, S. A. Biggs and W. H. Harris, *J Arthroplasty*, 16 (2001).
45. V. Saikko, O. Calonius and J. Keranen, *J. Biomed Mater Res Part B: Appl Biomater*, 69B (2004).
46. V. Saikko, Proc Instn Mech Engrs Part H, 220 (2006).
47. M. Turell, A. Wang and A. Bellare, *Wear*, 255 (2003).
48. M. E. Turell, G. E. Friedlaender, A. Wang, T. S. Thornhill and A. Bellare, *Wear*, 259 (2005).
49. D. Mazzucco and M. Spector, *Wear*, 254 (2003).
50. Y. Sawae and T. Murakami, in Pro. International Tribology Conference Nagasaki, 2000, (JAST, 2001).
51. Y. Sawae and T. Murakami, in *Tribology in Biomechanical Systems, Science and Applications*, Ed. F. Franek (Expert Verlag , 1997).
52. Y. Sawae, T. Murakami, T Sawano, I. Noda and T. Shimotoso, *Key Engineering Materials*, 218-220 (2002).
53. Y. Sawae, T. Murakami and T. Sawano, in *Life Cycle Tribology,* Ed. D. Dowson (Elsevier, 2005).

54. T. Sawano, T. Murakami and Y. Sawae, in *Life Cycle Tribology,* Ed. D. Dowson (Elsevier, 2005).
55. H. A. McKellop, P. Campbell, S. H. Park, T. P. Schmalzried, P. Grigoris, H. C. Amstutz and A. Sarmiento, *Clinical Orthopaedics and Related Research.* **311** (1995).
56. H. Oonishi, E. Tsuji and Y. Y. Kim, *J. Mater. Science: Materials in Medicine.* **9** (1998).
57. R.M. Hall, A. Unsworth, P. Siney and B.M. Wroblewski, *Proc Instn Mech Engrs Part H*, 210 (1996).
58. D. Dowson, *Wear*, 190 (1995).
59. J. R. Cooper, D. Dowson and J. Fisher, *Clinical Materials*, 14 (1993).
60. V. Saikko, O. Calonius and J. Keranen, *J Biomed Mater Res*, 57 (2001).
61. L. Costa, K. Jacobson, P. Bracco and E. M. Brach del Prever, *Biomaterials*, 23 (2002).
62. F. W. Shen and H. A. McKellop, *J Biomed Mater Res.* **61** (2002).
63. V. Medhekar, R. W. Thompson, A. Wang and W. G. McGimpsey, *J Appl Polym Sci*, 87 (2003).
64. P. Eyerer and Y. C. Ke, *J Biomed Mater. Res*, 18 (1984).
65. M. Kurth, P. Eyerer, R. Ascherl, K. Dittel and U. Holz, *J Biomater Applications*, 3 (1988).
66. L. Costa, M. P. Luda, L. Trossarelli, E. M. Brach del Prever, M. Crova and P. Gallinaro, *Biomaterials*, 19 (1998).
67. E. Brach del Prever, M. Crova, L. Costa, A. Dallera, G. Camino and P. Gallinaro, *Biomaterials*, 17 (1996).
68. L. Ambrosio, G. Carotenuto, G. Marletta, L. Nicolais and A. Scandurra, *J Mater science Mater in Medicine*, 7 (1996).
69. Y. Sawae, T. Murakami and S. Doi, in *Thinning Films and Tribological Interface*, Ed. D. Dowson (Elsevier, 2000).
70. S. P. James, S. Blazka, E. W. Merrill, M. Jasty K. R. Lee, C. R. Bragdon and W. H. Harris, *Biomaterials*, 14 (1993).
71. F. W. Shen, Y. J. Yu and H. McKellop, *J Biomed Mater Res*, 48 (1999).
72. H. Higaki, T. Murakami, Y. Nakanishi, H. Miura, T. Mawatari and Y. Iwamoto, *Proc Instn Mech Engrs Part H*, 212 (1998).
73. V. Saikko, *J Tribology*, 125 (2003).
74. A. Wang, A. Essener and G. Schmidig, *J biomed Mater Res Part B: Appl Biomater*, 68B (2004).
75. Y. Sawae, T. Murakami and J. Chen, *Wear*, 216 (1998).
76. M. P. Heuberger, M. R. Widmer, E. Zobeley, R. Glockshuber and N. D. Spencer, *Biomaterials*, 26 (2005).
77. S. C. Scholes and A. Unsworth, *Proc Instn Mech Engrs Part H*, 220 (2006).
78. L. Costa, P. Bracco, E. Brach del Prever, M. P. Luda and L. Trossarelli, *Biomaterials*, 22 (2001).

79. A. Aurora, J. D. DesJardins, P. F. Joseph and M. LaBerge, *Proc Instn Mech Engrs Part H*, 220 (2006).
80. V. Saikko and T. Ahlroos, *Wear*, 207 (1997).
81. J. Bell, J. L. Tipper, G. Ingham, M. H. Stone and J. Fisher, *Proc Instn Mech Engrs Part H*, 215 (2001).
82. E. S. Greenbaum, B. B. Burroughs, W. H. Harris and O. K. Muratoglu, *Biomaterials*, 25 (2004).
83. Y. Sawae, A. Yamamoto and T. Murakami, *Tribology International*, 41 (2008).

CHAPTER 7

BIOPOLYMER TRIBOLOGY

Tom Joyce

School of Mechanical and Systems Engineering, Newcastle University
Stephenson Building, Newcastle upon Tyne, NE1 7RU, UK
E-mail: t.j.joyce@ncl.ac.uk

This chapter focuses on the tribology of biopolymers used in total joint replacements. These biopolymers include ultra high molecular weight polyethylene, various types of cross-linked polyethylene and, for historical completeness, poly tetra fluoro ethylene and polyacetal. The main applications of these polymers, and where they have shown the greatest success, are in replacement hip and knee joints, though these same polymers have also been used in other prostheses implanted around the human body such as the shoulder, elbow, wrist, finger, ankle and toe. The key tribological concern regarding these biopolymers is their wear and this area is examined in detail including consideration of why wear is so critical, details of polymeric wear debris, the important types of wear, and laboratory testing which aims to replicate the wear rates which are found in the body. The study of wear necessarily leads on to a consideration of friction and lubrication. While most replacement joints generally have one component of a polymeric material, consideration is given to the potential of all-polymer prostheses. The chapter ends with a look at future possible developments in biopolymers and current challenges for tribologists concerned with biopolymers.

1. A Brief History of Biopolymers in Total Hip Replacements

Historically, poly tetra fluoro ethylene (PTFE) was the first biopolymer to be implanted as an acetabular cup, where it articulated against a stainless steel femoral head in the design of total hip replacement first proposed by the late Sir John Charnley.[1] PTFE was chosen due to its low friction against stainless steel. However, it was found to show excessive wear in the body. This provoked an inflammatory response in the patients fitted with the device and over 99% of the implants had to be revised within two to three years of implantation.[2] An average linear wear rate of 2.26mm/year was reported for PTFE which corresponded to an average volumetric wear of 859mm^3/year.[1] A later study on explanted PTFE acetabular cups suggested that a typical wear factor was of the order of 37 x 10^{-6}mm^3/Nm.[3]

Therefore PTFE was soon replaced by ultra high molecular weight polyethylene (UHMWPE) by Charnley in 1962 and this biopolymer has shown positive clinical results leading to the widespread success of most total hip replacements which utilize a polymeric acetabular cup. As an example, one worldwide review of 5,089 Charnley total hip replacements employing UHMWPE acetabular cups showed that the probability of implant survival at 20 years was 83%.[4] The mean patient age at operation was 63 years which is significant as younger patients tend to be more active and therefore put greater demands on their implants. Despite this reality, in a study of patients with a mean age at operation of 41, a 74% survival rate for Charnley hip prostheses at 20 years has been reported.[5] In both these cases survivorship refers to implants which have not undergone revision surgery. Due to such long-term survival, total hip replacements have been hailed as the major success in orthopedics in the 20th century and as one of the most successful surgical interventions ever developed.[6] A key factor in this success has been the clinical performance of the biopolymer UHMWPE.

The Charnley hip prosthesis is cemented in place with the stainless steel femoral component fixed into the femur and the UHMWPE acetabular cup into the acetabulum. Other designs of hip prosthesis are 'non-cemented' and achieve fixation by encouraging bone ingrowth through the use of controlled topography on the outer aspects of the prostheses. As will be seen, it is now recognized that the wear of UHMWPE serves as a limitation on the life of the hip prosthesis. Essentially it is widely accepted that UHMWPE wear debris provokes a negative cascade of events within the body which can eventually lead to loosening of the prosthesis, in turn leading to pain for the patient and the need for a revision operation.

In an effort to improve the wear resistance of UHMWPE, new types of polyethylene have been developed. Those which currently appear to show the most promising clinical results are the cross-linked polyethylenes (XLPE). For many years sterilization of UHMWPE components was achieved by irradiating at a dose of 2.5-4 Mrad. It should be noted that the radiation dose is typically expressed in kilogray [kGy] or megarads [Mrad] where 10 kGy = 1 Mrad.[7] By increasing the dose of radiation, increased cross-linking can be achieved and with it a concomitant increase in wear resistance has been found.

Currently most types of XLPE are made by irradiating UHMWPE with electron beam or gamma irradiation and then eliminating any free radicals produced during irradiation by applying a subsequent melting step.[8] As a specific example, one method of producing XLPE is to heat compression-molded UHMWPE to 125°C before irradiating the material with a 9.5Mrad dose using a 10-MeV electron beam accelerator. After irradiation the material is heated to 150°C for 2 hours to eliminate detectable free radicals and is then slowly cooled. After machining into acetabular cups the components are sterilized with gas plasma.[9] XLPE can also be manufactured using a different dose, for example of 5Mrad.[9] The method of producing highly cross-linked polyethylene acetabular cups has also been comprehensibly explained by McKellop *et al.* who in addition showed that wear rate fell with increasing radiation dose.[10]

The advantages of such types of XLPE include extremely high wear resistance, excellent oxidative resistance, low susceptibility to third body wear, forgiveness in conditions such as component malpositioning and edge loading, material and handling properties similar to the existing UHMWPE and low cost.[8]

2. The Use of Biopolymers in Other Prostheses

In turn, UHMWPE has been employed in other prostheses around the human body, once again as a component generally articulating against a harder counterface of stainless steel, cobalt chrome molybdenum (CoCrMo) or alumina.

In the UK and Sweden total hip replacement is more common than that of the knee, although this situation is reversed in the US and Australia. Total knee replacements show a similar degree of success to hip replacements. One recent meta-analysis of the long-term survival of 5,950 total knee replacements indicated that the majority showed survivorship rates of greater than 90% at 10 to 11 years.[11] Another huge study of 11,606 total knee arthroplasties offered a 91% survivorship at 10 years and 78% at 20 years.[12] Total knee arthroplasty refers to replacement of the articulation between the femur and the tibia, as well as between the femur and the patella.[2] Unicondylar knee arthroplasty refers to replacement of the articulation between a single condyle of the knee and the tibia. In both total and unicondylar knee replacement an UHMWPE tibial component is usually articulated against a polished CoCrMo femoral component.

There are a range of reasons for revision (i.e. replacement of an implant) knee surgery but one of the most common is loosening[13] which again is associated with UHMWPE wear debris. A form of fatigue failure known as delamination has been said to be the primary cause of most polyethylene related total knee failures.[7] Such delamination clearly causes the release of significant volumes of polymeric wear debris. It will therefore be appreciated that the performance of UHMWPE plays a crucial role in the long-term success of knee arthroplasty.[2] Due to the fact that wear and damage of the UHMWPE insert compromises the longevity of knee arthroplasty,[2] as with hip replacements improved wear performance of the biopolymer is still sought.

After the hip and the knee, replacement joints are far less common but where they do occur UHMWPE is often used. For example in total shoulder prostheses the glenoid component is often of UHMWPE[14] while in ‘reverse’ shoulder prostheses it is the humeral component which is made of UHMWPE.[15] In both cases CoCrMo usually forms the counterface material. In the elbow UHMWPE is again used[16] and here too there are concerns about the wear of this material.[17]

A range of designs of different finger prostheses have been proposed over the years.[18] These have included two-piece, articulating designs which have used UHMWPE for one of the bearing surfaces. One such design has a part-hemispherical UHMWPE shell on the metacarpal component and has shown low wear *in vitro*.[19]

Regarding the various designs of toe prosthesis which are available,[19] metal-on-polymer designs of metatarsophalangeal implant include the Kenetik Great Toe Implant and the Bio-Action.[20] Both of these implants articulate a CoCrMo metatarsal component against an UHMWPE phalangeal component.

The wear of UHMWPE has also been a concern when it has been used in artificial ankle,[21] wrist,[22] spinal,[23] and jaw[24] joints. The wear of UHMWPE and other biopolymers will now be discussed in greater detail.

3. Biopolymer Wear and Wear Debris

As has been seen, wear is the most crucial component of biopolymer tribology and it is therefore beneficial to begin by considering the wear factors. Firstly, it is worth recalling the mean wear factor of 37 x $10^{-6}mm^3/Nm$ for failed PTFE acetabular cups used in the first total hip prostheses.[3] In contrast, failed Charnley hip prostheses which employed an UHMWPE acetabular cup have been calculated to have a wear factors for UHMWPE of 2.1 x $10^{-6}mm^3/Nm$.[25] The superiority of UHMWPE over PTFE in this application is therefore clear.

In addition, functioning UHMWPE acetabular cups which have been retrieved at autopsy (and have therefore been operating successfully) typically show a wear rate of 45% to 69% of failed implants.[26] Combining this data with that of Hall *et al.*[25] permits a wear factor of 0.95 to 1.45 x $10^{-6}mm^3/Nm$ to be calculated for functioning UHMWPE acetabular cups. Another paper offered data comparing autopsy with revision cups.[27] Here, autopsy all-polyethylene acetabular cups showed a mean wear of $35mm^3$/year compared with revision all-polyethylene cups which showed $62mm^3$/year. In other words, autopsy cups showed 56% of the wear of revision cups. Taking this 56% value and the wear factor of 2.1 x $10^{-6}mm^3/Nm$ offered for revision cups[25] implies that functioning cups would show a wear factor of 1.2 x $10^{-6}mm^3/Nm$. These wear factors for functioning UHMWPE will be returned to when *in vitro* testing of this material is discussed.

Wear can also be quantified in other ways. The scientific literature suggests that functioning UHMWPE acetabular cups typically show linear wear rates of the order of 50-250μm/year[28] or volumetric wear rates of 40-130mm^3/year.[29] As an aside it is useful to compare these values with those for PTFE acetabular cups, which are 2,260μm/year and 859mm^3/year respectively. While for most patients such UHMWPE wear rates are acceptably low, for younger patients who might place greater demands on their replacement joints, wear can be such that the total volume of wear debris generated can produce a negative cascade of biological reactions within the body leading to osteolysis (bone death) and the eventual need for a revision operation.[30-32] Indeed the wear rate of UHMWPE acetabular components has been correlated with the severity of bone destruction and the resultant loss of component fixation.[33] Therefore, understanding and minimizing the wear processes of biopolymers such as UHMWPE is crucial to extending the longevity of implants using such materials.

What are the important wear mechanisms related to biopolymers in orthopedic applications? There are multiple wear mechanisms affecting UHMWPE *in vivo* including abrasion,[34] three-body abrasion,[35] adhesion,[36] and fatigue,[7] the latter being especially associated with UHMWPE knee components.

Many explanted prostheses show scratch damage which helps to indicate the importance of abrasive wear. One way to minimize such wear is to offer a smooth, hard, scratch resistant counterface. Therefore femoral heads made of CoCrMo, zirconia and alumina have been introduced. Given their hardness these materials show a high degree of scratch resistance and so should be less easily damaged in the body. One study which supports this line of argument compared retrieved titanium femoral heads with those made of CoCrMo. While the surface finish of the CoCrMo heads had been maintained in the body with at average roughness of around 0.05μm Ra, that of the titanium heads had risen to approximately 0.20μm Ra.[37] Crucially, the wear rates in the patients with CoCrMo femoral heads was significantly less than those who had titanium femoral heads implanted.

It has been shown that it is not only the volume of the polyethylene wear debris, but also the size of the wear debris that can influence the intensity of the biological reaction.[38] In total hip joints, most UHMWPE wear debris has been shown to be of the submicron size.[39,40] In hip simulator testing using bovine serum as the lubricant, similar size wear debris can be produced.[41] This fact serves as one means of validating such a test device. Unfortunately such submicron sizes have been shown to constitute the most biologically active particles[42] i.e. the most effective size at initiating osteolysis.

In terms of maximum volume it has been suggested that a total wear volume of UHMWPE above $600mm^3$ is associated with the failure of total hip replacements.[25] By taking a typical UHMWPE particle size of 0.5μm and a wear volume of $30mm^3$/year for an UHMWPE acetabular cup it can be calculated that this equates to 40,000 wear particles being generated per step.[43] For research groups interested in such work an international standard on the isolation, characterization and quantification of wear debris from implant materials is available.[44] In a hip simulator study it was shown that XLPE debris was of a similar size and shape to that of UHMWPE.[45] UHMWPE wear debris from total knee replacements has also been analyzed. Shanbhag *et al.* offered a mean UHMWPE particle size of 1.7μm ± 0.7μm (standard deviation) and this larger size was thought to be one of the reasons why osteolysis is less of a problem in knee prostheses than in hip prostheses.[46]

The truest test of any biopolymer is the clinical situation and here some significant reductions in wear have been reported when XLPE has been used. An 81% reduction in wear when XLPE acetabular liners were used, as opposed to conventional UHMWPE liners, was reported in one short-term clinical study.[47] Another clinical study, at 31 months mean follow-up, of XLPE acetabular cups reported a steady state wear rate of 0.007mm/year as opposed to 0.174mm/year for a conventional UHMWPE group.[48]

Oonishi and co-authors have reported clinical and laboratory results of cross-linking UHMWPE with irradiation doses of up to 100Mrad and greater.[49-52] One paper undertook *in vitro* wear tests on samples which had received irradiation doses of between 0 and 200Mrad and concluded that 200Mrad was the optimum dose to be applied.[49] Clinical results of cups irradiated with 100Mrad showed linear wear rates of 0.05mm/year compared with 0.28mm/year for conventional UHMWPE.[50] Most recently Oonishi *et al.* reported clinical and laboratory results of 6Mrad acetabular cups.[52] Their hip simulator study ran to 5.4 million cycles and the wear of the XLPE cups was said to be 3% of that of the conventional UHMWPE cups. Clinical wear results, at greater than one year follow up, were reported as 0.006mm/year and 0.098mm/year respectively.[52] In the clinical study, all of the XLPE cups articulated against alumina heads of 28mm diameter. Another recent report offered a mean wear rate of 0.01mm/year for cross-linked acetabular liners compared with 0.19mm/year from conventional UHMWPE liners at a mean follow-up of 5.7 years.[53] Bragdon *et al.* reported on electron beam irradiated XLPE liners at an average of 45 months follow-up. They stated that the average penetration was 25μm/year compared with 144μm/year for a control group of UHMWPE components.[54] However, not all *in vivo* results have been positive, with wear and cracking damage to some XLPE hip explants having been reported.[55]

Polyethylene has also been cross-linked by the silane method. A small number of acetabular cups manufactured from this material were produced and implanted. After initial bedding in, these implants have shown low wear against an alumina ceramic counterface.[56,57] The most recent report stated that, at 17 years follow-up, after an initial bedding-in of up to 0.41mm, the mean penetration was 0.02mm/year.[58] This same type of XLPE has also been used for a novel all-polymer finger prosthesis.[59,60]

It should also be noted that a modified type of polyethylene has been tried as an acetabular cup material but was found to be less successful than conventional UHMWPE. Polyethylene with the trade name Hylamer™ was treated by the hot isostatically pressed process.[61] However, wear rates *in vivo* were found to be of the order of 0.49mm/year when rubbing against 28mm femoral heads.[62] This is approximately twice the wear rate of conventional UHMWPE, a distinction also reported by Livingston *et al.*[63] Wear rates for these and other polymers used in acetabular cups have been summarized in Table 1. A carbon fiber reinforced polyethylene known as Poly-II™ was implanted as a tibial tray material in artificial knee joints. Unfortunately, clinical results were poor with many patients presenting with osteolysis and mechanical failure of the inserts.[61]

Table 1. Wear rates of biopolymers used as the acetabular cup material in total hip replacements.

Biopolymer	Reference	*In vivo/ in vitro*	Wear factor ($x10^{-6}mm^3/Nm$)	Linear wear (mm/year)	Volumetric wear	Gravimetric wear (mg/mill cycles)	Notes
PTFE	Charnley[1]	*In vivo*		2.26	$859mm^3$/year		Revision
PTFE	Dowson[3]	*In vivo*	37				Revision
UHMWPE	Hall[25]	*In vivo*	2.1				Revision
UHMWPE	Jasty[27]	*In vivo*			$62mm^3$/year		Revision
UHMWPE	Jasty[27]	*In vivo*			$35mm^3$/year		Autopsy
UHMWPE	ASTM[29]	*In vivo*			40-130 mm^3/year		
UHMWPE	Clarke[64]	*In vitro*			$23mm^3$/mill cycles		22mm heads
UHMWPE	Edidin[65]	*In vitro*			$90mm^3$/mill cycles		28mm heads
UHMWPE	Smith[66]	*In vitro*			$52mm^3$/mill cycles		28mm heads
UHMWPE	Scott[62]	*In vivo*		0.49			Hylamer™
UHMWPE	Rieker[28]	*In vivo*		0.050-0.250			
UHMWPE	Manning[48]	*In vivo*		0.174			
UHMWPE	Engh[53]	*In vivo*		0.190			
UHMWPE	Bragdon[54]	*In vivo*		0.144			
UHMWPE	Oonishi[52]	*In vivo*		0.098			
UHMWPE	Saikko[67]	*In vitro*	0.57			15.5	28mm heads
UHMWPE	Hermida[68]	*In vitro*				15	28mm heads
XLPE	Manning[48]	*In vivo*		0.007			
XLPE	Engh[53]	*In vivo*		0.010			
XLPE	Bragdon[54]	*In vivo*		0.025			
XLPE	Oonishi[52]	*In vivo*		0.006			
XLPE	Hermida[68]	*In vitro*				1.5	
XLPE	Estok[9]	*In vitro*				0.01	9.5MRad
XLPE	Estok[9]	*In vitro*				3.8	5 MRad

4. Wear Testing of Biopolymers

As the wear of biopolymers in the human body has been recognized to be a critical concern, so a great deal of emphasis has been focused on laboratory wear testing of these polymers. This wear testing can be divided into two broad areas: the wear testing of biopolymers in the form of implants; and the wear testing of the biopolymer as a material.

4.1. *Implant Wear Testing*

4.1.1. *UHMWPE Acetabular Components Tested In Vitro*

A large amount of testing of UHMWPE hip prostheses has been undertaken over many years.[64-67,69] Testing has been developed to such an extent that guidance is now published in the form of international standards.[70] However, such standards can inevitably lag slightly behind the latest developments in research and therefore debate exists around aspects of their use related to the testing of total hip prostheses.[71] It should also be noted that in most cases wear is measured gravimetrically and guidelines on this issue is given in international standards.[72,73]

Clarke *et al.* reported wear rates of 23mm^3/million cycles for UHMWPE acetabular cups articulating with 22mm diameter femoral heads. He suggested that this value was somewhat less than the value measured *in vivo*.[64] Edidin *et al.* reported mean wear rates of UHMWPE acetabular cups articulating with 28mm diameter femoral heads as 90mm^3/million cycles.[65] One explanation for the difference in these results is related to the size of the femoral head, as it is recognized that wear of UHMWPE increases with head size.

As will also be seen, another important factor is the protein content within the bovine serum lubricant. Good *et al.* tested UHMWPE acetabular cups and varied the protein content of the serum lubricant. Very low wear was found when the lubricant had zero protein content, but over the wide range of 3mg/ml to 63mg/ml protein, wear rates between 22 and 44mm^3/million cycles were measured against 22mm diameter femoral heads.[69] For 28mm diameter femoral heads articulating against UHMWPE cups in the presence of 25% bovine serum, a mean UHMWPE wear rate of 52mm^3/million cycles was reported.[66]

Saikko tested 28mm CoCrMo femoral heads against UHMWPE cups in the presence of bovine serum diluted 1:1 with distilled water and offered a mean wear rate of 15.5mg/million cycles which was said to correspond to a mean wear factor of 0.57 x $10^{-6}mm^3/Nm$.[67] It has also been shown that the method of sterilization of the UHMWPE cups can have an influence on the wear of this biopolymer, with greater than a two fold difference in wear rates being reported.[74]

It should be recognized that wear rates can vary between hip simulators. Factors that cause this variation include the upright or inverted position of the hip prosthesis, the motion pattern of the simulator, the type of lubricant involved, the temperature control, the load curve, the frequency of cycles, and the motion profile.[9]

It has been seen that average *in vivo* wear rates of UHMWPE acetabular cups of $35mm^3$/year for autopsy cups and $62mm^3$/year for revision cups have been offered.[27] Another source offered a range of values of volumetric wear rates of 40-$130mm^3$/year, depending on the size of femoral head.[29] To permit a comparison between *in vitro* and *in vivo* results, it is generally accepted that on average one million steps are taken per year. This average value masks a wide range of activity levels, with a 45-fold difference between the least and most active patients fitted with hip or knee implants.[75] This piece of information helps to explain one of the sources of variability in rates of UHMWPE wear *in vivo*.

4.1.2. *XLPE Acetabular Components Tested In Vitro*

An early report of *in vitro* wear tests of total hip replacements with XLPE acetabular cups showed zero wear after 20 million cycles of testing in a hip simulator in the presence of a bovine serum lubricant.[76] Conventional UHMWPE acetabular cups were also tested and were reported as showing mean wear rates of 360mg after 20 million cycles when rubbing against 22mm femoral heads. It was recognized by the authors that such overly positive results were influenced by the XLPE test components absorbing a greater amount of lubricant than the soak controls[76] but despite this caveat the wear results were still impressive.

Bowsher *et al.* tested XLPE acetabular cups in a hip simulator and investigated the effects of activity levels and roughened femoral heads.

As might be expected both situations led to increased wear of XLPE and the greatest wear was found by combining a roughened femoral head with the highest simulated activity level.[77]

In a comparison of XLPE acetabular liners with those of UHMWPE, both tested in 90% bovine serum at 37°C in a hip simulator, it was reported that wear rates were 1.5 and 2.6mg/million cycles for 28mm and 32mm head sizes for XLPE compared with 15 and 17mg/million cycles for 28mm and 32mm head sizes for UHMWPE.[68] Again the superiority of XLPE over UHMWPE from a wear point of view was seen.

In a study of radiation XLPE acetabular cups rubbing against 28mm CoCrMo femoral heads in a hip simulator, two types of third-body particulate were added to a bovine serum lubricant. Parallel tests were undertaken with UHMWPE cups. In the presence of aluminium oxide particles wear rates were 37mg/million cycles for the XLPE and 149mg/million cycles for the UHMWPE. When bone cement particles were added, wear rates were 0.5mg/million cycles for the XLPE and 19mg/million cycles for the UHMWPE. In both cases, the use of radiation XLPE was seen to significantly reduce the wear compared with conventional UHMWPE.[78]

Recently, Estok *et al.* compared two different types of XLPE in a hip simulator study. They reported that wear was reduced with greater cross-linking.[9] Specifically, when rubbing against 32mm diameter femoral heads the average weight losses were 0.01mg/million cycles for 9.5 Mrad XLPE and 3.8mg/million cycles for 5 Mrad XLPE. In both cases, test components were compensated by 'motion soak' controls.[9]

4.1.3. *UHMWPE Tibial Components Tested In Vitro*

Using the knee simulator he helped to design, Burgess reported UHMWPE wear rates of 2.6mg/million cycles (2.8mm^3/million cycles) with tests done in 30% serum at 37°C. It was noted that the equivalent wear factor was 0.05 x 10^{-6} mm^3/Nm[79] which, when compared with typical wear factors for UHMWPE in functioning hip prostheses of 1 x 10^{-6} mm^3/Nm, appears to be on the low side. Using the same test device, again with 30% serum at 37°C, an average wear rate of 3-4mg/million cycles was reported for two types of total knee replacement.[80]

In their AMTI knee simulator study, Young *et al.* compared UHMWPE tibial inserts which had low and high levels of oxidation. In the latter delamination occurred whereas in the former only burnishing of the articulating surfaces was seen after five million cycles of testing.[81] Wear rates were said to be 2.8mg/million cycles for the low oxidation group and 8.3mg/million cycles for the high oxidation group. Barnett *et al.* reported mean UHMWPE wear rates of 18mm^3/million cycles for knee prostheses tested in a ProSim simulator, using a 25% bovine serum lubricant.[82]

DesJardins *et al.* compared the wear of UHMWPE tibial inserts with two different lubricants. They used an Instron/Stanmore knee simulator and the test ran to 5 million cycles. With 50% bovine serum with a protein content of 35g/l and heated to 37°C, wear rates of 9.4mg/million cycles were reported. When hyaluronic acid was added to the same lubricant, wear increased to 65mg/million cycles.[83] As with the hip, models of joint loading have been developed and international standards now exist to offer guidance for the testing of knee prostheses.[84]

A number of recent papers have reported further testing of knee prostheses. Using a lubricant of 25% serum Jennings *et al.* reported UHMWPE wear in the range 5 to 17mm^3/million cycles without and with condyle lift-off respectively.[85] In a mainly finite element study, auxiliary knee simulator tests using a bovine serum lubricant at 37°C in a Stanmore knee simulator produced UHMWPE wear rates of the order of 16mg/million cycles.[86]

4.1.4. *XLPE Tibial Components Tested In Vitro*

In a knee simulator study, aged XLPE tibial inserts were compared with aged UHMWPE tibial inserts and unaged UHMWPE tibial inserts. Wear rates were reported for these three groups as 1mm^3/million cycles, 10mm^3/million cycles and 9mm^3/million cycles respectively.[87] Importantly, the aged UHMWPE components showed delamination, whereas the aged XLPE components did not. A later study by the same authors, again using a knee simulator, compared XLPE with conventional UHMWPE tibial inserts in cruciate-retaining designs of total knee replacements. Undiluted bovine serum was used as the lubricant. Wear rates, including backside wear, were reported as 23mm^3/million cycles and 5mm^3/million cycles for UHMWPE and XLPE respectively.[88] XLPE tibial inserts were introduced clinically in the US in 2001.[2] Unlike in the hip, no widely accepted techniques for quantifying *in vivo* wear rates of knee prostheses exist. The most effective way to quantify their wear performance is through an analysis of explanted components.[2] As far as this author is aware, such explant studies of XLPE tibial implants have not yet been reported in the scientific literature.

4.1.5. *Other Simulators*

After devices for testing artificial hips and knees, there are far fewer joint simulators which have been validated against data from explants. Spinal simulators have recently been developed by several companies and these devices are likely to become more widely used in the near future. One *in vitro* set of tests of two different designs of UHMWPE-on-metal total disc replacements offered wear rates of 18 and 21mm^3/million cycles respectively.[89] Wear and failure of UHMWPE disc replacements have recently been reported[23,90] and it can therefore be argued that more appropriate *in vitro* testing of such devices should have been undertaken prior to implantation. An American Standard exists related to the testing of spinal implants[91] and it is expected that a similar ISO standard will appear in the near future.

For testing an artificial ankle joint, a knee simulator has been employed.[21] For finger prostheses a device has been developed which has matched clinical type failures of these implants.[92,93] This test device has been used for *in vitro* tests of UHMWPE and poly ether ether ketone (PEEK) capped finger prostheses, with the UHMWPE implant showing lower wear.[94]

4.2. *Biopolymer Wear Testing*

It is recognized that hip and knee simulators are expensive devices, available at only a small quantity of research establishments and they offer only a limited number of test stations.[36] Therefore less expensive wear screening devices are often used to try and assess potential biomaterials for use in total joint replacements. Once again a key aim is to be able to replicate the *in vivo* wear rates using a screening test machine. For UHMWPE, the most important biopolymer, only a limited number of test devices have been shown to have such a capacity.[95] Some of these are listed in Table 2.

Table 2. Wear of UHMWPE in wear screening devices which apply multi-directional motion to test pins in the presence of a bovine serum based lubricant.

Ref.	Wear	Comments
Bragdon[96]	9.3mg/million cycles	0.1mg/million cycles - reciprocation only
Saikko[97]	0.8×10^{-6} mm^3/Nm	0.005μm Ra counterface
Jones[98]	0.122×10^{-6} mm^3/Nm	0.013×10^{-6} mm^3/Nm - reciprocation only
Joyce[99]	1.1×10^{-6} mm^3/Nm	0.08×10^{-6} mm^3/Nm - reciprocation only
Rieker[100]	1.5×10^{-6} mm^3/Nm	
Mazzuco[101]	$0.3\text{-}1.8 \times 10^{-6}$ mm^3/Nm	AMTI OrthoPOD, range of contact stresses
Saikko[36]	1.63×10^{-6} mm^3/Nm	

From these machines, the key issues for matching *in vivo* wear factors have been shown to be: to apply multi-directional motion to the test samples; and to use a protein containing lubricant, most often dilute bovine serum. If only unidirectional motion, such as applied in a typical rotational pin-on-disk device, or reciprocating motion, as seen in a conventional pin-on-plate device, is applied then this results in wear factors which are approximately 1 to 2 orders of magnitude less than comparable rates *in vivo* associated with metal-on-polymer total hip replacements.[102] As a single example, reciprocating pin-on-plate wear tests in presence of bovine serum were undertaken on UHMWPE test pins. For counterface roughness values between 0.005 and 0.04μm Ra, wear factors between 7.4 and 16.5×10^{-9} mm^3/Nm were reported.[103] Indeed, it is not only reciprocating pin-on-plate devices that can produce such low wear of UHMWPE. It was reported that by applying reciprocating motion in a hip joint simulator, no detectable wear of UHMWPE acetabular components was found after a seven million cycle test.[104]

In general a constant load is applied to the test sample, as clearly this is one important way of maintaining a low cost of the wear screening rig. For guidance on wear screening tests of biopolymers an ASTM standard exists[29] as does guidance on gravimetric measurements.[105] Such screening test rigs have a varying number of test stations. From single station,[106] through four stations,[99] to six stations.[101] However, more recently a 12-station device[97] has been developed which was then modified to 100 stations.[36]

Wear testing of silane XLPE in a multi-directional wear screening device against a metallic counterface and in the presence of a dilute bovine serum lubricant was reported in 1999.[107] Under such conditions the wear of XLPE (k = $1.2 \times 10^{-6} mm^3/Nm$) was found to be similar to that of UHMWPE (k = $1.1 \times 10^{-6} mm^3/Nm$). Later testing of silane XLPE suggested that its wear was lower than that of UHMWPE at lower values of counterface roughness but greater at higher values of roughness.[108] Silane XLPE acetabular cups have shown low wear *in vivo*, perhaps in part because they have articulated against scratch resistant alumina femoral heads.[57]

Some screening wear testing has also been reported for radiation XLPE, where this material has been subject to multi-directional motion. Saikko *et al.*[109] tested Durasul® XLPE against CoCrMo plates of various roughness values in the presence of a bovine serum lubricant diluted 1:1 with distilled water. Against polished CoCrMo counterfaces, wear factors of the order of $2 \times 10^{-9} mm^3/Nm$ were reported for XLPE compared with $2 \times 10^{-6} mm^3/Nm$ for conventional UHMWPE. This is a remarkable difference in wear factors. In a separate test of XLPE in a multi-directional pin-on-plate wear test device, undiluted bovine serum was used as the lubricant.[110] Results showed that XLPE wore at significantly lower rates than UHMWPE but, frustratingly, specific values of wear rate were not offered in the text. More recently, in a test in a multi-directional pin-on-plate device with 50% bovine serum as the lubricant, XLPE pins were reported as showing a wear rate of 0.405mg/million cycles compared with 2.53mg/million cycles for unirradiated XLPE.[111]

It is also worth recognizing that clinical wear data from other polymers is available and can serve as a benchmark against which wear screening rigs for biopolymers can be validated. These include, for example, PTFE[112] and polyacetal.[113]

5. Influence of Counterface Roughness on Wear

A number of wear screening studies under reciprocation motion have provided evidence of a strong correlation between wear of UHMWPE and the roughness of the counterface it articulates against.[103,114,115] This is one reason why femoral heads with a high hardness, such as alumina and CoCrMo have often been preferred to softer and less scratch resistant materials such as stainless steel and titanium. In the body such scratch damage can be caused by bone fragments or the radio-opaque agents used in bone cement.[43] Explant studies have shown that femoral heads tend to roughen *in vivo*. As a couple of examples, roughness values between 0.06 and 0.19μm Ra have been reported[25] while another study offered a range of 0.01 to 4.3μm Ra.[116]

When multi-directional motion has been applied to polyethylenes, again an increase in wear with increasing roughness has been reported. Saikko *et al.* tested UHMWPE and XLPE in the presence of bovine serum diluted 1:1 with distilled water.[109] For UHMWPE these authors offered the power equation

$$k = 5.87\, x\, (R_a)^{0.91} \tag{1}$$

where k is the wear factor (mm^3/Nm) and Ra is the arithmetic mean surface roughness value. For XLPE the power equation offered was

$$k = 7.87\, x\, (R_a)^{2.49} \tag{2}$$

Although Saikko *et al.* found that the regression line for XLPE was steeper than that for UHMWPE, the wear rate of XLPE was much lower. Specifically they reported wear factors for XLPE of 2 x 10^{-9}mm^3/Nm compared with 2 x 10^{-6}mm^3/Nm for UHMWPE.[109] Clearly this is a remarkable reduction in wear.

6. Influence of Lubricant on Wear

When considering an 'ideal' lubricant for wear testing biopolymers, human synovial fluid might be suggested as the lubricant of choice as this is the lubricant in which implanted artificial joints must function. The key components of human synovial fluid are considered to be proteins, hyaluronic acid and phospholipids.[83] Of course, there are practical and ethical difficulties to obtaining human synovial fluid, and it is recognized that the properties of synovial fluid are influenced by diseases such as osteoarthritis and rheumatoid arthritis.[117]

In addition it has been shown that the viscosity of rheumatoid synovial fluid is less than that of osteoarthritic synovial fluid which in turn is less that that of 'normal' synovial fluid.[118] As synovial fluid is mainly a filtrate of plasma plus hyaluronic acid[117] bovine serum is commonly used as a lubricant in wear tests. Such a lubricant is the one recommended by various international standards related to the wear testing of biopolymers.[29,70,84]

Bovine serum is generally diluted with distilled water. One of the key reasons for dilution is to bring the protein content of the serum down to a level matching that seen in synovial fluid in human joints. For example, protein concentrations of human synovial fluid have been offered as 20-35g/l[119] and 21g/l.[69] In addition it has been stated that disease has an influence on the protein content of human synovial fluid. For osteoarthritic synovial fluid a value of 25g/l was offered and for periprosthetic fluid the value was 28g/l.[83] From a group of patients Liao *et al.* reported a range of 15-55g/l with a mean of 34g/l.[120]

Another paper reviewed previous data and offered the following values for protein content of synovial fluid: normal 18g/l and 20g/l; traumatic 26g/l; osteoarthritic 31g/l and 32g/l; rheumatoid 45g/l and 35g/l; and total hip arthroplasty 34g/l.[121] When human periprosthetic synovial fluid was examined it was reported that the total protein content was about 30g/l and that approximately 70% of the proteins were albumin.[122] Another reason for dilution of serum is to match the recommendations given in the relevant international standards.[29,70,84]

One example will hopefully suffice to show why bovine serum should be used as a lubricant for the wear testing of biopolymers. A hip simulator study comparing the influence of distilled water and bovine serum on UHMWPE wear found that the use of serum resulted in wear particles of a similar size to those found *in vivo* as well as no evidence of a transfer film on the femoral head.[41] From explant studies on hip prostheses it is known that no polymeric transfer film occurs in the body. When distilled water was used as the lubricant wear particles were measured to be significantly larger than those from explant studies and a transfer film was found on the femoral heads.[41] These authors therefore concluded that serum lubrication reproduced clinical wear mechanisms in total hip replacement while water lubrication did not.[41] Many other studies have shown polyethylene transfer films when using water or saline as a lubricant.[123-126]

Disadvantages of bovine serum include its cost and the fact that it degrades over time, therefore it needs to be changed regularly with the possibility that wear particles which may contribute to third-body wear could be removed from the test environment. There are also some safety issues in using biological fluids, while the addition of chemicals such as sodium azide, which aim to retard microbial growth, carry their own safety risk. It should also be recognized that different batches of serum can have different protein contents, anywhere between 33g/l and 52g/l being the experience of this author. Values up to 70g/l[69] and 80g/l have been reported by others.[127]

The batch protein content can be critical if a percentage dilution is undertaken. For example a lubricant of 25% serum is recommended by one standard for the testing of hip prostheses.[70] Therefore it would be most useful if both dilution and protein content of serum based lubricants were offered in reports of wear testing of biopolymers. Protein content is important because during testing of PTFE acetabular cups in hip simulators, it was seen that wear increased with increasing protein content.[128] However, this situation was not repeated with UHMWPE. Instead, above a threshold, the wear rate of UHMWPE tended to decline with protein content.[129]

The optimum protein content is still a source of debate, between 3 and 30g/l having been suggested by one group[69], 20g/l having been suggested by another[74] while others suggest it should not be below 20g/l[130], and other groups appear happy to use undiluted bovine serum.[88] To complicate matters further, in a hip simulator study involving PTFE acetabular cups it was shown that the albumin to globulin ratio had a critical impact on the wear measured, as did whether alpha serum or bovine calf serum was used.[74]

Given these issues, it should also be accepted that bovine serum is not synovial fluid. Due to this fact many studies have been undertaken which have aimed to add or remove elements of the lubricant so that its properties more closely match those that are considered to be important in human synovial fluid. For example, a lubricant that diluted bovine serum with Ringer solution, with the aim of comparing well with synovial fluid, has been employed.[131] Saline and Ringer solution contain salts equivalent to those found in the body. Soybean lecithin and soybean protein have also been tried as additives.[132]

More recently some work has investigated the influence of the addition of hyaluronic acid to bovine serum in wear tests. Hyaluronic acid is one of the important constituents of synovial fluid. In a knee simulator study 1.5g/l was added to 50% dilute bovine serum.[83] This lubricant was said to give biochemical and rheological properties which were not statistically different to aspirated total knee arthroplasty joint fluids. A five million cycle wear test was reported as causing 6.88 times more UHMWPE wear than the 50% dilute bovine serum alone. Further it was claimed that the addition of hyaluronic acid led to surface corrosion on the CoCrMo counterface, which in turn was said to be seen on CoCrMo femoral component retrievals.[83]

It has been stated that the four results that a lubricant for biopolymer wear testing should provide are: no transfer of polymer to the metal counterface; a wear factor of the order $1 \times 10^{-6} mm^3/Nm$; most of the wear debris consisting of particles smaller than $1 \mu m$; and the worn polyethylene surface should look polished.[132] It is felt that these benchmarks should still be applied today.

All of the preceding comments refer to the influence of lubricant on the **wear** of biopolymers. If the issue were not complex enough, it has been shown that the choice of lubricant has an equally strong influence on the **friction** of biopolymers. Using a standard pin-on-disk tribometer, Yao *et al.* reported that changing from distilled water to diluted bovine calf serum as the lubricant led to increased friction and that bovine synovial fluid did not simulate closely human synovial fluid, despite both fluids having comparable rheological properties.[133]

As can be seen, the issues regarding lubricants and the testing of orthopedic biopolymers is an important and intriguing one, and much useful work in this area is still to be undertaken.

7. Soak Controls

Apart from PTFE, the wear rates of most biopolymers tend to be so low that the influence of fluid uptake by test samples can have a crucial impact on the accuracy of gravimetric wear measurements. Most wear tests therefore employ what may be termed a 'soak control' or simply a 'control' component. Any increase in weight of the soak control would generally be assumed to be equal to that of the test component. In hip and knee simulators one or more stations may be set aside as such soak controls. Here, the test components would tend to be loaded but no motion would be applied. These control components would be subject to the same cleaning and weighing protocol as the test specimens.

In wear screening devices, such as a pin-on-plate test rig, an unloaded pin would generally serve as a soak control. It should be recognized that there could be important differences between soak controls in simulators and those in wear screening devices. In wear screening devices the test components are usually subject to a static load. In simulators they are usually subject to dynamic loads. In addition the surface area of an acetabular cup, which will generally be totally submerged in lubricant, is much greater than that of a pin in a pin-on-plate test device where generally only a portion of the pin is submerged.

The importance of soak controls in the wear testing of biopolymers has been recognized and several recent papers have attempted to aid understanding of this area.[9,110,111,134-137]

For UHMWPE it has been shown that both the surface area of the component and the sterilization method employed can influence lubricant uptake.[134] It has also been reported that soak acetabular cups subject to a load increased in weight more than those subject to no load.[135] Later work by the same authors reported that hydration in undiluted bovine calf serum prior to wear testing significantly increased the wear of UHMWPE.[110] In a test of total knee replacements, one study reported a six fold difference in weight gain of the control components when two different bovine serum based lubricants were used.[127] The amount of increase in weight due to lubricant uptake has been seen to increase with temperature, at 37°C it has been reported as being approximately twice that at room temperature (25°C).[136]

With the introduction of low-wearing XLPE the importance of soak controls has become more critical still. When wear testing this material papers have reported that test samples, even when compensated for fluid uptake by the soak controls, have increased in weight.[10,76] Clearly this 'zero wear' situation cannot be the case. Such results have pointed to the need to have soak controls which more closely replicate the situation encountered by the test components, while not actually causing these control components to undergo any wear.

One method proposed by a very recent paper has been termed 'motion soak controls'.[9] Here, based on the fact that linear reciprocating motion applied to test components in a hip simulator was known to produce no wear of UHMWPE,[104] it was argued that applying such motion to XLPE control components would serve to more accurately reproduce interface temperatures and various loading conditions, while not leading to weight loss from wear.[9] Estok *et al.* claimed that motion soak controls were more accurate than load soak corrections as they increased in weight by a greater amount than the conventional load soak controls. They also noted that lubricant uptake by the XLPE samples varied depending on whether they had been irradiated to 5-Mrad or 9.5-Mrad.[9]

Similarly, another recent paper has highlighted the importance of both lubricant composition and load profile on fluid uptake by control components.[137] UHMWPE cylindrical samples showed a significantly higher weight gain under dynamic load than when subject to no load or static load. In addition, two different dilute bovine serum based lubricants, which varied only in the chemicals added to them to minimize bacterial growth, caused fluid uptake to change by between 2.4 and 4.2 times, depending on the load condition applied.[137]

The lubricant uptake of silane XLPE finger prostheses has been investigated. In a test in bovine serum lubricant heated to 37°C and lasting over 160 days it was found that lubricant uptake of the unloaded samples was greater than that of the loaded samples.[60] Tests were repeated at lubricant temperatures of 30°C and 40°C, but no significant different in uptake was measured. When the tests were repeated with distilled water as the lubricant the results revealed that weight increase due to soaking in bovine serum was several times that of soaking in distilled water.[60]

In summary the area of soak control is one that still lacks full understanding and it is a topic that would benefit from more research.

8. Theoretical Lubrication Analysis

A theoretical lubrication analysis can be undertaken for a range of total replacement joints. What follows is an example related to total hip replacements, as these are arguably the most studied type of implant.

Modeling the 'ball and socket' hip prosthesis as an equivalent ball-on-plane model and employing elastohydrodynamic theory[138] allows the minimum effective film thickness (h_{min}) to be calculated from:

$$\frac{h_{min}}{R_x} = 2.80\left(\frac{\eta u}{E^* R_x}\right)^{0.65}\left(\frac{w}{E^* R_x^2}\right)^{-0.21} \tag{3}$$

where R_x is the equivalent radius (m), η is the viscosity of the lubricant (Pa s), u is the entraining velocity (m/s), E^* is the equivalent elastic modulus (Pa), and w is the load (N). In turn, given that R_a is the surface roughness and assigning subscript 1 to the ball (femoral component) and subscript 2 to the socket (acetabular component) of the hip prosthesis under consideration, then the lambda ratios can be calculated from:

$$\lambda = \frac{h_{min}}{\left[(R_{a1})^2 + (R_{a2})^2\right]^{1/2}} \tag{4}$$

This allows the lubrication regime to be identified, as $\lambda<1$ indicates boundary lubrication, $\lambda>3$ designates fluid film lubrication, and between these values mixed lubrication is indicated.[139]

Before these calculations can be undertaken, the equivalent radius (R_x) should be calculated from:

$$\frac{1}{R_x} = \frac{1}{R_1} - \frac{1}{R_2} \tag{5}$$

where R refers to the radius of the component and subscript 1 refers to the ball and subscript 2 to the socket of the hip prosthesis. The equivalent modulus of elasticity can be determined from the equation:

$$\frac{1}{E^*} = 0.5\left(\frac{1-\nu_1^2}{E_1} + \frac{1-\nu_2^2}{E_2}\right) \quad (6)$$

Again, E refers to the Young's modulus of the component and again subscript 1 refers to the ball and subscript 2 to the socket of the hip prosthesis, similarly for the two Poisson's ratios (ν).

Clearly, the natural hip joint will encounter a range of loads during a lifetime and similarly the joint will move at a range of speeds. However, for this calculation single values will be taken. In terms of loading, an average value of 1kN has been suggested[140] while an entraining velocity of 15mm/s was taken.[141] A radial clearance of 50μm was taken for the metal-on-polymer combination.[142] For a cobalt chrome femoral head roughness values as smooth as 3nm have been reported[143] while an UHMWPE cup can have a roughness of 1.29μm.[144] A viscosity of the synovial fluid lubricant of 0.005Pa s was assumed.[145] The Young's moduli of CoCrMo and UHMWPE were taken as 210GPa[146] and 1GPa[147] respectively, while their Poisson's ratios were taken as 0.3[146] and 0.4.[145]

Using these values, R_x can be calculated to be 3.93m and E^* to be 2.37GPa. In turn these quantities permit a lambda ratio of 0.23 to be calculated, indicating that such a metal-on-polymer hip prosthesis would operate in the boundary lubrication regime ($\lambda<1$).

Clearly the evidence from explants, that of scratches on the surface, together with volumes of wear debris, supports the theoretical conclusion that metal-to-polymer contact occurs, as would be expected under boundary or mixed lubrication. It also needs to be acknowledged that any *in vivo* wear will likely lead to a change in the values of the surface roughness of the articulating surfaces. Data from explanted artificial hip joints indicates that the femoral heads tend to roughen while the articulating region of the acetabular cup shows a decrease in roughness.[148] One set of data describes an increase in femoral head roughness from 3nm to 10nm and a change in UHMWPE socket roughness from 0.247μm to 0.047μm.[148] Taking these 'worn' values of 10nm and 0.047μm and repeating the lubrication regime calculations shows that a lambda ratio of 6.1 can be achieved, indicating that under similar 'worn' conditions fluid film lubrication is possible with metal-on-polymer combinations. However, before celebrating the positive side effect of such wear, a separate paper examined a different set of explanted UHMWPE acetabular cups and, while roughness values did decrease from their original values, an average worn roughness of 0.266μm Ra was measured for six cups subject to 'adhesive/abrasive' wear.[149] Taking this value plus the 10nm worn head roughness offered by Elfick *et al.*[148] corresponds to a lambda ratio of 1.1, just within the mixed lubrication regime. Therefore, 'bedding-in' wear of polyethylene components may not always be sufficient to shift the lubrication regime from boundary to fluid film.

The above calculations have been applied to a total hip prosthesis but a similar analysis has been applied in other joints such as the finger,[150] the wrist[151] and the toe.[152]

Given that the boundary lubrication regime dominates then, as would be expected, total wear volumes would increase with sliding distance, as suggested by the wear equation.

$$k = \frac{V}{LD} \tag{7}$$

Here, V is the volume of material lost (units mm^3), L is the load (N) and D is the sliding distance (m).

Indeed it has been shown that wear increases with the femoral head diameter of the hip prostheses, both for PTFE and UHMWPE acetabular cups when used in a hip simulator.[64] The same result has been shown clinically by an analysis of components retrieved at autopsy and revision.[27] As another, more recent example, at a follow up of 9-21 years, replacement hip joints with 32mm diameter femoral heads showed a revision rate 2.8 times higher than those with 22mm diameter heads.[153]

9. Friction

Charnley's original choice of PTFE was based on the fact that it would offer a low friction counterface to a stainless steel femoral head. Indeed Charnley's design aim has resulted in his implant being sometimes known as a 'low friction arthroplasty' and this term has been used as a surrogate descriptor of Charnley's design.[4,154]

Values of friction can be converted to a friction factor which is defined as

$$f = \frac{T}{rL} \tag{8}$$

where T is the frictional torque generated between the bearing surfaces, r is the femoral component radius, and L is the load applied.[155]

By performing tests at various lubricant viscosities, Stribeck analyses can be undertaken. In a Stribeck plot, friction factor is plotted against Sommerfeld number z which is defined as

$$z = \frac{\eta u r}{L} \tag{9}$$

where η is the lubricant viscosity and u is the entraining velocity of the bearing surfaces. A rising trend in the friction factor with increasing Sommerfeld number indicates full fluid film lubrication whereas a falling trend indicates mixed lubrication.[155] Such an analysis helps to identify lubrication regimes from *in vitro* testing.

Although wear has been seen to be a key concern in the choice of biopolymers, their friction is important too. Tests of UHMWPE acetabular sockets in hip friction simulators have been undertaken by several centres.[141,156] Friction measurements of biopolymers have also been undertaken using pin-on-disk tribometers.[133,157-159]

Sawae *et al.* undertook tribological studies using a uni-directional pin-on-disk device.[157] They tested UHMWPE against stainless steel and alumina in the presence of several lubricants, namely saline, saline plus albumin, saline plus hyaluronic acid and 30% bovine serum. They reported that the lowest friction was found with the hyaluronic acid solution, with a coefficient of friction of approximately 0.02. With 30% bovine serum as the lubricant the coefficient of friction was closer to 0.06.[157] In another study using a standard pin-on-disk tribometer, the friction of UHMWPE rubbing against CoCrMo counterfaces was investigated in the presence of several lubricants.[133] Lubricant was found to have a significant influence on friction. Changing from distilled water to dilute bovine serum led to an increase in friction while the use of human periprosthetic synovial fluid was associated with the lowest friction. It was also found that XLPE had a similar coefficient of friction to that of conventional UHMWPE.[133]

In their pin-on-disk friction tests involving several lubricants Gispert *et al.* also reported that the addition of hyaluronic acid tended to decrease the values of friction coefficient.[158] These authors argued that the addition tended to move the lubrication regime from mixed towards hydrodynamic conditions. The addition of proteins to the lubricant was seen to prevent a transfer film (a result reported by many other researchers) leading to the friction coefficient remaining approximately constant in their tests of protein-including lubricants.[158]

The previous three papers all used friction measuring devices which applied uni-directional motion to their test samples. In wear tests this has produced wear rates which are not clinically relevant. Recently, Saikko modified his multi-directional wear test device so that one station could measure friction.[159] Using 50% bovine serum as the lubricant he reported coefficients of friction in the range 0.26 to 0.34 for UHMWPE rubbing against CoCrMo at a nominal contact pressure of 1.1MPa. As an explanation for such high values of friction it was suggested that the values would fall over time with increasing protein degradation.[159]

Friction tests in hip simulators have employed a number of lubricants including water, carboxymethyl cellulose (CMC) solution, silicone fluids, bovine serum and synovial fluid.[155] With CMC fluids, friction factors have been reported for UHMWPE against CoCrMo of the order of 0.05, with synovial fluid they were of the order of 0.05, and with bovine serum they were of the order of 0.06.[141] In another recent hip friction paper it was reported that metal-on-polymer hips gave a slightly higher friction factor than a ceramic-on-polymer combination.[156] Tests were done in water, 25% serum and 100% serum. In general it was found that friction increased with protein concentration, water (0% serum) offering the lowest friction.[156]

New and explanted total hip replacements were tested in a hip friction simulator and said to operate in the mixed lubrication regime.[160] In this study, most explanted joints were shown to have a similar friction factor to that that of new joints, though some showed higher friction factors. In another frictional study of explanted hip prostheses,[161] friction factors of 0.040 were offered for the explanted Charnley prostheses compared with 0.025 for explanted Porous Coated Anatomic (PCA) prostheses. The friction tests of explanted PCA prostheses were said to indicate a mixed lubrication regime.[161]

Friction of knee prostheses has also been measured. Ash *et al.* showed that the friction of a polyurethane compliant layer knee (see below) was less than that of a conventional UHMWPE knee.[162]

There are still challenges in regard to linking friction and wear in relation to orthopedic biopolymers. For example it should be recalled that, against polished metallic counterfaces and in the presence of a bovine serum based lubricant, PTFE shows lower friction than UHMWPE, but PTFE also shows much greater wear than UHMWPE.

10. Other Polymers (Nonpolyethylene)

Other polymers have been used in total hip and other joint replacements. PTFE has already been mentioned but polyacetal (Delrin®) was used in the Christiansen design of hip prosthesis.[163,164] Clinical results were poor, but the Christiansen design placed the femoral head on a cylindrical trunion bearing, also of polyacetal, on the femoral stem. In effect therefore there were two articulations, one at the trunion and one between the femoral head and the acetabular cup. In addition, due to other design features, the acetabular liner was relatively thin.[65] These design details may have influenced clinical results more than the choice of polyacetal as the biopolymer.

Tests of polyacetal against stainless steel under multi-directional motion and in the presence of a lubricant of dilute bovine serum resulted in a mean wear factor of $3.8 \times 10^{-6} mm^3/Nm$.[95] Lower wear factors from different blends of polyacetal may be possible, as these materials are still being developed, and as manufacturers of implants employing polyacetal have not always divulged the specific blend they have used.

Poly ether ether ketone (PEEK) has been used as an implant material for a proposed shoulder prosthesis[165] and also in a potential two-piece finger prosthesis. For the latter, after five million cycles of testing in a clinically validated test rig, the PEEK finger prostheses showed greater wear than comparable UHMWPE prostheses.[94] When PEEK was compared with UHMWPE in a hip simulator tests it was found that PEEK showed six times the wear rate of the UHMWPE.[166]

Different types of carbon fiber reinforced (CFR) PEEK were wear tested against clinically relevant hard counterfaces in the presence of a bovine serum lubricant.[166] In hip simulator tests the various types of CFR PEEK showed between 1/10 and 1/30 the wear of conventional UHMWPE components. However when the same materials were tested in a ring-on-flat wear tester, they all showed significantly greater wear than UHMWPE.[166] The authors therefore suggested that CFR PEEK was unsuitable for high stress conditions, as reproduced in the ring-on-flat wear tester.

One type of PEEK, with the trade name Optima®, has been carbon fiber reinforced and then articulated against ceramic counterfaces in a wear screening rig which applied multi-directional motion to the test pins and employed a lubricant of dilute bovine serum.[167] Wear factors of approximately $0.2 \times 10^{-6} mm^3/Nm$ were reported for the CFR PEEK Optima® polymers,[167] which compares with $1.1 \times 10^{-6} mm^3/Nm$ for conventional UHMWPE tested on the same rig under similar test conditions.[95]

11. All-Polymer Articulations

Some fascinating work has been undertaken regarding the potentially intriguing situation of all polymer artificial joints. McKellop wear tested polyacetal against UHMWPE in a hip simulator and reported 40% less wear than when UHMWPE was tested against CoCrMo.[168] For the knee, a prosthesis of polyacetal rubbing against UHMWPE was proposed.[169,170] Clinical results of this polyacetal against UHMWPE total knee replacement were reported as showing no instances of failure due to wear.[171] This is remarkable for such a heavily loaded joint as the knee. The result implies that there could be a massive potential for such all-polymer articulations which would be much less expensive than conventional metal-on-polymer implants.

Independent tests of polyacetal against UHMWPE in a wear screening device which applied multi-directional motion and used a lubricant of dilute bovine serum revealed an average wear factor for this material combination of 4.1 x $10^{-6}mm^3/Nm$.[172] This value seems on the high side compared with Moore *et al.*'s clinical result. Two possible explanations include the fact that precise details of the polyacetal polymer were not offered by the manufacturer and that the degree of multi-directional motion applied may have been on the high side for the situation at the knee. The potential exists for more work to be done in this area.

Novel work which concerned a two-piece finger prosthesis, with each component manufactured from silane XLPE (the same material as used for acetabular cups of hip prostheses as reported by Wroblewski) has been undertaken. Tests of the implant, and of the material rubbing against itself in screening wear tests, under a range of biotribological conditions, have been reported.[59,173-175]

In wear screening tests under reciprocating motion, the lowest wear was found with bovine serum as the lubricant (k = 0.6 x $10^{-6}mm^3/Nm$) compared with a mean wear factor of 5.8 x $10^{-6}mm^3/Nm$ when distilled water was used. Intriguingly, with bovine serum and under multi-directional motion, wear factors increased to 3.4 x $10^{-6}mm^3/Nm$ but they fell to 0.7 x $10^{-6}mm^3/Nm$ with distilled water under multi-directional motion.[175]

In a clinically validated finger simulator wear factors of the XLPE prosthesis were found to depend on the lubricant. With Ringer solution the total wear factors were 0.98 x $10^{-6}mm^3/Nm$, with distilled water they were 0.60 x $10^{-6}mm^3/Nm$ and with dilute bovine serum they were 0.07 x $10^{-6}mm^3/Nm$.[174] Later tests studied in greater detail the influence of the bovine serum based lubricant on the wear of the finger prosthesis. From five separate tests totaling over 45 million cycles of flexion-extension, it was found that the weight increase of the test components exceeded that of the control components.[60] Therefore, as well as the wear test results, a comprehensive investigation of fluid uptake by silane XLPE finger prostheses was also reported.[60]

In the hope of gaining more information regarding the wear of biopolymers against themselves, wear tests of all-UHMWPE articulations have been undertaken.[176] Tests included pin-on-plate studies under reciprocating and multi-directional motion, while a pseudo-prosthesis was also manufactured and tested in a finger wear simulator. In all cases UHMWPE wear was high, with wear factors as great as 91 x $10^{-6}mm^3/Nm$ being seen. It was also found from the pin-on-plate studies that the lubricant and the motion applied to the test pins had a significant effect on the wear measured.[176]

12. Future Developments in Biopolymers

Clearly UHMWPE has proved to be a major success in many ways and the developments in XLPE point to a bright future. However there is always room for improvement in producing lower wear polymers or polymers whose debris provokes a more benign reaction within the body. A wide range of suggested improvements to polymers has been proposed and some of the most recent and promising will be briefly mentioned here.

The use of hyaluronan - a well-known natural lubricant present in synovial fluid - has been proposed, in an UHMWPE-hyaluronan microcomposite.[177] This microcomposite was compared with UHMWPE, and with XLPE which had been irradiated to 100kGy with an electron beam. The materials were tested on a wear screening device which applied multi-directional motion to the test samples. Zhang *et al.* reported that XLPE showed the lowest wear but that the wear of the microcomposite was less than that of the unmodified UHMWPE. In addition it was found that the wear rates for the microcomposite increased with time, suggesting that the benefits of the surface modification were lost due to the removal of the hyaluronan film. The authors concluded that future work would investigate whether the hyaluronan treatment could similarly improve the wear resistance of XLPE.[177]

Some remarkable results have been reported from the grafting of a surface layer of biocompatible polymer to XLPE and its subsequent influence on wear.[178] The graft is a phospholipid polymer 2-methacryloyloxyethyl phosphorylcholine (MPC). Results showed a four fold reduction in wear compared with XLPE acetabular cups and a forty fold reduction compared with conventional UHMWPE in a hip simulator study. A separate study showed that the surface graft also significantly reduced the friction of the material compared with XLPE and conventional UHMWPE. In addition, wear particles were shown to be much less reactive than those of conventional UHMWPE.[178] A very recent paper by the same group is also available reporting wear rates as high as 4mg/million cycles for unmodified XLPE cups and 0.16mg/million cycles for XLPE cups with the MPC graft.[179]

One area of modification of UHMWPE concerns the addition of Vitamin E.[180,181] The aim of adding vitamin E is to remove the need for post-irradiation re-melting (which exists to decrease the concentration of residual free radicals) by diffusing vitamin E (α-tocopherol) which is an anti-oxidant, into irradiated UHMWPE.[181] Oral *et al.* reported the testing of vitamin E stabilized UHMWPE cups in a hip simulator with a 100% bovine serum lubricant heated to 37°C and stated that wear rates were reduced to between a quarter and a tenth of those seen with conventional UHMWPE acetabular cups.[181]

One of the most interesting developments in biotribology has been to try and replicate, in part at least, the natural synovial joints. Instead of two relatively rigid surfaces articulating together, such as UHMWPE against CoCrMo, the UHMWPE has been replaced with a relatively soft polymer such as polyurethane. The aim has been to use the inherent compliance of the 'soft layer' or 'compliant layer' to encourage elastohydrodynamic lubrication and thus the two surfaces to be separated by a film of lubricating fluid, as occurs with cartilage in natural synovial joints.[182,183] In such a situation both friction and wear should be minimized. One concern is at 'start up', when the relatively soft surface will be in contact with the harder counterface. Despite this anxiety, simulator studies have shown reductions in friction and wear to approximately a third of the friction and wear seen with conventional UHMWPE acetabular cups.[184] Positive results from a 48-month study in sheep have recently been reported with the authors concluding that the use of such implants in humans remains an intriguing prospect.[185] A very recent paper reported a new design of multi-directional wear screening device. This rig was used to compare the wear of UHMWPE with that of an elastomeric polyurethane, and it was reported that the wear rate of the polyurethane was half that of the UHMWPE.[186]

13. Future Challenges

What then are some of the challenges for researchers concerned with biopolymer tribology for orthopedic applications? As has been seen, one key area concerns the lubricant to be used for biopolymer wear testing. While bovine serum is widely regarded as the most appropriate lubricant currently available, it is also recognized that it differs from synovial fluid in certain respects. Challenges such as determining the optimum protein content and appropriate additives exist, as does a search for a lubricant which is safer and less expensive than bovine serum and the properties of which remain constant over the test duration.

As has also been seen efforts have been made to improve on UHMWPE and, most recently, to study XLPE more fully. These efforts will no doubt continue. Important issues still remain related to soak controls, while all-polymer couples in artificial joints could potentially offer a relatively inexpensive step forward. Equally, moves away from relatively rigid and homogeneous materials such as UHMWPE will also persist, in an aim to reproduce more closely the superb tribological properties of articular cartilage.

14. Summary

The use of UHMWPE in replacement joints has been a major factor in their success and the use of XLPE promises to continue this success into the future, for the benefit of patients fitted with such implants. Biotribology is crucial to understanding the success of the majority of current joint replacements and to making future implants even more successful.

References

1. J. Charnley, A. Kamangar and M. D. Longfield, *Med Biological Engng*. 7, 31 (1969).
2. S. M. Kurtz. *The UHMWPE Handbook. Ultra-high molecular weight polyethylene in total joint replacement*. (Elsevier Academic Press, San Diego, 2004).
3. D. Dowson and N. C. Wallbridge, *Wear*. 104, 203 (1985).
4. J. Older, *J. Arthroplasty*. 17, 6, 675 (2002).

5. B. Wroblewski, P. Siney and P. Fleming, *J. Bone Joint Surg. Br.* 84-B, 4, 540 (2002).
6. D. J. Berry, W. S. Harmsen, M. E. Cabanela and B. F. Morrey, *J. Bone Joint Surg. Am.* 84, 2, 171 (2002).
7. O. K. Muratoglu, A. Mark, D. A. Vittetoe, W. H. Harris and H. E. Rubash, *J. Bone Joint Surg. Am.* 85, suppl_1, S7 (2003).
8. M. Jasty, H. E. Rubash and O. Muratoglu, *J. Arthroplasty.* 20, Supplement 2, 55 (2005).
9. D. M. Estok, B. R. Burroughs, O. K. Muratoglu and W. H. Harris, *J. Arthroplasty.* 22, 4, 581 (2007).
10. H. A. McKellop, F. Shen, B. Lu, P. Campbell and R. Salovey, *J. Orthop. Res.* 17, 2, 157 (1999).
11. M. C. Forster, *J. Arthroplasty.* 18, 3, 265 (2003).
12. J. A. Rand, R. T. Trousdale, D. M. Ilstrup and W. S. Harmsen, *J. Bone Joint Surg. Am.* 85, 2, 259 (2003).
13. O. Robertsson, K. Knutson, S. Lewold and L. Lidgren, *Acta Orthopaedica Scandinavica.* 72, 503 (2001).
14. G. R. Williams and J. A. Abboud, *J. Should. Elbow Surg.* 14, 1, Supplement 1, S122 (2005).
15. P. Boileau, D. Watkinson, A. M. Hatzidakis and I. Hovorka, *J. Shoulder Elbow Surg.* 15, 5, 527 (2006).
16. M. Radford and A. Carr, *Current Orthop.* 16, 5, 325 (2002).
17. B. Lee, R. Adams and B. Morrey, *J. Bone Joint Surg. Am.* 87, 5, 1080 (2005).
18. T. J. Joyce, *Expert Rev. Med. Devs.* 1, 2, 193 (2004).
19. T. J. Joyce, *Expert Rev. Med. Devs.* 2, 4, 453 (2005).
20. K. Olms and A. Dietze, *Intl. Orthop.* 23, 240 (1999).
21. C. J. Bell and J. Fisher, *J. Biomed. Mat. Res. B.* 81B, 1, 162 (2007).
22. D. Groot, T. Gosens, N. C. M. v. Leeuwen, M. v. Rhee and H. J. L. J. M. Teepen, *J. Hand Surg.* 31, 10, 1615 (2006).
23. S. M. Kurtz, A. van Ooij, R. Ross, J. de Waal Malefijt, J. Peloza, L. Ciccarelli and M. L. Villarraga, *Spine.* 7, 1, 12 (2007).
24. J. P. Van Loon, G. J. Verkerke, L. G. M. de Bont and R. S. B. Liem, *Biomaterials.* 20, 1471 (1999).
25. R. M. Hall, A. Unsworth, P. D. Siney and B. M. Wroblewski, *J. Engng. Med.* 210, 197 (1996).
26. C. J. Sychterz, K. H. Moon, Y. Hashimoto, K. M. Terefenko, C. Anderson and T. W. Bauer, *J. Bone Jt. Surg.* 78-A, 8, 1193 (1996).
27. M. Jasty, D. D. Goetz, C. R. Bragdon, K. R. Lee, A. E. Hanson, J. R. Elder and W. H. Harris, *J. Bone Joint Surg. Am.* 79, 3, 349 (1997).
28. C. B. Rieker, R. Schon and P. Kottig, *J. Arthroplasty.* 19, 8, Supplement 1, 5 (2004).
29. ASTM-F732-00. (2000).

30. W. H. Harris, *Clin. Orthop. Rel. Res.* 311, 46 (1995).
31. E. Ingham and J. Fisher, *J. Engng. Med.* 214, 21 (2000).
32. J. J. Jacobs, N. J. Hallab, R. M. Urban and M. A. Wimmer, *J. Bone Joint Surg. Am.* 88, suppl_2, 99 (2006).
33. J. H. Dumbleton, M. T. Manley and A. A. Edidin, *J. Arthroplasty.* 17, 5, 649 (2002).
34. A. Wang, D. C. Sun, C. Stark and J. H. Dumbleton, *Wear.* 181-183, 241 (1995).
35. V. C. Jones, I. R. Williams, D. D. Auger, W. Walsh, D. C. Barton, M. H. Stone and J. Fisher, *J. Engng. Med.* 215, 171 (2001).
36. V. Saikko, *J. Engng. Med.* 219, H, 309 (2005).
37. K. Brummitt, C. S. Hardaker, P. J. McCullagh, K. J. Drabu and R. A. Smith, *J. Engng. Med.* 210, 3, 191 (1996).
38. T. R. Green, J. Fisher, M. H. Stone, B. M. Wroblewski and E. Ingham, *Biomaterials.* 19, 2297 (1998).
39. P. Campbell, S. Ma, H. A. McKellop, T. P. Schmalzried and H. C. Amstutz, *J. Biomed. Mat. Res.* 29, 127 (1995).
40. A. P. Elfick, S. M. Green, S. Krikler and A. Unsworth, *J. Biomed. Mat. Res.* 65A, 95 (2003).
41. A. Wang, A. Essner, C. Stark and J. H. Dumbleton, *Biomaterials.* 17, 865 (1996).
42. J. Fisher, J. Bell, P. S. M. Barbour, J. L. Tipper, J. B. Matthews, A. A. Besong, M. H. Stone and E. Ingham, *J. Engng. Med.* 215, 127 (2001).
43. R. M. Hall, M. J. K. Bankes and G. Blunn, *Curr. Orthop.* 15, 4, 281 (2001).
44. BS_ISO_17853. (2003).
45. M. D. Ries, M. L. Scott and S. Jani, *J. Bone Joint Surg. Am.* 83, 2_suppl_2, S116 (2001).
46. A. S. Shanbhag, H. O. Bailey, D. S. Hwang, C. W. Cha, N. G. Eror and H. E. Rubash, *J. Biomed. Mat. Res.* 53, 1, 100 (2000).
47. C. Heisel, M. Silva, M. Dela Rosa and T. P. Schmalzried, *J. Bone. Jt. Surg.* 86-A, 4, 748 (2004).
48. D. W. Manning, P. P. Chiang, J. M. Martell, J. O. Galante and W. H. Harris, *J. Arthroplasty.* 20, 7, 880 (2005).
49. H. Oonishi, M. Kuno, E. Tsuji and A. Fujisawa, *J. Mat. Sci: Mat. Med.* 8, 1, 11 (1997).
50. H. Oonishi and Y. Kadoya, *J. Orthop. Sci.* 5, 3, 223 (2000).
51. H. Oonishi, I. C. Clarke, S. Masuda and H. Amino, *J. Arthroplasty.* 16, 8, Supplement 1, 129 (2001).
52. H. Oonishi, S. C. Kim, Y. Takao, M. Kyomoto, M. Iwamoto and M. Ueno, *J. Arthroplasty.* 21, 7, 944 (2006).
53. J. C. A. Engh, A. S. Stepniewski, S. D. Ginn, S. E. Beykirch, C. J. Sychterz-Terefenko, J. R. H. Hopper and C. A. Engh, *J. Arthroplasty.* 21, 6, Supplement 1, 17 (2006).

54. C. R. Bragdon, S. Barrett, J. M. Martell, M. E. Greene, H. Malchau and W. H. Harris, *J. Arthroplasty*. 21, 7, 935 (2006).
55. L. Bradford, D. A. Baker, J. Graham, A. Chawan, M. Ries and L. Pruitt, *J. Bone Joint Surg. Am.* 86A, 6, 1271 (2004).
56. B. M. Wroblewski, P. D. Siney, D. Dowson and S. N. Collins, *J. Bone Jt. Surg.* 78-B, 280 (1996).
57. B. M. Wroblewski, P. D. Siney and P. A. Fleming, *J. Bone Jt. Surg.* 81-B, 54 (1999).
58. B. M. Wroblewski, P. D. Siney and P. A. Fleming, *J. Bone Joint Surg. Br.* 87-B, 9, 1220 (2005).
59. T. J. Joyce, H. E. Ash and A. Unsworth, *J. Engng. Med.* 210, 11 (1996).
60. T. J. Joyce and A. Unsworth, *J Applied Biomaterials Biomechanics*. 2, 136 (2004).
61. S. M. Kurtz, O. K. Muratoglu, M. Evans and A. A. Edidin, *Biomaterials*. 20, 1659 (1999).
62. D. L. Scott, P. A. Campbell, C. D. McClung and T. P. Schmalzried, *J. Arthroplasty*. 15, 1, 35 (2000).
63. B. J. Livingston, M. J. Chmell, M. Spector and R. Poss, *J. Bone Jt. Surg. Am.* 79, 10, 1529 (1997).
64. I. C. Clarke, V. Good, L. Anissian and G. A. Gustafson, *J. Engng. Med.* 211, 25 (1997).
65. A. A. Edidin and S. M. Kurtz, *J. Arthroplasty*. 15, 3, 321 (2000).
66. S. L. Smith and A. Unsworth, *J. Engng. Med.* 215, 61 (2001).
67. V. Saikko, *J. Engng. Med.* 219, 437 (2005).
68. J. C. Hermida, A. Bergula, P. Chen, C. W. Colwell and D. D. D'Lima, *J. Bone. Jt. Surg.* 85A, 12, 2325 (2003).
69. V. D. Good, I. C. Clarke, G. A. Gustafson, B. Downs, L. Anissian and K. Sorenson, *Acta Orthop. Scand.* 71, 4, 365 (2000).
70. ISO14242-1, Geneva, ISO. (2000).
71. C. Kaddick and M. A. Wimmer, *J. Engng. Med.* 215, 429 (2001).
72. ISO14242-2, Geneva, ISO. (2000).
73. ASTM-71714-96, (2002).
74. A. Essner, G. Schmidig and A. Wang, *Wear*. 259, 7-12, 882 (2005).
75. T. P. Schmalzried, E. S. Szuszczewicz, M. R. Northfield, K. H. Akizuki, R. E. Frankel, G. Belcher and H. C. Amstutz, *J. Bone Joint Surg. Am.* 80, 1, 54 (1998).
76. O. Muratoglu, C. Bragdon, D. O'Connor, M. Jasty and W. Harris, *J. Arthroplasty*. 16, 2, 149 (2001).
77. J. G. Bowsher and J. C. Shelton, *Wear*. 250, 1-12, 167 (2001).
78. C. R. Bragdon, M. Jasty, O. K. Muratoglu, D. O. O'Connor and D. Harris, *J. Arthroplasty*. 18, 5, 553 (2003).
79. I. C. Burgess, M. Kolar, J. L. Cunningham and A. Unsworth, *J. Engng. Med.* 211, 37 (1997).
80. H. E. Ash, I. C. Burgess and A. Unsworth, *J. Engng. Med.* 214, 437 (2000).

81. S. K. Young, A. S. Keller, K. W. Greer and M. C. Gorhan, *J. Tribology*. 122, 323 (2000).
82. P. I. Barnett, H. M. J. McEwen, D. D. Auger, M. H. Stone, E. Ingham and J. Fisher, *J. Engng. Med.* 216, 51 (2002).
83. J. DesJardins, A. Aurora, S. L. Tanner, T. Pace, K. B. Acampora and M. LaBerge, *J. Engng. Med.* 220, 609 (2006).
84. ISO14243-2, Geneva, ISO. (2000).
85. L. M. Jennings, C. J. Bell, E. Ingham, R. D. Komistek, M. H. Stone and J. Fisher, *J. Engng. Med.* 221, 305 (2007).
86. L. A. Knight, S. Pal, J. C. Coleman, F. Bronson, H. Haider, D. L. Levine, M. Taylor and P. J. Rullkoetter, *J. Biomech.* 40, 7, 1550 (2007).
87. O. K. Muratoglu, C. R. Bragdon, M. Jasty, D. O. O'Connor, R. S. Von Knoch and W. H. Harris, *J. Arthroplasty*. 19, 7, 887 (2004).
88. O. K. Muratoglu, H. E. Rubash, C. R. Bragdon, B. R. Burroughs, A. Huang and W. H. Harris, *J. Arthroplasty*. 22, 3, 435 (2007).
89. W. Nechtow, M. Hintner, M. Bushelow and C. Kaddick, *52nd Annual Meeting of the Orthopaedic Research Society*, 0118 (2006).
90. R. Ross, A. H. Mirza, H. E. Norris and M. Khatri, *J. Bone Joint Surg. Br.* 89-B, 6, 785 (2007).
91. ASTM-F2423-05, (2006).
92. T. J. Joyce and A. Unsworth, *J. Engng. Med.* 214, 519 (2000).
93. T. J. Joyce, R. H. Milner and A. Unsworth, *J. Hand Surg.* 28B, 1, 86 (2003).
94. T. J. Joyce, C. Rieker and A. Unsworth, *BioMed. Mat. Engng.* 16, 1, 1 (2006).
95. T. J. Joyce, *Trib. Mat. Surf. Interfaces*. 1, 2 In Press, (2007).
96. C. R. Bragdon, D. O. O'Connor, J. D. Lowenstein, M. Jasty and W. H. Harris, *Fifth World Biomaterials Congress*, 788 (1996).
97. V. Saikko, *J. Biomed. Mat. Res.* 41, 58 (1998).
98. V. C. Jones, D. C. Barton, D. P. Fitzpatrick, D. D. Auger, M. H. Stone and J. Fisher, *BioMed. Mat. Engng.* 9, 189 (1999).
99. T. J. Joyce and A. Unsworth, *Biomed. Mat. Engng.* 10, 241 (2000).
100. C. B. Rieker, S. Schaffner, R. Schon and R. Konrad, *J. Bone Joint Surg. Br.* 84-B, SUPP_I, 19 (2002).
101. D. Mazzucco and M. Spector, *Wear*. 254, 514 (2003).
102. C. R. Bragdon, D. O. O'Connor, J. D. Lowenstein, M. Jasty, S. A. Biggs and W. H. Harris, *J. Arthroplasty*. 16, 5, 658 (2001).
103. J. G. Lancaster, D. Dowson, J. Fisher and G. H. Isaac, *J. Engng. Med.* 211, 17 (1997).
104. C. R. Bragdon, D. O. O'Connor, J. D. Lowenstein and W. D. Synivta, *J. Engng. Med.* 210, 157 (1996).
105. ASTM-F2025-06, (2006).
106. H. Marrs, D. C. Barton, R. A. Jones, I. M. Ward, J. Fisher and C. Doyle, *J. Mat. Sci: Mat. Med.* 10, 333 (1999).

107. T. J. Joyce, A. Unsworth, T. M. Cartwright and D. Monk, *J. Bone. Jt. Surg.* 81B, Supplement I, 85 (1999).
108. H. Sakoda, A. M. Voice, M. J. McEwan, G. H. Isaac, C. S. Hardaker, B. M. Wroblewski and J. Fisher, *J. Arthroplasty.* 16, 8, 1018 (2001).
109. V. Saikko, O. Calonius and J. Keranen, *J. Biomed. Mat. Res.* 57, 506 (2001).
110. J. Q. Yao, T. A. Blanchet, D. J. Murphy and M. P. Laurent, *Wear.* 255, 1113 (2003).
111. K. Laraia, N. Leone, R. MacDonald and T. A. Blanchet, *Tribology Transactions.* 49, 3, 338 (2006).
112. T. J. Joyce, P. Thompson and A. Unsworth, *Wear.* 255, 1030 (2003).
113. T. J. Joyce, D. Monk, P. Thompson, P. Chiu, A. Unsworth and S. M. Green, *Seventh World Biomaterials Congress*, 537 (2004).
114. D. Dowson, M. Diab, B. Gillis and J. R. Atkinson, *Am Chem Soc.* 171 (1985).
115. B. Weightman and D. Light, *Biomaterials.* 7, 20 (1986).
116. J. L. Tipper, E. Ingham, J. L. Hailey, A. A. Besong, J. Fisher, B. M. Wroblewski and M. H. Stone, *J. Mat. Sci: Mat. Med.* 11, 2, 117 (2000).
117. P. Geborek and F. A. Wollheim *Synovial fluid.* In: V. Wright and E. Radin *Mechanics of human joints: physiology, pathophysiology and treatment.* (Marcel Dekker, New York, 1993).
118. A. F. Cooke, D. Dowson and V. Wright, *Engng. Med.* 7, 2, 66 (1978).
119. H. Saari, S. Santavirta, D. Nordstrom, P. Paavolainen and Y. T. Konttinen, *J. Rheumatol.* 20, 87 (1993).
120. Y. S. Liao, P. D. Benya and H. A. McKellop, *J. Biomed. Mat. Res.* 48, 465 (1999).
121. A. Wang, A. Essner and G. Schmidig, *J. Biomed. Mat. Res. B.* 68B, 45 (2004).
122. J. Q. Yao, M. P. Laurent, L. N. Gilbertson, C. R. Blanchard, R. D. Crowninshield and J. J. Jacobs, Dallas, Texas, USA. (2002).
123. H. A. McKellop, *Engng. Med.* 10, 3, 123 (1981).
124. P. Kumar, M. Oka, K. Ikeuchi, K. Shimizu, T. Yamamuro, H. Okumura and Y. Koloura, *J. Biomed. Mat. Res.* 25, 813 (1991).
125. J. R. Cooper, D. Dowson and J. Fisher, *Clin. Matl.* 14, 295 (1993).
126. B. Derbyshire, J. Fisher, D. Dowson, C. S. Hardaker and K. Brummitt, *Med. Engng. Phys.* 16, 229 (1994).
127. T. Schwenke, C. Kaddick, E. Schneider and M. A. Wimmer, *J. Engng. Med.* 219, 457 (2005).
128. I. C. Clarke, F. W. Chan, A. Essner, V. Good, C. Kaddick, R. Lappalainen, M. P. Laurent, H. A. McKellop, W. McGarry, D. Schroeder, M. Selenius, M. C. Shen, M. Ueno, A. Wang and J. Q. Yao, *Wear.* 250, 188 (2001).
129. A. Wang, V. K. Polineni, A. Essner, C. Stark and J. H. Dumbleton, *45th Annual Meeting Orthopaedic Research Society*, 52 (1999).
130. V. Saikko, *Trans ASME J Tribology.* 125, 638 (2003).
131. R. M. Streicher, M. Semlitsch, R. Schon, H. Weber and C. Rieker, *J. Engng. Med.* 210, 223 (1996).

132. T. Ahlroos and V. Saikko, *Wear.* 211, 1, 113 (1997).
133. J. Q. Yao, M. P. Laurent, T. S. Johnson, C. R. Blanchard and R. D. Crowninshield, *Wear.* 255, 1-6, 780 (2003).
134. C. Klapperich, S. Niedzwiecki, M. Ries and L. Pruitt, *J. Biomed. Mat. Res.* 53, 73 (2000).
135. J. Q. Yao, M. P. Laurent, L. N. Gilbertson and R. D. Crowninshield, *Wear.* 250, 140 (2001).
136. T. A. Blanchet, S. L. Peterson and K. D. Rosenberg, *J. Trib.* 124, 1 (2002).
137. T. Schwenke, E. Schneider and M. A. Wimmer, *J. ASTM Intl.* 3, 8, 1 (2006).
138. B. J. Hamrock and D. Dowson, *Trans. ASME. J. Lubn. Tech.* 100, 236 (1978).
139. K. L. Johnson, J. A. Greenwood and S. Y. Poon, *Wear.* 19, 91 (1972).
140. V. Saikko and T. Ahlroos, *J. Engng. Med.* 213, 301 (1999).
141. S. C. Scholes and A. Unsworth, *J. Engng. Med.* 214, 49 (2000).
142. V. Saikko, T. Ahlroos, O. Calonius and J. Keranen, *Biomaterials.* 22, 1507 (2001).
143. S. L. Smith, D. Dowson and A. A. J. Goldsmith, *J. Engng. Tribology.* 215, 5, 483 (2001).
144. S. C. Scholes, A. Unsworth, R. M. Hall and J. T. Scott, *Wear.* 241, 209 (2000).
145. D. Jalali-Vahid, M. Jagatia, Z. M. Jin and D. Dowson, *J. Biomech.* 34, 2, 261 (2001).
146. I. J. Udofia and Z. M. Jin, *J. Biomech.* 36, 4, 537 (2003).
147. Z. M. Jin, D. Dowson and J. Fisher, *J. Engng. Med.* 211, 247 (1997).
148. A. P. D. Elfick, R. M. Hall, I. M. Pinder and A. Unsworth, *J. Biomed. Mat. Res.* 48, 712 (1999).
149. A. A. Edidin, C. M. Rimnac, V. M. Goldberg and S. M. Kurtz, *Wear.* 250, 1-12, 152 (2001).
150. T. J. Joyce, *Med. Engng. Phys.* 29, 1, 87 (2007).
151. T. Pylios and D. E. T. Shepherd, *J. Biomech.* 37, 405 (2004).
152. T. J. Joyce, *IFMBE Proc.* 15, 17, 651 (2007).
153. S. Tarasevicius, U. Kesteris, O. Robertsson and H. Wingstrand, *Acta Orthopaedica.* 77, 706 (2006).
154. O. A. Nercessian, G. Martin, R. P. Joshi, B. W. Su and N. S. Eftekhar, *J. Arthroplasty.* 20, 2, 162 (2005).
155. S. C. Scholes and A. Unsworth, *J. Engng. Med.* 220, 687 (2006).
156. C. Brockett, S. Williams, Z. Jin, G. Isaac and J. Fisher, *J. Biomed. Mat. Res. B.* 81B, 2, 508 (2007).
157. Y. Sawae, T. Murakami and J. Chen, *Wear.* 216, 213 (1998).
158. M. P. Gispert, A. P. Serro, R. Cola and B. Saramago, *Wear.* 260, 1-2, 149 (2006).
159. V. Saikko, *J. Engng. Med.* 220, 723 (2006).
160. A. Unsworth, R. M. Hall, I. C. Burgess, B. M. Wroblewski, R. M. Streicher and M. Semlitsch, *Wear.* 190, 2, 226 (1995).
161. A. P. Elfick, R. M. Hall, I. M. Pinder and A. Unsworth, *J. Engng. Med.* 212, 5, 395 (1998).

162. H. E. Ash, S. C. Scholes, A. Unsworth and E. Jones, *Phys. Med. Biol.* 49, 15, 3413 (2004).
163. L. I. Havelin, N. R. Gjerdet, O. D. Lunde, M. Rait and E. Sudmann, *Acta Orthop. Scand.* 57, 419 (1986).
164. A. Ohlin and P. G. Persson, *J. Arthroplasty*. 4, 3, 207 (1989).
165. D. Muller, A. Marti and P. Herzig, *47th Annual Meeting, Orthopaedic Research Society*, 0782 (2001).
166. A. Wang, R. Lin, C. Stark and J. H. Dumbleton, *Wear*. 225-229, 724 (1999).
167. S. C. Scholes and A. Unsworth, *J. Engng. Med.* 221, 281 (2007).
168. H. A. McKellop, T. Rostlund and G. Bradley, *Clin. Mat.* 14, 2, 117 (1993).
169. G. Bradley, M. A. R. Freeman, M. Tuke, S. G. Elias and H. A. McKellop, *Am. J. Knee Surg.* 5, 1, 3 (1992).
170. G. W. Bradley, M. A. R. Freeman, M. Tuke and H. A. McKellop, *Clin. Mater.* 14, 2, 127 (1993).
171. D. J. Moore, M. A. R. Freeman, P. A. Revell, G. W. Bradley and M. Tuke, *J. Arthroplasty*. 13, 388 (1998).
172. T. J. Joyce, *J. App. Biomat. Biomech.* 3, 3, 141 (2005).
173. T. J. Joyce and A. Unsworth, *J. Engng. Med.* 210, 297 (1996).
174. T. J. Joyce and A. Unsworth, *Wear*. 250, 199 (2001).
175. T. J. Joyce, C. Vandelli, T. Cartwright and A. Unsworth, *Wear*. 250, 206 (2001).
176. T. J. Joyce and A. Unsworth, *J Applied Biomaterials Biomechanics*. 2, 29 (2004).
177. M. Zhang, P. Pare, R. King and S. P. James, *J. Biomed. Mat. Res. A*. 82A, 1, 18 (2007).
178. T. Moro, Y. Takatori, K. Ishihara, T. Konno, Y. Takigawa, T. Matsushita, U.-I. Chung, K. Nakamura and H. Kawaguchi, *Nature Materials*. 3, 11, 829 (2004).
179. M. Kyomoto, T. Moro, T. Konno, H. Takadama, N. Yamawaki, H. Kawaguchi, Y. Takatori, K. Nakamura and K. Ishihara, *J. Biomed. Mat. Res. A*. 82A, 1, 10 (2007).
180. E. Oral, K. K. Wannomae, N. Hawkins, W. H. Harris and O. K. Muratoglu, *Biomaterials*. 25, 24, 5515 (2004).
181. E. Oral, S. D. Christensen, A. S. Malhi, K. K. Wannomae and O. K. Muratoglu, *J. Arthroplasty*. 21, 4, 580 (2006).
182. A. Unsworth, M. J. Pearcy, E. F. T. White and G. White, *J. Engng. Med.* 17, 3, 101 (1988).
183. D. Dowson, J. Fisher, Z. Jin, D. D. Auger and B. Jobbins, *J. Engng. Med.* 205, 2, 59 (1991).
184. S. L. Smith, H. E. Ash and A. Unsworth, *J. Biomed. Mat. Res.* 53, 6, 710 (2000).
185. A. Carbone, D. W. Howie, M. McGee, J. Field, M. Pearcy, N. Smith and E. Jones, *J. Arthroplasty*. 21, 6, 899 (2006).
186. C. J. Schwartz and S. Bahadur, *Wear*. 262, 3-4, 331 (2007).

CHAPTER 8

FRICTIONAL BEHAVIOR OF MINIATURE JOURNAL POLYMER-ON-POLYMER BEARINGS

Monika Kwacz and Zygmunt Rymuza

Warsaw University of Technology, Department of Mechatronics
Institute of Micromechanics and Photonics
sw.A.Boboli 8, 02-525 Warsaw, Poland
E-mail: z.rymuza@mchtr.pw.edu.pl

The frictional behavior of miniature polymer-on-polymer bearings is discussed. The transition from static to kinetic friction was simulated based on material parameters, surface topography and loading conditions. Therefore the effect of rheological behavior of contacting polymeric materials on friction was investigated. The elaborated model was used for prediction of friction coefficient (torque) in the analyzed bearings. The friction force was considered to be a sum of adhesive and mechanical (deformation) components. The extended experiments then have been carried out to compare the results from computer simulation with the results obtained in the experimental part of the studies. Very good agreement between simulation and experimental results was found. The results of the studies can be used in design process of miniature components to predict their frictional behavior and to select optimum combination of rubbing polymeric materials to obtain a required coefficient of friction.

1. Introduction

The problems of friction during transition from static to kinetic friction in polymer tribosystems and prediction of friction in particular miniature polymer-on-polymer journal bearings are essential in microtribology.[1-3] Such bearings are often applied in miniature mechanisms (e.g. quartz clocks) driven by stepping micromotors supplied by microbatteries. Such mechanisms operate in start-stop mode so the very high friction in starting phase reduces the lifetime of operation of microbatteries. Also bearings and other components of the mechanisms are heavily loaded by friction resistance forces.[4-5] Polymeric small mechanisms are often applied including also MEMS (Micro Electro Mechanical Systems) devices.[6-9] Polymeric components are used in medical technology (heart valves, endoprotheses, heart pumps etc.). Such polymer-on-polymer systems are very useful since they can operate without lubrication. Polymer-on-polymer journal bearings are most often used, but also miniature polymeric gears and guides are commonly embodied in small mechanisms. The frictional behavior of such systems is very important information for a designer of miniature devices.[10-14]

Advances in experimental studies in tribology are not correlated enough with generalization which can be applied for predicting the behavior of tribosystems. The importance of friction on energy state of tribosystem as well as on its reliability makes it necessary to explain frictional processes. One possible method is modeling and computer simulation of tribological processes to predict e.g. coefficient of friction.

During friction of polymer-on-polymer components mechanical and adhesive interactions play important role between rubbing surfaces.[29-36] According to molecular-mechanical theory of friction proposed by Kragelski[15] the friction force is a sum of deformation (mechanical) and adhesive components. Deformation component is connected with the necessity of deformation of micro-asperities of surface layer and adhesive component results from the effect of molecular adhesive bonds between contacting surfaces. When the surface roughness is small the contribution of adhesive component is high. The ratio of adhesive and mechanical component depends among others on applied load, roughness of contacting surfaces, their mechanical properties, adhesive characteristics of friction contacts and ambient conditions.

Very important for tribological contacts is an estimation of real area of contact (RAC) as the surface on which direct discrete contact occurs. The friction is localized on the RAC where heat is generating and wear appears. For polymer-on-polymer tribosystems it is necessary to take into consideration the rheology of stress and deformation of both contacting polymers. The estimation of RAC during sliding is a more complex problem as compared with static contact. The analysis of transition from static contact to sliding needs consideration of process of deformation of micro-asperities.

We present new procedure of modeling and computer simulation of the friction process in transition from static contact to kinetic contact (sliding) in polymer-on-polymer tribosystem. In modeling we apply theories describing mechanical and molecular interactions and descriptions of rheological processes occurring in polymers. The results of experimental studies by the use of Atomic Force Microscope (AFM) of surface topography and mechanical properties of thin surface layer are applied in the theoretical analysis of the system studied.

2. Experimental Studies

Frictional behavior of miniature polymer-on-polymer bearings depends on operational conditions and combination of materials. In particular, time of contact significantly influences the maximum value of friction coefficient of tested bearings.

For modeling contact and frictional behaviors of studied bearings the rheological as well as adhesive behaviors are important. The elaborated methods enable us to estimate elasticity modulus of a thin surface layer of used polymeric components as well as its surface energy.

2.1. *Investigation of Friction Coefficient and Time Needed to Start to Operate of Bearing*

2.1.1. *Test Rig and Procedure of Measurements*

The miniature journal polymer-on-polymer bearing shown in Fig. 1 was investigated.

Thermoplastic polymers such as PS, PA, POM and others were used to mold journals and bearing bushes. The bearing was not lubricated. The starting phase of operation of bearing was studied. The transition from static to kinetic friction was observed. The sliding speeds were selected from 1 μm/s to 100 μm/s at loads from 5 to 30N. The time of contact of the journal and bearing bush under load before starting the operation was varied from 1 to 900s. The test rig is shown in Fig. 2 and it is described

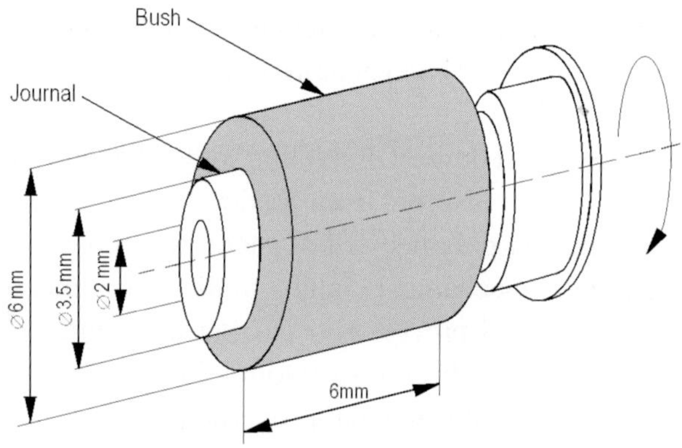

Fig. 1. Investigated bearing.

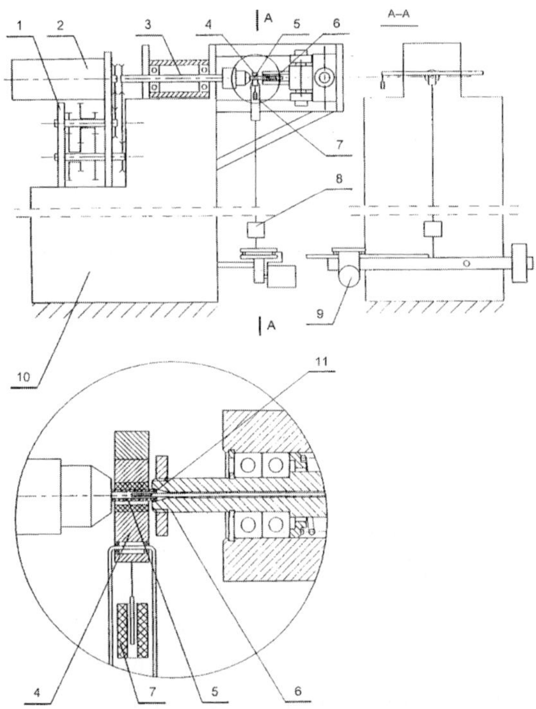

Fig. 2. Schematics of test rig for investigation of journal bearings. 1 – belt gear transmissions, 2 – DC motor, 3 – Driving shaft, 4 – Tested bearing, 5 – Journal, 6 – Support of journal, 7 – Displacement sensor, 8 – Load transducer, 9 – System for loading, 10 – concrete base, 11 – thermo element.

elsewhere.[16] The friction torque (friction coefficient), wear, the temperature inside a hole in the journal were measured.

2.1.2. *Results and Discussion*

An example of characteristics of friction coefficient vs. P (load), v (sliding speed) and t_p (time of contact under load before operation of the bearing) is presented in Fig. 3.

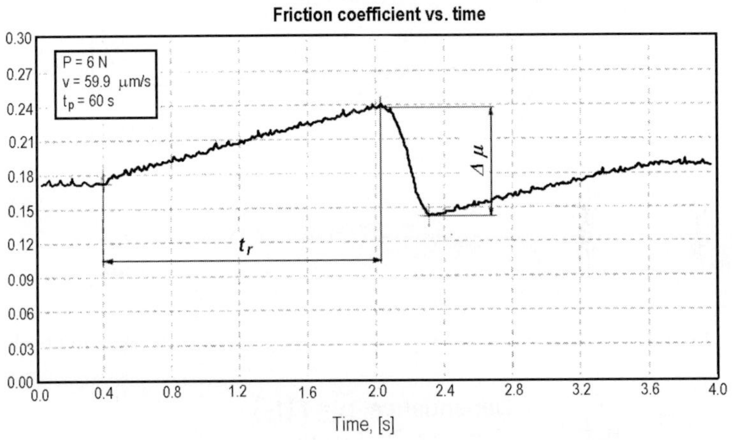

Fig. 3. Example of results of friction measurements: friction coefficient μ vs. time t at P – load, v – sliding speed and t_p – time of contact under load before test. Characteristic parameters: μ_{max} – maximum friction coefficient at phase to transit to slide, t_r – time needed to start to slide. Journal PA6 – bearing bush PA6.

As characteristic parameters were chosen:

- μ_{max} - maximum friction coefficient at start,
- t_r - time needed to start to slide.

The results of studies of friction showed that:

- increase of sliding speed (v) yields a decrease of time needed to slide (t_r) but does not influence on μ_{max} (Fig. 4),
- increase of time of contact (t_p) before operation yields a significant increase of time needed to slide (t_r) and increase of maximum friction coefficient (μ_{max}) (Fig. 5),

- increase of load P slightly affects the time needed to transit to sliding t_r and causes decrease of maximum friction coefficient μ_{max} in the range of relatively low loads (up to 13N) (Fig. 6).

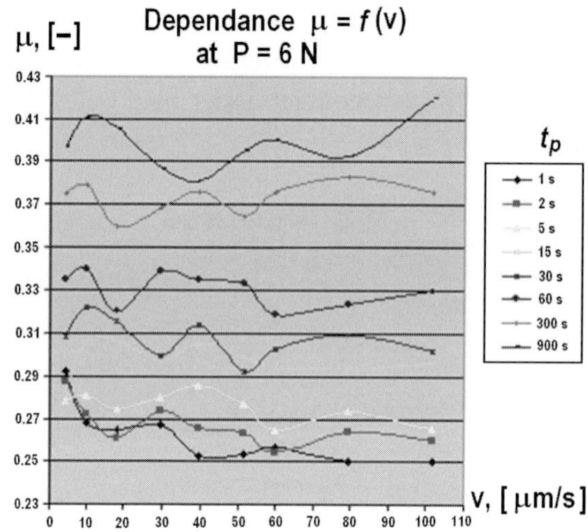

Fig. 4. Friction coefficient μ = f(v) for load P = 6N. Journal PA6 – bearing bush PA6.

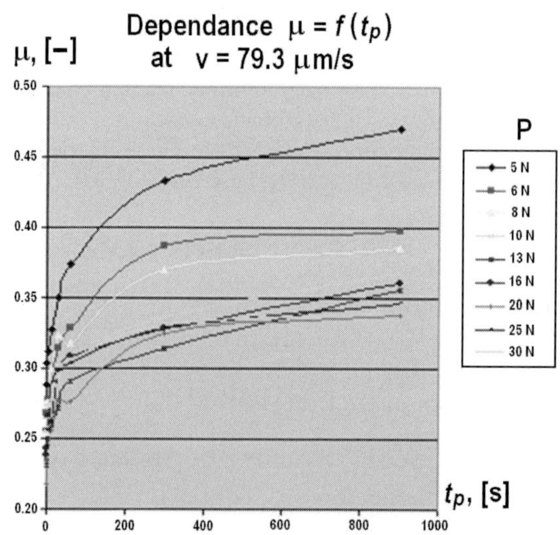

Fig. 5. Friction coefficient μ = f(t_p) at sliding speed v=79.3μm/s. Journal PA6 – bearing bush PA6.

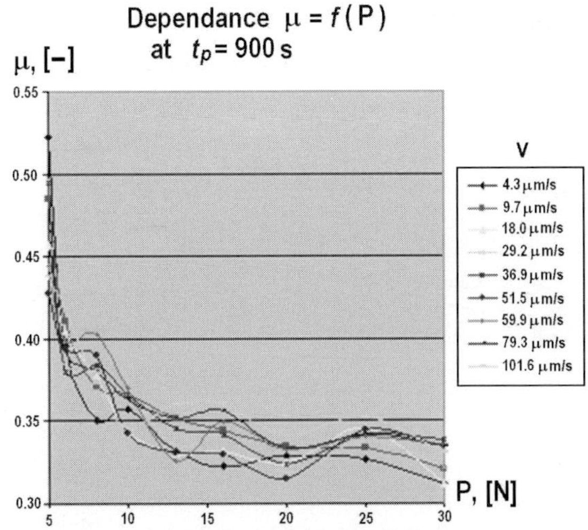

Fig. 6. Friction coefficient μ = f(P) at time of contact before test t_p=900s. Journal PA6 – bearing bush PA6.

The comparison of maximum friction coefficients μ_{max} for the tested bearings is shown in Fig. 7.

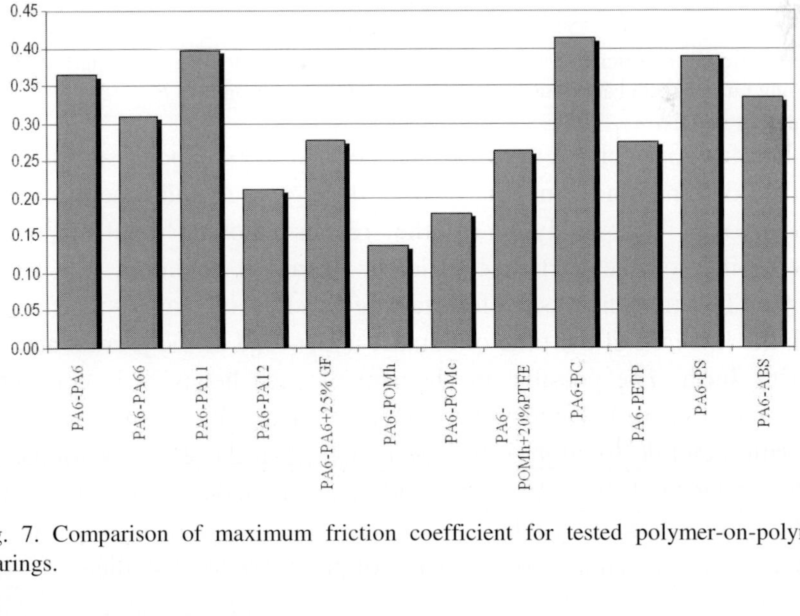

Fig. 7. Comparison of maximum friction coefficient for tested polymer-on-polymer bearings.

An example of friction coefficient vs. time of contact under load t_p for various sliding speeds and loads is shown in Fig. 8.

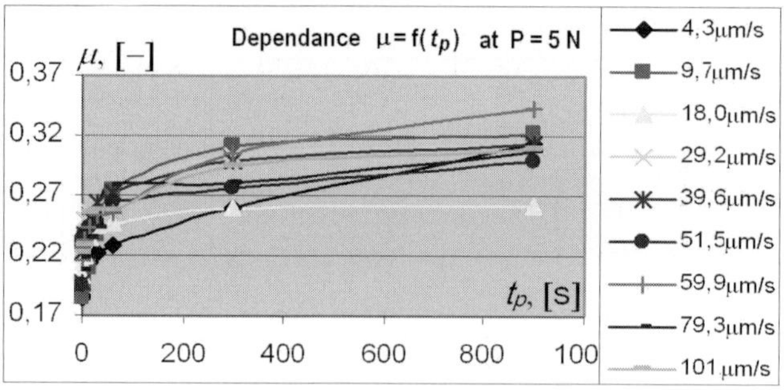

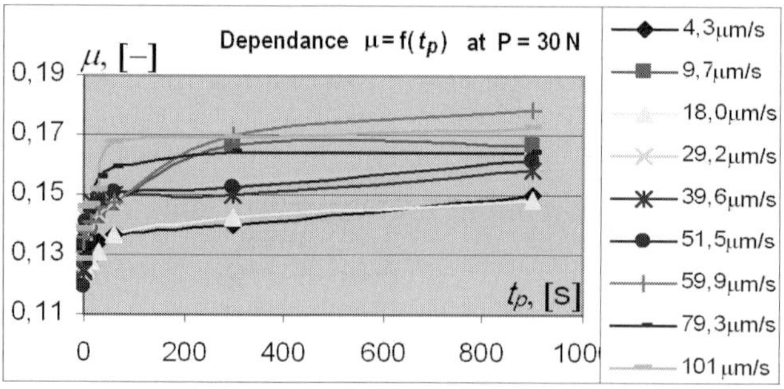

Fig. 8. Friction coefficient vs. time of contact before test for PA6 (journal) – POMc (bearing bush) bearing at various speeds and load (a) P = 5N and (b) P = 30N.

The frictional characteristics of tested bearings depend significantly on the combination of materials, load, sliding speed and time of contact under load. The analysis of the results showed relatively high repeatability of the friction coefficient (standard deviation was below ±10%). The time necessary for starting to slide increases significantly with the increase of the time of contact under load of the journal and the bearing bush. The possible mechanism of such behavior is connected with rheological behavior of polymeric materials in the phases of formation and destruction of frictional contact in the tested bearings. To better understand these processes and also for modeling of contact, the surface topography of contacting surfaces, the mechanical properties of surface layer as well as surface energy of polymers were studied.

2.2. *Investigation of Surface Topography of Rubbing Components by Use of AFM*

2.2.1. *Test Instrument and Procedure*

The use of standard profilometer to study roughness of rubbing surfaces of the bearing components was not reasonable because of high load and large radius of the stylus. The application of AFM operating in tapping mode[17-19] was reasonable. Such non-contact testing was performed in different regions of working surfaces of the journal and bearing bush. The resolution in such measurements was very high[20]: ±0.01nm in z axis, ±0.2 nm along x, y axes. The scan size was 18×18μm. The data about surface topography were registered with resolution 256×256 pixels and stored in tables. The step of scanning was $\Delta x=\Delta y=78.4$ nm.

2.2.2. *Results and Discussion*

The visualization of the results of AFM scanning was presented as 2D and 3D images of the surface. Examples of the results are shown in Fig. 9. The maximum height of asperities and the root mean square (RMS) parameter of roughness have been estimated. The studied surfaces demonstrated different roughnesses. The scanning with the use of AFM appeared very useful to obtain precise images of the polymeric surfaces of real components of the studied bearings. The 3D AFM images of the working surfaces of the journals and bearing bushes were useful in modeling of contact of the journal and bearing bush to estimate mechanical and adhesive components of friction force.

2.3. *Experimental Investigation of Elasticity Modulus of Surface Layers of Polymeric Bearing Materials*

The process of friction in the investigated bearings depends on visco-elastic behavior of bearing materials under load during contact. The process of formation of contact occurs as follow: in the moment of application of load the visco-elastic flow of material begins, which according to laws of relaxation processes, is possible to interpret as

Polymer	AFM images		Roughness parameters
	2D	3D	
1	2	3	4
1. Polystyren KM	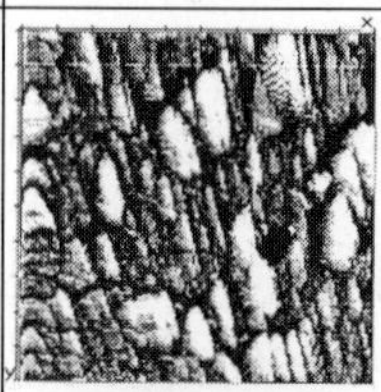	9.9 x 9.9 μm 	H – maximum height H = 355.6 nm RMS = 36 nm
2. Tarnamid T-27		3.8 x 6.2 μm 	H = 163.6 nm RMS = 18.7 nm

Fig. 9. Topography and roughness parameters for surfaces of PS (Polystyren KM) and PA6 (Tarnamid T27).

decrease of elasticity modulus $E_{(t=i)} > E_{(t=i+1)}$. As a result of decrease of elasticity modulus, the contact which was initially in equilibrium state becomes unstable. To balance of external constant load it is necessary to increase (additionally) deformation of the material which leads to a new degree of the surface deformation. This additional deformation of the material yields an increase of supporting area.[21-23] These processes are relaxation phenomena. Both chemical structure of macromolecules as well as physical state of polymers effect substantially on relaxation processes. The important parameter in the analysis of the creep processes is retardation time (t_{ret}) i.e. time needed for deformation $\varepsilon = \varepsilon_\infty \left(1 - \frac{1}{e}\right) = 0{,}632\ \varepsilon_\infty$, where: $\varepsilon, \varepsilon_\infty$ – deformation in time t and at $t \to \infty$ respectively. Such retardation time is essential in rheological models of visco-elastic materials. The preliminary results of rheological behavior of polymeric bearing materials showed that for micro-contacts the variation of elasticity modulus versus time is different from macro-scale behavior. Such characteristic $E(t)$ for the real polymeric bearing micro-components should be therefore studied. We performed an experimental investigation for the bearing materials and respective model functions of $E(t)$ were found to describe the observed visco-elastic behavior of the studied polymers in micro-contact. The variation of elasticity modulus vs. time was estimated based on creep characteristics. The tests were carried out by use of special test rig. The studies concerning the application of selected rheological models of visco-elastic materials for the description of variation of elasticity modulus in micro-contact for various polymeric materials were performed.

2.3.1. *Test Rig and Procedure*

The creep behavior was studied for real micro-components of the bearings. The following materials were tested: PA66 - Ultramid A3K (BASF, Germany), PA6 – TarnamidT27 (Z.A., Tarnów, Poland), PC – Makrolon (Bayer, Germany), POMh+20%PTFE – Delrin AF (Du Pont, USA), POM – Delrin 500NC10 (Du Pont, USA), POM – Tarnoform300 (Z.A., Tarnów, Poland).

The micro-components were uniaxially compressed under constant load. It was a creep test with long time of loading. The deformations were measured at room temperature of 20°C. The tests have been carried out by use of specially designed test rig as shown in Fig. 10.

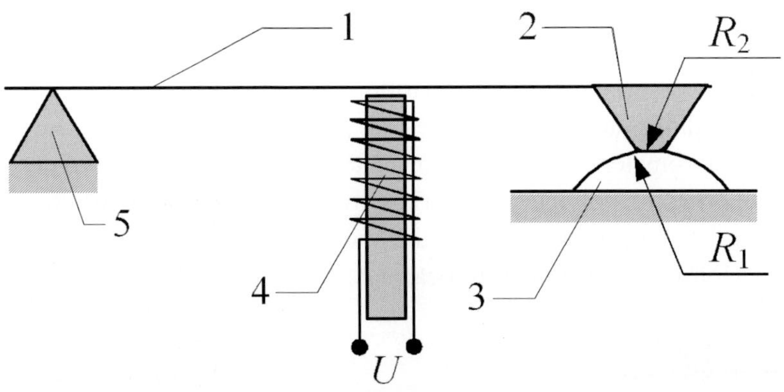

Fig. 10. Schematics of test rig for investigation of elasticity modulus. 1 – lever, 2 – indenter, 3 – specimen, 4 – electromagnet, 5 – support.

The investigated sample (part of real journal (radius R_1=1.75mm) of bearing) was fixed on the table. On one end of the lever (1) the steel spherical indenter was attached with radius R_2=1.5mm. The load was applied by electromagnet 4 and lever 1 as it is shown in Fig. 11.

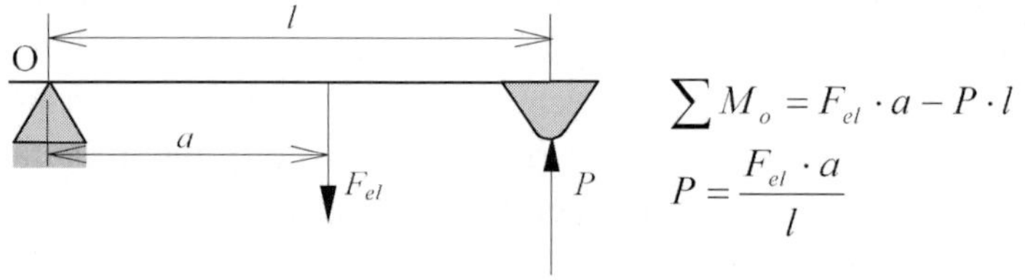

Fig. 11. Schematics of application of load in test rig (see Fig. 18).

The test were carried out at five loads: 0.01N, 0.02N, 0.03N, 0.04N and 0.05N. The time of contact was 380s. The displacement of indenter $h(t)$ vs. time was measured (Fig. 12).

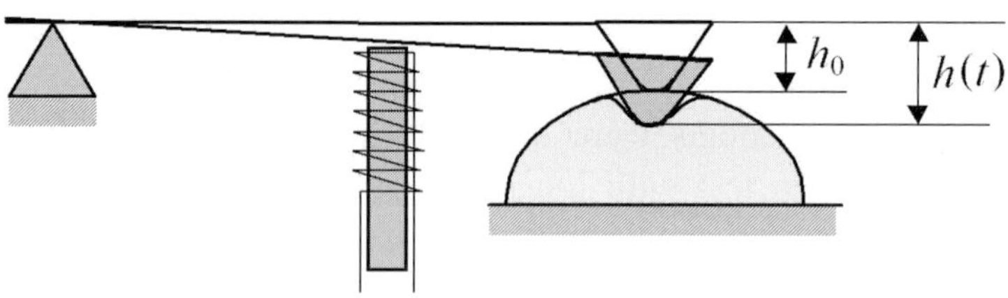

Fig. 12. Test procedure – recording of displacement $h(t)$ at constant load P.

The time of measurement (380s) was divided into 128 periods of time. The measurement of deformation was carried out after each 3s. The measured deformations of the tested polymers were in the range of several micrometers. The results of measurements can therefore describe the properties of polymeric materials in thin surface layer. Such results are useful to understand and to the model friction behavior of studied polymer-on-polymer miniature bearings.

2.3.2. *Results*

As a result of the creep test under constant load many characteristics of displacement of the indenter versus time $h(t)$ have been obtained. The plots of $h(t)$ for various load P are presented in Fig. 13. The characteristics presented in Fig. 13 are similar to typical creep characteristics of visco-elastic material.

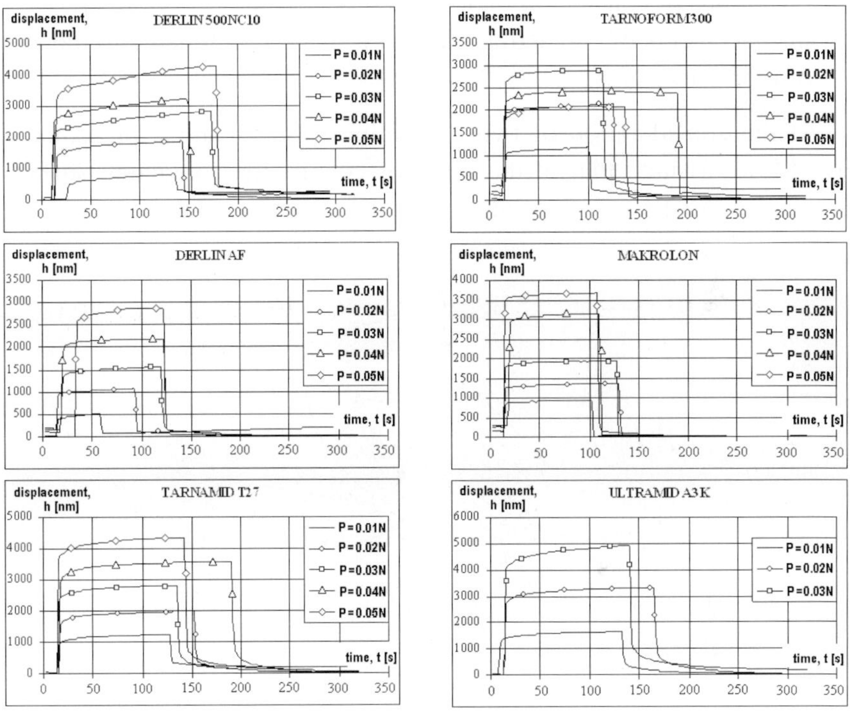

Fig. 13. Displacement of indenter vs. time $h(t)$ for various materials at loads P.

To estimate elasticity modulus (E_1) of the tested polymer at certain moment of time (t) related to known displacement of indenter ($h(t)$) the equations of contact mechanics according to Hertz theory were applied.[24] The following assumptions were taken into consideration:

- two contacting solids are completely elastic (when time needed for single measurement of displacement is very short, $t \to 0$; during this time polymer does not demonstrate visco-elastic properties but behaves as elastic solid),

- in the contact area the bodies are limited by smooth surfaces with regular curvature,
- the area of contact created by the application of load has smaller size then the surface of contacting bodies,
- on surface of contact only normal stresses occur,
- displacements of contacting solids are small as compared to the radii of their curvatures.

At the consideration of two contacting solids 1 and 2 presented in Fig. 14 we assume that in the point of contact the contacting surfaces have one normal z and perpendicular to it a tangential plane.

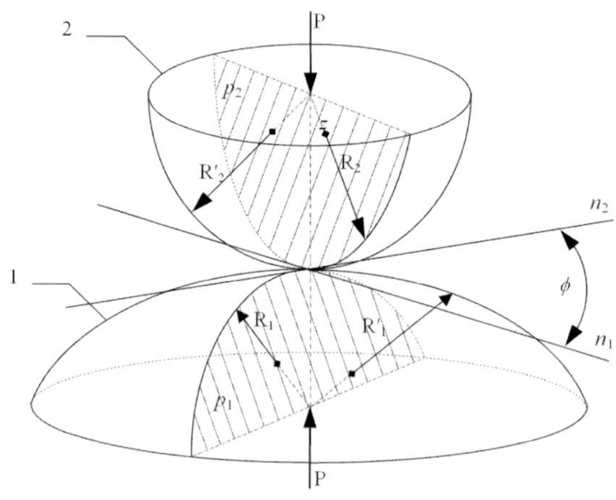

Fig. 14. Schematics of contact of two solids. R_1 – minimum radius of curvature of solid 1, R_1' – maximum radius of curvature of solid 1, R_2 – minimum radius of curvature of solid 2, R_2' – maximum radius of curvature of solid 2, p_1 – plane of maximum curvature of solid 1, p_2 – plane of maximum curvature of solid 2, n_1, n_2 – normal to maximum curvature of solid 1 and 2 in contact point, ϕ – angle between normal n_1 and n_2.

In the contact area the contacting surfaces are defined by principal radii of curvatures: minimum R_1 and maximum R'_1 for surface 1 and minimum R_2 and maximum R'_2 for surface 2. The main curvatures (main curvature for solid 1 is defined as $\dfrac{1}{R_1}$ and $\dfrac{1}{R_1'}$ and for solid 2 as $\dfrac{1}{R_2}$ and $\dfrac{1}{R_2'}$). The elastic properties of materials 1 and 2 are defined by elasticity moduli E_1, E_2 and Poisson ratios ν_1, ν_2. As the results of load of the contact by force P the local displacement of solids occurs and the approaching of the centers of curvatures of two bodies according Hertz can be defined as:[25]

$$y = \lambda \sqrt[3]{\frac{P^2}{K^2 \delta}} \tag{1}$$

where: y – approach of center of curvatures of contacting solids,

P – load (in the experiments: $P = 0.01$N, 0.02N, 0.03N, 0.04N, 0.05N),

$$K = \frac{8}{3} \frac{E_1 E_2}{E_2 \left(1 - \nu_1^2\right) + E_1 \left(1 - \nu_2^2\right)} \tag{2}$$

$$\delta = \frac{4}{\dfrac{1}{R_1} + \dfrac{1}{R_2} + \dfrac{1}{R_1'} + \dfrac{1}{R_2'}} \tag{3}$$

λ – shape coefficient (parameter given in tables) depending on angle Φ estimated from the Eq. (4):

$$\Phi = \arccos \frac{1}{4} \delta \sqrt{\left(\frac{1}{R_1} - \frac{1}{R_1'}\right)^2 + \left(\frac{1}{R_2} - \frac{1}{R_2'}\right)^2 + 2\left(\frac{1}{R_1} - \frac{1}{R_1'}\right)\left(\frac{1}{R_2} - \frac{1}{R_2'}\right) \cos 2\varphi} \tag{4}$$

where angle φ is defined in Fig. 14 as the angle between normal n_1 and n_2 to the planes of maximum curvatures of contacting bodies in the area of contact.

In Table 1 the values of shape coefficient λ as a function of angle Φ are listed.[25]

Table 1. Shape coefficient λ vs. angle Φ according [25].

Φ	0°	10°	20°	30°	35°	40°	45°	50°
λ	–	0.851	1.220	1.453	1.550	1.637	1.709	1.772
Φ	55°	60°	65°	70°	75°	80°	85°	90°
λ	1.828	1.875	1.912	1.944	1.967	1.985	1.996	2.00

We analyzed the contact of the steel indenter with spherical shape and polymeric journal (cylindrical shape). As contacting body 1 is considered the journal (cylinder: radii R_1=1.75mm and $R'_1=\infty$) and as body 2 – indenter (spherical shape with curvature radius $R_2=R'_2$=1.5mm) (Fig. 14). Then

$$\delta = \frac{4}{\frac{1}{R_1}+\frac{1}{R_2}+\frac{1}{R_1'}+\frac{1}{R_2'}} = \frac{4R_1R_2}{R_2+2R_1} = 0{,}002470\,\mathrm{mm}$$

For the contact of sphere and cylinder the angle $\varphi = 0$. The angle Φ can be calculated from Eq. (4) as

$$\theta = arccos\frac{1}{4}\delta\sqrt{\left(\frac{1}{R_1}-\frac{1}{R_1'}\right)^2+\left(\frac{1}{R_2}-\frac{1}{R_2'}\right)^2+2\left(\frac{1}{R_1}-\frac{1}{R_1'}\right)\left(\frac{1}{R_2}-\frac{1}{R_2'}\right)cos\,2\phi} = arccos\frac{1}{4}\delta\frac{1}{R_1} = 79{,}83°$$

For the calculated value $\Phi = 79.83°$ from the Table 2 the shape coefficient is $\lambda \cong 1.985$.

After introduction of these data into Eq. (1) we receive

$$y = \lambda\sqrt[3]{\frac{P^2}{K^2\cdot\delta}} = \lambda\sqrt[3]{\frac{P^2}{\frac{64}{9}\frac{E_1^2E_2^2}{\left[E_2\left(1-\nu_1^2\right)+E_1\left(1-\nu_2^2\right)\right]^2}\frac{4R_1R_2}{R_2+2R_1}}} \quad (5)$$

From Eq. (5) we extract the modulus of elasticity of the polymer material (E_1) for known displacement of indenter

$$E_1(y) = \frac{1-\nu_1^2}{\frac{8}{3}\left(\frac{4R_1R_2}{R_2+2R_1}\right)^{1/2}\cdot\lambda^{-3/2}\cdot P^{-1}\cdot y^{3/2}-\frac{1-\nu_2^2}{E_2}} \quad (6)$$

where index 1 relates to polymeric material and index 2 to steel.

By use of Eq. (6) the modulus of elasticity $E_1(t)$ in every moment of measurement was calculated at $y(t) = h(t) - \Delta h$ (where Δh – normalizing correction).

As a result of the calculations, the experimental rheological characteristics of variation of the elasticity modulus of polymeric material $E_1(t)$ in micro-contact at load have been found. The plots are presented in Fig. 15.

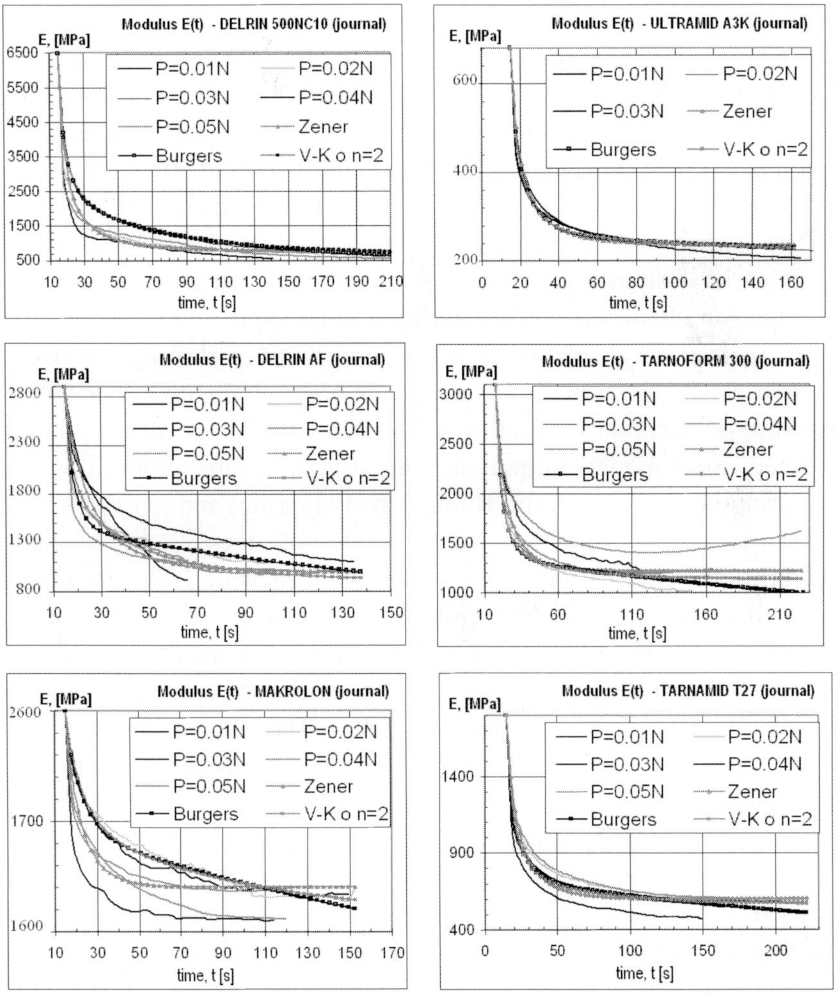

Fig. 15. Rheological variation of elasticity modulus $E(t)$ for various polymers.

2.3.2. *Discussion*

In Fig. 15 it is possible to observe exponential shape of the characteristics $E_1(t)$. Comparing the curves for the same material obtained at various loads, it is possible to see small differences resulting probably from inaccuracy of measurement method for the estimation of the displacement of the indenter or insufficient relaxation of the deformation before next series of measurements. Very big differences were found for characteristics $E_1(t)$ for various polymers. It is possible to observe rapid decrease of elasticity modulus for polyacetals POM (Delrin, Tarnoform) as compared with polyamides PA (Tarnamid, Ultramid). The reasons for that is probably higher stiffness of POM macromolecules.

The experimental testing of rheological variation of elasticity modulus of polymer is time consuming and expensive. Also the experimental characteristics are not used to be applied in modeling. The use of mathematical relationships for description of rheological models of visco-elastic materials is advantageous because simulation models of visco-elastic materials describe effectively the properties of tested polymers. Such models for polymers are usually combinations of models describing behavior of elastic Hooke's body (spring) and plastic Newton body (damper).[22] For all combinations of these models linear relationships between stress, deformation and their derivates have been found. It creates possibility to find analytically the relationship $E=f(t)$. Because of it, a trial of definition of parameters of selected rheological models to describe precisely experimental relationship elasticity modulus vs. time in the micro-contact was undertaken. The criteria for selection of these parameters are also defined. For the analysis the following rheological models of visco-elastic materials have been selected: Voigt-Kelvin with n=1 (n – number of simple Voigt elements used to compose complex model), Zener, Burgers, Voigt-Kelvin with n=2. Such models are used for polymers in visco-elastic state.[21-23] Because of the complex structure we will not use the Voigt-Kelvin model with unlimited number of n elements. The schematics of the selected models and the equations describing the rheological variation of elasticity modulus $E(t)$ for these models are presented in Table 2.

Table 2. Rheological models of viscoelastic materials according [22].

Type	Rheological model	Creep formula (deformation vs. time $\varepsilon(t)$ at constant stress σ)	Selected parameters t_{ret} – retardation time	Parameter E_a (see Table 3)
1.	2.	3.	4.	5.
Voigt-Kelvin with n=1	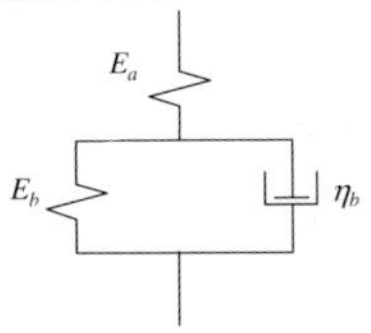 $t_{ret}=\dfrac{\eta_b}{E_b}$	$\dfrac{\varepsilon(t)}{\sigma}=\dfrac{1}{E(t)}=\dfrac{1}{E_a}+\dfrac{1}{E_b}\left(1-e^{-t/t_{ret}}\right)$	E_b t_{ret}	PA6 (Tarnamid T27): E_a=1800 MPa PA66 (Ultramid A3K): E_a=600 MPa
Zener	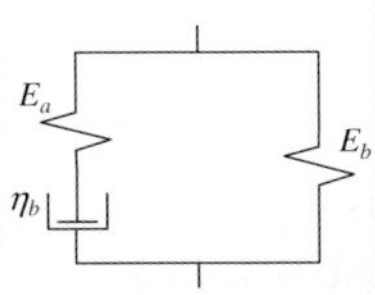 $t_{ret}=\dfrac{\eta_b}{E_b}$	$\dfrac{\varepsilon(t)}{\sigma}=\dfrac{1}{E(t)}=\dfrac{1}{E_a}e^{-t/t_{ret}}+$ $+\dfrac{1}{E_b}\left(1-e^{-t/t_{ret}}\right)$	E_b t_{ret}	PC (Makrolon): E_a=2600 MPa POMh+20%PTFE (Delrin AF): E_a=2800 MPa
Burgers	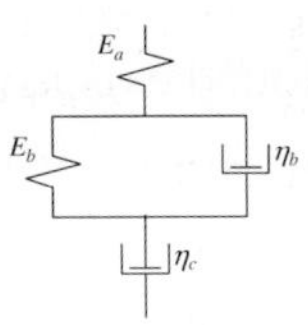 $t_{ret}=\dfrac{\eta_b}{E_b}$	$\dfrac{\varepsilon(t)}{\sigma}=\dfrac{1}{E(t)}=\dfrac{1}{E_a}+$ $+\dfrac{1}{E_b}\left(1-e^{-t/t_{ret}}\right)+\dfrac{t}{\eta_c}$	E_b t_{ret} η_c	POMh (Delrin 500NC10): E_a=6500 MPa
Voigt-Kelvin with n=2	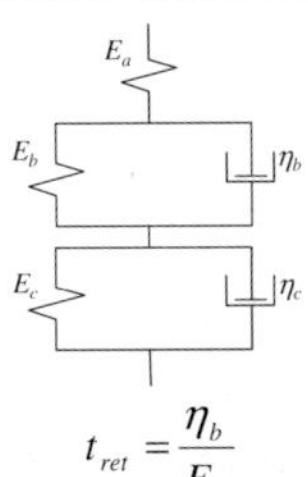 $t_{ret}=\dfrac{\eta_b}{E_b}$ $t_{ret}=\dfrac{\eta_c}{E_c}$	$\dfrac{\varepsilon(t)}{\sigma}=\dfrac{1}{E(t)}=\dfrac{1}{E_a}+$ $+\dfrac{1}{E_b}\left(1-e^{-t/t_{ret1}}\right)+$ $+\dfrac{1}{E_c}\left(1-e^{-t/t_{ret2}}\right)$	E_b t_{ret1} E_c t_{ret2}	POMc (Tarnoform 300): E_a=3100 MPa

The assumed parameter was E_a interpreted as initial elasticity modulus of polymeric material for $t = t_{pocz}$ equal to elasticity modulus at compression. This modulus was estimated experimentally by the use AFM (Table 3) according the procedure described by us elsewhere.[26]

Table 3. Elasticity modulus of selected polymers estimated by AFM.

Polymer	POMh+20%PTFE	PC	PA11	PA6
E_a[MPa]	2800	2600	2000	1800
Polymer	ABS	PS	PA66	PETP
E_a[MPa]	1400	1200	600	450

The parameters selected for the models are collected in column 4 in Table 2. After every assumption of new values of selected parameters of the analyzed model the coincidence between theoretical and experimental characteristics of elasticity modulus by use of the χ^2 test was investigated. The statistics χ^2 was calculated by use of Eq. (7):

$$\chi^2 = \sum_{t=t_{pocz}}^{t=t_{kon}} \frac{\left(E_1^{ex}(t) - E_1^{theoret}(t)\right)^2}{E_1^{teoret}(t)} \tag{7}$$

where: $E_1^{ex}(t)$ – characteristics of elasticity modulus found experimentally,
$E_1^{theoret}(t)$ – characteristics of elasticity modulus found from model equation with selected parameters.

The parameters of Voight-Kelvin model with $n = 2$ for various polymers are given in Table 4.

Table 4. Parameters of Voight-Kelvin model with $n = 2$ for different polymer materials.[38]

Polymer	E_b [MPa]	t_{ret1} [s]	E_c [MPa]	t_{ret2} [s]
Tarnamid T27 (PA6)	1470	9.9	1850	65.0
Ultramid A3K (PA 66)	721	11.8	920	160.0
Makrolon (PC)	7600	9.4	4500	110.0
Delrin AF (POM)	7500	3.8	3000	44.4
Delfin 500 NC10 (POM)	7250	5.0	735	136.4
Tarnoform 300 (POM)	2990	6.6	4550	42.5

2.4. *Experimental Studies of Surface Free Energy of Polymer*

2.4.1. *Test Rig and Experimental Procedure*

Experimental studies of surface free energy were carried out by the measurement of contact angle (Fig. 16).

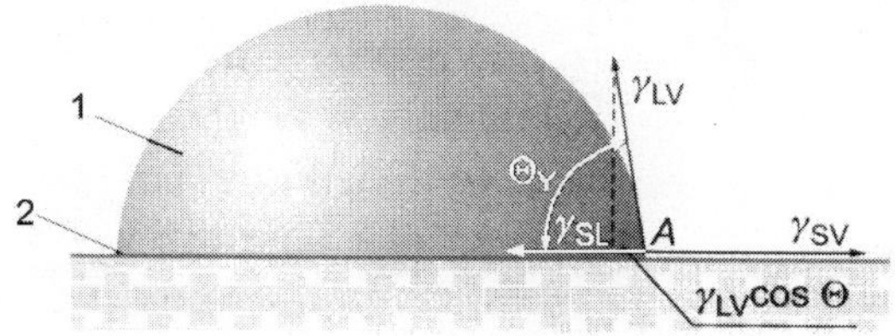

Fig. 16. Graphic presentation of Young's equation at wetting. 1 – droplet of test liquid, 2 – tested material.

The Young's equation was used to estimate the surface free energy.

$$\gamma_{SV} = \gamma_{SL} + \gamma_{LV}\cos\Theta_Y \tag{8}$$

where: γ_{SV} – interfacial energy at solid – vapor phases,
γ_{SL} – interfacial energy at solid – liquid phases,
γ_{LV} – interfacial energy at liquid – vapor phases,
Θ_Y – contact angle (Fig. 16).

The estimation of the surface free energy from Eq. (8) is not possible directly because γ_{SV} and γ_{SL} are unknown. If the adsorption at the interfacial surface is neglected the following equation is useful:

$$\gamma_S = \gamma_{SL} + \gamma_L\cos\Theta \tag{9}$$

where: γ_S – surface free energy of solid,
γ_L – surface free energy (surface tension) of liquid,
Θ – contact angle.

The measurements were carried out by the use of special test rig constructed also to study of wetting of components of micro-electromechanical systems (MEMS) (Fig. 17).[27]

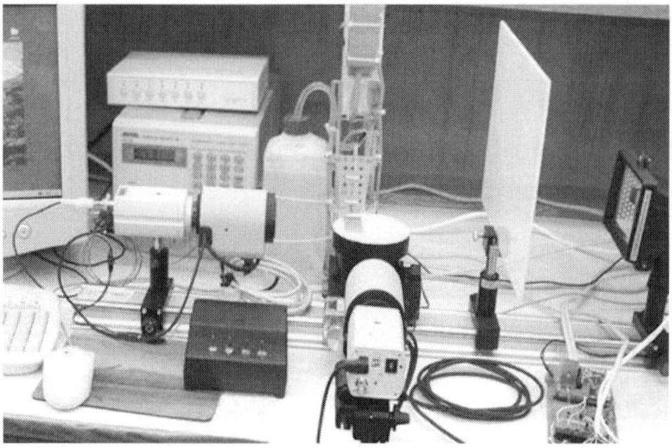

Fig. 17. Test rig to study surface energy by measurement of contact angle.

Owens-Wendt method[28] was used to estimate the surface free energy of tested polymers. The total surface free energy γ_S is composed of two components: dispersion γ_S^d and polar γ_S^p:

$$\gamma_S = \gamma_S^d + \gamma_S^p \tag{10}$$

Owens and Wendt proposed the Eq. (11) to find the interfacial energy:

$$\gamma_{SL} = \gamma_S + \gamma_L - 2(\gamma_S^d \gamma_L^d)^{0.5} - 2(\gamma_S^p \gamma_L^p)^{0.5} \tag{11}$$

From equations (10) and (11) results that:

$$\gamma_L \frac{1+\cos\Theta}{2} = (\gamma_S^d \gamma_L^d)^{0.5} + (\gamma_S^p \gamma_L^p)^{0.5} \tag{12}$$

To estimate γ_S there were measured contact angles Θ for two liquids with known γ_L^d and γ_L^p and the system of two equations was solved. In the studies as test liquids water and diiodomethane were applied with the following parameters: $\gamma_L = 72.8$ mJ/m^2, $\gamma_L^d = 21.8$ mJ/m^2, $\gamma_L^p = 51.0$ mJ/m^2 for water and $\gamma_L = 50.8$ mJ/m^2, $\gamma_L^d = 48.5$ mJ/m^2, $\gamma_L^p = 2.3$ mJ/m^2 for diiodomethane. Taking into account dispersion component γ_S^d and polar component γ_S^p of the surface of the used liquids the dispersion and polar components of the surface energy of tested polymer were found from the following equation:

$$(\gamma_S^d)^{0.5} + 1.53(\gamma_S^p)^{0.5} = 7.80(1+\cos\Theta_1)$$
$$(\gamma_S^d)^{0.5} + 0.22(\gamma_S^p)^{0.5} = 3.65(1+\cos\Theta_2) \quad (13)$$

where: Θ_1 – contact angle for water,
Θ_2 – contact angle for diiodomethane.

2.4.2. *Results and Discussion*

In Table 5 are listed results of measurements for selected polymers. The used method appeared to be sensitive enough to find relatively small differences between the surface free energies of tested polymers.

The estimated values of the surface free energy are useful in recognizing adhesive properties of the tested polymers and in further modeling of adhesive interactions between rubbing surfaces for predicting the coefficient of friction.

Table 5. Surface energy estimated by measurement of contact angle.

Polymer	Polar component of surface energy, γ_S^p [mJ/m^2]	Dispersion component of surface energy, γ_S^d [mJ/m^2]	Surface energy, γ_s [mJ/m^2]
PA6 (Tarnamid T27)	33.8	38.3	72.1
PA66 (Ultramid)	32.8	30.6	63.4
POMc (Tarnoform)	32.2	32.7	64.9
POMh (Delrin 500NC10)	33.7	26.8	60.5
POMh+20% PTFE (Delrin AF)	39.6	8.8	48.4

3. Model of Contact and Predicting of Friction Coefficient

The computer simulation of the process of formation of RAC between rubbing surfaces of analyzed miniature bearing was carried out. The rheological process is taken into consideration. The elaborated method enabled the identification of visco-elastic deformations in the area of contact of polymeric materials as well the effect of these deformations on the friction coefficient. The model was projected to predict friction coefficient as a function of selected parameters of frictional process. These parameters are: load P, time of contact under load before test t_p, roughness of contacting surfaces (studied by AFM) and selected mechanical properties of applied polymers (v.e. elasticity modulus E, Poisson's ratio) and surface free energy.

3.1. *Model of Force Interactions between Contacting Polymeric Surfaces — Static Contact*

3.1.1. *Model of Formation of Real Area of Contact between Elastic Rough Surface and Ideal Flat Stiff Surface*

We assume that the contact is under constant external load P, distance between contacting surfaces is d and topography of rough surface can be described by the model of elastic Winkler's layers (Fig. 18).

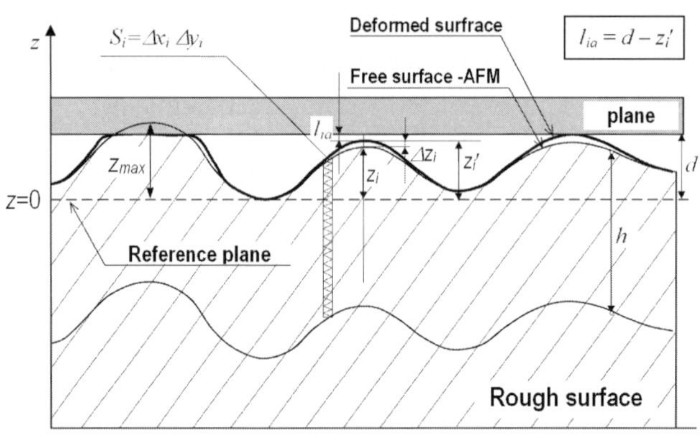

Fig. 18. Model of elastic Winkler's layers.

According to this model in the case of free surface, every single spring simulates column of material with height h. Height h relates to such thickness of surface layer of material in which processes of elastic deformations occur. We assume that height h of all modeling springs is the same and depends on mechanical properties of surface layer of material. We assume that there are no interactions between modeling springs. Because in modeling the AFM images of the surface topography are used in the Winkler's model we assume that the base area of single spring relates to elementary discrete area S_i of the AFM image (where $S_i=\Delta x \cdot \Delta y$; and Δx and Δy – step of AFM scanning along x and y axes). For description of topography of the free surface and process of its deformation the idea of base reference plane ($z = 0$) is introduced which is defined as the plane in which the lowest point of topography of the free surface is located. In relation to reference plane are defined the distance between contacting planes d and height of topography of free surface (z_i) as well as deformed plane (z_i').

After loading of the contact, some of the springs modeling the rough surface deform elastically by Δz_i. Elastic deformation Δz_i depending on type of contact (mechanical or adhesive) can be interpreted as compression (in the case of mechanical contact) or elongation (in the case of adhesive contact).

The springs on which direct mechanical contact with flat surface occurs (i.e. when the height of topography of free plane z_i is higher then distance d) are compressed with the elastic deformation $\Delta z_i=d-z_i$. According to the principle of similarity the unit force P_i which reacts against elastic deformation of single spring can be expressed as follows:[24]

$$P_i(\Delta z_i) = \frac{E}{kh} \cdot \Delta z_i \tag{14}$$

where: $k = \dfrac{(1+\nu)\cdot(1-2\nu)}{(1-\nu^2)}$

(ν - Poisson's ratio of material of rough surface)

Summarizing forces P_i estimated for all compressed springs being in mechanical contact with the plane we obtain the total contact force of mechanical interaction P_s as

$$P_s(d) = \frac{E}{kh} \sum_{i=1}^{n} \Delta S_i (d - z_i) \tag{15}$$

where: n – number of compressed modeling springs being in mechanical contact with plane, i.e. the following condition is fulfilled $d\text{-}z_i<\varepsilon$ (where ε–molecular distance according Lenard-Jones equation).

For the same steps of scanning by use of AFM the force P_s can be described as

$$\boldsymbol{P}_s(\boldsymbol{d}) = \frac{\boldsymbol{E}}{\boldsymbol{kh}} \cdot \Delta \boldsymbol{x} \cdot \Delta \boldsymbol{y} \cdot \sum_{i=1}^{n} (\boldsymbol{d} - z_i) \tag{16}$$

The modeling springs for which adhesive contact occurs (i.e. when the heights of topography of free surface z_i is lower than d) are elongated under forces of adhesive interactions. It was assumed that the force of molecular interaction between modeling spring and plane, which are separated by distance l_i=d- z_i, is described by the relationship (17):

$$F(l_i) = \frac{8}{3} \cdot \frac{\Delta\gamma}{\varepsilon} \cdot \left[\left(\frac{\varepsilon}{l_i} \right)^3 - \left(\frac{\varepsilon}{l_i} \right)^9 \right] \tag{17}$$

where: ε – molecular distance according Lenard-Jones equation.[24,37]

In Eq. (17) the component $(\varepsilon/l_i)^9$ characterizing van der Waals interactions, is relatively small as compared with components $(\varepsilon/l_i)^3$. Therefore at the distance higher then intermolecular distance during estimation of total force of intermolecular interactions the above mentioned component can be omitted. We receive finely the following equation:

$$\boldsymbol{F}(\boldsymbol{l}_i) = \frac{8}{3} \cdot \frac{\Delta\boldsymbol{\gamma}}{\boldsymbol{\varepsilon}} \cdot \left(\frac{\boldsymbol{\varepsilon}}{\boldsymbol{l}_i} \right)^3 = \frac{8}{3} \cdot \Delta\boldsymbol{\gamma} \cdot \frac{\boldsymbol{\varepsilon}^2}{\boldsymbol{l}_i^3} \tag{18}$$

At distance l_i>1.5ε the introduced error is smaller than 10%. The force $F(l_i)$ is attractive force and it effects on elongation of spring by Δz_i. The value of Δz_i can be estimated from the equilibrium of forces

$$F(l_i) = \frac{E}{kh} \cdot \Delta z_i \tag{19}$$

We obtain therefore

$$\Delta z_i = \frac{8}{3E} \Delta\gamma \cdot k \cdot h \cdot \varepsilon^2 \cdot \frac{1}{(l_i)^3} \tag{20}$$

We have to remember that in the assumed simplifications the value Δz_i is correct at l_i>1.5ε. For l_i<1.5ε in the condition of equilibrium of forces the complete formula (17) should be applied.

After estimation of elastic deformation Δz_i we can define the height of topography of rough deformed surface z_i' for every elongated spring as

$$z_i' = z_i + \Delta z_i \tag{21}$$

Finely, the adhesive attraction force between single elongated spring and plane can be expressed by Eq. (22)

$$\boldsymbol{F}_i = \frac{8}{3}\Delta\gamma\varepsilon^2 \frac{1}{(\boldsymbol{d} - z_i')^3} \tag{22}$$

By summation of force F_i for elongated modeling springs being in adhesive contact with plane (for these springs $d-z_i'$>1.5ε) we obtain total non-contact attractive (adhesive) force F_s as

$$\boldsymbol{F}_s(\boldsymbol{d}) = \frac{8}{3}\sum_{i=1}^{m} \Delta\gamma \cdot \varepsilon^2 \cdot \Delta \boldsymbol{S}_i (\boldsymbol{d} - z_i')^{-3} \tag{23}$$

where: m – number of elongated modeling springs being in adhesive contact with plane, i.e. the condition $d-z_i'$>1.5ε is fulfilled.

For scanning points of surface topography the force F_s can be expressed as

$$\boldsymbol{F}_s(\boldsymbol{d}) = \frac{8}{3} \cdot \Delta x \cdot \Delta y \cdot \sum_{i=1}^{m} \Delta\gamma \cdot \varepsilon^2 \cdot (\boldsymbol{d} - z_i')^{-3} \tag{24}$$

Non-contact attractive force F_s and contact force P_s do not balance because randomly selected distance d is not related to the equilibrium situation. General equation is therefore written as

$$F_s(d) - P_s(d) = P \tag{25}$$

In Eq. (25) depending on minus or plus acting forces the resulting force P relates to external force which is necessary to approach surfaces to the distance d, or external force acting into direction of separation of surfaces, which is necessary to keep given distance between surfaces.

If the external force P is known it is possible to find distance $d=d_0$ to obtain equilibrium. The estimated distance is the real distance between reference plane of deformed rough surface and flat in the case of loaded contact. At this distance the supporting area of contact A_r, called real area of contact (RAC) can be treated as a sum of Winkler's layers, which are compressed $A_r = \sum_{i=1}^{n} \Delta S_i$ or as area related to uniformly distributed scan points of surface topography $A_r = \Delta x \cdot \Delta y \cdot n$.

3.1.2. *Model of Formation of Real Area of Contact between Two Elastic Rough Surfaces*

It was assumed that at contact an external load N is applied, contacting surfaces are separated by certain distance d and topographies of rough surfaces can be described by models of elastic Winkler's layers. The distance d is measured between reference planes of both contacting rough surfaces.

If the values of heights of contacting surfaces z_1 and z_2 in defined scanning points are known, therefore by application of the modeling procedure afore-described, we can easily find areas of mechanical and adhesive contact, which will appear during approaching the distance d. To carry-out computer analysis of contact it is necessary to fulfill two conditions: first one needs the same size of scan range of both analyzed surfaces and second one the application of the same scan steps Δx and Δy in x and y directions respectively.

By introduction of the equivalent rough surface z^*, the height of topography is equal to sum of heights of topography of surfaces 1 and 2 in respective scanning points $z_i^* = z_{1i} + z_{2i}$, it is possible to assume that the problem of modeling of the contact of two rough surfaces is reduced to modeling of contact of the equivalent rough surface z_i^* with the flat surface.

3.1.3. *Model of Formation of Real Area of Contact of Two Rough Surfaces with Taking into Account their Visco-elastic Properties*

The process of formation of real area of contact in the case of visco-elastic materials is a continuous process. It begins in the moment of the application to contacting surfaces of external load P. Visco-elastic deformation (creep) of material results from general rules for relaxation processes and it is connected with decrease of elasticity modulus of loaded polymeric material $E(t_{i+1})<E(t_i)$. The decrease of the elasticity modulus results in a decrease in the stiffness of material and load capacity of the contact and as a result the equilibrium at beginning of contact becomes unstable. To equilibrate the contact again, additional deformation of microasperities leading to increase of real area of contact (load carrying surface) is necessary. The result of this process is decrease of the distance d between contacting surfaces. As a result the process of formation of visco-elastic contact can be characterized by the time sequences of positions of contacting surfaces.

In the elaborated algorithm of computer modeling we assumed the relationship describing time dependence of elasticity modulus of polymeric materials as the conditions of their visco-elastic deformation in the form:[22]

$$\frac{1}{E(t)}=\frac{1}{E_a}+\frac{1}{E_b}\left(1-e^{-t/t_{ret1}}\right)+\frac{1}{E_c}\left(1-e^{-t/t_{ret2}}\right) \tag{26}$$

where: E_a – elasticity modulus of polymeric material estimated on free surface by the AFM,

E_b, E_c, t_{ret1}, t_{ret2} – parameters of Voigt-Kelvin model with n=2 (Table 2),

t – time.

Numerical calculations of the visco-elastic formation of the contact of rough surfaces have been carried out according the following algorithm described below.

For the elasticity modulus E_a and initial topography of equivalent surface z^* is calculated, then initial distance d_0 between surfaces and real area of contact A_{r0} related to external load N, and finally the topography $z^{*\prime}(t_0=0)$ is found which describes the equivalent deformed surface for the time $t_0=0$.

The time step Δt is the time after which the next calculations are performed. By this way the discretization of the continuous process is introduced in which $t_i=t_{i-1}+\Delta t$.

The function of time variation of elasticity modulus is defined according the Eq. (13).

The new value of elasticity modulus E_i in i period of time is then found. For the new value of E_i the new deformed topography $z^{*\prime}(t_0=t_i)$ describing equivalent surface in i-th time period is identified. The calculations of the distance between surfaces d_i and real area of contact A_{ri} for the plane described by the topography $z^{*\prime}(t_0=t_i)$ with taking into account new value of elasticity modulus E_i and constant external force N are then performed.

In the case of consideration of the process of formation of contact during longer period of time we come back to paragraph 2 of the algorithm. In every next calculation step the initial topography $z^{*\prime}(t=t_i)$ is assumed as the result of the last step.

On the base of the afore-mentioned algorithm the computer program was written which enables the realization of calculations and visualization of obtained results of modeling of process of formation of visco-elastic contact.

3.2. *Model of Force Interactions Occurring between Polymeric Rubbing Surfaces — Phase of Transition from Stationary State to Movement*

It is assumed that during relative tangential movement of rubbing surfaces the shearing process in elastic regime occurs on all real area of contact. Final output of the real area of contact can be described for any combination of materials with known topographies and physical properties by the computer simulation.

3.2.1. *Model of Force Interactions between Elastic Rough and Ideally Flat and Stiff Plane*

For the estimation of total friction force T necessary to initiate relative displacement (slip) between contacting surfaces on the area related to scan area of both contacting surfaces, it is necessary to separately analyze phenomena occurring during contact for:

- elementary areas $S_i=\Delta x\cdot\Delta y$ of real area of contact for which direct contact does not occurs but the distance l_{ia} (Fig. 1) is close to intermolecular distances – estimation non contact (adhesive) component of friction force.
- elementary areas $S_i=\Delta x\cdot\Delta y$ of real area of contact on which direct contact occurs – estimation of contact component of friction force.

Estimation of non-contact component of friction force

Taking into account the movement of non-deforming plane relatively to rough surface it can be notice that for elementary areas of surface $S_i=\Delta x\cdot\Delta y$ of real area of contact, for which direct contact does not occur but the distance l_{ia} is in the range of intermolecular adhesive forces, bonding of moving surfaces occur and shearing of elastic Winkler's layer appears. The model of Winkler's layer during shearing is presented in Fig. 19 [38].

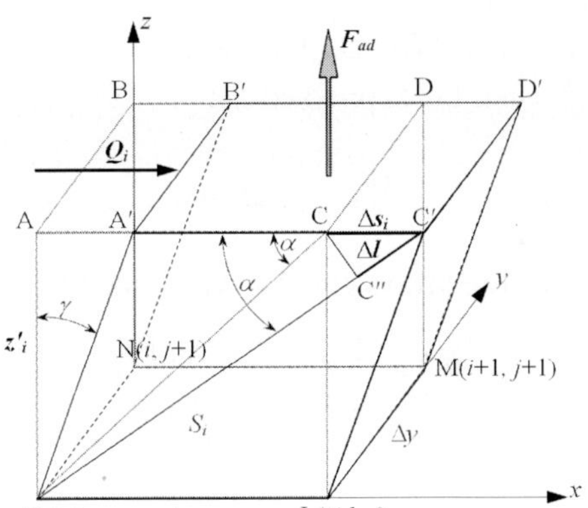

Fig. 19. Model of shearing of elastic Winkler's layer.

After consideration of deformation of element relating to single Winkler's layer it is possible to arrive finely to relationship of lateral shearing load Q_i with displacement Δs_i:

$$\Delta s_i = \frac{1}{2CG}\cdot\frac{Q_i}{S_i}\cdot z_i' \tag{27}$$

where: $C = \dfrac{tg\alpha}{tg^2\alpha + 1}$ and $tg\alpha = \dfrac{z_i'}{\Delta x}$

z_i' – height of deformed Winkler's layer,

Δx – scanning step along sliding direction,

$G = \dfrac{E}{2(1+\nu)}$ – shear modulus of polymeric material,

$S_i = \Delta x \cdot \Delta y$ – area of single Winkler's layer (Δy – scanning step along perpendicular direction to sliding direction).

Relationship (27) is true only for isotropic material. For polymers, this can be applied in nanometer scale assuming that for every separate elastic Winkler's layer material parameters (elasticity modules G, E and Poisson's ratio ν) are constant and that Winkler's layer is isotropic. If these assumptions cannot be fulfilled on all real area of contact it is necessary to investigate distribution of these parameters in nanoscale (e.g. by the use of AFM).

Assumption that shear force Q is acting statically the work of this force on distance Δs_i can be expressed as:

$$W_i = \frac{1}{2} Q_i \Delta s_i \tag{28}$$

By introducing to Eq. (28) the value of Δs_i according (27) we obtain

$$W_i = \frac{1}{4} Q_i^2 \cdot z_i' \cdot \frac{1}{CGS_i} \tag{29}$$

For relative displacement between contacting surfaces (slip) the value of work of shear realized by force W_i should be over the energy of molecular interactions E_1 then destruction of Winkler's layer occurs or if energy of molecular interaction is over $E_{1\text{-}2}$ then destruction of adhesive bond appears. This condition can be expressed as follows

$$W_i > En \tag{30}$$

where: $En = \min(E_1, E_{1\text{-}2})$.

By assuming that the friction process of polymeric tribosystem occurs without generation of wear particles (assumption of elastic deformation of Winkler's layer) for sliding remains the conditions as follows

$$W_i > E_{1\text{-}2} \tag{31}$$

Energy $E_{1\text{-}2}$ needed to break of contact can be estimated as the work of van der Waals forces on the distance from l_{ia} to ∞. Considering that for polymeric materials van der Waals forces are described by Eq. (4) and the initial gap can be described as

$$l_{ia} = d - z_i' \tag{32}$$

where: d – distance between contacting surfaces measured between surface of non-deformed flat surface and reference plane of rough surface,

z_i' – height of deformed Winkler's layer.

En can be described as follows:

$$En = E_{1-2} = \int_{l_{ia}=d-z_i'}^{\infty} \frac{8}{3}\frac{\Delta\gamma}{\varepsilon}\left[\left(\frac{\varepsilon}{l_i}\right)^3 - \left(\frac{\varepsilon}{l_i}\right)^9\right] dl_i \tag{33}$$

By introduction (29) and (33) to the condition (31) we obtain

$$\frac{1}{4} Q_i^2 \cdot z_i' \cdot \frac{1}{CGS_i} > \int_{l_i=d-z_i'}^{\infty} \frac{8}{3}\frac{\Delta\gamma}{\varepsilon}\left[\left(\frac{\varepsilon}{l_i}\right)^3 - \left(\frac{\varepsilon}{l_i}\right)^9\right] dl_i \tag{34}$$

From Eq. (34), at known parameters of deformed topography of rough surface (z_i', Δx, Δy) and known material parameters of both contacting surfaces (G, $\Delta\gamma$, ε), it is possible to estimate minimum shear force Q_i necessary to deform (slip) tangentially contacting surfaces on the area S_i relating to area of basement of single Winkler's layer

$$Q_i^2 > \frac{1}{z_i'} \cdot 4CGS_i \cdot \int_{l_{ia}=d-z_i'}^{\infty} \frac{8}{3}\frac{\Delta\gamma}{\varepsilon}\left[\left(\frac{\varepsilon}{l_i}\right)^3 - \left(\frac{\varepsilon}{l_i}\right)^9\right] dl_i \tag{35}$$

Summation of forces Q_i , estimated from (35), for all Winkler's layer for the condition of absence of direct contact, when l_{ia} is in the range of several intermolecular distance ($l_{ia} \sim \varepsilon$, $l_{ia} \geq \varepsilon$), it gives the component of non-contact friction force T_{ad} resulting from adhesive interaction of contacting polymeric surfaces. It can be written that:

$$T_{ad} = \Sigma\, Q_i \text{ for all Winkler's layers for which } l_{ia} = d - z_i' \geq \varepsilon$$

Estimation of contact component of friction force

To estimate contact component of friction force the destruction of contact for elementary areas of contact $S_i = \Delta x \cdot \Delta y$ of real area of contact for which direct contact occurs i.e. $l_i = d - z_i < \varepsilon$. In this case between contacting surfaces are acting normal forces of mechanical interaction P_i effecting in appearance of contact pressure. The contact pressure for one Winkler's layer (P_i/S_i) can be defined according the algorithm elaborated for computer simulation of formation of stationary contact. When normal force of mechanical interaction (P_i) is known the distance l_{im} can be estimated at which the force of intermolecular interaction $F(l_i)$ (van der Waals force) defined by Eq. (17) will form the equilibrium with this interaction. It can be written that $P_i = F(l_{im})$ i.e.:

$$\boldsymbol{P_i} = \frac{8}{3} \cdot \frac{\Delta\gamma}{\varepsilon} \cdot \left[\left(\frac{\varepsilon}{l_{im}} \right)^3 - \left(\frac{\varepsilon}{l_{im}} \right)^9 \right] \tag{36}$$

From Eq. (36) by iteration the l_{im} was estimated.

For sliding i.e. relative displacement between contacting surfaces the work done by shear force W_i defined by Eq. (29) must be equivalent to energy of intermolecular interaction i.e. energy needed to change the distance from l_{im} to ε ($E_{1\text{-}1}$) and lead to break adhesive contact for initial gap $l_i = \varepsilon$ ($E_{1\text{-}2}$) i.e. change of distance between molecules from ε to ∞. This condition can be expressed us:

$$W_i > E_{1\text{-}1} + E_{1\text{-}2} \tag{37}$$

where:

$$E_{1-1} = \int_{l_{im}}^{\varepsilon} \frac{8}{3} \frac{\Delta\gamma}{\varepsilon} \left[\left(\frac{\varepsilon}{l_i} \right)^3 - \left(\frac{\varepsilon}{l_i} \right)^9 \right] dl_i \tag{38}$$

and

$$E_{1-2} = \int_{\varepsilon}^{\infty} \frac{8}{3}\frac{\Delta\gamma}{\varepsilon}\left[\left(\frac{\varepsilon}{l_i}\right)^3 - \left(\frac{\varepsilon}{l_i}\right)^9\right] dl_i \tag{39}$$

Therefore

$$W_i > \int_{l_{im}}^{\infty} \frac{8}{3}\frac{\Delta\gamma}{\varepsilon}\left[\left(\frac{\varepsilon}{l_i}\right)^3 - \left(\frac{\varepsilon}{l_i}\right)^9\right] dl_i \tag{40}$$

Introducing into (40) the value W_i estimated from (29) in which z_i' must be treated as d, we will receive:

$$\frac{1}{4} Q_i^2 \cdot d \cdot \frac{1}{CGS_i} > \int_{l_{im}}^{\infty} \frac{8}{3}\frac{\Delta\gamma}{\varepsilon}\left[\left(\frac{\varepsilon}{l_i}\right)^3 - \left(\frac{\varepsilon}{l_i}\right)^9\right] dl_i \tag{41}$$

From Eq. (41), for well-known parameters of deformed topography of rough surface (z_i'=d, Δx, S_i) and known material properties of contacting surfaces (G, $\Delta\gamma$, ε), it is possible to estimate minimum shear force Q_i necessary to separate contacting surfaces on area S_i relating to the area of single Winkler's layer as

$$Q_i^2 > \frac{1}{z} \cdot 4CGS_i \int_{l_{im}}^{\infty} \frac{8}{3}\frac{\Delta\gamma}{\varepsilon}\left[\left(\frac{\varepsilon}{l_i}\right)^3 - \left(\frac{\varepsilon}{l_i}\right)^9\right] dl_i \tag{42}$$

Summation of forces W_i, found from (42), for all Winkler's layers for the conditions of indirect contact, i.e. $l_i = d - z_i < \varepsilon$ it gives the contact component of friction T_{mech} resulting from mechanical interaction of contacting polymeric surfaces. It can be written therefore that contact component of friction force is $T_{mech} = \Sigma\, Q_i$ for Winkler's layer, for which $l_i = d - z_i < \varepsilon$.

Estimation of total friction force

Total friction force T relates to shear force necessary for lateral displacement (sliding) between contacting surfaces on the area of contact (scanning are). Total friction force is composed of two components: adhesive component T_{ad} and mechanical component T_{mech}:

$$T = T_{ad} + T_{mech} \tag{43}$$

3.2.2. *Model of Force Interactions between Two Rough and Elastic Surfaces*

Definition of discrete contact of two rough surfaces is very complex problem for statistical modeling. It reduces to the definition of the parameters of certain equivalent parameters or to the definition of probability function of microasperity contacts. The problem simplifies when there is known numerical function of heights of topography of contacting surfaces $z_1(x, y)$ and $z_2(x, y)$. Such functions are set of 3D images of surface obtained by scanning by use of AFM microscope. For computer analysis of contact of two rough surfaces it is necessary to select areas of two surfaces with same size and scan the surface by AFM with the same step Δx and Δy along x and y axes respectively. By assuming that the distance between two reference planes of two surfaces is d it is possible to predict the mechanical compression that takes place at contacts at scanning points (i, j) in which the following inequality:

$$z_1(x, y) + \Delta z_1(x, y) + z_2(x, y) + \Delta z_2(x, y) > d \tag{44}$$

where: $\Delta z_1(x, y)$, $\Delta z_2(x, y)$ – elastic deformation of surface 1 and 2 respectively

The contact between two rough surfaces can be interpreted as contact between rough equivalent surfaces with topography $z^*(x, y) = z_1(x, y) + z_2(x, y)$ and absolutely flat plane. Further modeling of the contact of two rough surfaces can be realized as it is described in Section 3.2.1.

3.2.3. *Model of Force Interactions between Two Rough Visco-elastic Surfaces*

The process of formation of real area of contact between two polymeric rough surfaces with the consideration of their visco-elasticity in the case of static contact loaded by normal force has been already described in Section 3.1.3. During analysis of the phenomena occurring at the initiation of relative displacement of rubbing surfaces, i.e. the process of transition from static to kinetic friction, it is assuming that tangential load of contact effects in visco-elastic deformations of microasperities. The sequence of visco-elastic displacement versus time is described with the use of rheological function of variation of elasticity modulus of polymer. Rheological processes occurring during formation of real area of contact results in an increase in the coefficient of friction. Basing on Boltzman's superposition we assume that during transition from stationary state to movement both lateral and normal visco-elastic deformations of polymeric materials are additive.

3.3. *Results of Modeling*

The final result of calculations is used to predict friction force between the journal and bearing bush. The plots presented below demonstrate the contribution of adhesive and mechanical components of friction force (Fig. 20). The plots can be applied to study the effect of stationary state of the bearing under load on starting characteristics (Fig. 21).

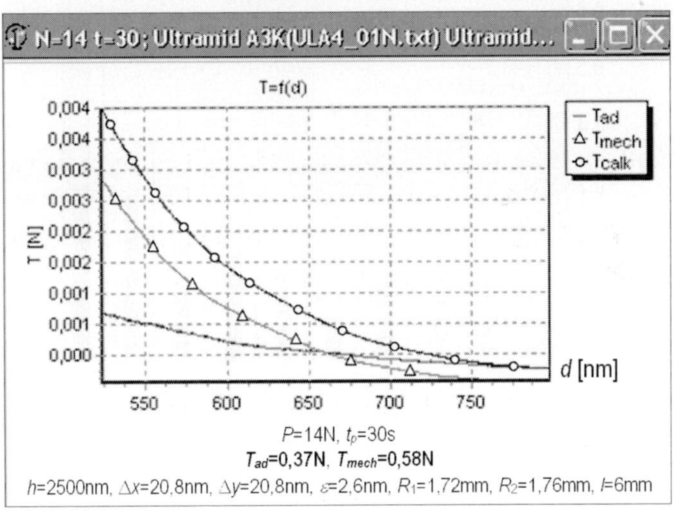

Fig. 20. Friction force T (adhesion T_{ad} and mechanical T_{mech} components) in bearing as function of distance d. Journal: PA66 (Ultramid A3K), bearing bush: Ultramid A3K, load of bearing P=14N, time of contact under load t_p=30s, thickness of Winkler's layer h=2500nm, step of scanning $\Delta x=\Delta y$=20.8nm, equilibrium intermolecular distance ε=2.6nm, radius of journal R_1=1.72mm, radius of bearing bush R_2=1.76mm, length of bearing l=6mm.

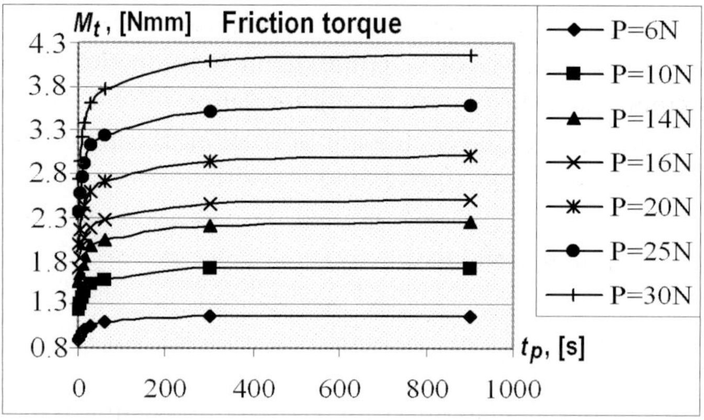

Fig. 21. Starting torque M_t in bearing vs. time of contact under load t_p for various loads P. Journal PA66 (Ultramid A3K), bearing bush PA66 (Ultramid A3K).

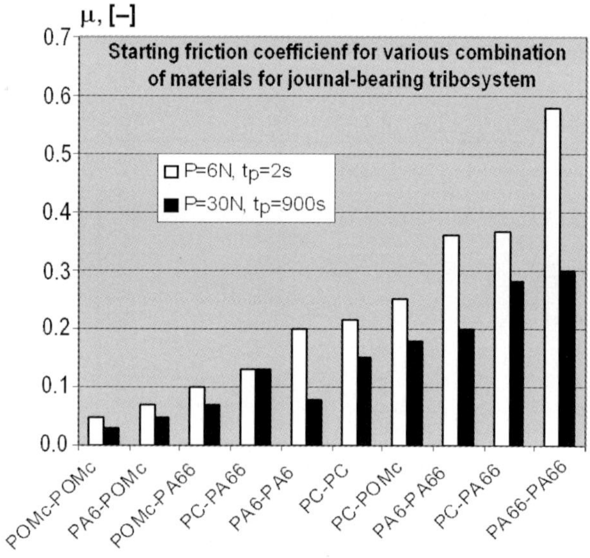

Fig. 22. Comparison of starting friction coefficient μ for various materials of journal and bearing bush.

The numerical calculations can be carried out for any combination of materials of journal and bearing bush. For realization of calculations it is necessary to know the nano-topography of both contacting surfaces, parameters described in Section 3.1.1 as input data and geometrical dimensions of the bearing and applied load. It is possible therefore to realize optimum selection of materials of journal and bearing bush to obtain low torque. In Fig. 22 comparison of initial friction coefficient μ (defined as ratio of total friction force $T = T_{ad} + T_{mech}$ to load of the bearing P) predicted for various combinations of materials of journal and bearing bush is presented.

The friction coefficients were calculated for selected loads P and selected times of contact under load before operation of the bearing t_p.

The realization of numerical calculations for wide range of loads and time of contact under load enables to obtain a family of characteristics describing starting phase of operation of a microbearing. Examples are presented in Fig. 23 for PA6-PA6 and PA6-POMc bearings.

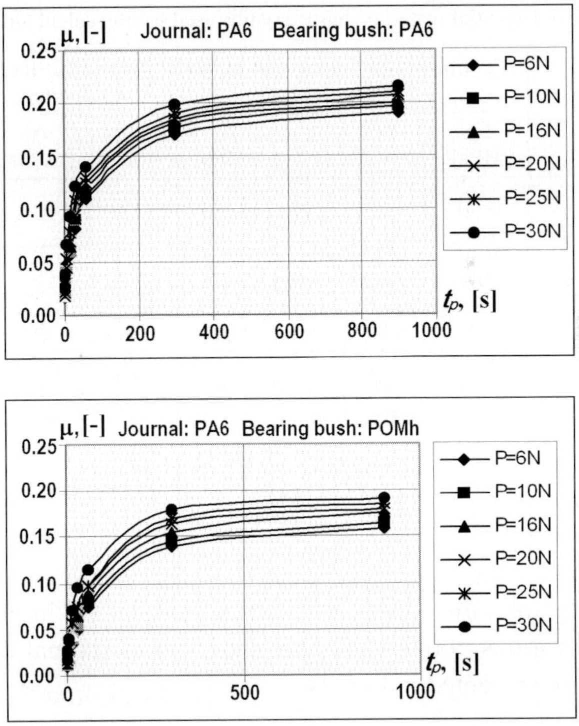

Fig. 23. Comparison of characteristics of starting friction coefficient μ in microbearing vs. time of contact under load t_p for combinations PA6 – PA6 and PA6 – POMh.

The simulation analysis of the behavior of various polymeric tribosystems at the following conditions: load P from 5N to 30N and time of contact (under load) t_p from 1s to 15min were performed. The calculations have been carried out for the following combinations of materials for journal-bearing bush tribosystem: PA66 (Ultramid A3K) – PA66 (Ultramid A3K), PC (Makrolon) – PA66 (Ultramid A3K), PC (Makrolon) – PC (Makrolon), PC (Makrolon) – POM (Delrin AF), PC (Makrolon) – PA6 (Tarnamid T27), POM (Delrin AF) – POM (Delrin AF), POM (Delrin AF) – PA66 (Ultramid A3K), PA6 (Tarnamid T27) – PA66 (Ultramid A3K), PA6 (Tarnamid T27) – POM (Delrin AF), PA6 (Tarnamid T27) – PA6 (Tarnamid T27). The essential material characteristics used for the calculations of the frictional properties are listed in Table 6.

Table 6. Essential material characteristics used for the calculations.[38]

Polymer	Parameters of Voight-Kelvin model with n=2					Surface energy	Poisson's ratio
	E_a [MPa]	E_b [MPa]	t_{ret1} [s]	E_c [MPa]	t_{ret2} [s]	$\Delta\gamma$ [mJ/m^2]	ν [-]
Tarnamid T27 (PA6)	1800	1470	9.9	1850	65.0	72.1	0.35
Ultramid A3K (PA 66)	600	721	11.8	920	160.0	63.4	0.34
Makrolon (PC)	2600	7600	9.4	4500	110.0	52.5	0.33
Delfin AF (POM)	2800	7500	3.8	3000	44.4	48.4	0.35
Delfin 500 NC10 (POM)	6500	7250	5.0	735	136.4	60.5	0.35
Tarnoform 300 (POM)	3100	2990	6.6	4550	42.5	64.9	0.34

The results of simulations of starting torque M_t in microbearing versus t_p for various combinations of materials of journal bearing bush tribosystem are presented in Fig. 24.

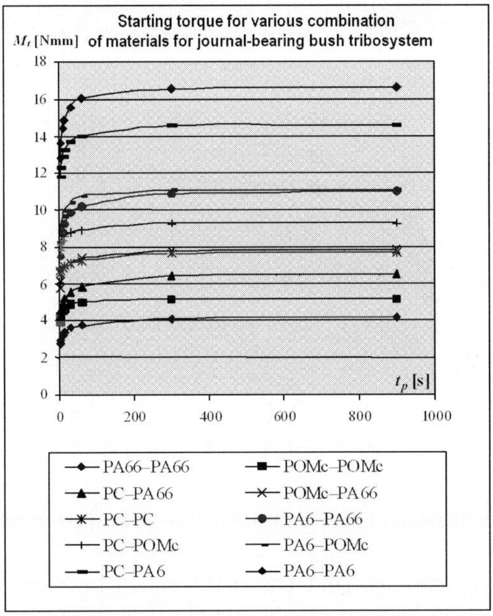

Fig. 24. Starting torque M_t in microbearing vs. time of contact under load t_p for various combinations of materials for journal – bearing bush tribosystem. Load P=30N.

The analysis of obtained characteristics shows that variation of load has small effect on the friction coefficient but increase of time of contact under load yields significant increase of the friction coefficient. The effect of time of contact on friction coefficient was observed in the range of 1s to 900s. The further increase of time of contact does not influence dramatically on friction coefficient. This predicted behavior agrees perfectly with the experimental results (see below).

4. Comparison of Predicted and Experimentally Determined Frictional Behaviors of Studied Bearings

The results of simulation of friction characteristics in the phase of starting to operation of the studied microbearings have been compared with the results of experimental studies. In Fig. 25 are presented examples of such comparison for the material combination PA6 – POMh at load P=6N and P=25N.

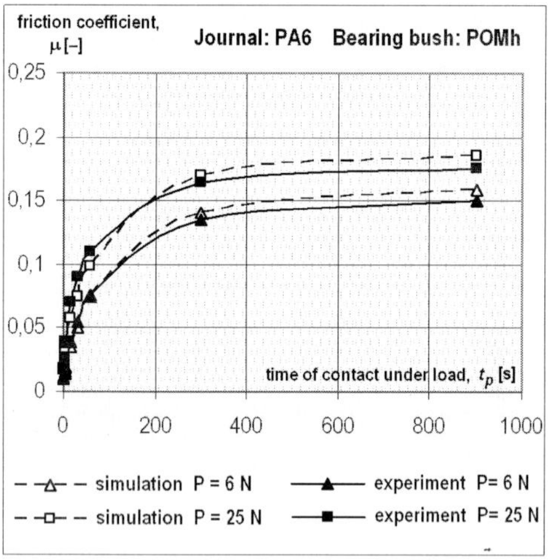

Fig. 25. Comparison of results of simulation and experimental studies.

It is possible to remark that the differences between simulation and testing results are very small (below 10%). These differences can result from inaccuracies of measurements and varied test conditions (applied load, presence of vibrations, effect of temperature etc.). Also in the model several parameters are not well-defined (Poisson's ratio, intermolecular distance ε, inaccuracies of identification of rheological variation of elasticity modulus etc.). Also, it was assumed that the properties of polymers in the area of contact are the same. The 2D (and 3D) maps of distribution of elasticity modulus is interesting problem: the data from such studies would be interesting to be used in the proposed simulation procedures.

The comparisons of simulations and experimental results are also carried-out for other combinations of materials (see Fig. 24). The differences between the results are also very small. We can conclude therefore that the elaborated simulation procedure is correct.

5. Conclusions

In this chapter we presented a new model of contact of rubbing polymeric components, description of the program or prediction of friction coefficient (or torque) for miniature polymer-on-polymer journal bearings. The simulation results were compared with experimental ones. Very good agreement was found.

The presented model is based on the assumption of the combined effect of mechanical and adhesive components of friction force. The values of both components depend strongly on real area of contact. At modeling we took into consideration applied load, roughness of both surfaces, surface energetics and rheological processes occurring in the contact. For modeling 3D AFM images of contacting surface have been used.

The computer simulation enables rapid calculation of frictional resistances. The original elaborated software open perspectives for optimization of the analyzed bearings to minimize energy losses during starting phase which is critical for proper operation of the studied microbearings. The elaborated software is very important tool in designing of micromechanisms constructed with the use of polymeric materials.

We believe that the presented theory can be applied to any other sliding polymer-on-polymer contacts (e.g. guides) in particular when starting phase is considered. The estimation of the area of contact is then crucial. The proposed methodology can be also applied for steel-polymer sliding pairs or any other combinations of materials. If the material does not demonstrate visco-elastic properties then the constant (versus time) value of elasticity modulus should be used.

The proposed theory is applicable to any size (area of contact) of studied bearings. It is necessary to remember that the increase of the area of contact can lead to:

- increase of time of computations, in particular in the case of journal bearings (cylinder-on-cylinder contact)
- errors of calculations because of the lack of homogeneity of the properties of material in the area of contact (in presented theory it was assumed that the elasticity modulus and surface energy do not change in various points of contact).

References

1. Blau P.: *Friction Science and Technology*, Marcel Dekker, New York (1996).
2. Blau P.: *Friction mechanisms and modeling on the macroscale*, in.: Fundamentals of Tribology and Bridging the Gap Between the Macro- and Micro/Nanoscales, B.Bhushan (ed.), Kluwer Academic Publishers, Dordrecht (2001), 241-260.
3. Blau P.: *Experimental aspects of friction research on the microscale*, in: see. Ref 2, 261-278.
4. Karapetyan A.N., Rymuza Z., Kusznierewicz Z.: *Tribological properties of polymer-on-polymer pairs*, Trenie and Iznos 25 (2004), 292-303, in Russian.
5. Rymuza Z., Kusznierewicz Z., Solarski T., Kwacz M., Chizhik S.A., Goldade A.V.: *Static friction and adhesion in polymer-polymer microbearings*, Wear 238 (2000), 56-69.
6. Mehregany M., Senturia S.D., Lang J.H., Nagarkar P.: *Micromotor fabrication*, IEEE Trans. Electron Devices 39 (1992), 2060-2069.
7. Senturia S.: *Microsystem Design*, Kluwer Academic Publishers, Boston (2001).
8. Dhuler V.R., Mehregany M., Philips A.M.: A comparative study of bearing designs and operational environments for harmonic side-drive micromotors, IEEE Trans. Electron Devices 40 (1993), 1985-1989.
9. Wallrabe U., Blej P., Krevet B., Menz W., Mohr J.: *Design rules and test of electrostatic micromotors made by the LIGA process*, J. Micromech. Microeng. 4 (1994), 40-50.
10. Belyi V.A., Sviridenok A.I. *et al.*: *Friction and Wear of Polymer Based Materials*, Nauka and Tekhnika, Minsk (1976), in Russian.
11. Rymuza Z.: *Tribology of Miniature Systems*, Elsevier, Amsterdam (1989).
12. Ainbinder S.B., Tiunina E.L.: *Introduction in Theory of Friction of Polymers*, Zinaite, Riga (1978), 224, in Russian.
13. Rymuza Z.: *Tribology of Anti-Friction Polymers*, WNT, Warszawa (1986), in Polish.
14. Yamaguchi. Y.: *Tribology of Plastic Materials*, Elsevier, New York (1990).
15. Kragielski I.V.: *Friction and Wear*, Mashgiz, Moskva (1962), in Russian.
16. Rymuza Z., Kusznierewicz Z., Mańturzyk G.: *Testing miniature, in particular polymer- polymer journal bearings*, Wear 174 (1994), 39-46.
17. Meyer E., Hug H.J., Bennewitz R.: *Scanning Probe Microscopy*, Springer Berlin (2004).
18. Kaneko R., Nonaka K., Yasuda K.: Summary abstract: scanning tunneling microscopy and atomic force microscopy for microtribology, J. Vac. Sci. Technol. A6 (1988), 291-292.
19. Bhushan B., Fuchs H., Hosaka S. (eds): *Applied Scanning Probe Methods*, Springer Berlin (2004).
20. Bhushan B. (ed): Nanotribology and Nanomechanics - An Introduction, Springer Berlin (2005).

21. Rymuza Z.: *Plastics*, in: Surface Modification and Mechanisms, G.E.Totten and H.Liang eds., Marcel Dekker, New York (2004).
22. Laczynski B.: *Mechanics of High-Molecular Materials*, Wydawnictwa Politechniki Warszawskiej, Warszawa (1977), in Polish.
23. Tonell A.E.: Polymers From the Inside Out: An Introduction to Macromolecules, Wiley, Chichester (2001).
24. Johnson K.L: *Contact Mechanics*, Cambridge, UK, (1985).
25. Roark R.J.: *Formulas for Stress and Strain*, McGraw-Hill Book Company 53030, (1965).
26. M.Kwacz, S.Chizhik, Z.Rymuza, Z.Kusznierewicz: Investigation of mechanical properties of surface laser of polymeric components of miniature journal bearing with use of atomic force microscope, Polimery 7/8 (2004), 551-557, in Polish.
27. Wyrobek J.T., Wrzesinska H., Strom A., Rymuza Z., Nowek A.: *Nanowear of ultrathin TiN/NbN superlattices deposited on silicon,* Tribology Letters 18 (2005), 165-172.
28. Żenkiewicz M.: Adhesion and Modification of Surface Layer of High-molecular Materials, WNT Warszawa (2000), in Polish.
29. Rymuza Z., *Tribology of Polymers,* Archives of Civil and Mechanical Engineering 7 (2007), 177-184.
30. Erhard G.: Sliding friction behavior of polymer-on-polymer material combinations, Wear 84 (1983), 167-181.
31. Lavielle L.: Polymer- on polymer friction: relation to adhesion, Wear 151 (1991), 63-75.
32. Chen N.H., Maeda N., Tirrell M., Israelachvili J.N.: *Adhesion and friction mechanisms of polymer-on-polymer surfaces*, Science 297 (2002), 379-386.
33. Bahadur S.: The development of transfer layers and their role in polymer tribology, Wear 245 (2000), 92-99.
34. Jia B.B., Li TS. Liu X.J., Cong P.H.: Tribological behaviors of several polymer-on-polymer sliding combinations under dry friction and oil-lubricated conditions, Wear 262 Issue 11-12 (2007), 1353- 1359.
35. Chen N.H., Maeda N., Tirrell M., Israelachvili J.N.: *Adhesion and friction of polymer surfaces: the effect of chain ends*, Macromolecules (2005), 3491-3503.
36. Eiss N.S., Mcann: Frictional instabilites in polymer-polymer sliding, Tribol. Trans. 36 (1993), 686-692.
37. Lennard-Jones J.E.: Proc. Roy. Soc. Ser. A196 (1924), 463
38. Kwacz M.: *Minimization of Energy Losses in Starting Phase of Polymer Tribological Microcontacts*, PhD Thesis, Warsaw University of Technology, Oficyna Wydawnicza PW (2005), Warszawa, in Polish

CHAPTER 9

STATE-OF-THE-ART OF RUBBER TRIBOLOGY

Si-Wei Zhang

School of Mechanical and Electronic Engineering
China University of Petroleum, P.O. Box 902, Beijing, 100083, P.R. China
E-mail: swzhang99@sina.com

Along with the developments and applications for tribological purposes of rubber and rubber-like materials, remarkable progress has been made in rubber tribology over the years. This chapter is mainly aimed at reviewing the recent advances in rubber tribology in the last five years from the following five aspects: rubber friction, rubber lubrication, rubber wear, wear of metal by rubber, and tribology of rubber assemblies. This comprehensive material could served as a valuable source of references for the future tribologists. Up to now, rubber tribology seems still in its infancy in the strict sense. However, increasing the understanding of rubber tribology will put the tribological application of elastomers forward. Conversely, spreading the tribological application of rubbers will certainly bring about a great advance in rubber tribology. It is expected that a new developing prospect of the rubber tribology would be opened up in the near future.

1. Introduction

Rubber tribology, a branch of tribology, might be defined as "the science and technology of investigating the regularities of the emergence, change and development of various tribological phenomena occurred in rubber and rubber-like materials and their applications". According to the accepted usage, in this chapter, the two terms 'rubber' and 'elastomer' are interchangeable since the word 'rubber' is referred to the broad concept of 'rubber' more often than not.

Nowadays, as compared with biotribology and nanotribology, rubber tribology is not a topic of general interest in the field of tribology. However, rubber possesses many unique properties which metals and other polymeric materials lack in addition to its high elasticity. This has led the ever more extensive application thereof to a great variety of tribological components in machinery and to the fact that it recently occupies a prominent position amongst the contemporary engineering materials. Moreover, encouraging progress is being made in rubber tribology in recent years. This situation is merited attention.

In order to provide a recent and complete understanding of the rubber tribology, a comprehensive work entitled "Tribology of Elastomers" was published by the present author recently.[1] It covers the main facets of rubber tribology, such as rubber friction, rubber abrasion, fatigue wear, frictional wear, abrasive erosion and its surface mechano-chemical effects, wear of metal by rubber, and lubrication of rubber seals. This book fills a gap in the currently available literature of rubber tribology. Therefore, as a supplement to the book, this chapter is emphatically to review the new developments in rubber tribology in the last five years.

2. Rubber Friction

Rubber friction differs in many aspects from the frictional properties of most other solids, which is accounted for the very low elastic modulus of rubber and the high internal friction exhibited by rubber over a wide frequency region.[2] During the last decade, tribologists were still followed the friction of elastomers with great interest. Studies in rubber friction were much extensive and deep-going. Most of them have obvious engineering background.

2.1. *Sliding Friction on Self-Affined Rough Surface*

For type-road friction, Palasantzas and co-workers[3-6] have systematically investigated rubber friction during sliding of a rubber body on a rough self-affined surface in the recent years.

The term "self-affinity" represents a generalization of the term "self-similarity" that denotes invariance under isotropic dilation[2]. Hence, by self-affine surface (SA) we mean a surface possessed of surface invariance under isotropic dilation. This surface is characterized by three parameters as follows: rms (root mean square) roughness amplitude w, namely, the root mean square fluctuations around the mean height; in-plane correlation length ξ, which is the correlation length parallel to the surface: roughness (or Harst) exponent H ($0 \leq H \leq 1$),[3,4] that is a measure of the degree of surface irregularity at short length scales.

2.1.1. *Effect of SA Surface Roughness on Rubber Friction*[3,4]

The adhesive friction coefficient of a rubber body sliding on a solid substrate was shown to be proportional to the roughness amplitude w, and to be depended strongly on the roughness exponent H. Moreover, in any case, understanding of the adhesive friction should take properly into account of the precise roughness nature both at all lateral roughness wavelengths[3]. It has been found that with increasing surface roughening at short and/or long length scales, the coefficient of friction, μ_f, is increasing in all cases if the roughness exponent H is decreasing and/or the ratio w/ξ is increasing.[4]

The maximum μ_f shifts to low sliding speed for small contact length scales λ ($<\xi$), *i.e.* under incomplete contact. However, in all cases, the maximum μ_f attains for sufficiently large contact length scales λ ($>>\xi$).[4] These results are shown on Fig. 1.

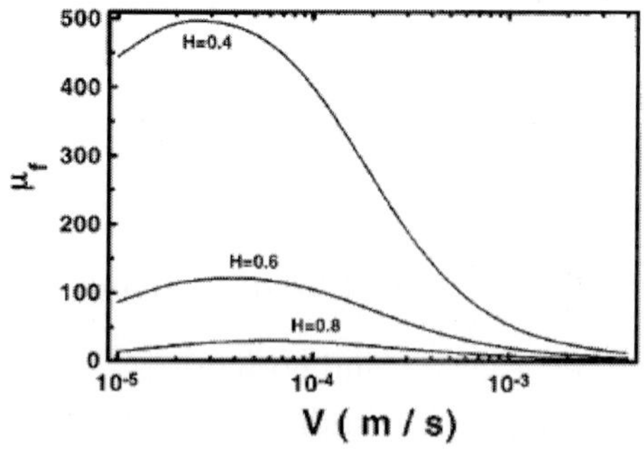

Fig. 1. Friction coefficient μ_f vs. sliding velocity v for relaxation time $\tau = 10^{-3}$ s, λ = 30nm, the ratio of Young modulus $E_1 = E_\infty$ to applied macroscopic load σ, $E_1/\sigma = 1000$, $w = 3nm$, $\xi = 100nm$ and various roughness exponents H.[4]

As shown, as a function of roughness exponent H, the velocity distribution becomes sharp for small H (<0.5). Obviously, H has a strong influence on μ_f. Moreover, the influence of H is increased with increasing contact length λ up to sufficiently large contact length scales. (Fig. 2)

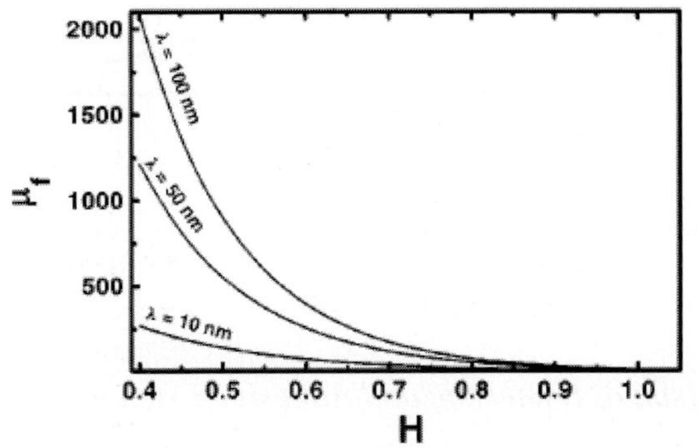

Fig. 2. Friction coefficient μ_f vs. roughness exponent H for sliding velocity $v = 2 \times 10^{-4}$ *m/s*, $\tau = 10^{-3}$ *s*, various contact lengths λ, $E_1/\sigma = 1000$, $w = 5nm$, $\xi = 100nm$.[4]

2.1.2. *Effect of SA Surface Roughness on Rubber Friction in the Presence of a Liquid Interlayer*[5]

When a thin liquid layer was existed between the rubber and solid substrate, it has been founded that the liquid layer will reduce the energy dissipation from the small surface asperities and cavities of lateral sizes smaller than the healing length ζ, as a result, the friction coefficient μ_f decreases. The meaning of healing length is that roughness fluctuations of the substrate/liquid interface at length scales smaller than healing length are smoothened out, and thus they are not further replicated onto the rubber/liquid interface. However, μ_f could be attained significant values for small healing lengths ($\zeta << \xi$), the effect of H becomes less important. It is shown more clearly in Fig. 3.

The theoretical results as stated above would be interesting to be tested by rubber sliding experiments on well-defined random rough surface in the presence of a liquid interlayer between rubber and substrate.

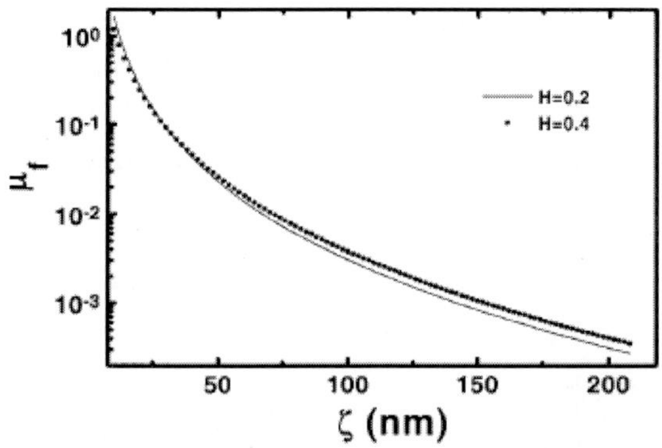

Fig. 3. Friction coefficient μ_f vs. healing length ζ for sliding velocity $v = 2 \times 10^{-4}$ *m/s*, $\xi = 100nm$, contact length scale $\lambda = 1000nm$ ($>> \xi$), $E_1/\sigma = 1000$, $\tau = 2 \times 10^{-3}$ *s*, $w = 5nm$, $\zeta = 30nm$ and various roughness exponents H.[5]

2.1.3. *Comparison of Hysteretic and Adhesive Friction*[6]

It has been observed that for large roughness exponent H ($H \approx 1$) and/or small roughness ratio $w/\xi < 10^{-2}$, the adhesive friction becomes significant at relatively high sliding speeds beyond the maximum of the hysteretic component.

With increasing the roughness exponent H, say up to 0.8 as shown in Fig. 4, the influence of adhesive friction becomes evident at low sliding velocity. Therefore, in order to gauge the contributing mechanisms of friction phenomena at various sliding speeds, the influence of H should be taken into account carefully.

2.1.4. *Sliding Friction on Self-Affine Road Traces*[2]

In order to predict the likely level of rubber friction in engineering practice more economical and time saving, Klüppel and Heinrich[2] advanced a basic formulation of rubber friction on rough, rigid surfaces that relates the frictional force to hysteretic energy losses of the rubber, i.e., the dissipated energy of the rubber during sliding stochastic excitations on a broad frequency scale which can be determined in the laboratory. They adopted the view of Persson[7] that assumed hysteretic energy losses arising from the deformation by surface asperities as the

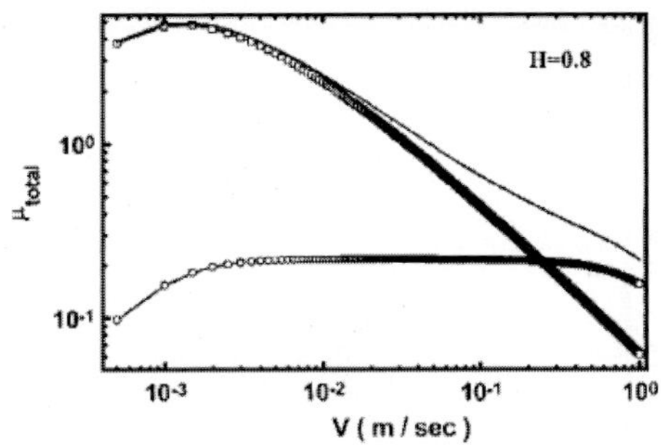

Fig. 4. Total friction coefficient (solid line) μ_{total} vs. sliding velocity v for w = 10nm, ξ = 200nm and H=0.8 (the squares show the hysteretic friction coefficient μ_{HF}, and the circles show the adhesive friction coefficient μ_{HAF}).[6]

dominant factor of rubber friction and further considered the typical scaling invariant morphology of many rough as well as macroscopically appearing smooth surfaces. Moreover, the effect of the surface roughness of road tracks was described by the concept of fractal geometry. Figure 5 shows a correlation between laboratory rating of four different tire tread compounds and tire test results based on the newly developed formulation of rubber friction on rough surfaces during the ABS dry braking process. The tire tests were performed on asphalt tracks (tire size: 225/55 ZR16). As shown, the physics of rubber friction supports well the prediction of tire performance.[2] It means that a prediction of the likely level of friction of tread compounds on the basis of viscoelastic data is possible.

When rubber is slid on a hard and rough substrate, the surface asperities of the substrate exert oscillating forces on the rubber surface leading to energy dissipation via the internal friction of rubber. Persson[7] estimated this contribution to the friction force. He showed that the adhesion force will deform the rubber at the rubber/substrate interface, and the rubber completely followed the short-wavelength surface roughness profile. This gave an additional contribution to the friction force.

Recently, Persson[8] investigated the effect of the local heating on the rubber friction. The local heating is resulted from the viscoelastic deformations when a rubber block is sliding on a hard rough surface. It has been found that the temperature increase results in a decrease in

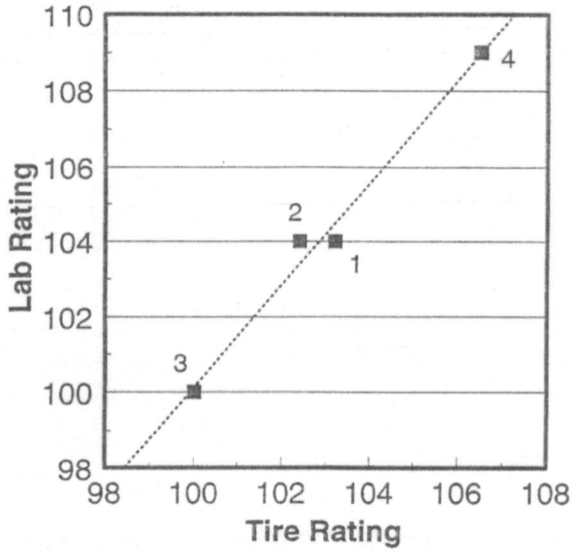

Fig. 5. Correlation between lab rating and tire rating.[2]

rubber friction with increasing sliding velocity for $v > 0.01 m \cdot s^{-1}$. This might result in stick-slip instabilities, and is very important in many practical applications, such as for tyre-road friction and in particular for ABS braking systems.

Gal *et al.* [9-10] presented a physically based model to describe the dynamic contact and the sliding friction of elastomers on self-affine rough surfaces more recently. The hysteresis friction which originates from the energy losses due to the local rubber deformation from the surface asperities during dynamic contact can be modelled. Predications of the load-dependent hysteresis friction coefficient were formulated and gave a deeper insight into the physical mechanisms involved in the material viscoelastic properties, the surface roughness and the friction behaviour.

2.2. *Friction of Polydimethylsiloxane (PDMS) Elastomers*

Recently, Bistac and Galliano[11] have made a quantified comparison between the macroscopic and microscopic friction of PDMS elastomers using a tribometer and atomic force microscopy (AFM), respectively.

The objectives are to verify if frictional properties are comparable at macroscopic and nanoscopic scales, and to propose molecular mechanisms to be able to explain the observed behaviour. However, no quantitative correlations between the macro and micro friction were obtained by their experimental work.

Based on the experimental results, they proposed the interfacial molecular mechanisms involved in both macrofriction and nanofriction, and displayed a schematic illustration as shown in Fig. 6.

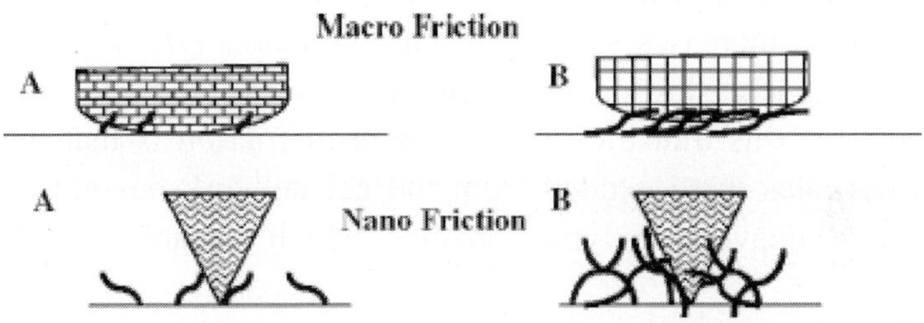

Fig. 6. Schematic representation of the interface for PDMS A and B, in macrofriction (PDMS hemisphere in contact with glass plate) and nanofriction (AFM tip on the PDMS surface).[11]

Both samples, PDMS A and B, are varied by their initial molecular weight M_W, being equal to 6000*g/mol* and 17200*g/mol* respectively.

In Fig. 6, black lines (in bold) represent free and pendant chains emerging from the PDMS surface (square and short chains for PDMS A, and long and numerous chains for PDMS B). Bistac and Galliano proposed that long and numerous free and pendant chains of PDMS B can be oriented acting as a sliding layer during sliding, as a result, the macroscopic friction decreases. However, the short and sparse free and pendant chains of PDMS A are strongly adsorbed and stick to the AFM tip during sliding, which are able to increase the nanoscopic friction.

Casoli *et al.*[12] investigated experimentally the friction behaviour of compatible polymers sliding past each other at model interfaces of controlled molecular parameters. A crosslinked PDMS drop was moved on solid substrates coated with grafted or adsorbed PDMS "monolayers". It has been found that systematic variation of the grafting density of long connector molecules which tend to penetrate into the network yielded two opposing trends. The same connectors resulted in higher friction at lower velocities and led to a reduction of friction with respect to a surface without connectors, respectively.

The surface sliding friction of chemically crosslinked PDMS and swollen with linear PDMS as an oligomer and its transition phenomenon were studied.[13] It has been found that the friction force increases with the normal pressure in a power-law relation, where the exponent changes in a range of 0-1, depending on the degree of polymerization, N poly, of the linear PDMS oligomer. When N poly is in a range of 240-320, a dramatic decrease in friction force was observed at a critical normal pressure, which increases with decreasing networks size W_{net} of the gel and also with increasing polymer length related to N poly. A possible explanation for this transition phenomenon in friction is that the linear PDMS molecules were exuded from the gel network beyond a certain pressure and behave as polymer brushes, which are able to reduce the friction.

2.3. *The Other Topics*

In addition, some interesting and representative topics related to rubber friction were presented more recently. The effect of surface polarity on friction coefficient of silicone rubber/poly (acrylic acid) hydrogel composite was investigated.[14] The friction was found to be strong dependence on polarity of the composite surface.

Deladi *et al.*[15] have developed a static friction model between rubber-like material and rigid asperities (metal surface) taking into account the viscoelastic behaviour of rubber. A single asperity contact was modelled in such a way that the asperities stick together in a central region and slip over an annulus at the contact edge. This model can be used to calculate the traction distributions, contact area, tangential and normal displacement of two contacting asperities. Moreover, it has been extended to the multi-asperity contact case by using a statistical approach. The effect of viscoelasticity in a displacement-controlled case consists in a decrease of the static friction force in time; the limiting displacement remains the same. In a load-controlled case, an increase in the limiting displacement with respect to time is obtained by using a viscoelastic model, but the static friction force being unaffected.

Thermoplastic polyurethane elastomers (TPU) that combine the mechanical properties of vulcanized rubber with the process ability of thermoplastic polymers are used in many fields because of their moderate cost, excellent mechanical properties and high resistance to tear, oxidation and humidity. Elleuch *et al.*[16] have studied its tribological behaviours under reciprocating sliding conditions. The studied material was a polycaprolactone-copolyester based TPU. In the monotonous friction tests, it was founded that the friction coefficient between the steel and the TPU increased rapidly at the beginning until reaching the full sliding contact (static friction), and then reached a constant value 0.5 (dynamic friction) (Fig. 7).

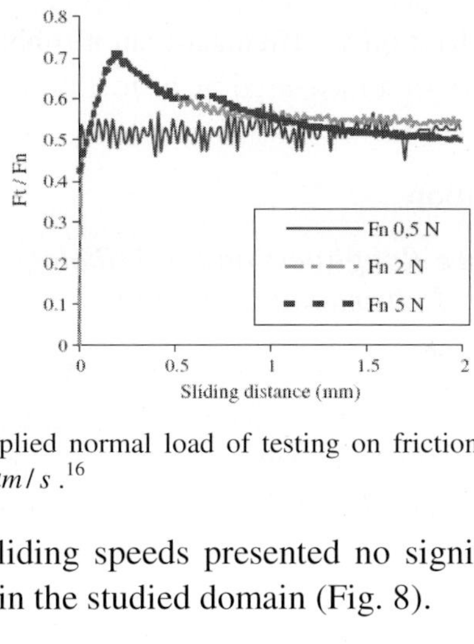

Fig. 7. Influence of applied normal load of testing on friction with sliding distance at sliding speed $v = 250\mu m/s$.[16]

However, the sliding speeds presented no significant impact on the friction coefficient in the studied domain (Fig. 8).

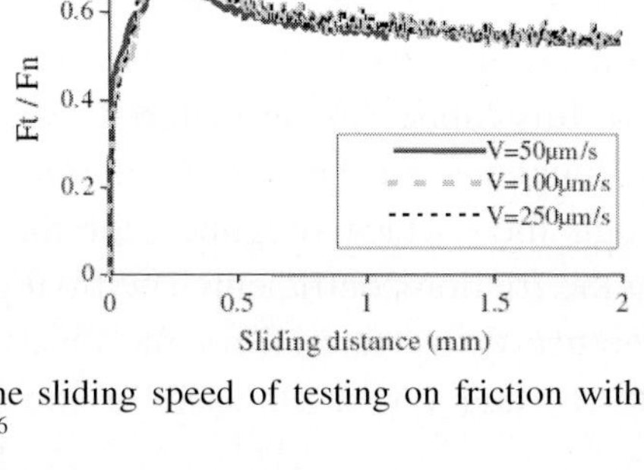

Fig. 8. Influence of the sliding speed of testing on friction with sliding distance under normal load $F_N = 5N$.[16]

The role of interface friction in the formation of mechanical behaviour of the debonded structural cell of particulate rubber composites was examined.[17] It has been found that the effective properties of cells are dependent on the interface friction.

Recently, Gee *et al.*[18] developed a novel friction measurement system that can be used to evaluate the frictional feel of materials. The test system based on the use of strain gauged flexure elements can provide the necessary information about contact forces and position of contact. The use of this soft friction system was demonstrated by reference to the preliminary experimental results on the finger-tip-surface frictional behaviour for a series of thermoplastic elastomers that show how the test system discriminates between them. It has been found that touch by finger gave a higher friction coefficients than a rubber probe that was in turn higher than the friction measured by a steel probe.

3. Rubber Lubrication

3.1. *Effects of Surface Roughness on the Tribological Behaviours of Rubber under Lubrication Conditions*

In order to prove the fact that the microscopic roughness of the footwear soling surface is a major determinant of slip-resistance on lubricated surface, Manning *et al.*[19] conducted an experiment in which five pairs of shoes were soled with the same rubber compound. Four of the pairs were abraded by different grades of grit to produce a range of roughness values. The friction coefficient of the five solings was measured repeatedly by the walking traction method on wet surfaces including glazed wall tiles, vinyl asbestos coated with the wax floor polish, and both sides of a sheet of float glass. The Kruskal-Wallis statistical test proved that the soling roughness is a major factor in determining the friction coefficient of this rubber soling material. Nearly all of the grips were due to surface roughness of the soling material on these atypically smooth floors, though the surface roughness of the floors also had a significant effect on the friction coefficient. The float glass was shown to be a promising reference floor material for the measurement of friction coefficient of footwear and could be used in the development of a standard friction coefficient test method. Manning and co-workers pointed out that any specification of flooring by measuring the friction coefficient based on dry surface could lead to an increase in the number of injuries caused by slipping on the wet surface.

More recently, Chang[20,21] investigated the effect of surface roughness and contaminant on the dynamic friction of porcelain tile, under the conditions with different footwear materials and different mixtures of glycerol and water as contaminant at the interface. It was shown that the dynamic friction decreased as the interface sliding speed and glycerol content in the contaminant increased due to the hydrodynamic lubrication effect. The correlation coefficient between the surface roughness parameters and the measured friction was affected by the glycerol ratio. However, the skewness of surface heights (R_{sk}), the arithmetical average and root mean square of surface slope (D_a and D_q) and the kernel roughness depth (R_k) had the highest correlation with dynamic friction.

Cossin *et al.*[22] have studied the effect of surface roughness on the lubrication properties of a high-molecular-weight polysaccharide (guar gum) against the background of bio-tribological applications. The test contact was formed between a stainless steel and a soft silicone rubber ring mounted on a spinning disk.

The polysaccharide guar gum was not found to adsorb onto the surface. Its lubrication properties and Stribeck curves obtained for water and various guar solutions are shown in Fig. 9. Although there shows high friction in boundary lubrication, these solutions still significantly reduced friction in the mixed regime. The Stribeck curves obtained for water and a 0.05% guar solution using roughened silicone surface R_{a2} as shown in Fig. 10. In Figs. 9 and 10, R_a and R_b are the root-mean-squire surface finishes of the interacting surfaces between silicone rubber ring and stainless steel ball. It was found that in the boundary regime, the friction coefficient increased with the speed due to the increase in the shear stress with sliding velocity. The direct effect of the average surface roughness is to displace the speed range corresponding to the mixed lubrication regime towards high speeds.

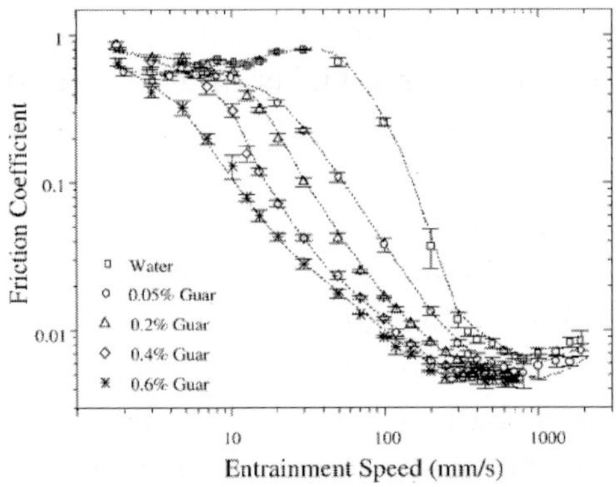

Fig. 9. Stribeck curves obtained for water and guar gum solutions, using untreated rubber surface ($R_a = 0.17\mu m$), T=35°C, W=3N.[22]

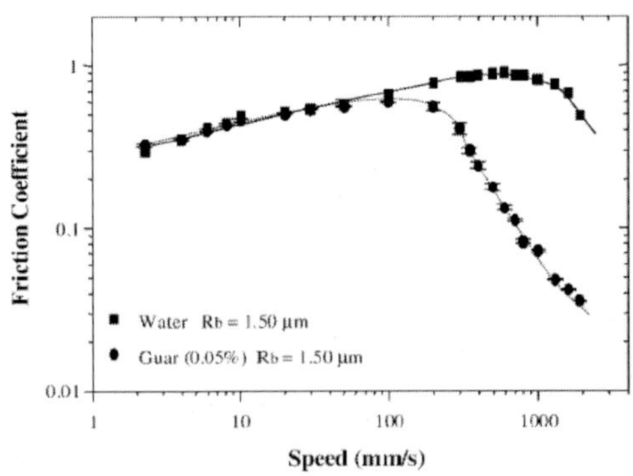

Fig. 10. Stribeck curves obtained for water and a guar gum solution, using roughened rubber surface ($R_{a2} = 0.17\mu m$), T=35°C, W=3N.[22]

3.2. *Wet Sliding Friction of Elastomers*

X.-D. Pan[23-25] conducted a series of studies in wet sliding friction of elastomer compounds over the years. More recently, in order to improve the capability of rational design for enhanced driving safety, he further investigated the wet sliding friction of four compounds made of a tin-coupled styrene-butadiene (SBR) on a Portland cement concrete surface under either ethanol or water lubricated conditions.[26] It has been found that the wet sliding friction is markedly reduced under ethanol lubrication in comparison to that under water lubrication. Additionally, the silica-reinforced compounds by comparison with the carbon black-filled compounds exhibit higher wet sliding friction in water. Moreover, it was indicated that in addition to bulk viscoelastic hysteresis, interfacial interactions between filler particles and the road surface were also a significant contributor to wet sliding friction of elastomer compounds on a rough surface.

The friction of butyl rubber sliding on a hard abrasive in unidirectional sliding was investigated under lubricated conditions with water and oil at room and elevated temperatures.[27] Experiments were conducted in an X-ray microscope chamber using a pin on disk apparatus. It was observed that the presence of lubricant reduced the frictional coefficient.

3.3. *Robust Molecular Lubrication Layers*

Luzinov *et al.*[28] have fabricated the thermoplastic elastomeric molecular lubrication layers with organized micro-domain structure and superior nanotribological properties from the functionalized tri-block copolymer, poly [styrene-b-(ethylene-co-butylene)-b-styrene] (SEBS). This polymer was chemically attached to a functionalized silicon oxide surface via the interfacial epoxy-terminated monolayers, which correspond to a molecular lubrication regime where a single macromolecule separates mating surfaces. The frictional, adhesive, shear and wear properties of a series of molecularly reinforced polymer layers chemically tethered to a solid substrate were investigated. The grafting density and the thickness of the SEBS films were optimized to assure the development of organized microphrase structure within the molecularly thick (<10nm) films. Tethered SEBS monolayers, composed of a rubber matrix reinforced by a two-dimensional net of glassy polystyrene (PS) micro-domains, possessed a friction coefficient as low as 0.02 and shear strength in the range 0.15-1.5 GPa. Moreover, chemically tethered SEBS monomolecular films are much more stable under shear stresses than conventional molecular coatings.

The friction coefficients for SEBS films with various thicknesses are illustrated in Fig. 11. Although the friction coefficient for the bare silicon is about 0.1, the friction coefficients drops to values close to 0.02-0.03

for SEBS films of 8-9 nm thickness. By contrast, a relatively high friction coefficient was observed for tethered PS films of the same thickness (Fig. 11). However, the adhesive forces increase after tethering a very thin SEBS film to the silicon surface (Fig. 11) and are higher than ones for tethered PS layers of comparable thickness.[29] In addition, it was observed that the chemically tethered SEBS film remained intact under loading conditions that led to a complete destruction of the conventional SEBS film with the same chemical composition and microstructure.

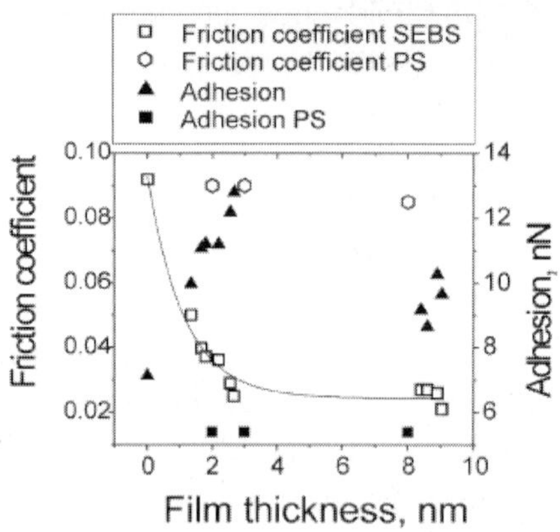

Fig. 11. Friction coefficients and adhesive forces for SEBS film with different thicknesses in comparison with PS films and silicon oxide surface. The line is a guide for the eyes.[28]

4. Rubber Wear

4.1. *Wear Behaviour of Elastomers*

The understanding of wear mechanism of rubber is still arouses general concern. In situ sliding wear tests of butyl rubber on a hard abrasive under both dry and lubricated conditions with water and oil at room and elevated temperatures in an X-ray microscope chamber were performed.[27] It has been found that during sliding contact, dry sliding led to thermal decomposition of rubber by elevated frictional temperatures, and the wearing contact area on rubber changed in shape and location (Fig. 12), which gave rise to non-uniform wear of rubber. Moreover, lubrication by mineral oil of sliding rubber specimens at elevated temperatures led to accelerated wear at higher normal loads due to chemical deterioration of rubber. It is possible that the abrasive grains reacted with the rubber at high temperatures to form new compounds during sliding.

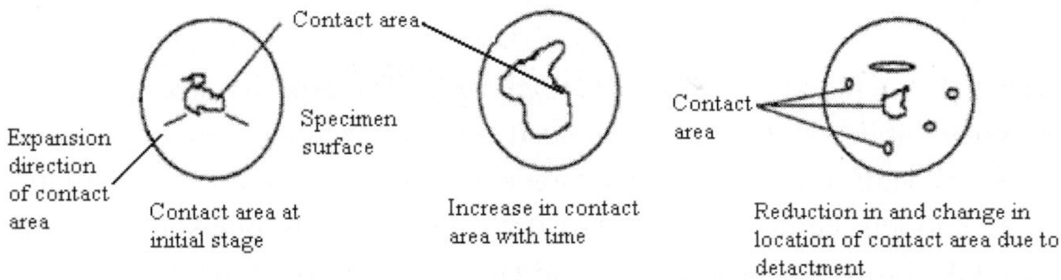

Fig. 12. Model of wearing contact in rubber, which shows changes in contact area location with time during sliding of rubber.[27]

Time-dependent wear process between lubricated soft materials was studied using an apparatus with a half-cylinder sliding on a flat silicon rubber plate.[30] It was shown that severest wear was at the starting region, while slight wear was at the end of the trace. No wear was observed in the middle zone where complete fluid film existed. Obviously, at start-up, lubricated compliant solids were subject to wear until complete fluid film was formed due to soft-EHL effect.

Ridge formation in the course of pattern abrasion was investigated using a rubber wheel rubbed against a steel cylinder. Distribution maps of the friction coefficient and worn surface profile along the circumference of the rubber wheel were made throughout the experiment by monitoring friction and wear during sliding.[31] The movement rate of ridge was calculated from the gradient of each stripe-like variation in the maps. It has been found that the ridges remained until the steady state moved faster than the combined ridges before reaching the steady state, and the difference between ridge movement rates was the source of the ridge combination phenomenon.

Gerrard and Padovan[32] have offered an analytical model of the micro-mechanical processes during intrinsic wear. While analyzing the processes surrounding the removal of an "average" intrinsic nodule, it was shown that nodules experience fatigue crack propagation rates in the constant, linear and power regions of Fatigue Crack Propagation (FCP). It was found that the wear rate was a function of the local frictional work into the nodule raised to a power exponent that was simply based on the distribution of loading across the full range of FCP. This proposed analytical description of wear was tested against the early friction and wear data published by Schallamach. It was shown that the wear relationship offered in the analytical study can be used to predict rubber wear over a wide range of pressures and abrasive asperity sizes.

Since rubber shows a pronounced non-linear behaviour in the strain range relevant for tires. Many practical tribological problems require rubber models being incorporated the non-linear viscoelastic properties of rubber. More recently, Popov and Geike[33] proposed a new constitutive model for rubber. It was introduced the concept of elastic moduli as order parameters to incorporate the strain amplitude dependence. The model is characterized by one kinetic equation for each spring, i.e. the moduli are no longer constants but variables. This model is able to describe both linear and non-linear behaviours of rubber as well as the failure of partial spring and thus the failure mechanisms of rubber. Moreover, it can be incorporated into numerical simulation models for simulating both friction and wear in tire applications.

A dry sand-rubber wheel abrasion test (DSRW) was used to investigate the wear behaviour of polyurethanes (PU).[34] The wear performance of PU was assessed for the range of Shore hardness 85A to 65D, and a correlation between the wear rate and the sample hardness was identified. It has been found that PU can be separated into three classes based on their hardness and wear performance, and each class shows a different dependence on the specimen temperature. It might be useful for predicting the field performance of PU by using the DSRW test.

Beck and Truss[35] have studied the effect of chemical structure on the wear behaviour of a series polyurethane-urea elastomers of two hardness levels (83/90 Shore A) using two different test methods. It was observed that in the high-speed slurry erosion test, wear was higher in the harder elastomers but in DSRW test, the softer rubbers underwent greater wear. This change in rankings of the elastomers was associated with a change in the wear mechanism from cutting and gouging in the slurry test to a fatigue mechanism in the DSRW test. Moreover, it has been found that in both wear tests and for the two levels of hardness tested, elastomers cured with MBOCA (methylene bis-o-chloroaniline) were more wear-resistant than those cured with ETHACURE which was linked to a higher level of crosslinking in the MBOCA.

Zhang *et al.*[36] investigated the surface physical and chemical effects of abrasive erosion of PU in three kinds of fluids (water, polyacrylamide and NaOH) containing quartz particles respectively. During the abrasive erosion process, the surface physical effects were found to be microcutting, high-cycle fatigue and plastic fracture, while the surface chemical effects turned out to be mechanical rupture of macromolecular chains, thermal decomposition, hydrolysis, and surface oxidative degradation.

More recently, El-Tayeb and Nasir[37] have examined the effect of soft carbon black on the tribological properties of deproteinised natural rubber (DPNR) and synthetic cis-1,4-polyisoprene rubber(IR). As was expected, the addition of carbon black has significantly affected the wear resistance of both DPNR and IR. It was found that the abrasion weight loss was reduced by more than 70% compared to unfilled rubber, depending on the conditions and the concentration of carbon black.

4.2. *Wear Behaviour of Rubber Composites and Coatings*

Bijwe *et al.*[38] were investigated the wear behavior of polyetherimide (PEI) and its two composites, one containing only short glass fibers (GF) 20% and the other, commercially established bearing material containing 25% GF, in four types of wear modes, i.e., adhesive wear, abrasion, fretting and erosive wear. It has been proved that a bearing grade material is extremely good in adhesive and fretting wear modes, but the same composite is rather poor in abrasive and erosive wear modes. Neat PEI performed best in these two wear modes.

Recently, against a background of application in aeronautical construction, a study of the fretting behaviour of glass-reinforced polypropylene composite against 2024-T6 aluminium alloy was conducted.[39] The particular attention was given to the effects of relative amplitude displacement, temperature, relative humidity and surface treatment by anodisation of the Al alloy. To analyze the influence of each factor, friction force and displacement were used to establish the fretting cycles for every condition tested. Figure 13 shows the fretting cycles for two pairs of materials, namely composite being paired up with the anodized and non-anodized Al alloy respectively. It has been shown that variations in the shape of the cycles enabled three typical fretting regimes to be identified: stick, slip and partial slip, but most effective way to characterize the transition between regimes was based on energy dissipation by friction. It was found that surface treatment by anodisation led to lower wear value for smaller displacement amplitude, whereas for higher displacement amplitudes the wear values were larger, for the case of room temperature and humidity. Moreover, the fretting behaviour of the polypropylene composite/Al alloy pair was much affected by temperature change. An increase in temperature resulted in a rise in wear volumes, especially for non-anodized aluminium. However, variation in humidity did not greatly influence the wear behaviour of the pairs of materials studied. In addition, it was observed that in the test performed at high temperature, considerable amounts of composite matrix were transferred to the pin surface.

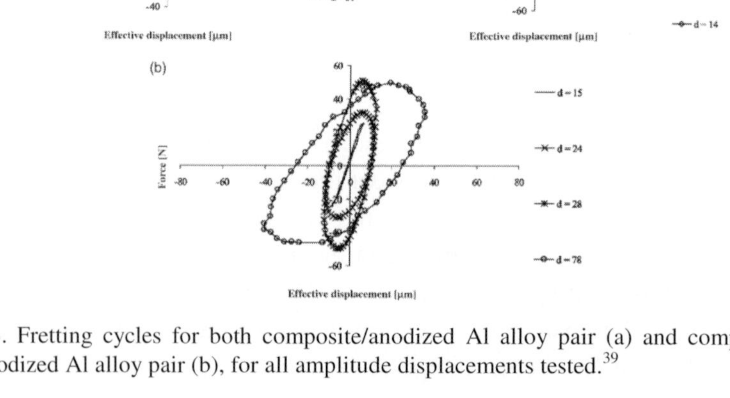

Fig. 13. Fretting cycles for both composite/anodized Al alloy pair (a) and composite/non-anodized Al alloy pair (b), for all amplitude displacements tested.[39]

The wear behaviour of thin film formed from plasma polymerization of perfluoro(methylcycohexane) on an EPDM (ethylene-propylene-diene terpolymer) substrate under cold plasma process operated at 13.56MHz was examined using scanning electron microscopy (SEM).[40] It has been observed that the coated substrate shows superior abrasion resistance characteristics with respect to the neat elastomer, and the adhesion between the substrate and the plasma polymer coating appears to be excellent and remains strong after weathering.

5. Wear of Metal by Rubber

5.1. *A Brief Historic Background*

It is felt the necessity of giving a brief historic background of the study in wear of metal by rubber as this is a special topic in rubber tribology. In a common sense, the soft material is always worn out by the hard one. Hence, wear of metal by rubber became an unexpected and much interesting wear phenomenon in rubber-metal frictional pairs. Unfortunately, it escaped the attention all along, even though this special wear mode was discovered as early as the 1960's.[41] Until late in the 1970's, Gent and his group carried out a series of studies with varieties of rubber formulations under dry friction conditions.[42-45] However, they did not get obvious breakthrough after years of research and broke off this study to the mid 1980's. During this time, AB-Malek and Stevenson[46] examined the wear of alloy-steel by rubber. Particularly, King and Lancaster[47] investigated the wear of metal by rubber in the presence of a clean fluid or in a dispersion of abrasives. Entering 1990's, Zhang and his group[48,49] focused attention on the wear mechanisms of steel by rubber in liquid medium. More recently, an overview of the developments of the study in wear of metals by polymers was made by the present author.[50]

5.2. *Wear of Steel by Rubber in the Presence of Liquid Media*

The wear mechanisms for steel T10 (a carbon tool steel) being worn by nitrile rubber (NBR) in the presence of paraffinic mineral oil were investigated[48]. It has been found that under the action of shear forces, the steel lattice was disordered and new dislocations formed, as a result, the chemical activity of steel increased. In the meantime, the macromolecular chains of NBR and the alkane chains of the mineral oil were broken at the stress concentration sites. These broken macromolecular chains produced active macromolecular free radicals, which reacted with the activated iron atoms and iron oxides, and generated the metal-polymer film on the steel surface. This metal-polymer film was destroyed and removed by the action of the iron oxide particles on the sliding surface, thus, the rubber worn away the steel. Therefore, the mechanism of wear of steel by rubber is mainly a cyclic and alternating process of occurrence and removal of the metal-polymer films.

Both the wear process and the wear mechanisms of steel 45 (a carbon steel) by natural rubber (NR) in water medium were studied.[49] Two kinds of effects were observed in the wear process, *i.e.*, surface physical effect and surface mechano-chemical effect. The former is mainly microcutting action of steel debris and particulates of rubber additive remained in the sliding interface. The latter are the fracture and oxidization of macromolecular chains of NR, and the reaction between the macroradical of NR and iron. Moreover, it was found that the process of wear of steel 45 by NR in water medium involved two stages. Firstly, under the action of frictional force, mechano-chemical reactions were generated among the surfaces of both steel and rubber as well as the water medium, namely, the macro-radicals produced by the rupture of molecular chains of rubber were reacted with the steel surface and generated the metal-polymer complexes based on X-ray photoelectron spectroscopy (XPS) analysis (Fig. 14), as a result, a chemical reaction film (CRF) was formed on the steel surface. Secondly, the CRF was peeled off by microcutting of the debris on the sliding surface, as a result, the steel surface worn off. The wear mechanism might be mainly attributed to the chemical-mechanical self-catalytic destruction of steel being rubbed by NR.

In fact, both the process and mechanisms of wear of steel by rubber are much complicated. Some influence factors must be considered and investigated further, such as, friction heat, swell of rubber, hydrogen brittleness and fatigue of steel, and so forth.

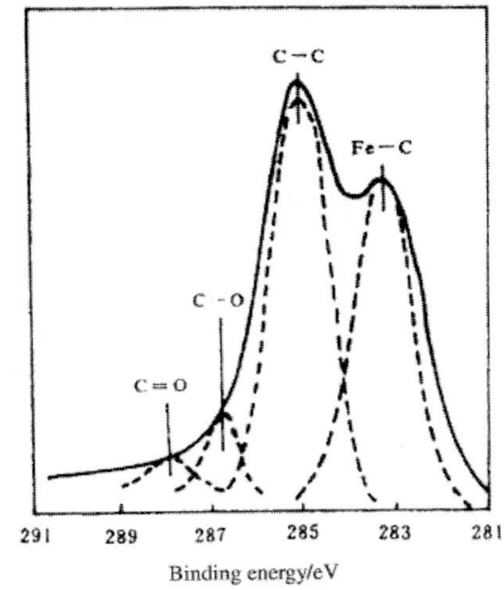

Fig. 14. XPS spectrum of C element on worn surface of steel 45 after 5 min sputtering.[49]

In addition, Wang *et al.*[51] investigated the formation of transfer film during the process of wear of steel AlSl 1020 by NR in air. It has been found that the formation of transfer film on the worn surface of steel involved two stages. Firstly, an adhesive layer emerged on the steel surface owing to the macro-molecular chains of rubber adhere to the steel surface by van der Waals' force of the intermolecular chains. Secondly, the iron atom and metal oxide reacted with the macro-radicals of NR in the adhesive layer and produced Fe-polymer compound. Consequently, the transfer film was formed on the worn surface of the steel by the chemical bonds and electrostatic force.

6. Tribology of Rubber Assemblies

6.1. *Tires*

Pavement surface texture plays a vital role in the development of both tire-pavement friction and tire wear. In light of that the available relationships quantify tire wear and tire-pavement friction simply in terms of empirical pavement texture parameters, Gunaratre *et al.*[52] developed the correlations of tire wear and tire-pavement friction to texture of concrete pavements, with which the tire wear rate and tire-pavement friction on smooth concrete pavement can be predicted using actual pavement texture properties. Moreover, these correlations can be useful in the design of pavements with acceptable tire wear rates with a sufficient level of skid resistance, and also useful for predicting the tire wear rate for a given pavement based on conventional Mu-meter measurements and grease patch test results. Figure 15 shows the variation of predicated values of the gravimetric wear rate Kw versus their corresponding observed values.

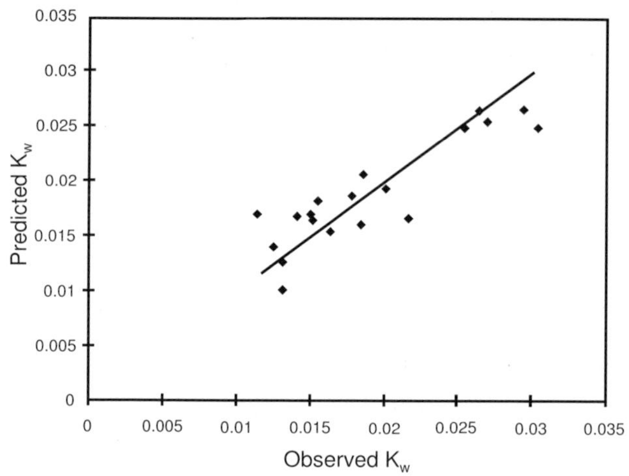

Fig. 15. Plot of Predicted Kw versus Observed Kw.[52]

Recently, Grosch[53] designed a road test computer simulation program, which took account of the major influence in road wear and using distribution functions to reproduce the large range of variables encountered during a road test. Moreover, using the equations obtained from laboratory abrasion experiments in the program produces tire lives which are close to practical experience. It has been shown that this computer simulation program is able to simulate a very wide range of road wear testing severities and the effect which they have on achievable mileage and wear rating of tread compounds. More recently, Hofstetter *et al.*[54] established a simulation procedure for sliding tread blocks under consideration of thermal effects and abrasion. They developed a contact model suited for the contact pairing rubber and rough road. Subsequent changes of the model geometry in the course of the simulation allowed for the reproduction of abrasion effects. The simulations were carried out by means of the Finite Element Method in a thermo-mechanically coupled manner. By incorporation of sipes and by consideration of parallelogram-shaped bases, a close approach to realistic tread patterns was achieved.

6.2. *Elastomeric Seals*

Rotary shaft lip seals are widely accepted as efficient and cost-effective bearing seals. The trend towards are mainly increasing shaft speed, maximizing seal life and minimizing seal leakage either for cost or environmental reasons. However, higher speed and lower leakage result in higher seal friction and under-lip temperatures as well as limiting seal life. In this connection, Embury and Armour[55] investigated the influence of elastomer compound design on the performance of rotary shaft lip seals and the development of a new elastomer compound with inherent heat dissipation properties and superior wear resistance without prohibitive increase in friction and cost.

Acrylonitrile-butadiene rubbers (NBR) are widely used for lip seals of ball-bearing in the automotive industry. In order to reduce the contact friction and improve the seal life, Degrange *et al.*[56] investigated the relationship between the viscoelastic and tribological behaviours of carbon black filled nitrile rubber (NBR). Two rubbers, mainly differing by their glass transition temperature (T_g) values were studied, it has been found that the wear resistance of the rubber with a high T_g values is better than that of the rubber with a low T_g values while the difference in their mechanical properties is very limited. Moreover, the experimental results and comments addressed the difference of wear mechanism by relating it to the different characteristics of the rubber sample surface: viscoelastic, themo-mechanical and physicochemical (obtained by XPS analysis). Later on, Thomine *et al.*[57] also investigated the relations between viscoelastic properties and tribological behaviours of two filled nitrile rubbers (NBR), which mainly differed in their glass transition temperature (T_g) values. It has been found that the sample with high T_g values showed an oxidized protective layer formed on the surface, which prevented its wear, but the sample with lower T_g values did not have such a protective layer.

In recent years, Nikas[58-63] developed a numerical model to study the rectangular elastomeric seals used in linear hydraulic actuators. Elastomeric seals often operate in extreme environments of pressures up to 35 MPa and temperatures in the range of -60 to +140°C. This type of seal is of great engineering significance in industrial and aviation hydraulic assemblies. A theoretical modelling of rectangular elastomeric seals has been addressed[58] and a simple version was made later on.[59] Recently, Nikas and Sayles[63] updated the model to account for finite strains by using the popular Mooney-Rivlin model of rubber elasticity with a suitable modification to include thermally induced strains into the equations that were initially meant for isothermal loading. This new stress-strain model was then used with the elastohydrodynamic (EHL) model in Ref[59] to produce fundamental numerical results for the seal performance (contact pressure and width, film thickness, leakage rate and friction).

For the typical assembly of a hydraulic actuator, it has been found that the operating temperature plays the major role in the stress-strain behaviour of the seal elastomeric material, which shows a nonlinear behaviour at high temperature. Moreover, increasing the sealed pressure for an outstroke causes a linear increase of the contact pressure and the film thickness, as well as a linear increase in the mass leakage rate at low and medium temperature and a nonlinear increase at high temperature (Fig. 16), and a nonlinear decrease in both the contact width and the friction force on the seal. In addition, the nonlinear model was compared with the classic Hookean model of linear elasticity in order to establish the models and the limits of applicability and validity of each one for typical operating conditions. Certainly, the nonlinear model stated above is only a rough approximation of a complex problem.

Further on, Nikas and Sayles[64] expanded the numerical model to study the system of two identical seals in a row at some distance from each other, namely tandem (or dual) elastomeric rectangular seals for reciprocating motion and to compare their sealing performance with the typical one-seal arrangement. The computational model based on a linear

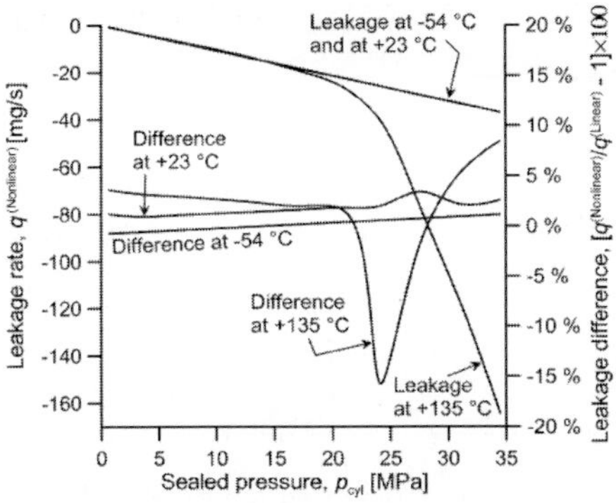

Fig. 16. Mass leakage rate (the negative sign implies fluid loss from the actuator) of the seal.[63]

hydraulic actuator as used in many industrial and aviation applications is covered the full typical range of operating conditions in terms of sealed pressure (0-35MPa), stroking velocity (0.05-0.5m/s), and operating temperature (-54-+135°C).

It has found that the system leakage-per-cycle is reduced with increasing temperature (Fig. 17 and Fig. 18) but is not significantly affected by the sealed pressure (Fig. 17), however, the system leakage-per-cycle is markedly increased with increasing stroking velocity (Fig. 18).

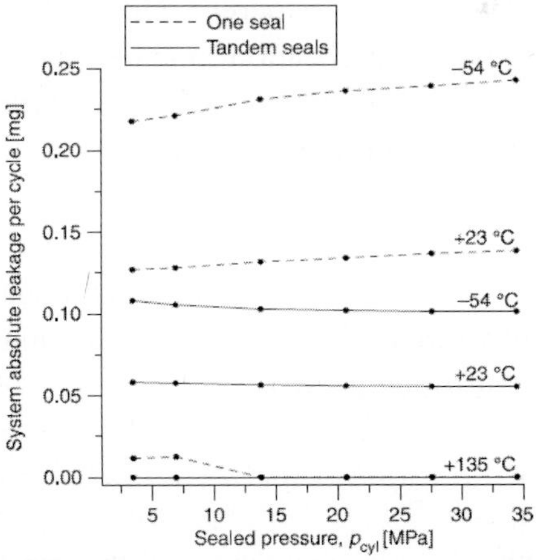

Fig. 17. Effect of the sealed pressure on leakage for stroking velocity |v|=0.2m/s.[64]

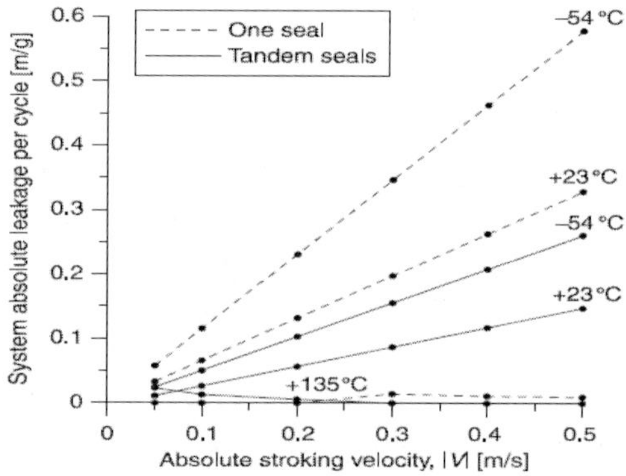

Fig. 18. Effect of the stroking velocity on leakage for sealed pressure Pcyl =13.8MPa.

The theoretical simulation showed that in comparison with the single-seal configuration, the tandem seal arrangement offered a substantial reduction of leakage (in the order of 50% or more) at all but very high temperatures, where leakage was very small or zero anyway. Therefore, through the simulation, the tandem seal solution appears to be a worthy one in a number of applications with strict leakage constraints although there are some design problems associated with this seal arrangement.

More recently, Salant[65] developed a new analysis method for elastomer reciprocating seals, in which, both surface roughness and mixed film lubrication were considered. This method is used a fluid pressure/density function and cavitation function to calculate the pressure distribution under the seal. The effect of the surface roughness of the seal lip was taken into account through the use of flow factors.[66] These are functions of the ratio of the film thickness to the roughness amplitude and the roughness geometry of the seal. This new model appears to provide improved predictions that are closer to practical experience.

However, the above-mentioned numerical simulations of seals are still remained to be confirmed further.

6.3. *Brakes*

In order to increase the damping behaviour of the cured friction material, styrene-butadiene-styrene (SBS), poly[styrene-b-(ethylene-co-butylene)-

b-styrene)] (SEBS) and nitrile rubber/polyvinyl chloride (NBR/PVC) blend system were used as thermoplastic elastomers (TPE) materials, namely as viscoelastic polymeric material into the formulation, and the damping characteristics were investigated.[67] It was found that NBR/PVC and SEBS were much more effective in damping behaviour. Obviously, the damping characteristics of commercial friction materials were strongly affected by the TPE ingredients. In addition, it was shown that the specimens with high TPE content had low noise propensity.

More recently, Shojaei *et al.*[68] investigated the thermal conductivity of rubber-based composite friction materials used in railroad vehicles. Based on a commercially available railroad friction material, various friction materials containing different thermally conductive fillers (Cu, brass, Al, Al_2O_3 and talc) were fabricated and their thermal conductivities were measured at various filler content. It has been observed that addition of the thermally conductive fillers caused an increase in thermal conductivity of the friction material from 0.48 up to 5.8 w/m·k, depending on the type and content of the filler. It was found that the thermal conductivity of the friction material can be influenced by the indirect effects including mainly the shape and size of the filler. Fillers with large size and platelet shape were more effective in increasing the thermal conductivity of the friction material. In addition, in order to estimate the thermal conductivity of the friction materials, a new semi-empirical model was proposed for multiphase systems based on the geometric mean model, in which, the filler may be regarded as thermally conductive or non-conductive one. The agreement between model predications and experimental measurements was satisfactory.

6.4. *Rubber Journal Bearing and Rubber Acetabular Bearing Materials*

Cabrera *et al.* [69] measured the fluid film pressures on water-lubricated rubber journal bearing and indicated that the film pressure profile were very different from those of conventional rigid bearing. It has been found that the relatively low film pressures caused significant rubber defections but were too low to produce changes of lubricant viscosity. Moreover, the integration of the pressure in these bearings showed that the water-lubricated rubber bearings were operated in mixed lubrication regime.

The polyurethame(PU) elastomer Corethane 80A(Corvita) is being considered as the acetabular bearing material in a novel total replacement hip joint. Its bio-stability was investigated in vitro together with three other commercially available biomedical PUs, namely Pellethane 2363-80A (DOW Chemical), a polyhexamethylene oxide based PU, PHMO-PU (CSIRO) and ChronoFlex AL-80A (Cardio Tech).[70] It has been found that Corethane 80A displayed the best overall resistance to hydrolysis, ESC (environmental stress-cracking), MIO and calcification, followed by ChronoFlex 80A and PHMD-PU, with Pellethane 80A being the least stable. Building on the in vitro study, the follow-up in vivo investigation assessed Corethane 80A as the bearing layer in a prototype compliant layer acetabular cup, in a fully functioning ovine total hip arthoplasty (THA) model. It was shown that the Corethane 80A functioned well in the THA model, with the bearing surfaces of the retrieved hip cups showing no significant evidence of biodegradation or wear damage after 3 years in vivo. These findings provided compelling evidence for the bio-stability and effectiveness of acetabular cups incorporating a Corethane 80A compliant bearing layer.

7. Concluding Remarks

Taking a general view of the rubber tribology, it seems still in its infancy in the strict sense. The understanding of rubber friction is rather sophisticated, in contrast, that of the rubber lubrication and rubber wear is much inadequate or rather poor even though a large number of experimental results are available. As for the wear process of metal by rubber, there are still a number of key problems needed to be addressed. Obviously, increasing the understanding of rubber tribology will promote the tribological application of elastomers. Vice versa, along with the extensive application of rubbers for tribological purposes, especially the new developments of elastomeric materials, rubber tribology will face new challenges and opportunities. For instance, a nature's supper-rubber was produced by using a gene from fruit flies.[71] It is the most resilient and stretchy rubber known. This supper-rubber could be the starting point for a wide variety of applications, from industry to medicine. Therefore, we would optimistically expect a new developing prospect of the rubber tribology will be opened up in the present decade.

Although it is intended to provide an overview of the recent advances in rubber tribology as inclusive as possible, this chapter only traced a few interesting and important topics owing to limited space.

Acknowledgments

The author is profoundly grateful to the publishers and authors for their kind permission to quote some of the illustrations to this article. Acknowledgements are presented in the figure captions and in the list of references. Special acknowledgement is due to Miss Xiao Yuqi for her assistance in the preparation of this article.

References

1. S. W. Zhang, *Tribology of Elastomer,* B. J. Briscoe, Ed. Tribology and Interface Engineering Series, 47 (Amsterdam, Elsevier, 2004).
2. M. Klüppel and G. Heinrich, *Rubber Chem. Technol.* **73**(2000), 578-606.
3. G. Palasantzas, *Surf. Sci.* **565**(2004), 191-196.
4. *idem, J. Chem. Phys.* **120**(2004), 2889-2892.
5. G. Palasantzas and J. Th. M. De Hosson, *J. Appl. Phys.* 95(2004), 389-392.
6. G. Palasantzas, *ibid.* **97**(2005), 03906-1~034906-4.
7. B. N. J. Persson, *Surf. Sci.* **401**(1998), 445-454.
8. *idem, J. Phys.: Condensed Matter* **18**(2006), 7789-7823(35).
9. A. Le Gal *et al., Applied and Computational Mechanics* (Lecture notes) 2006, 27.
10. A. Le Gal *et al., Wear* 2007, In Press, Available online 20 June.
11. S. Bistac and A. Galliano, *Tribol. Lett.* **18**(2005), 21-25.
12. A. Casoli *et al., ibid.*, **8**(2000), 249-253.
13. T. Tada *et al., ibid.*, **17**(2004), 505-511.
14. S. Kheirandish and E. Jabbari, *Colloid and Polym. Sci.* **284**(2006), 1411-1417.
15. E. L. Deladi *et al., Tribol. Int.* **40**(2007), 8-594.
16. R. Elleuch *et al., Materials and Design* **28**(2007), 24-830.
17. A. S. Koltsov *et al., Int. J. Solids and Structures* **40**(2003), 1299-1310.
18. M. G. Gee *et al., Wear* **259**(2005), 437-1442.
19. D. P. Manning, C. Jones, F. J. Rowland and M. Roff, *J. Safety Res.* **29**(1998), 275-283.
20. W.-R. Chang, *Appl. Ergonomics* **32**(2001), 173-184.

21. *idem.*, *Safety Sci.* **40**(2002), 577-591.
22. G. Cassin, E. Heinrich and H. A. Spikes, *Tribol. Lett.* **11**(2001), 95-102.
23. X.-D. Pan, *J. Polym. Sci., Part B: Polym. Phys.* **42**(2004), 2467-2478.
24. *idem.*, *Rheol. Acta* **40**(2005), 379-395.
25. *idem.*, *Trbol. Lett.* **20**(2005), 209-219.
26. *idem.*, *Wear* **262**(2007), 707-717.
27. M. Chandrasekaran and A. W. Batchelor, *Wear* **211**(1997), 35-43.
28. I. Luzinov *et al.*, *Tribol. Int.* **34**(2001), 327-333.
29. *idem.*, *Macromolecules* **42**(2000), 2267-2274.
30. K. Ikenchi *et al.*, *Wear* **225-229**(1999), 656-659.
31. T. Iwai and Y. Vchiyama, *Wear* **259**(2005), 669-675.
32. D. P. Gerrard and J. Padovan, *Rubb. Chem. Technol.* **76**(2003), 101-121.
33. V. L. Popov and T. Geike, *Tribol. Int.*, **40**(2007), 1012-1016.
34. D.J. T.Hill *et al.*, *Wear* **208**(1997), 155-160.
35. R. A. Beak and R.W.Trass, *Wear* **218**(1998), 145-152.
36. Zhang *et al.*, *J. Mater. Sci.* **36**(2001), 5037-5043.
37. N. S. M. El-Tayeb and R. Md. Nasir, *Wear* **262**(2007), 30-361.
38. J. Bijue *et al.*, *Wear* **249**(2001), 715-726.
39. P. V. Antunes and A. Ramalho, *Tribol. Int.* **38**(2005), 363-379.
40. N. D. Tran *et al.*, *Polymer Degradation and Stability* **91**(2006), 1052-1063.
41. D. I. James, Ed., *Abrasion of Rubber* (McLaren & Sous, Ltd., London, 1967).
42. A. N. Gent and C. T. R. Pulford, *Wear* **49**(1978), 135-139.
43. *idem.*, *J. Mater. Sci.* **14**(1979), 1301-1307.
44. A. N. Gent and W. R. Rodgers, *J., Polym. Sci.* **23**(1985), 829-841.
45. J. M. Charrier *et al.*, *J. Elastom. Plast.* **18**(1986), 200.
46. K. AB-Malek and A. Stevenson, *J. Mater. Sci.* **19**(1984), 585.
47. R. B. King and J. K. Lancaster, *Wear* **61**(1980), 341-352.
48. S. W. Zhang *et al.*, *Lubr. Sci.* **13**(2001), 167-180.
49. S. W. Zhang *et al.*, *Wear* **256**(2004), 226-232.
50. S. W. Zhang, *J. Mater. Sci.* **39**(2004), 5625-5632.
51. D. G. Wang *et al.*, in *Proc. of the* 2nd *Asia International Conference on Tribology* (Korea Society of Tribologists and Lubrication Engineers, Seoul, 2002) p. 223.
52. M. Gunaratne *et al.*, *J. Mater. In Civil Eng.* **12**(2000), 46-54.
53. K. A. Grosch, *Rubb. Chem. Technol.* **77**(2004), 791-813.
54. K. Hofstetter *et al.*, *Computer and Structure* **84**(2006), 1151-1163.
55. P. Embury and J. Armour, *Sealing Technol.* **2004**(2004), 7-10.
56. J.-M. Degrange *et al.*, *Wear* **259**(2005), 684-692.
57. M. Thomine *et al.*, *Tribol. Int.* **40**(2007), 405-411.
58. G. K. Nikes, Determination of polymeric sealing principles for end user high reliability, Technical Report DOW-08/01 (Donoty project), Mechanical Engineering Department, Tribology Section (Impirical College London, London, 2001).
59. *idem.*, *J. Tribol.* **125**(2003), 60-69.

60. *idem.*, *Tribol. Int.* **37**(2004), 689-699.
61. *idem.*, *J. Eng Tribol.* **217**(2003), 365-373.
62. *idem.*, *ibid.* **217**(2003), 461-473.
63. G. K. Nikes and R. S. Sayles, *Tribol. Int.* **37**(2004), 651-660.
64. *idem.*, *ibid.* **39**(2006), 622-634.
65. R.F. Salant, *Sealing Technol.* **1**(2007), 7-11.
66. N.Patir and H.S. Cheng, *J. Lubr. Technol.* **100**(1978), 12-17.
67. M. Masoomi *et al.*, *Appl. Composite Mater.*, 2006, published online first.
68. A. Shojaei *et al.*, *Composite Sci. and Technol.*, 2007, In Press, Available online 20 March.
69. D. L. Cabrera *et al.*, *Proc. IMechE, Pt.J, J Eng Tribol.* **219**(2005), 125-132.
70. I. Kham *et al.*, *Biomaterials* **26**(2005), 633-643.
71. J. Bohannon, *Science Now*, 10/12/2005, p. 1-2.

PART II

Reinforced Polymers

CHAPTER 10

WEAR OF POLYTETRAFLUOROETHYLENE AND PTFE COMPOSITES

Thierry A. Blanchet

Rensselaer Polytechnic Institute, Troy NY, United States

Polytetrafluoroethylene (PTFE) was discovered by chance in 1938 by Roy Plunkett at the DuPont company, and within a decade establishment of initial commercial production facilities brought PTFE's widespread availability. In addition to this polymer's high melting point and chemical resistance, its unusually low friction behavior was noted early on,[1] leading to a great deal of activity over the half-century since developing a basic understanding of its tribological capabilities as well as its useful application as a self-lubricating bearing material. This chapter surveys the molecular and crystalline structure of PTFE and its related transfer film and friction behavior, the severe wear mechanism often accompanying such transfer behavior and its prevention via hard micro-fillers within a PTFE composite, as well as alternate approaches towards wear resistance which avoid the abrasive countersurface damage such hard filler particles also typically present. Finally, the recent development of extremely wear-resistant PTFE nano-composites is presented.

1. Processing and Structure

PTFE, $-(CF_2-CF_2)_n-$, is polymerized into a powder form from tetrafluoroethylene gas[2] under pressure in an aqueous medium via a free radical-initiated addition process to extremely high molecular weight chains of up to $M_n>10^8$. This polymerization creates linear chains of smooth profile, lacking branches or bulky side groups. As opposed to the planar zigzag of the carbon backbone of polyethylene chains, where each CH_2 group has effectively rotated 180° from its neighbor, the carbon backbone of PTFE must instead adopt a helical conformation to accommodate the larger fluorine along its length (Fig. 1). At temperatures of 19°C or less, each CF_2 group is only rotated 166.2° from its neighbor, requiring 13 groups to attain 6 twists of the chain, in other words a 13_6 helical structure.[3] For temperatures above 19°C the PTFE molecule instead transforms to a 15_7 helix.[4]

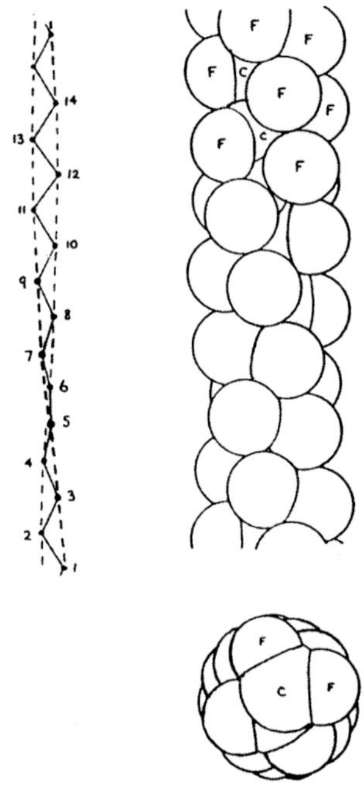

Fig. 1. Helical structure of PTFE molecule (side-views above and end-view below). Below 19°C, thirteen CF_2 groups are required for the helix to complete six full twists.[3] Reprinted by permission from Macmillan Publishers Ltd: Nature.

With such a smooth and linear profile, the molecules within as-polymerized powder are able to organize themselves to levels of crystallinity as high as 92-98%. The melt of such as-polymerized powder occurs at approximately 342°C, however as a result of its extremely high molecular weight PTFE does not truly liquefy under such heating. Thus bulk PTFE must be consolidated by other than typical melt flow processing techniques such as screw extrusion. Press-sinter approaches are instead commonly employed, where powder is cold-pressed into a mold at pressures of 10MPa or more then sintered above the melting point, typically at 380°C, into a bulk form that upon cooling may finally be machined to final shape. PTFE composites may be formed by simply blending filler particles into the PTFE powder prior to press-sinter.

Sintered PTFE is thought to form a 'banded' microstructure (Fig. 2), where the chains arrange parallel to one another with their axes falling within the plane of thin crystalline slices, with disordered regions separating slices as they stack to form bands.[5,6] Based on the spacing of striations along the length of these bands observed upon fracture surfaces, the hypothesized crystalline slices appear to have thickness in the range 20-30nm.[7]

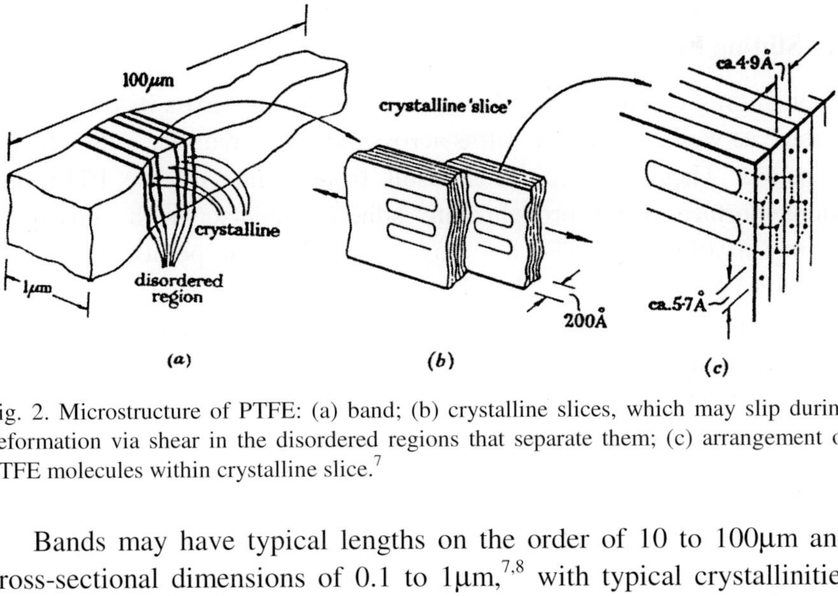

Fig. 2. Microstructure of PTFE: (a) band; (b) crystalline slices, which may slip during deformation via shear in the disordered regions that separate them; (c) arrangement of PTFE molecules within crystalline slice.[7]

Bands may have typical lengths on the order of 10 to 100μm and cross-sectional dimensions of 0.1 to 1μm,[7,8] with typical crystallinities ranging from 30 to 70%, depending upon the rate at which the PTFE is cooled after sintering with the larger values corresponding to slower cooling rates.[8,9] These reduced levels of attainable crystallinity relative to the very high values present in the as-polymerized powder result from the introduction of occasional CF_3 bulky side groups along the otherwise smooth PTFE molecules during sintering.[10]

At ambient pressures, below 19°C the 13_6 helical PTFE chains within the crystalline slices arrange themselves into a triclinic arrangement with 0.554nm separation between neighboring chains, referred to as Phase II. As temperature is increased above 19°C and transition to a 15_7 helix occurs, the slice crystal structure instead becomes hexagonal with chain separation increased to 0.566nm. Such a Phase I exists only up to 30°C, beyond which disorder in the 15_7 helix initiates and a resulting increased chain spacing in a pseudohexagonal crystal structure comprises a Phase IV at higher temperatures up to the melt point.[11,3-4,7] Following the initial 342°C melt of powder during sintering, any subsequent melt point of PTFE occurs more near to 327°C. Biswas and Vijayan[12] offer a detailed summary of this temperature-dependent structure of PTFE.

2. Sliding Friction

Upon sliding contact, PTFE forms 'running' films across its own sliding surface as well as 'transfer' films across the countersurface with which it is mated. These surface films result from points on the PTFE and countersurface being brought into adhesive contact with subsequent sliding, as fibrils of PTFE are drawn between these points maintaining adhesive contact while also accommodating relative motion. With increased sliding displacement the fibrils eventually rupture along their length, leaving fragments running along the PTFE surface as well as transferred to the countersurface, oriented along the sliding direction (Fig. 3). Such fragments fractionally cover the countersurface, and with increased sliding they fill in to form a transfer film. As the shear strengths of such running and transfer surface films can be an order of magnitude lower than that for bulk PTFE, unusually low friction for dry sliding contact is often observed, with friction coefficients as low as μ=0.005 cited.[13] As PTFE-on-countersurface sliding couples are transformed to PTFE-on-overlaying PTFE transfer film, the frictional behavior of PTFE becomes largely insensitive to countersurface composition or roughness.

Early studies of PTFE transfer films measured thicknesses near to the hypothesized 20-30nm slice thickness.[7,8] This suggested that PTFE

(a) (b)

Fig. 3. Development of PTFE surface films. (a) strands of running film along sliding direction (between lower-left and upper-right) over PTFE surface spanning machining mark recess persisting from surface preparation (reprinted from ref. 14 with permission from Elsevier), and (b) strands of transfer film over metal countersurface along sliding direction (between upper-left and lower-right) spanning pore.[15]

surface films form by a destruction of the banded structure, whereby shearing of the disordered regions between neighboring crystalline slices allowed sliding displacement to be accommodated by a spreading out of the slices from their stacks into the drawn fibrils and resultant surface films. However, measurements of transfer films with thicknesses considerably less than the proposed slice thickness have since been made.[16-19] Transfer film thickness measurements even approaching the 0.56nm thickness of the PTFE molecule[20-21] instead suggest that these sliding-induced surface films form via the drawing and extension of individual molecules along the sliding direction. Such a surface film formation mechanism would result in molecular orientation along the sliding direction, which has been confirmed by electron diffraction,[22,16] birefringence measurements,[22,17] and atomic force microscopy.[23] This sliding-induced molecular orientation is aided by the high degree of lateral mobility of PTFE chains, which results from not only their smooth linear molecular profile (also found in high-density polyethylene) but also importantly from the ability of the electron clouds of the large fluorine ions to screen bonding forces between neighboring molecules[22].

With regards to corresponding friction behavior, upon initial sliding the coefficients of friction are first high $\mu>0.2$, reflective of the non-oriented molecules on the PTFE surface and absence of transfer film upon the countersurface. It is not until the oriented running and transfer surface films are established that PTFE's more well-known low dry-sliding friction coefficients $\mu<0.1$ become apparent. In applications where the sliding direction on the PTFE surface continuously changes, molecular orientation coincident with the sliding direction may never be attained and friction coefficients will remain at the high initial values. In other applications where one extended period of unidirectional sliding or reciprocation is followed by a period of sliding or reciprocation along an alternate direction, friction will become momentarily high until molecular re-orientation along the new sliding direction restores the previous low steady-state values.[16]

At room temperature Tanaka *et al.*[8] observed the steady-state coefficient of friction of PTFE in unidirectional sliding to monotonically increase with sliding speed, from roughly $\mu=0.07$ at V=1mm/s up to values as high as $\mu=0.4$ at V=1m/s (Fig. 4). In other words, PTFE's often-cited low-friction $\mu<0.1$ behavior does not exist universally, as friction coefficients can actually become quite high under some operating conditions. Tanaka *et al.*[8] empirically described the speed-dependent variation in friction behavior by a proportionality to V^n where the power-law exponent is approximately n=0.26. Smurugov *et al.*[24] also employ such power law description of friction, with a comparable speed exponent of 0.3. As relative motion of the two contacting surfaces is being accommodated by the displacement of oriented PTFE molecules past one another within drawn fibrils and their resultant surface films, and such polymeric deformation is viscoelastic, increased sliding speed would prescribe higher rates of deformation and therefore the observed increase in frictional resistance. As the viscosity of this chain-drawing deformation process would decrease with increasing temperature, at any constant speed Tanaka *et al.*[8] indeed found friction to decrease as temperature was increased in the range 296K to 373K.

As shown in Fig. 4, at any test temperature in this range the shape of the friction behavior maintained the same power law relationship to

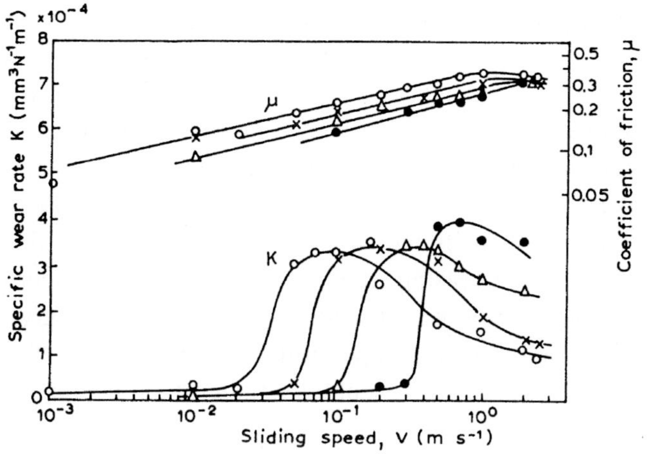

Fig. 4. Wear rate and coefficient of friction of unfilled PTFE, as a function of sliding speed at various temperatures ∘ 23°C, × 50°C, Δ70°C, • 100°C. Reprinted from ref. 8 with permission from Elsevier.

sliding speed, though shifted on the speed axis by a temperature-dependent factor a_T, such that friction coefficient proportionality to $(a_T V)^n$ could describe both its sliding speed and temperature dependence. This temperature dependence was shown to have an Arrhenius form

$$a_T = \exp\left(\frac{\Delta H}{R}\left(\frac{1}{T} - \frac{1}{T_0}\right)\right) \qquad (1)$$

where T_0 is a reference temperature (commonly room temperature) at which the speed-dependent data set is not shifted (i.e. $a_T=1$), and ΔH was found to be approximately 7kcal/mol. Thus, the combined speed- and temperature-dependent behavior of PTFE sliding friction was found to be described as

$$\mu \propto V^n \exp\left(\frac{n\Delta H}{RT}\right) \qquad (2)$$

Blanchet and Kennedy[14] measured PTFE friction as a function of sliding speed at four different temperatures within the range investigated by Tanaka[8] and found the same speed and temperature dependencies, with comparable values of n and ΔH. At a single constant sliding speed of 0.05m/s, Burris[25] showed the behavior of decreased PTFE friction with increasing temperature to maintain such an Arrhenius dependence over an expanded range of test temperatures from 249K to 416K. The activation energy E_a=1.2kcal/mol they determined over this broader range is comparable to the effective activation energy $n\Delta H$=1.8kcal/mol in the preceding description of the temperature dependence of friction from Tanaka[8]. It should be stated that McLaren and Tabor[26] reported prior observation of such increases in PTFE friction with increasing sliding speed or decreasing temperature, though no corresponding mathematical description was provided. Additionally, at a given value of environmental background temperature, the increase of friction with increasing sliding speed will eventually flatten then decrease as a result of increased frictional heating.

3. Wear

Countering its ability to provide low-friction in dry sliding, it is also commonly held that PTFE has very poor wear resistance with severe wear rates reaching towards ~10^{-3} mm^3/Nm. However such poor wear resistance does not prevail universally for PTFE. For example, considering the wear measurements in Fig. 4 by Tanaka *et al.*,[8] at room temperature PTFE's wear rate remains mild at more near to $2*10^{-5}$ mm^3/Nm for sliding speeds up to 0.02m/s before suddenly transitioning to the more well-known severe values. At 100°C these mild wear rates are maintained to speeds as high as 0.3m/s before finally transitioning to severe rates at higher speeds. Blanchet and Kennedy[14] similarly showed PTFE to possess mild wear behavior before transitioning to severe wear as sliding speed was increased beyond a critical value, and that the critical speed threshold up to which mild wear would prevail increased with increasing temperature. The temperature-dependent shifting of speeds at which wear phenomena such as mild-severe transition occur is described by the same factor a_T as the friction behavior, such that combinations of speed and temperature conditions which produce the same product $a_T V$ result in comparable wear behavior.

In a much earlier room-temperature study corroborating the limitation of $\mu<0.1$ low friction PTFE behavior to low sliding speeds ~1mm/s, with higher speeds V>1m/s instead producing high friction

$\mu > 0.3$, Makinson and Tabor[7] correspondingly noted at lower speeds that the PTFE transfer on the countersurface was limited to thin (<40nm) films while at the higher speed lumps and slabs of PTFE transfer with thicknesses up to 5μm appeared. While at low speed mild wear proceeded by the cyclic deposition and eventual detachment of thin transfer films, Blanchet and Kennedy[14] noted that the transition to severe wear was accompanied by the onset of a mechanism producing large plate-like PTFE debris (Fig. 5) with thicknesses on the order of micrometers and in-plane dimensions of several hundreds of micrometers.

Fig. 5. Large plate-like wear debris particle formed by PTFE when increased sliding speed induces severe wear conditions.[19]

The severe wear mechanism producing such plate-like debris appeared to be a fracture process, where cracks nucleate in the subsurface at a depth of several micrometers, with subsequent propagation preferentially on planes parallel to the sliding surface as directed by sliding-induced molecular orientation along the same planes within a surface-affected zone. The enabling of such a severe wear mechanism only at higher speeds is thought to be related to the higher friction coefficient and therefore higher shear stresses being imposed on the PTFE at the sliding surface, driving the hypothesized subsurface cracking mechanism.[14]

4. PTFE Composite Wear Reduction Mechanisms

The ability of many hard filler materials to dramatically reduce its wear has been known since some of the earliest investigations of PTFE tribology, for example.[27,28] As compared to a typically severe $0.46*10^{-3}$ mm^3/Nm value measured for unfilled PTFE in room-temperature sliding at 2.7m/s, Lancaster[29] found that adding any of a broad variety of chopped fiber or other hard particulate filler reduced wear rate for the PTFE composite down to mild levels of at least $0.71*10^{-5}$ mm^3/Nm, or in some instances lower by up to another order of magnitude. Similarly, Tanaka and Kawakami[30] showed over an expanded range of high speeds 0.1 to 2.5m/s that many additional particulate fillers similarly reduced the wear rate of unfilled PTFE from severe values approaching 10^{-3} mm^3/Nm by two or more orders of magnitude down to a mild range of 10^{-5} to10^{-6} mm^3/Nm, and similarly reduced wear rates as shown in the example of Fig. 6 have since been repeatedly reported.

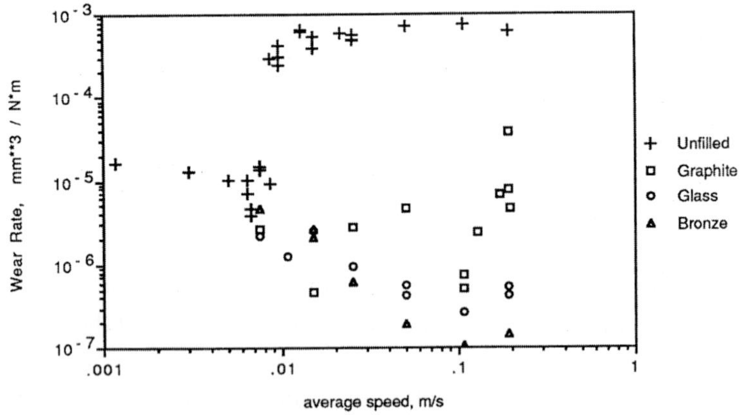

Fig. 6. Wear rate of PTFE composites (15% graphite-filled, 25% glass-filled, 40% bronze-filled by weight), as well as unfilled PTFE, as a function of sliding speed.[15]

Numerous wear reduction mechanisms of these particulate fillers in PTFE have been hypothesized and debated. Considering unfilled PTFE wear to proceed by a cyclic process whereby eventual removal of previously transferred PTFE from the countersurface is followed by replenishment of transfer from the PTFE body, the various hypothesized wear reduction mechanisms can be divided between those concentrating on the initial removal of material from the PTFE body to form the transfer as the rate-limiting step, and others such as augmented transfer film adhesion that consider the subsequent removal of transferred material from the countersurface as rate-limiting.

4.1. *Transfer Film Adhesion*

With regards to the augmented transfer film adhesion hypothesis, Cadman and Gossedge[31,32] and Gao *et al.*[33] conducted X-ray Photoelectron Spectroscopic (XPS) investigations of interactions between PTFE and various metallic countersurface as well as filler materials, and did note chemical interaction as evidenced by the production of metal fluorides. However, given fluorine's monovalency such metal fluorides are not able to directly provide the chemical bonding between a countersurface and the carbon backbones of PTFE chains in the transfer film that could increase their adhesion. Nonetheless, it does serve as indirect evidence of unsatisfied reactive carbon sites along the PTFE backbone from which the fluorine was abstracted, provided such metal fluoride production indeed occurs under tribological conditions and is not a damage artifact from incident radiation of the spectroscopic technique such as noted by Wheeler and Pepper.[34]

The field ion microscopy and the Auger emission spectroscopy studies of Brainard and Buckley[35] and Buckley and Brainard[36] suggested that chemical bonding could occur directly to carbon in the PTFE backbone, provided that the metallic countersurface is sputtered clean of its oxides, with abrasion by the hard fillers in a PTFE composite potentially providing removal of such contaminant countersurface films. However, using metal fluoride as a proxy indication of direct chemical interaction, Gong *et al.*[37] cast doubt on the importance of transfer film adhesion to the wear resistance of PTFE composites, since XPS analysis of steel countersurfaces with transfer films deposited under ambient conditions indicated metal fluoride formation even in the case of unfilled PTFE.

4.2. *Preferential Load Support*

Focusing instead on the initial removal of material from the PTFE body as the rate-limiting step of the transfer wear process, Lancaster[29] suggests as one important role that hard filler particles simply help to support the normal load while also possessing a wear rate K_f less than that of the PTFE matrix K_m. If it is initially assumed that a uniform and equal pressure (EP) condition is exerted over both filler- and matrix-occupied regions of the sliding surface, the wear rate of the composite K_c adopts a linear rule-of-mixtures on the volume fractions of the filler x_f and the matrix $x_m=1-x_f$.

$$K_c = x_f K_f + x_m K_m = K_m - (K_m - K_f)x_f \tag{3}$$

While such a simplistic expression does describe a composite wear rate that decreases with increasing volume fraction of wear-resistant filler, it can not describe the observed orders-of-magnitude decreases in composites whose majority is comprised of the PTFE matrix.

More precisely, Lancaster[29] suggested filler particles support the normal load preferentially, with contact pressures σ_f exceeding the average level of contact pressure p over the composite surface. Tanaka and Kawakami[30] described this filler contact pressure for fillers of aspect ratio l/r oriented normal to the sliding surface with shear stress τ acting along their side surfaces and average pressure p existing uniformly in the far-field of the composite acting on their back surfaces as

$$\sigma_f = p + 2\tau\left(\frac{l}{r}\right) \tag{4}$$

It can be shown that this preferential load support by the filler (Fig. 7) shields the less wear-resistant PTFE matrix from load, reducing its contact pressure to

$$\sigma_m = p - 2\tau\left(\frac{l}{r}\right)\frac{x_f}{1 - x_f} \tag{5}$$

The resulting expression for composite wear rate indeed possesses a steeper decrease with increasing filler volume fraction.

$$K_c = K_m - \left(1 + 2\frac{\tau}{p}\frac{l}{r}\right)\left(K_m - K_f\right)x_f \tag{6}$$

However, such a model has several inadequacies, such as the ability of flake-shaped fillers with negligible aspect ratio *l/r* and therefore no preferential load support capability such as mica, graphite and MoS_2 to still provide multiple order-of-magnitude wear reductions[29,30,38]. Additionally, as the preferential load support term $\tau(l/r)$ increases, the expression for matrix contact pressure would eventually become zero, implying no matrix wear, then negative. The actual upper limit on filler preferential load support is the equal wear (EW) condition, $\sigma_f K_f = \sigma_m K_m$, where the filler and matrix each experience the same rate of increasing wear depth.

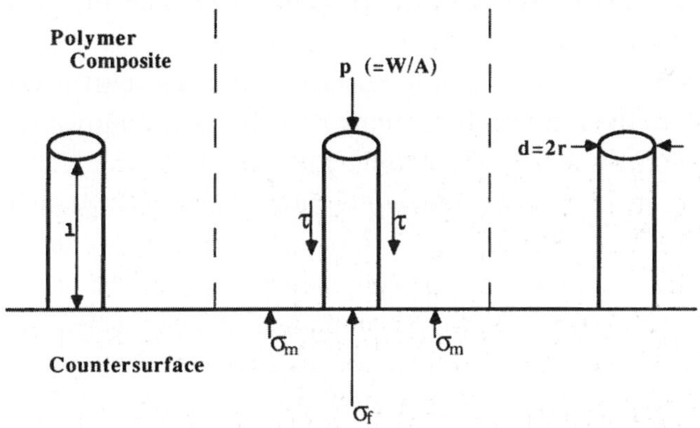

Fig. 7. Composite supporting load W over apparent contact area A. Cylindrical fillers of length *l* and radius r oriented normal to contacting countersurface, subject to a contact pressure σ_f greater than nominal contact pressure p due to shear stress τ at interface with matrix. This preferential load support by the filler reduces matrix contact pressure to $\sigma_m < p$.[39]

A well-known drawback of PTFE structurally is its propensity for creep deformation under sustained load, and as such it is also unlikely here to sustain its application of interfacial shear stresses τ onto the fillers. A contact situation much more near to uniform pressure on filler and matrix would instead exist, but other events must also correspond at the sliding surface. Specifically, since the PTFE matrix is less wear-resistant than the filler, under more uniform pressure its rate of wear depth increase would instantaneously exceed that of the filler, leaving the filler incrementally protruding and thus subject to a higher load. This process can not progress indefinitely for a PTFE composite of initially uniform filler volume fraction, instead the heightened contact pressure on the filler leads to it simultaneously sinking back into the matrix via PTFE cold flow. With as yet unexposed filler also in the underlying bulk, this process leads to accumulation of filler in the near-surface region and development of a volume fraction profile varying from x_{fb} in the bulk to x_{fs} at the sliding surface.

Such filler surface accumulation in PTFE has been noted in several studies[40-44], and an example is shown in Fig. 8a. As the wear-resistant filler gradually becomes more present at the sliding surface, and thus increasingly supports the normal load, the wear rate of the composite gradually drops throughout this transient run-in process. This increasing presence of filler on the sliding surface continues until a condition is reached where the ratio of the volume rates of wear removal of filler and matrix from the surface is equal to the ratio of the volume fractions of the filler and matrix in the composite bulk. Han and Blanchet[43] have modeled the steady-state surface filler volume fraction x_{fs} that will result under this scenario, and have also shown for such a mechanism of filler preferential load support that the composite wear rate K_c will instead follow

$$\frac{1}{K_c} = x_f \frac{1}{K_f} + x_m \frac{1}{K_m} = \frac{1}{K_m} + x_f \left(\frac{1}{K_f} - \frac{1}{K_m} \right) \tag{7}$$

Such an inverse rule-of-mixtures description of composite wear rate K_c, equivalently a linear rule-of-mixtures description of composite wear resistance $1/K_c$ as observed experimentally in the example of Fig. 8b, would also result under assumption of equal wear (EW) conditions

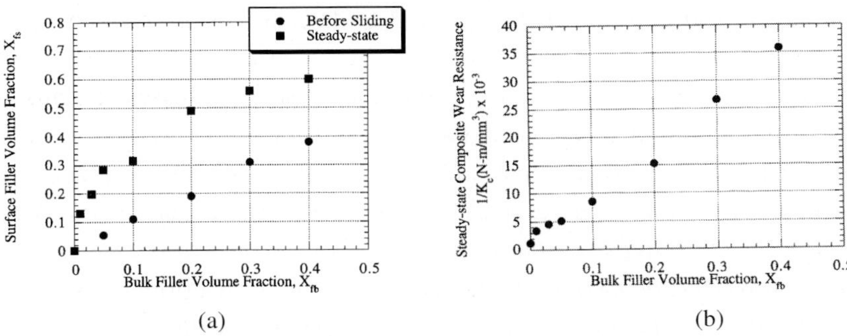

Fig. 8. Wear behavior of copper-filled PTFE as a function of bulk filler volume fraction x_{fb}. (a) Volume fraction x_{fs} of fillers measured at the surface of the composite before sliding as well as increased values due to surface accumulation under steady-state wear conditions, and (b) steady-state composite wear resistance, $1/K_c$.[43]

previously discussed. Such a model more closely reflects the precipitous orders-of-magnitude decrease in PTFE composite wear rate that occurs with increasing filler bulk volume fraction even at low filler contents.

4.3. *Debris Size Control*

While each of the two previously discussed wear reduction mechanisms of fillers may still have some pertinence to PTFE composite wear, neither addresses the foremost observation that unfilled PTFE itself is already capable of wear at mild rates ~10^{-5} mm^3/Nm with transfer in the form of thin films, provided that sliding speed is kept sufficiently low. It is not until a critical speed threshold is exceeded that unfilled PTFE wear rates become severe ~10^{-3} mm^3/Nm as the production of much larger plate-like debris becomes enabled. While augmented transfer film adhesion may indeed be an active wear reduction for secondarily decreasing wear rate towards ~10^{-6} mm^3/Nm or lower after transfer wear has already been limited to the form of thin films, the primary wear mechanism must address the initial reduction of wear rate to ~10^{-5} mm^3/Nm through prevention of large plate-like debris production under sliding conditions that would otherwise be severe for unfilled PTFE.

In his 1977 review, Ricklin[45] pointed out that in PTFE one of the primary wear-reducing functions of filler particles, through its shape and size, must be the prevention of large wear particle production. Bahadur and Tabor[46] similarly add that the reduced wear rate of PTFE must be primarily attributed to a filler mechanism controlling the size of the wear debris, since PTFE composites produce fine particles as opposed to the large fragmented sheets released by unfilled PTFE. With respect to the previously described process of large plate-like debris production, occurring by cracks at several micrometers depth within the PTFE propagating hundreds of micrometers parallel to its sliding surface and activated by sliding conditions providing sufficient frictional shear stress to drive this propagation, Blanchet and Kennedy[14] simply hypothesized that filler particles interrupted such crack tips and controlled debris size by limiting propagation distances, such as seen in the example of Fig. 9. By such a mechanism the filler particle must be of sufficient size on at least the scale, of micrometers, of the subsurface depth at which such cracks exist to effectively interfere with their propagation, as opposed to smaller scales where fillers particles would instead be encapsulated within the large plate-like debris the PTFE would otherwise continue to produce.

With such a debris size control primary wear reduction mechanism in place, others may now act secondarily. For example, descriptions above for decreasing composite wear rate with further increases in filler volume fraction via filler load support may be employed with K_m already more nearly reduced to 10^{-5} mm^3/Nm rather than ~10^{-3} mm^3/Nm. Or such a primary and secondary wear reduction may couple, for example filler surface accumulation related to the filler load support mechanism may lead to decreased filler separation in the near-surface and in turn further decreased PTFE debris size.

Similar to the above wear reduction mechanism where filler particles interfere with a cracking mechanism that would otherwise liberate large plate-like debris, Tanaka *et al.*[47] proposed filler particles interfere with the large-scale destruction of PTFE's banded structure that occurs during transfer wear. In Tanaka and Kawakami's[30] study of a variety of fillers they similarly point out that a sufficient particle size is required to effectively interfere with this destruction process, with smaller filler particles instead simply transferred with the wearing PTFE onto the countersurface, such as for 0.3μm TiO_2 observed to have a lesser capability to reduce PTFE wear.

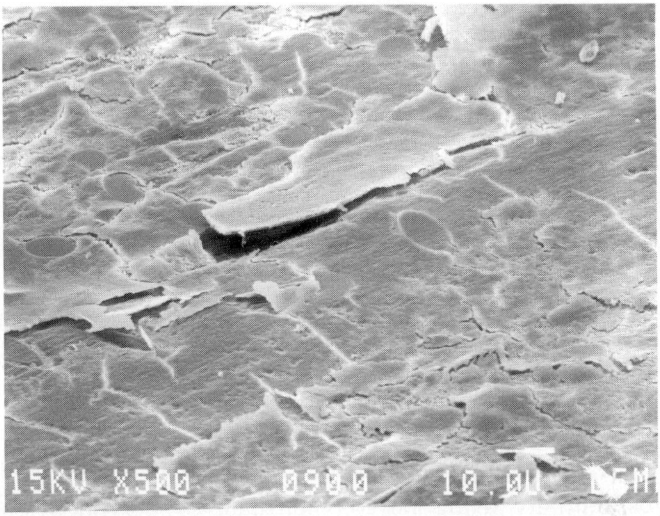

Fig. 9. Glass fiber fillers (appearing as circles upon intersecting composite wear surface) limiting PTFE debris to ~10μm scale, preventing the formation of larger plate-like debris typical of severe wear.[15]

5. Polymer-Filled PTFE Composites

When using hard-particle fillers to resist severe wear of PTFE, it is of concern whether such fillers compromise PTFE's attractive low-friction ($\mu<0.1$) properties. It should be kept in mind, however, that sliding conditions posing a severe wear risk to PTFE and thus demanding wear-resisting fillers are already at speeds that are sufficiently high to raise the friction coefficient of even unfilled PTFE to typical values of $\mu=0.2$ or higher. Under such high speed conditions the friction of hard particle-filled PTFE often may not differ significantly from unfilled PTFE, such as in Lancaster[29] where most friction coefficients for PTFE composites as well as unfilled PTFE are measured to be near $\mu=0.25$. Tanaka and Kawakami[30] made a similar observation about a broad insensitivity of the friction coefficient of PTFE and its composites on filler type, and attributed this behavior to the ability of PTFE at surface positions adjacent to filler particles to draw running films over the filler particles, effectively lubricating them. Exceptions to this behavior, where friction coefficients may instead increase to $\mu>0.4$ or higher, were associated with larger filler particles whose size (>50μm) the PTFE surface running films may not be able to span.

The primary compromise hard-particle fillers actually present to a tribosystem while reducing the wear of the PTFE body is the abrasive wear damage they may introduce to the mating countersurface.[41,48] In one interesting approach Burroughs[49] was able to obtain two-order-of-magnitude reductions of PTFE wear using hard boric oxide fillers while additionally not inducing countersurface abrasion, relying on reaction of the filler with water vapor to continuously maintain boric acid lamellar solid lubricant layers covering the filler contact surface. Of course, such a novel approach is quite limited to applications in environments of sufficient humidity.

Arkles[50] addressed countersurface abrasion via polymer-filled PTFE composites, initially using polyoxybenzoate or polyphenylene sulfide (PPS) fillers. These composites provided three-order-of-magnitude reductions in wear equivalent to that provided by glass fiber filler at 15 wt%, while reducing steel countersurface abrasive wear ten-fold, towards negligible levels experienced with unfilled PTFE. Friction coefficients of these composites were elevated relative to unfilled PTFE, however, presumably due to large polymeric filler particle size and the inability of PTFE running films to fully lubricate the composite surface. Additionally, the melt point of the PPS filler is 60°C lower than the 327°C value of PTFE, potentially imposing some slight compromise in thermal capabilities of such PTFE composites.

Using instead a 5μm particle size polyetheretherketone (PEEK) with melt point greater than PTFE, Burris and Sawyer[51] produced polymer-filled PTFE composites which at weight fractions of 10-30% and a sliding speed of 0.05m/s displayed wear rates sharply reduced by up to five orders-of-magnitude compared to the $0.6*10^{-3}$ mm^3/Nm value measured for unfilled PTFE, while maintaining comparable or lower friction. Using PEEK filler of unspecified particle size, Briscoe[52] had previously demonstrated similarly dramatic five-order-of-magnitude reductions in wear rate and non-compromised friction performance for polymer-filled PTFE at high 2m/s speed using 20wt.% PEEK, with wear rate continuing to drop modestly with further increases in PEEK content. Over similar ranges of filler content Hager and Davies[53] also observed reduced wear for PEEK-filled PTFE, but only by two orders-of-magnitude. Such presence of PEEK as a wear-resistant filler within PTFE can furthermore be graded with subsurface depth towards a fully PEEK bulk, thus creating a multifunctional body of superior stiffness and strength yet still possessing a self-lubricating surface region.[54] This multifunctional body can additionally be provided higher thermal conductivity to withstand greater 'PV' frictional heating challenges by also including an interconnected metallic network within the PEEK bulk.[55]

While wear resistances provided by PEEK fillers are certainly impressive, this large variation in extent of wear reduction is unclear. As a most unusual example of the ability of a broad variety of fillers to provide wear resistance to PTFE, Burris *et al.*[56] found that even the polymeric protein comprising human beard hair when chopped into a short-fiber filler was capable of multiple order-of-magnitude reductions of wear rate.

Hong and Pyun[57] produced non-abrasive polymer-filled PTFE composites that were entirely fluoropolymer by employing fluorinated ethylene propylene (FEP) filler particles. Attainable wear resistances were much more modest than those for previously discussed polymer fillers, with the ten-fold wear reduction initially provided by a 5% weight fraction doubling to a twenty-fold wear reduction with increasing FEP weight fractions up to an optimum at 35%. As the wear resistance of bulk FEP can be greatly increased by radiation processing,[58,59] such irradiation of the filler enabled the twenty-fold reductions in the wear rate of FEP-filled PTFE to be increased up to two-order-of-magnitude reductions, again without compromising friction performance relative to unfilled PTFE.[58,60]

6. Radiation-Induced Wear Resistance in PTFE

As FEP is a copolymer of tetrafluoroethylene with fluorinated propylene, it is not unexpected that the radiation-induced wear resistance noted above has also been attained in pure PTFE, without need for fillers or concern for their abrasivity. Briscoe and Ni[61] showed that the severe wear rates in the range 0.5 to $0.7*10^{-3}$ mm^3/Nm measured for unfilled

PTFE gradually decreased by up to two orders-of-magnitude with gamma irradiation as dose was increased to 20Mrads, remaining relatively level at higher doses up to 100Mrad. Wear reduction also appeared to be achieved via a mechanism of debris size control, as the 'plate' form observed for unirradiated PTFE became finer and powdery. While Menzel and Blanchet[59] also found gamma irradiation to provide such a two-order-of-magnitude reduction, considerably greater wear resistances were attained using electron beam irradiation (Fig. 10) as wear rates gradually dropped by up to three orders-of-magnitude down into the ~10^{-6} mm^3/Nm range as dose was increased to 30Mrad.[62]

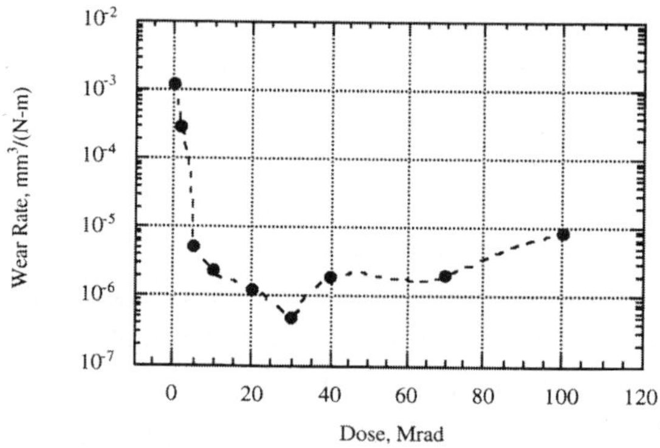

Fig. 10. Decreased steady-state wear rate of PTFE as a function of irradiation dose of electrons of 10MeV energy.[62]

To attain deep penetration into the PTFE as is inherently provided by gamma irradiation, and thus generate a wear resistance that will be long-lived even as wear depth progresses, a high electron energy of 10MeV was employed. For example, in the case of irradiation by 0.225MeV electrons the wear-resistant effect was limited to a ~0.5mm thick surface layer, with wear rate returning to the severe ~10^{-3} mm^3/Nm typical of unfilled PTFE once worn away.[63]

With regards to friction, McLaren and Tabor[64] observed a monotonic increase with increasing dose up to 50Mrad, asymptotically approaching μ=0.2 for tests conducted at sliding speeds of 0.01m/s or less. This however should not be viewed as a drawback for the radiation-based approach to wear resistance of PTFE since as previously noted at such low sliding speeds the wear rate of unfilled PTFE is already mild ~10^{-5} mm^3/Nm and without great need for further improvement, as seen for example in Fig. 6. In fact, at the very high speeds where the wear rate of PTFE is most in need of reduction via an approach such as irradiation the opposite trend is observed, with friction coefficient also monotonically decreasing with increasing dose from the high $\mu>0.3$ values for unirradiated PTFE towards a $\mu=0.24$ asymptote. At intermediate speeds between 0.1 and 1m/s the initial increase of friction with doses up to 5Mrad was thereafter followed by a decrease, asymptotically approaching levels of friction comparable to or better than the unirradiated PTFE.[64] Therefore, under sliding conditions of sufficient speed to cause unfilled PTFE to become severe, the 20 to 30Mrad doses found to be effective in greatly reducing PTFE wear rate will additionally leave PTFE friction at least relatively unaffected if not somewhat improved.

While in ultra-high molecular weight polyethylene similar radiation-induced wear resistance is well-known to result from crosslinking, for example,[65] PTFE's chemical resistance impedes the solvent extraction techniques typically employed to characterize the sol-gel state of the polymer and assess whether the prevailing molecular changes induced by irradiation are chain scission or crosslinking. Briscoe and Ni[61] did note large decreases in strain at rupture, and thus toughness, with increasing dose. The radiation-induced wear resistances they noted were thus limited to sliding against smooth counterfaces, with abrasive wear against rougher countersurfaces such as emery papers, or metallic wire gauzes,[66] instead increasing with increased dose. Additionally, Blanchet and Peng[58] found that PTFE's radiation-induced wear resistance against smooth countersurfaces could be annealed away if reheated to above its melt point.

7. Tomorrow's Challenges: Understanding Nano-Filled PTFE

The hypothesis that filler particles must be of sufficient size to effectively provide wear resistance to PTFE, and to be able to act by mechanisms such as debris size control via interference of subsurface crack propagation,[14] was founded on the observation of wear rates of sub-micrometer (0.3μm) TiO_2-filled composites exceeding those using fillers at sizes of several micrometers or more of materials such as chopped glass fibers, bronze and ZrO_2.[30] In that study, however, each filler material was tested at a single particle size, thus the effect of filler particle size was not isolated from any possible effect of filler material type. Despite wearing more rapidly than the glass, bronze and ZrO_2–filled composites, the sub-micrometer TiO_2 fillers in several instances did in fact provide wear rate reductions of two orders-of-magnitude compared to unfilled PTFE.

Li *et al.*[67] even showed 50nm ZnO nano-filler particles to provide wear resistance to PTFE, in some cases reducing wear volume after a fixed sliding distance of 776m by nearly 100-fold. Sawyer *et al.*[68] similarly found 40nm Al_2O_3 to provide two-order-of-magnitude reductions of PTFE wear rate when used at 10 to 20 wt.%, and in the same range of weight fractions Xiang *et al.*[69] measured comparable reductions in PTFE wear volume using a 50nm Al_2O_3. Xiang and Gu[70] also achieved reductions in PTFE wear volume approaching two orders-of-magnitude with 100nm kaolin particles at 10 wt.%. Using 10nm nano-diamond filler, Lee and Lim[71] found that 2 wt.% addition provided a more modest decrease in wear volume from PTFE coatings of not quite ten-fold, while Lai *et al.*[72] found that the reduction in wear rate of bulk PTFE upon adding 10 wt.% nano-diamond improved slightly, approaching twenty-fold. Regardless of extent achieved, each of the above examples involving particulates on the nano-scale clearly demonstrates that filler may still possess some mechanism to provide wear reduction, even if having size less than any micro-scale threshold, such as the debris which would otherwise be formed by the PTFE if unfilled.

Chen *et al.*[73] showed too that carbon nanotubes of 20-30nm diameter could reduce the wear rate of PTFE, by 290-fold when used at 20 wt.%. Similarly, Lai *et al.*[74,75] showed that a 5 wt.% addition of nano-attapulgite crystal of 10-25nm width could reduce wear rate of PTFE twenty-fold, from $0.6*10^{-3}$ mm^3/Nm down to $3*10^{-5}$ mm^3/Nm, with additional reductions down towards $5*10^{-6}$ mm^3/Nm if the attapulgite filler is also either acid- or thermally-treated. With larger diameters of a couple hundred nanometers, Shi *et al.*[76] showed carbon nanofibers to reduce the wear volume of PTFE by nearly two orders-of-magnitude even when used at a low weight fraction of 2%, with 400-fold reductions achieved with comparably sized potassium titanate whiskers.[77] However in each of these cases the filler is in the form of fibers, having longer lengths that extend into the microscale, and thus they do not demonstrated the ability of nano-fillers to provide PTFE wear resistance as clearly as the previous particulate examples.

For each of those particulate nano-filler examples, the studies did not include a micrometer-scale filler of the same material which could have served as a control to quantify variation in wear-reducing effectiveness as a function of filler particle size. Compared to their observation that carbon black filler reduced PTFE wear rate from $0.2*10^{-3}$ mm^3/Nm to less than 10^{-5} mm^3/Nm uniformly over particle sizes ranging from 13nm to 62nm, Takeichi *et al.*[78] showed this reduced wear rate doubled as particle size was increased to 122nm. But again, no micro-scale control was included to probe the extent of this trend. In their demonstration of the enhanced ability of an 80nm Al_2O_3 filler to reduce PTFE wear rate down as low as $\sim 10^{-6}$ mm^3/Nm, Burris and Sawyer[79] included a 0.5μm filler which in comparison was only able to reduce wear rate to $\sim 5*10^{-5}$ mm^3/Nm, however this study still lacked a control within the range from several micrometers to about 30μm conventionally deemed suitable for PTFE wear resistance.[30]

To more clearly assess effects of broadly varying filler size on PTFE wear resistance, McElwain *et al.*[80] studied two Al_2O_3 nano-fillers (40nm and 80nm), as well as four Al_2O_3 micro-fillers (0.5, 1, 2, and 20μm) all from the same commercial manufacturer and of the same alpha phase at a constant weight fraction of 5%. While the micro-fillers all provided a typical reduction of PTFE wear rate from $\sim 10^{-3}$ mm^3/Nm down to the $\sim 10^{-5}$ mm^3/Nm range, the nano-fillers not only demonstrated a wear-reducing effect, but also one that was superior to the micro-fillers with wear rates dropping another two orders-of-magnitude to $\sim 10^{-7}$ mm^3/Nm (Fig. 11). Furthermore, as gauged by energy-dispersive X-ray spectroscopic evidence of metallic debris embedded into the PTFE surface, the nano-composites were much less abrasive to their countersurfaces than the conventional micro-composites. Finally, as also

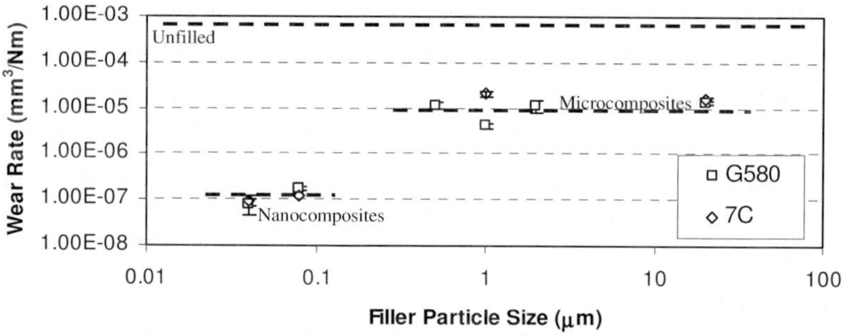

Fig. 11. Wear rate of alumina-filled PTFE as a function of filler particle size. Similar behavior of the filler is shown in both grades of PTFE matrix investigated (G580 and 7C). A datum wear rate of unfilled PTFE is provided for comparison of microcomposite and nanocomposite wear-reducing effectiveness.[80]

shown in many of the above studies,[67,71,72] under conditions which would produce severe wear in unfilled PTFE and thus justify the inclusion of wear-resistant fillers, the addition of nano-particles was typically not observed to compromise PTFE's frictional performance in any significant way.

While it may still hold that micro-fillers provide wear resistance by having a sufficient size to interrupt destructive processes such as subsurface crack propagation and delamination that would otherwise produce from unfilled PTFE the plate-like debris of micrometer-scale thickness, the wear resistance provided by nano-fillers in the examples provided must clearly originate from a different mechanism. In observing a wear rate that gradually decreased twenty-fold with increasing weight fractions of nano-diamond up to 10%, the differential scanning calorimetry (DSC) studies of Lai[72] indicate that the crystallinity of PTFE increased approximately from 40% up to 65%. However in the studies of McElwain[81] the unfilled, micro-filled (5%) and nano-filled (5%) PTFE materials had greatly differing wear rates of ~10^{-3}, 10^{-5} and 10^{-7} mm^3/Nm respectively, despite all having very similar PTFE crystallinities of roughly 30%. When Hu and Eiss[9] widely varied the crystallinity of unfilled PTFE from 25% up to 50% through use of different cooling rates during sintering they found that wear only decreased by approximately 10%. Thus it appears unlikely that any increase in PTFE crystallinity potentially induced by the inclusion of nano-fillers, if observed, could be solely credited with its multiple order-of-magnitude reduction in wear rate.

Schadler *et al.*[82] has alternately hypothesized that upon cooling a sintered composite, the nano-fillers may help to stabilize PTFE's high-temperature Phase I rather than the Phase II or IV that would normally appear near room temperature. Phase I is known to have higher fracture toughness than the lower temperature phases[83], with PTFE having an ability to draw fibrils spanning cracks, and such processes would also be helpful in countering the crack propagation and subsequent particle liberation in PTFE's severe wear mechanism of plate-like debris production. As precedent for the effect of phase on tribological properties, in tests involving unfilled PTFE both Flom and Porile[84] and Steijn[85] reported discontinuities in frictional behavior as temperature was varied through the range where transitions between Phases I, II and IV occur.

In probing potential wear reduction mechanisms of nano-fillers, McElwain *et al.*[80] also performed tests on PTFE having a mixture of 40nm and 20μm Al_2O_3, and in each case the resultant wear rate was equivalent to the ~10^{-5} mm^3/Nm values measured for micro-composites rather than ~10^{-7} mm^3/Nm more typical of nano-composites. Though the precise wear-reducing mechanism for nano-fillers may not yet be ascertained, it appears to be a separate and more effective mechanism than that for micro-fillers, which in this experiment seem to 'poison' the mixed composite by presenting wear mechanism pathways not otherwise present in a nano-composite. Once today's challenge is met of more fully understanding the role of nano-particles in these recent widespread observations of low wear rates, tomorrow's prospect of optimally designing nano-structured PTFE composites of extremely high wear resistance, negligible countersurface abrasivity, and non-compromised frictional performance may finally be realized.

Acknowledgments

The author acknowledges the support of an AFOSR-MURI grant FA9550-04-1-0367 during preparation of this summary. Any opinions, findings, and conclusions or recommendations expressed in this material are those of the author and do not necessarily reflect the views of the Air Force Office of Scientific Research.

References

1. K. V. Shooter and D. Tabor, *Proc. Of the Royal Society (B)*, **65**, 661 (1952).
2. B. C. Arkles and M. J. Schireson, *Wear*, **39**, 177 (1976).
3. C. W. Bunn and E. R. Howells, *Nature*, **174**, 549 (1954).
4. R. E. Moynihan, *J. Amer. Chem. Soc.*, **81**, 1045 (1959).
5. C. W. Bunn, A. J. Cobbold and R. P. Palmer, *Journal of Polymer Science*, **28**, 365 (1958).
6. C. J. Speerschneider and C. H. Li, *Journal of Applied Physics*, **33**, 1871 (1962).
7. K. R. Makinson and D. Tabor, *Proceedings of the Royal Society (A)*, **281**, 49 (1964).
8. K. Tanaka, Y. Uchiyama and S. Toyooka, *Wear*, **23**, 153 (1973).
9. T. Y. Hu and N. S. Eiss, *Proc. Int. Conf. Wear of Materials*, ASME, 636 (1983).
10. A. McNicol, D. Dowson and M. Davies, *Wear*, **181-183**, 603 (1995).
11. F. A. Quinn, D. E. Roberts and R. N. Work, *Journal of Applied Physics*, **22**, 1085 (1951).
12. S. K. Biswas and K. Vijayan, *Wear*, **158**, 193 (1992).
13. C. A. Sperati, Fluorocarbon Polymers, Polytetrafluoroethylene (PTFE), *Handbook of Plastic Materials and Technology*. I.I. Rubin, ed., John Wiley, 117 (1990).
14. T. A. Blanchet and F. E. Kennedy, *Wear*, **153**, 229 (1992).
15. T. A. Blanchet, The Wear Resistance Mechanism of Particulate Fillers in Polytetrafluoroethylene-based Polymer Composites, Ph.D. Thesis, Dartmouth College (1992).
16. C. M. Pooley and D. Tabor, *Proceedings of the Royal Society (A)*, **329**, 251 (1972).
17. J. L. Lauer, B. G. Bunting and W. R. Jones, *Tribology Transactions*, **31**, 282 (1988).
18. E. L. Yang, J. P. Hirvonen and R. O. Toivanen, *Wear*, **146**, 367 (1991).
19. T. A. Blanchet, F. E. Kennedy and D. T. Jayne, *Tribol. Trans.*, **36**, 535 (1993).
20. S. V. Pepper, *Journal of Applied Physics*, **45**, 2947 (1974).
21. D. R. Wheeler, *Wear*, **66**, 355 (1981).
22. R. P. Steijn, *Wear*, **12**, 193 (1968).
23. P. Dietz, P. K. Hansma, K. J. Ihn, F. Motamedi and P. Smith, *J. Mater. Sci.*, **28**, 1372 (1993).
24. V. A. Smurugov, A. I. Senatrev, V. G. Savkin, V. V. Biran and A. I. Sviridyonok, *Wear*, **158**, 61 (1992).
25. D. L. Burris, S. S. Perry and W. G. Sawyer, *Trib. Letters*, **27**, 323 (2007).
26. K. G. McLaren and D. Tabor, *Nature*, **197**, 856 (1963).
27. S. Ricklin and R. R. Miller, *Materials & Methods*, October, 112 (1954).
28. C. J. Speerschneider and C. H. Li, *Wear*, **5**, 392 (1962).

29. J. K. Lancaster, *Brit. J. Appl. Phys. (J. Phys. D)*, Vol. 1, 549 (1968).
30. K. Tanaka and S. Kawakami, *Wear*, **79**, 221 (1982).
31. P. Cadman and G. M. Gossedge, *Wear*, **51**, 57 (1978).
32. P. Cadman and G. M. Gossedge, *J. Mat. Sci.*, **14**, 2672 (1979).
33. J. Gao, J. Zhao and H. Dang, ASLE Proc. 3rd International Conference on Solid Lubrication, 308 (1984).
34. D. R. Wheeler and S. V. Pepper, *J. Vac Sci. Technol. A*, **8**, 4046 (1990).
35. W. A. Brainard and D. H. Buckley, *Wear*, **26**, 75 (1973).
36. D. H. Buckley and W. A. Brainard, *Advances in Polymer Friction and Wear*, L.H. Lee, ed., Plenum Publishing, 315 (1975).
37. D. Gong, Q. Xue and H. Wang, *Wear*, **148**, 161 (1991).
38. B. J. Briscoe, M. D. Steward and A. J. Groszek, *Wear*, **42**, 99 (1977).
39. T. A. Blanchet, *Tribol. Trans.*, **38**, 821 (1995).
40. K. Tanaka, *J. Lubr. Technol.,* **95**, 409 (1977).
41. M. N. Gardos, *Lubr. Engg.*, **37**, 641 (1981).
42. J. E and D. T. Gawne, *Wear*, **176**, 195 (1994).
43. S. W. Han and T. A. Blanchet, *J. Trib.*, **119**, 694 (1997).
44. H. Haiser and W. Haas, *Proceedings of the International Colloquium Tribology 2000-Plus*, Technische Akademie Esslingen, 1443 (2000).
45. S. Ricklin, *Lubr. Engg.*, **33**, 487 (1977).
46. S. Bahadur and D. Tabor, *Wear*, **98**, 1 (1984).
47. K. Tanaka, Y. Uchiyama, S. Ueda and T. Shimuzu, *Proc. of the JSLE-ASLE International. Lubrication Conf.*, pp. 110 (1976).
48. M. Nosaka, M. Oike, M. Kikuchi, K. Kamijo and M. Tajiri, *Lubr. Engg.*, **49**, 677 (1993).
49. B. R. Burroughs, J. H. Kim and T. A. Blanchet, *Tribol. Trans.*, **42**, 592 (1999).
50. B. Arkles, J. Theberge and M. Schireson, *Lubr. Engg.*, **33**, 33 (1977).
51. D. L. Burris and W. G. Sawyer, *Wear*, **261**, 410 (2006).
52. B. J. Briscoe, L. H. Yao and T. A. Stolarski, *Wear*, **108**, 357 (1986).
53. A. M. Hager and M. Davies, *Advances in Composite Tribology*, K. Friedrich, ed., Elsevier, 107 (1993).
54. D. L. Burris and W. G. Sawyer, *Wear*, **262**, 220 (2007).
55. D. L. Burris and W. G. Sawyer, *Wear*, in press (2008).
56. D. L. Burris, B. Boesl, G. R. Bourne and W. G. Sawyer, *Macromol. Mater. Eng.*, **292**, 387 (2007).
57. M. H. Hong and S. I. Pyun, *Wear*, **143**, 87 (1991).
58. T. A. Blanchet and Y. L. Peng, *Wear*, **214**, 186 (1998).
59. B. Menzel and T. A. Blanchet, *Lubr. Engg.*, **58**, 29 (2002).
60. B. Menzel and T. A. Blanchet, *Wear*, **258**, 935 (2005).
61. B. J. Briscoe and Z. Ni, *Wear*, **100**, 221 (1984).
62. T. A. Blanchet and Y. L. Peng, *Lubr. Engg.*, **52**, 489 (1996).
63. T. A. Blanchet, Y. L. Peng and S. V. Nablo, *Tribol. Lett.*, **4**, 87 (1998).

64. K. G. McLaren and D. Tabor, *Wear*, **8**, 3 (1965).
65. O. K. Muratoglu, C. R. Bragdon, D. O. O'Connor, M. Jasty and W. H. Harris, *Proc. 45th Ann. Mtg. of the Orthopaedic Research Society*, 817 (1999).
66. B. J. Briscoe and P. D. Evans, *Proc. International Conference on the Wear of Materials*, 449 (1989).
67. F. Li, K. Hu, J. L. Li and B. Y. Zhao, *Wear*, **249**, 877 (2002).
68. W. G. Sawyer, K. D. Freudenberg, P. Bhimaraj and L. S. Schadler, *Wear*, **254**, 573 (2003).
69. D. Xiang, K. Li, W. Shu and Z. Xu, *J. Reinforced Plastics and Comp.*, **26**, 331 (2007).
70. D. Xiang and C. Gu, *Mater. Lett.*, **60**, 689 (2006).
71. J. Y. Lee and D. S. Lim, *Surf. & Coat. Tech.*, **188-189**, 534 (2004).
72. S. Q. Lai, L. Yue, T. S. Li and Z. M. Hu, *Wear*, **260**, 462 (2006).
73. W. X. Chen, F. Li, G. Han, J. B. Xia, L. Y. Wang, J. P. Tu and Z. D. Xu, *Tribol. Lett.*, **15**, 275 (2003).
74. S. Q. Lai, T. S. Li, X. J. Liu and R. G. Lv, *Macromol. Mater. Eng*., **289**, 916 (2004).
75. S. Q. Lai, T. S. Li, X. J. Liu, R. G. Lv and L. Yue, *Tribol. Inter.*, **39**, 541 (2006).
76. Y. Shi, X. Feng, H. Wang, X. Lu and J. Shen, *J. Appl. Poly. Sci.*, **104**, 2430 (2007).
77. Y. J. Shi, X. Feng, H. J. Wang, C. Liu and X. H. Lu, *Tribol. Inter.*, **40**, 1195 (2007).
78. Y. Takeichi, A. Wibowo, M. Kawamura and M. Uemura, *Wear*, **3-4**, 308 (2008).
79. D. L. Burris and W. G. Sawyer, *Tribol. Trans.*, **48**, 1 (2005).
80. S. E. McElwain, T. A. Blanchet, L. S. Schadler and W. G. Sawyer, *Tribol. Trans.*, **51**, 247 (2008).
81. S. E. McElwain, *Wear Resistant PTFE Composites via Nano-scale Filler Particles*, M. Sc. Thesis, Department of Mechanical, Aerospace & Nuclear Eng., Rensselaer Polytechnic Institute, 83 (2006).
82. L. S. Schadler, L. C. Brinson and W. G. Sawyer, *J. Mineral. Metal. & Mat. Soc.*, March, 50 (2007).
83. E. N. Brown and D. M. Dattlebaum, *Polymer*, **46**, 3056 (2005).
84. D. G. Flom and N. T. Porile, *J. Appl. Phys.*, **26**, 1088 (1955).
85. R. P. Steijn, *ASLE Trans.*, **11**, 235 (1968).

CHAPTER 11

POLYMER COMPOSITES FOR TRIBOLOGICAL APPLICATIONS IN A RANGE BETWEEN LIQUID HELIUM AND ROOM TEMPERATURE

Klaus Friedrich[1], Geraldine Theiler[2] and Patrick Klein[3]

[1]*IVW GmbH, Erwin Schroedinger Str. 58, 67663 Kaiserslautern, Germany, and King Saud University, Riyadh, Saudi Arabia*

[2]*BAM, Unter den Eichen 87, 12205 Berlin, Germany*

[3]*Parker Hannifin GmbH & Co. KG, Prädifa-Packing Division Europe Arnold-Jäger-Str. 1, 74321 Bietigheim-Bissingen, Germany*

E-mail: [1]friedrich@ivw.uni-kl.de, [2]geraldine.theiler@bam.de [3]patrick.klein@parker.com

This study deals with the development of PEEK (polyetheretherketone) and PTFE (polytetrafluoroethylene) based composites, optimized for low friction and low wear performance under extreme environments. It is demonstrated that the incorporation of a harder polymer component into PTFE (such as PEEK particles), a short fibre reinforcement (e.g. carbon fibres CF), and internal lubricants (e.g. PTFE particles), helps to reduce the friction and to improve the wear resistance over a very wide temperature range.

1. Introduction to Cryotechnology

Low temperature technology has been mainly restricted to space applications and in superconductivity. However, new applications of cryotechnology are ready to take off in the near future. After the rocket technology showed the large potential of using liquid hydrogen (LH_2) for efficient and even pollution free power supply systems, the use of liquid hydrogen as an environmentally friendly energy supply and transportation system receives more and more interest.[1] Recent publications[2] reported many applications of hydrogen for aviation and automobile: Munich airport is equipped with hydrogen buses, BMW produced cars with hydrogen combustion (Fig. 1) and the former Deutsche Aerospace Airbus (now part of the EADS) has work on a prototype of an aircraft with a hydrogen engine.

Fig. 1. Challenge for new materials in automotive hydrogen technology: hydrogen storage tanks (upper right); filling equipment for the refuelling of hydrogen driven vehicles (lower right), such as the BMW Hydro 7 (lower left).

These applications involve new requirements of material properties, in particular regarding operability and reliability. Materials developed in the past cannot be transferred directly to new areas of cryogenic applications without more intense investigations.[3] Other examples regarding the hydrogen technology is described in[4]:

- Liquid hydrogen-fuelled propulsion units for space flights are constructed for a few minutes lifetime (Fig. 2). Propulsions units for aircraft have to work more than 40,000 hours. However, the long term behaviour of materials and components is still unknown.
- In the space shuttle, the main engines are powered by LH_2 and LOx, which are pressurized by turbopumps with rotational speeds of up to 50 000 min^{-1}, and with working temperatures ranging from $T = 20$ K to $T = 900$ K.
- For civilian applications, the safety of the equipment should be the prime requirement, since the technology is used by many people.

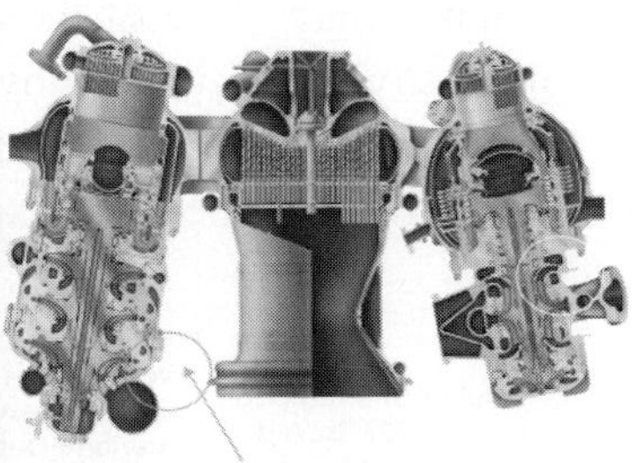

Fig. 2. Application of polymer composites as cages in steel roller bearings of turbo pumps for aerospace (see also Fig. 3).

Undetected leaks which can develop during operation can cause uncontrolled escape of gas with the consequence that ignitable gas mixtures may occur, and in the case of hydrogen, this can cause violent fire. Special attention must be paid to permeation, sealing and insulation. However, systematic investigations of materials at low temperatures are so far mainly focussed on thermal and mechanical characteristics.[5-8]

Stainless steels are well qualified for cryogenic applications because of the properties of their austenitic structure which enables high strength and sufficient ductility even at low temperatures. However, due to their high density, they tend to be replaced in many applications by polymers and polymer composites. Furthermore, in LH_2, the formation of martensite structure can lead to losses of ductility and changes in magnetic properties.[9]

At low temperatures, the Young's modulus and the hardness of polymers are much higher compared to their values at room temperature, whereas the already low heat conductivity continues to decrease.[10] Polymers are particularly suitable for cryogenic applications because of their low thermal and electrical conductivity as well as their permeation characteristics.[11] Fibre reinforced epoxy resins have been well investigated[12-13] and used in aerospace. In particular, the outstanding features of fibre composites favour their application in low temperature technology. The high specific strength and stiffness, the high fatigue life and the low density are some of the advantages in their use in cryogenic system as alternative to metals. In hydrogen technology, fibre-reinforced polymer composites are prime candidates and are used in lightweight pressure vessels for storage and transportation of liquefied gases.[14]

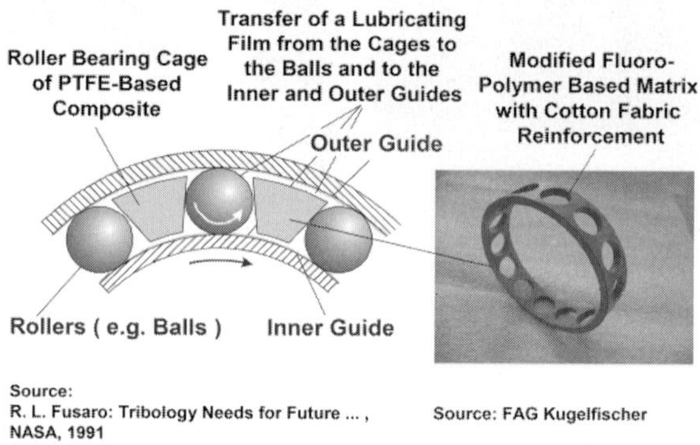

Fig. 3. Location of PTFE based composite cages and their function in ball bearings.

Components of cryotechnical machines and plants with interacting surfaces in relative motion such as bearings, seals or valves, tribosystems in general, are critical parts. Such tribosystems are characterized not only by low temperatures but also by special media. Environments are gaseous helium, cryogenic liquids or vacuum. Thus, materials of such systems have to be resistant to extreme conditions of simultaneous influences of low temperatures, mechanical deformation at the surface, special environments and the mutual interactions of these influences. In the case of tribological applications in hydrogen, possible chemical reactions due to mechanical activation must also be expected.[15]

Conventional lubricants cannot be used at very low temperatures. Similar technical solutions must be found to the conditions in vacuum, which are based on materials with good tribological characteristics. Self-lubricating polymers and polymer composites are therefore widely used as dry sliding materials in friction assemblies, where external supply of lubricants is impossible or not recommended (Figs. 3 and 4).

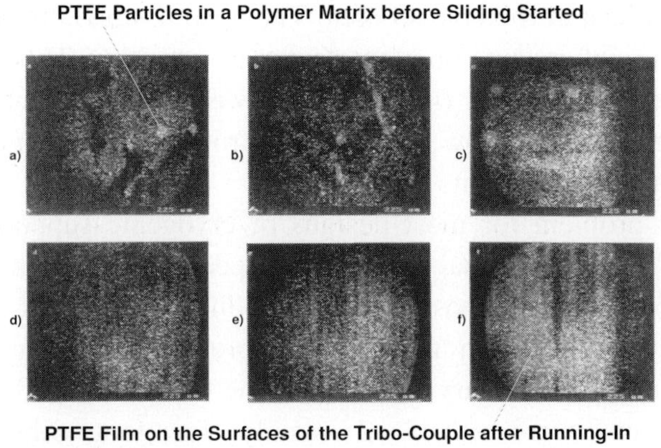

Fig. 4. Different steps (from a to f) in the development of a PTFE-transfer film on the surfaces of a tribocouple (after A. M. Häger and M. Davies, Adv. Compos. Tribol., Elsevier, 1993, p. 107).

Poly(tetrafluoroethylene) (PTFE) and Poly(etheretherketone) (PEEK) materials are good candidates due to their favourable characteristics at ambient and high temperature (Fig. 5). In particular, PTFE and PTFE-filled composites have been successfully used in rocket applications.[16-17] A list of various polymer- compounds, mostly based on PTFE, used as seals for extreme environments is given in Fig. 6.

One of the specific areas of interest is the use of hydrogen as a clean and renewable energy source instead of gasoline in the transportation industry. During refuelling, servicing and maintenance processes, safety has to be guaranteed. Insulation and sealing of components as well as the distribution of the medium (using pumps, valves, bearings etc.) are of increased importance. But tribological experimental data of materials at low temperatures and in hydrogen environment are hardly available due to cost and safety reasons. However, in order to satisfy the demand for new cryotechnologies, further material investigations need to be performed taking into account the following material related problems:

(a) Tribological performances of polymers depend significantly on the temperature at the friction contact. That applies also to very low temperature tribosystems, especially for polymers with low heat conductivity. This can be a problem due to bad evacuation of frictional heat at the contact area. The heat dissipation at the contact area is therefore an important factor to consider. This means, the tribological properties (including the wear processes) of a friction couple can differ at low temperatures from those known from room temperature experiments

(b) Another problem for most designs of cryogenic support structures comes from the thermal expansion, especially if different materials are involved. In composite materials, different thermal coefficients of expansion between matrix and fibres can lead to mechanical stresses in the boundary surfaces of the components. Such damage mechanisms were examined mainly for long fibre-reinforced polymers.[6,14,18]

(c) Experiments in low temperature environments also need to be carried out in order to consider the influence of its chemical reactivity on the performance of the materials.

(d) Finally, another specific cryogenic problem is due to the brittleness of some polymeric materials like epoxy resin at low temperatures. Precautions should be taken for the choice of the materials. Some polymers such as PTFE and PEEK possess, however, a relative ductility at low temperatures.[14]

The first major works on cryogenic self-lubricating composites were accomplished by NASA Lewis Research Center tribologists during the early 1960s.[19,20] They investigated steel bearings which operated in liquid hydrogen and liquid oxygen pumps in rocket engines. Various types of reinforced PTFE self-lubricating ball bearings were investigated. Other investigations for cryogenic friction and wear have been carried out for metal/insular parts in superconducting magnet windings. Most of the tribological studies for space applications were carried out in LN_2, which is a good compromise between LOx and LH_2. Tribological experiments in LHe are rare.

Different theories exist concerning the tribological behaviour of polymers at low temperatures. According to Gardos,[21] friction of PTFE should increases at low temperatures. To a significant extent, this increase was attributed to the generation of excessive amounts of large, abrasive but still counterface- adherent "transfer film" particles.

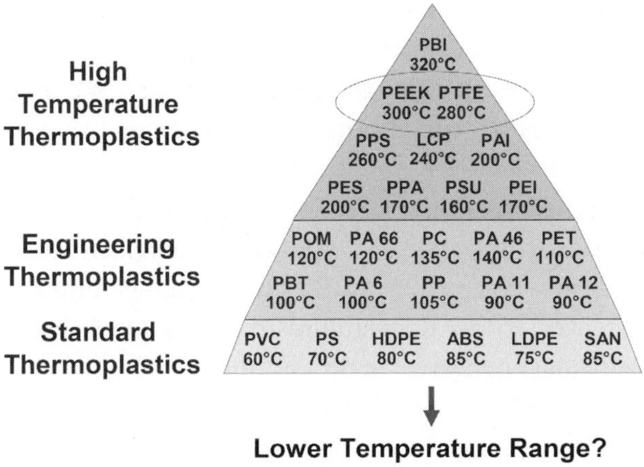

Fig. 5. Upper temperature limits of thermoplastic polymers.

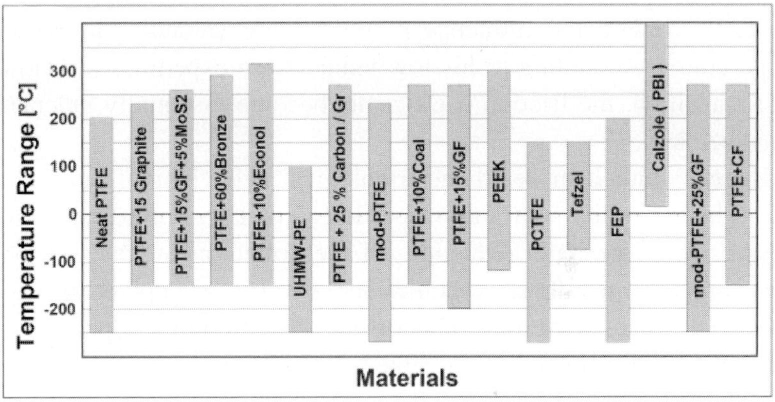

Source: Advanced Products: Seals for Extreme Environments, 2000

Fig. 6. Temperature limits for polymer sealing materials in use.

However, most of the other works found a better performance of polymer composites at low temperatures, even if the wear mechanism is still being discussed.

Wisander[19] investigated the friction and wear of PTFE composites in LN_2 for seals and bearings for missile power plants. He found good tribological properties for carbon-filled PTFE materials at low temperatures. Glaeser[16] studied the wear mechanisms of polymers at cryogenic temperatures. It was concluded from the surface analyses of the wear tracks that the wear process in pure PTFE at T = 77 K in LN_2 is totally different from that at room temperature. Wear at RT is a combination of gross plastic deformation or creep and shearing by prow formation. Wear at T = 77 K is closer to an abrasive wear mechanism producing fine powder debris.

Other work carried out by Michael[22] indicated that the wear mechanism remains constant from RT to low temperature and that the wear rate decreases at low temperatures due to a much larger hardness. He concluded that adhesive wear is the dominant polymer wear mechanism not only at room temperature, but also at cryogenic temperatures. Cross-plots of the polymers' wear coefficients versus the corresponding friction coefficients showed a good correlation and indicated a third- to forth- power functional relationship between friction and wear for both the unfilled and reinforced plastics. In a further work, Michael[23] studied the influence of the sliding velocities at cryogenic temperatures. According to his low temperature experiments, below the glass transition, the friction coefficients become essentially independent of the velocity.

Bozet[24] studied in detail the tribological behaviour of polyimide in LN_2. The reduction of friction coefficient in LN_2 compared to RT is caused by a sizeable reduction of deformation of the composite. According to him, adhesion should increase at low temperatures, and he concluded that adhesion phenomena are more important than deformation phenomena at cryogenic temperatures.

Hübner and Gradt[4] have investigated a wide range of materials at low temperatures. The behaviour of the materials indicated a significant influence of the temperatures as well as of the cryogenic environments. Experiments with different polymer materials indicated that the friction does not decrease homogeneously with temperature, instead a rise at T = 4.2 K in liquid helium was observed as compared to LN_2. It was also concluded that the molecular structure of the materials must be considered.

It is the aim of this chapter to demonstrate the tribological behaviour of various PTFE- and PEEK-matrix composites in a range between liquid helium and room temperature. In particular it should be verified if the good tribological behaviour of PTFE- and PEEK- based composites observed at room temperature (RT) is still existent at low temperatures. Due to the frictional heat produced at the contact, experiments were performed in different cryogenic environments. Suggestions are made regarding the composites and the conditions in which these materials can be used in cryotribosystems.

2. Materials and Tribological Characterization Methods

2.1. *Materials*

2.1.1. *Polymeric Matrices*

Despite its poor mechanical properties and high wear rate, PTFE is one of the most used polymers in tribological applications due to its low friction coefficient. On the other hand, PEEK has very good mechanical properties, but a higher friction coefficient. Since safety requirements are essential in low temperature technology, only composites with optimal performances should be employed in this condition. This is why in the current work, PTFE- and PEEK-matrix composites filled with different amounts of fillers and fibres were chosen to be investigated for tribological applications between room and low temperatures in various cryogenic environments.

Neat PTFE exhibits a very low coefficient of friction and retains useful mechanical properties at temperatures from T = 13 K (-260°C) to T = 533 K (260°C) for continuous use.[25] The crystalline melting point is T = 600 K (327°C), much higher than that of most other semi-crystalline polymers. Several transitions have been observed over the temperature range from T = 4 K to T = 600 K.[26] PTFE is nearly chemically inert and does not absorb water, leading to excellent dimensional stability. These characteristics of PTFE are very useful in the matrix of polymer-based composites which are used in sliding applications. The unique tribological properties of PTFE are due to its peculiar molecular and morphological structures.[25] On the other hand, PTFE is subjected to marked cold flow under stress and reveals the highest wear amongst the semi-crystalline polymers. This disadvantage is much improved by incorporating a filler into the PTFE.

PEEK itself is a tough aromatic thermoplastic polymer with properties which make it very attractive for use as high-quality engineering thermoplastic. It is a semi-crystalline polymer with a crystalline melting point around T = 613 K (340°C) and a glass transition at around T = 416 K (143°C). Its relative stiff backbone gives excellent high-temperature stability. It has a high continuous service temperature with the advantages of easy processability by injection moulding and other techniques common to thermoplastic polymers.[27]

Some of the mechanical properties of PTFE and PEEK at RT are shown in Table 1.

Table 1. Mechanical properties of PTFE and PEEK at RT.

Properties at RT	PTFE (Dyneon TF1750)	PEEK (Vitrex 450)
Density	2.16	1.3
Tensile strength [MPa]	42	103
Tensile modulus [GPa]	0.6	3.6
Coefficient of thermal expansion [10^{-5} K^{-1}]	13	5

2.1.2. *Reinforcing Fibres*

Carbon fibres (CF) are produced from organic fibres (rayon, acrylic) or from residues of petroleum or tar distillation. The first ones are called PAN-CF, the others, pitch-CF. Some of the mechanical properties are shown in Table 2.

For tribological applications, the addition of CF usually reduces the friction coefficient and increases the wear resistance. Their influence depends on the degree of graphitisation. High modulus CF with high graphitisation are used in tribological applications rather than high strength CF. Friedrich[28] exhibited (for a polyehternitrile (PEN) matrix) that some special pitch-based fibres from Idemitsu Kosan, Japan, can have a better wear reduction property than PAN-based fibres. The optimum CF content was found to be 20%. In another study of the same group[29], the friction and wear properties of other types of pitch – and PAN- based CF-composites with a PEEK- matrix showed, however, a different tendency. In this investigation, the short pitch-CF used (Kureha Co., Japan) have a diameter of about 20 μm, whereas PAN-CF usually have a diameter of about 6-8 μm. The coefficient of thermal expansion in the length direction of the fibres is -0.6 [10^{-6} K^{-1}] (for PAN-based CF) and -1.6 [10^{-6} K^{-1}] (for pitch-CF), respectively.

Table 2. Mechanical properties of carbon fibres.[30]

Characteristics	Fibres from PAN	Fibres from pitch
Strength (GPa)	1.8 - 7.0	1.4 - 3.0
Modulus (GPa)	230 - 540	140 - 820
Elongation at break (%)	0.4 - 2.4	0.2 - 1.3
Density (g/cm³)	1.75 - 1.95	2.0 - 2.2

2.1.3. *Other Fillers*

Ekonol® polyester (PAR) is a linear aromatic thermoplastic. The absence of aliphatic hydrogen contributes to its excellent thermal stability. It is a highly crystalline polymer. Flow and creep are virtually non-existent below T = 588 K (315°C).[31] When used as a filler in PTFE, it produces a composite material that has excellent temperature and wear resistance. The Ekonol® polyester/PTFE blend has also a high thermal conductivity, long-term thermal resistance, low thermal expansion and low moisture absorbance.

PEEK as a filler has been investigated in PTFE-matrix notably by Briscoe and Friedrich.[32,33] It reduces wear and increases strength.

On the other hand, PTFE as a filler is one of the mostly used solid lubricants. It reduces friction coefficient and stick/slip behaviour of pure PEEK.[32]

Bronze particles, filled into a polymeric matrix, are well known for their wear reduction properties and improvement of thermal conductivity.

2.1.4. *Compositions*

To get an idea about the tribological characteristics of PTFE- and PEEK-matrix materials at cryogenic temperatures, several composites have been selected from a list of materials prior tested at RT. Tables 3 to 5 give the compositions of all the PTFE-matrix composites.

Table 3. PTFE composites filled with bronze and CF.

PTFE-matrix	Bronze	CF
Dyneon TF1750	N.A	Pitch-CF
wt%	wt%	wt%
60	30	10

Table 4. PTFE composites filled with Ekonol® and CF.

PTFE-matrix	Aromatic polyester	CF
	Ekonol® T103, St Gobain	PAN-CF, Fortafil 383
wt%	wt%	wt%
75	10	15
75	20	5
65	20	15
60	20	20

Table 5. PTFE composites filled with PEEK and CF.

PTFE-matrix	PEEK	CF
Standard Dyneon	Victrex 150XF	Pitch-CF, Kureha M 201 S
wt%	wt%	wt%
76	9	15
71	14	15
66	19	15
60	25	15

The compositions of all the PEEK-based composites are listed in Table 6. Room temperature properties of these composites are described in detail in Ref [34].

Table 6. PEEK composites filled with PTFE and CF.

PEEK-matrix	PTFE	CF
Victrex 450G	Dyneon 9207	Pitch-CF, Kureha M-2007
wt%	wt%	wt%
100	0	0
80	5	15
70	15	15
65	20	15
75	10	15
65	15	20
55	15	30

2.1.5. *Processing*

The PTFE compounds were prepared via a sintering process at temperatures between T = 633 K (360°C) and T = 653 K (380°C) (Fig. 7). The mixing of the PEEK composites was achieved by a twin-screw-extruder at about T = 673 K (400°C) with a standard screw configuration. Afterwards, the materials were prepared as injection-moulded plates.

Samples were then cut into pins, discs or bars for the tribological, thermal or mechanical tests.

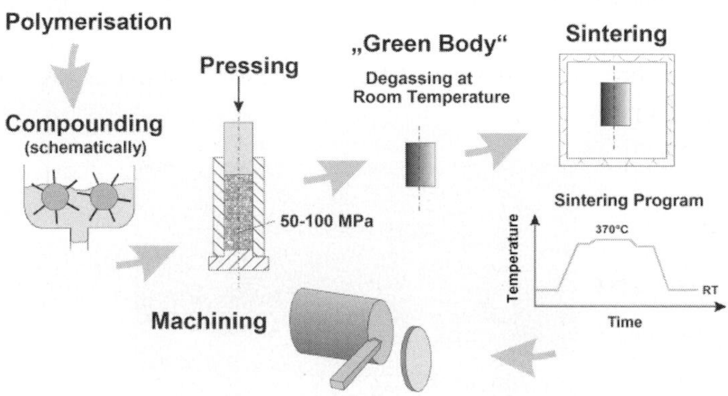

Fig. 7. Processing steps of PTFE and PTFE-compounds.

The chemical composition of the polymers described above is represented in Fig. 8, and two of the typical fillers are shown in Fig. 9.

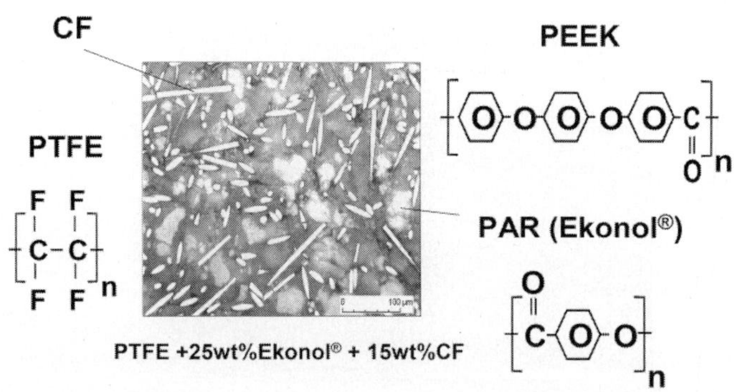

Fig. 8. Micrograph of a PTFE matrix, filled with PAR- or PEEK- particles and short CF.

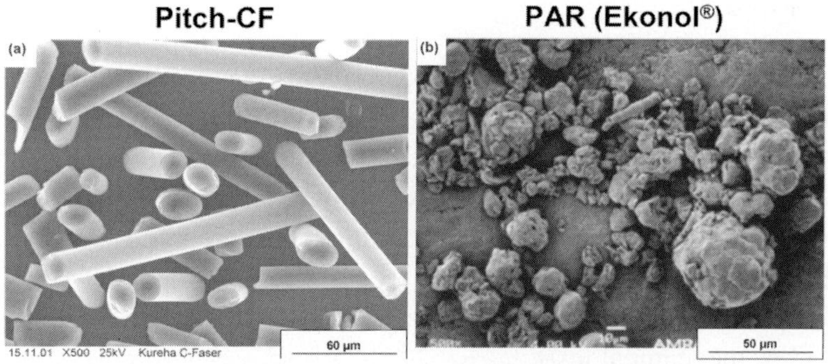

Fig. 9. Polymer fillers and short carbon fibers used in PTFE – composites.

Optical microscopy images of two of these composites are shown in Figs. 10 and 11.

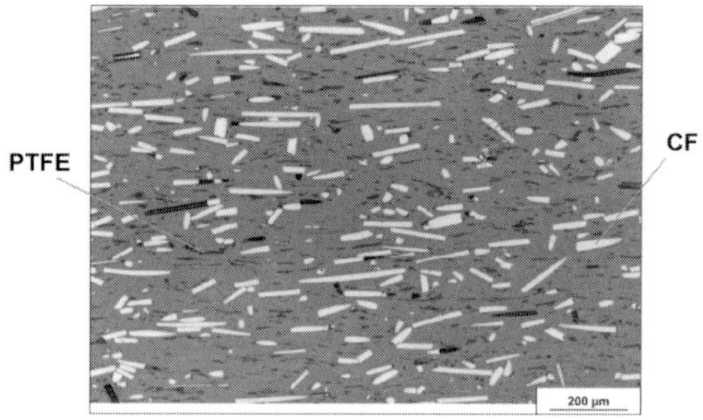

Fig. 10. Microstructure of a PEEK composite with 15% PTFE particles and 15% CF.

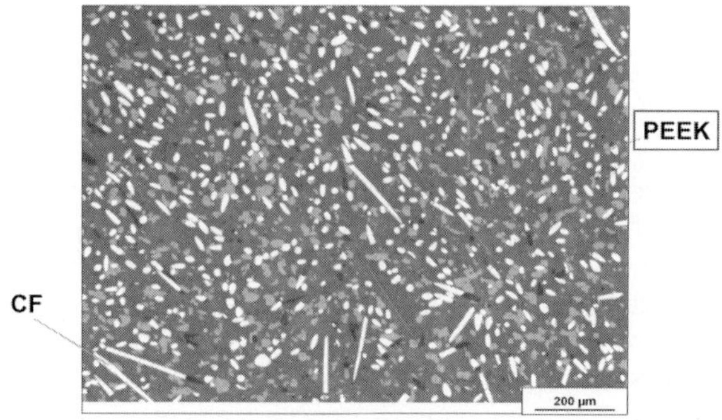

Fig. 11. Microstructure of a PTFE composite with 9% PEEK particles and 15% CF.

2.1.6. *Mechanical Properties*

Tensile tests with two characteristic types of composites (i.e. PTFE, filled with 15% CF and 9% PEEK, and PEEK, filled with 15% CF and 15% PTFE) were carried out on a Shimadzu Universal testing machine AG-10TC, using a reversing cage at room temperature and at T = 77 K in a bath of liquid nitrogen. For the experiments at T = 77 K, an isolated tank was used. The specimen dimensions were 10x2 mm² and 80 mm length. Table 7 gives the elongation rates. Tensile strength and elongation at break were recorded. An extensometer was used to calculate accurately the Young's modulus.

Samples were investigated after the experiments by means of Scanning Electron Microscopy (SEM).

Table 7. Elongation rates for tensile testing.

Material	Temperature	Elongation rate
PTFE composite	300 K	4 mm/min
PTFE composite	77 K	1 mm/min
PEEK composite	300 K	1 mm/min
PEEK composite	77 K	1 mm/min

The results of the tensile tests are summarised in Table 8.

Table 8. Average values of the tensile test results.

Composite	T	*E*-Modulus (GPa)	Tensile strength R_m (N/mm²)	Elongation at break (%)	Number of samples
PTFE	RT	1.4	9.7	146.3	3
PTFE	77 K	6.3	35	1.4	3
PEEK	RT	5.5	58	0.6	3
PEEK	77 K	7.4	96.2	0.6	4

The elongation at break was reduced significantly for the PTFE-matrix composite from 140% at RT down to 3% at T = 77 K (Table 8). On the other hand, the tensile strength increased from ca. 10 N/mm² at RT to 35 N/mm² at T = 77 K for this material. The improvement of the PEEK composite properties at low temperatures was relatively smaller. However, the tensile strength still increased from 60 N/mm² at RT to almost 100 N/mm² at T = 77 K.[3]

Corresponding fracture surfaces (Fig. 12) illustrate the measured differences in strength very drastically. While the PEEK surfaces look very similar at both temperatures, the PTFE surfaces are clearly different from each other. At room temperature, the PTFE-matrix looks highly fibrillized, with poorly bonded fibres in between. The poor bonding between the fibres and the PTFE-matrix becomes also obvious at 77K, although here the matrix has a typically brittle appearance. Higher magnifications (Fig. 13) show that there is also no adhesion between the PEEK-particles and the PTFE-matrix, so that it is not surprising that the measured tensile strength values are much lower than corresponding values of the neat PTFE, known from the literature. The same tendency is true for the PEEK- composites, which possess lower room temperature strength than neat PEEK, due to the poorly bonded, softer PTFE inclusions.

Figure 14 shows that the Young's modulus of each material increases at low temperatures from 1.4 GPa to 6.3 GPa for the PTFE composite filled with 15% CF and 9% PEEK and from 5.5 GPa to 7.4 GPa for the PEEK composite filled with 15% CF and 15% PTFE. It is obvious that the ductile-brittle transition in the PTFE – composites (from room temperature down to 77K) has a greater effect on the modulus than it is observed for the PEEK- composites, with their semi-brittle behaviour at both testing temperatures.

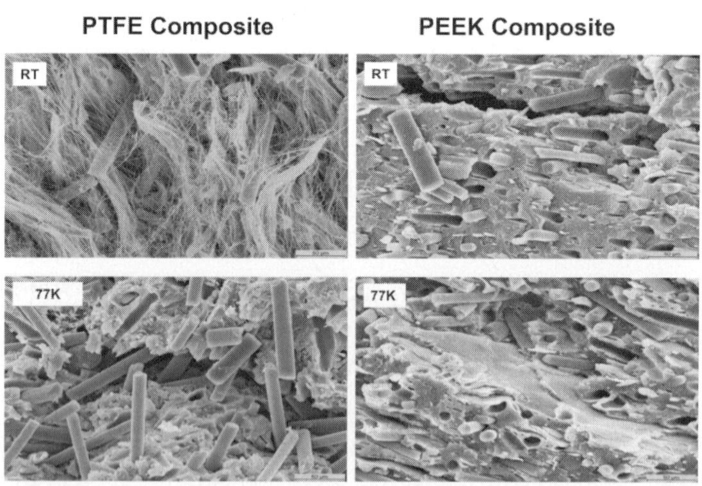

Fig. 12. Fracture surfaces of PEEK and PTFE composites at two different temperatures.

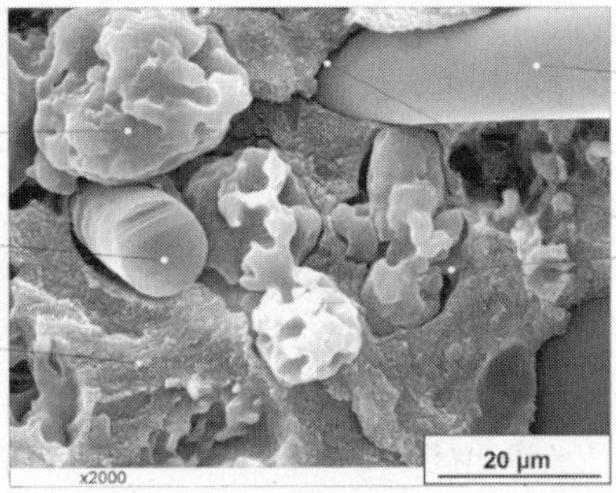

Fig. 13. Quality of fiber-matrix- and particle-matrix – adhesion, as seen on the fracture surface of PTFE/PEEK/CF (broken at –150°C).

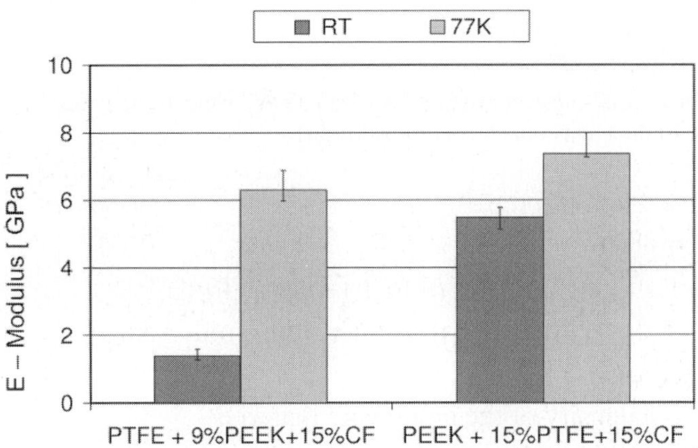

Fig. 14. Tensile modulus of two different composites at RT and 77 K.

2.2. *Tribotests at Room Temperature and Cryogenic Environments*

2.2.1. *Testing Procedures*

Unlubricated sliding wear tests were carried out on different pin-on-disc machines designed by IVW for room temperature experiments and by BAM for tests at lower temperatures. The IVW Equipment is illustrated in Fig. 15, and the measured data allowed determining the coefficient of friction μ and the specific wear rate w_s by the following equations:

$$\mu = F_R / F_N$$

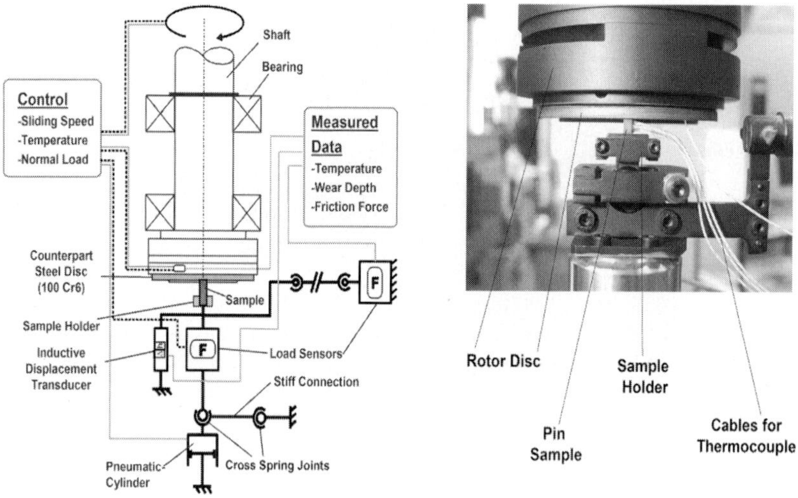

Fig. 15. Pin on disc- test configuration of an IVW- wear testing machine: schematic (left), photo of the local contact conditions (right).

where F_R = friction force

F_N = normal load

$$w_s = \Delta m / (d\, F_N\, L) = \Delta V / (F_N\, L),$$

usually in the dimensions $\{mm^3/(Nm)\}$

where Δm = mass loss

d = density of the composite material

ΔV = volume loss

L = sliding distance.

The cryotribometer used at BAM was thermally insulated by vacuum super-insulation and cooled directly by a bath of liquid cryogen (Fig. 16). Each experiment was repeated two or three times. More details about the equipment are described in Ref[3].

The pin-on-disc configuration (Fig. 17) consists of a fixed 2-pins holder continuously sliding against a rotating disc. The rotation is transmitted via a rotary vacuum feed-through to a shaft with the sample disc at the lower end. Loading is established by pressurized He-gas acting on a piston which presses a lever with the sample holder upwards against the lower face of the rotating disc. During the tribological tests, normal and friction forces are measured by strain gauges.

Fig. 16. Photo of the experimental set-up for friction and wear testing under cryogenic temperatures at BAM.

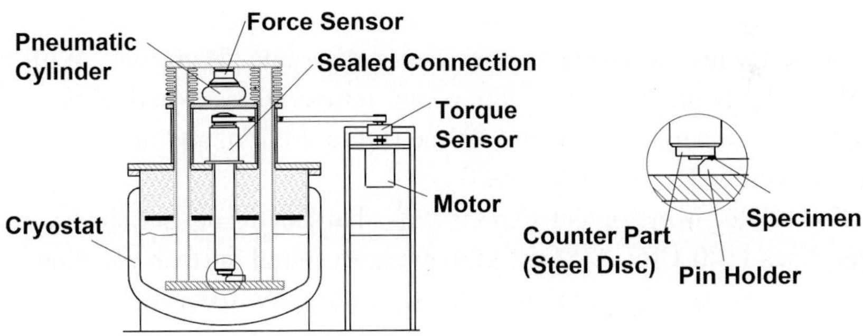

Fig. 17. Schematic illustration of the pin-on-disc configuration within the cryo-tribometer.

Before testing, the polymer composites were cut into pins (4 x 4 x 12 mm³). In the apparent contact area (4 x 4 mm²), the sliding direction was chosen to be parallel to the majority of the in plane oriented fibres, since under these conditions CF-based composites usually possess the lowest wear rates. In order to reduce the time of the running-in period, specimens were pre-worn with grinding paper (Grid 800) against the disc counterpart and then carefully cleaned with ethanol. With this method, the roughness of the specimens before testing was always the same, and so was the apparent area of contact with the steel disc due to a better parallelism between the two mating surfaces. In the measurements of the coefficient of friction with these samples, the normal load was kept constant at 50 N (leading to a nominal contact pressure of 3.125 MPa), the sliding distance amounted to 2000 m, and the following sliding speeds were applied: 1 m/s, 0.2 m/s, and (in very few cases) 0.06 m/s.

In the low temperature wear experiments, the sliding speed and the contact load were kept constant at 0.2m/s and 16 N, respectively, and the total sliding distance amounted to 2000m. Due to the fact that the amount of wear was very small in the low temperature tests, the tips of the pins (with the dimensions mentioned above) were rounded. This resulted in an initial contact pressure, which was clearly higher than the one used for the flat friction samples. Wear was planned to be measured by weight loss, in order to compare the results with the RT experiments performed at IVW.

For better comparison, also some room temperature studies were performed with the BAM facility under the same conditions. With the IVW equipment, however, the room temperature experiments were conducted over longer periods of time (up to 20 hrs), and also at various pressure and velocity conditions.[34]

For all the tribological experiments, hardened and polished carbon steel discs (100 Cr6; AISI 52 100) with an initial surface roughness of Ra = 50 nm served as counterparts. It is the most frequently used steel for bearing applications. It exhibits numerous advantages: high purity and tenacity, suitability for hardening without carburising, flexibility in the heat treatment. Chemical composition and some of its mechanical properties are indicated in Table 9. Steel discs with an inside diameter of 25 mm and an outside diameter of 42 mm were cleaned with ethanol before use.

Table 9. Composition and mechanical properties of the steel discs.

100 Cr 6 (AISI 52100)			
Chemical composition		Mechanical property	
C	0.98-1.10 %	Coefficient of thermal expansion	$\alpha_e = 12 \times 10^{-6}$ mm/mmK
Si	0.20-0.35 %	E-modulus	E = 205 000 N/mm²
Mn	0.25-0.45 %	Poisson ratio	0.3
Cr	1.3-1.6 %	Rockwell hardness	~ HRC 62

2.2.2. *Cryogenic Environments*

The tribological experiments were carried out in air, nitrogen, helium and hydrogen environments as indicated in Table 10. Nitrogen is often used as cryogenic medium since it is a cheap and inert coolant, which allows to carry out many low temperature experiments at T = 77 K. Due to the higher price of helium, fewer tests were performed at T = 4.2 K. Experiments in hydrogen were carried out only with selected composites for safety and cost reasons.

Table 10. Experimental environments.

Environment	Temperature	
Air	293 K (RT)	20°C
Nitrogen	77 K	-196°C
Helium	77 K 30 K 4.2 K	-196°C -243°C -268.8°C
Hydrogen	293 K (RT) 20 K	20°C -253°C

Table 11 gives some of the important thermal properties of the cryogenic liquids at boiling point T_b, and the properties of air at T = 293 K for comparison.

Typical information on the thermal conductivities of the three liquid cryogens can be found in Ref[35-37].

3. Friction and Wear of PTFE-based Composites at Room Temperature

Previous works[38] on the sliding wear of PTFE / PEEK – blends against smooth steel have shown that the low specific wear rate of neat PEEK can be further reduced by the addition of 5 to 15 vol.% PTFE particles (Fig. 18). At the same time, the PTFE improves remarkably the coefficient of friction of such blends (in comparison to the neat PEEK) due to the ability of PTFE to easily build-up a transfer film on the surfaces of both the polymer composite and the steel counterpart (cf. Fig. 4). At the other end of the diagram in Fig. 18, it is also evident

Table 11. Some properties of the cryogens and of air for comparison.

Property		LHe	LH_2	LN_2	Air
Boiling temperature T_b at 1 bar	K	4.2	20.4	77.3	80
Heat of evaporation at T_b	kJ/kg	20.9	446	199.1	205.2
Density					
at T_b	kg/m³	124.8	70.8	804.2	
at T = 273 K, 1 bar	kg/m³				1.29
Specific heat					
at T_b	kJ/kgK	4.4	9.3	2	
at T = 273 K, 1 bar	kJ/kgK				1.0
Thermal conductivity					
at T_b	10^{-3} W/mK	27.2	119	139.8	
at T = 273 K, 1 bar	10^{-3} W/mK				24.1
Viscosity					
at T_b	10^{-7} kg/ms	35	133	1650	
at T = 273 K, 1 bar	10^{-7} kg/ms				24.1

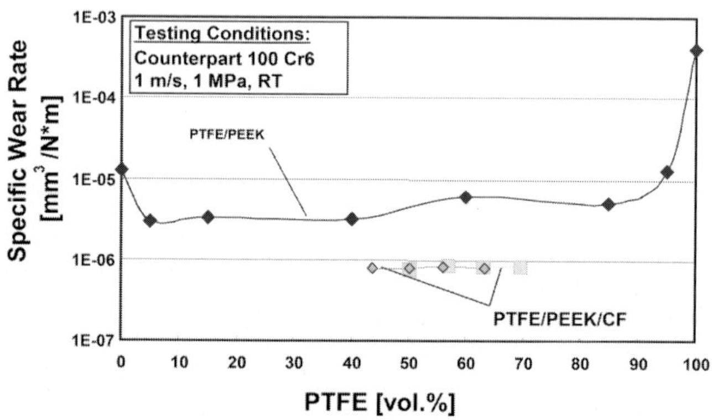

Fig. 18. Wear results of PTFE/PEEK/CF in comparison to PTFE/PEEK at room temperature.

that for the very high specific wear rate of the neat PTFE material (ca. 5×10^{-4} mm^3/Nm) an even more dramatic reduction (by ca. 2 orders of magnitude) can be achieved if ca. 15 vol.% of harder PEEK particles are incorporated into the PTFE- matrix. This is, however, associated with a slight increase in the coefficient of friction of the PTFE. In the intermediate range, i.e. between 15 and 85 vol.% PEEK in the blend, the values of both, coefficient of friction and specific wear rate remain on nearly the same low level.[39]

Adding short carbon fibres to such PTFE / PEEK – blends, results in a reduction in the specific wear rate level by nearly one order of magnitude (Fig. 18). This is true for both composites containing ca. 17 vol.% CF (squares; equivalent to 15 wt% CF) as well as ca. 24 vol.% CF (rhomboids; equivalent to 21 wt% CF). Within this small range, there is no remarkable effect of the amount of CF in these blends.

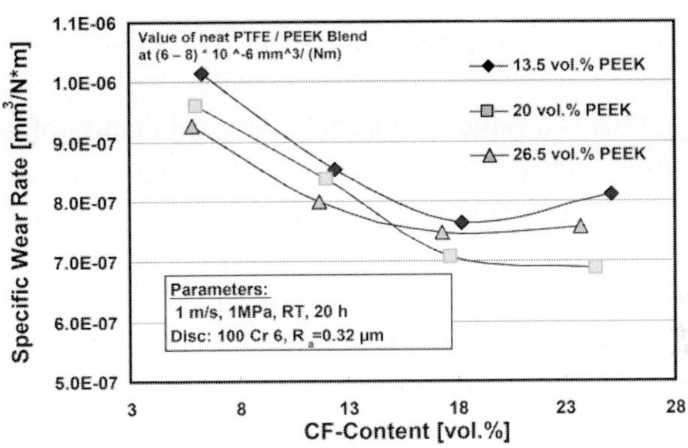

Fig. 19. Specific wear rate of PTFE composites as a function of CF content and PEEK particles.

More details of these materials are shown in Fig. 19. Here, the specific wear rate is plotted against the carbon fibre content for different amounts of PEEK- particles in the PTFE- matrix. Compared to the level of the unreinforced PTFE / PEEK –blend (ca. 7×10^{-6} mm^3/Nm), the short carbon fibres linearly reduce the specific wear rate down to a value of ca. 7.5×10^{-7} mm^3/Nm at ca. 18 vol.% CF. At higher CF content (up to 24 vol.%), the wear rate stabilized on this level or started to increase

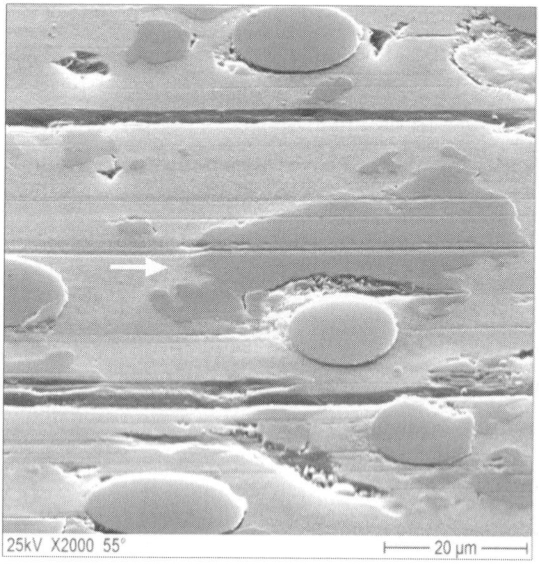

Fig. 20. Typical surface appearance of worn PTFE/PEEK/CF-specimens.

again slightly. This is a typical tendency, the explanation of which can be found in previous publications.[40] The general improvement in the wear resistance of the PTFE- matrix by the PEEK- particles and the CF can also be recognized on the worn surfaces of the pins, which exhibit that the harder CF and PEEK- particles are less severely worn than the softer PTFE- matrix in between (Fig. 20).[41]

The specific wear rate, mentioned above (also called "wear factor" of the material), can be considered as a kind of "material property", which allows to compare different materials with each other (when tested under equivalent system conditions). As long as this factor is not dramatically influenced by changes in the apparent contact pressure p or the sliding velocity v, a plot of the so called depth wear rate w_t (reduction in specimen height per testing time) against the pv-product should show a nearly linear relationship, according to

$$w_t = w_s\,(pv)$$

Schematically this is illustrated in Fig. 21. Above a critical pv-value, the wear factor k starts to deviate from being constant, which means it is

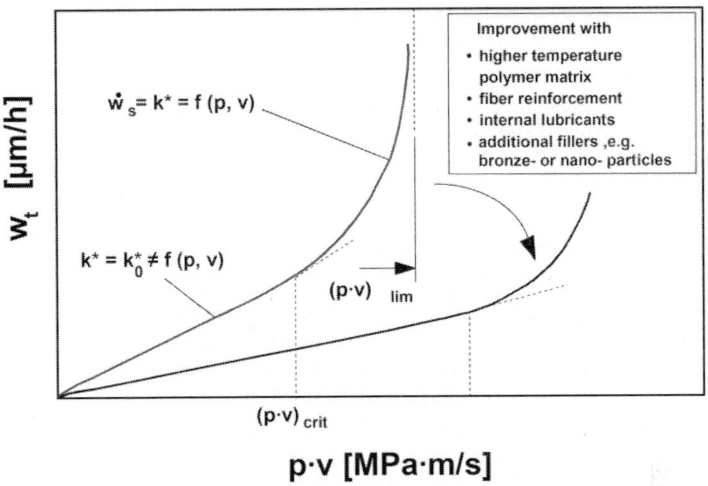

Fig. 21. Schematic illustration of the course of the depth wear rate vs. pv- product, and possible steps how to improve the wear behaviour.

not anymore independent on the pv-product. Finally the curve goes to infinity, indicating that the very upper limit of the material's load bearing capacity is exceeded. For the materials science people it is always the objective to develop materials with a reduced initial slope of this curve and to push the critical and limiting pv-levels to higher values, using different combinations of fillers and reinforcements in a given polymer matrix.[40]

The depth wear rate is a convenient term for the design engineer in order to estimate the height loss of a polymeric material at various pressure and velocity conditions (assuming that the wear rate of the steel counterpart is negligible). This, in turn, can help to calculate the development of a certain clearance in journal bearings with time, and thus to predict necessary maintenance intervals in order to prevent more catastrophic failures of complete machine units.

The PTFE- composites tested in this context, exhibited a nearly linear relationship between the depth wear rate and the pv-product within a range between 0.5 and 8 MPa m/s, using three different sliding speeds (Fig. 22). This is a good indication that these composites can be safely used within this broad pressure/velocity window.

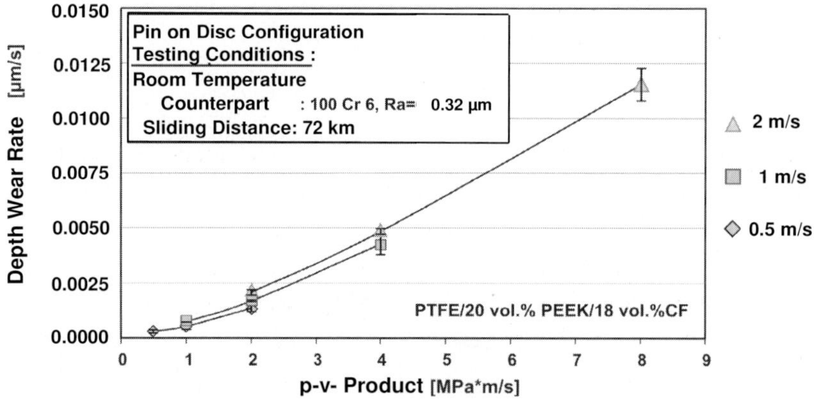

Fig. 22. Depth wear rate as a function of pv-product for a PTFE/PEEK/CF- composite.

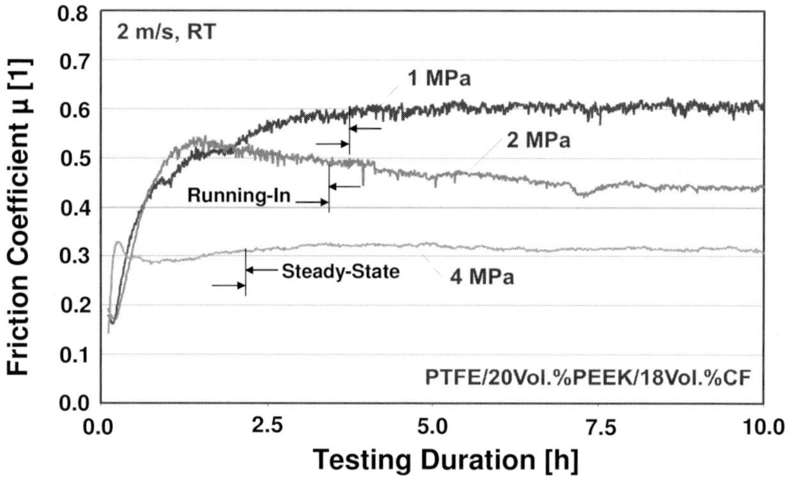

Fig. 23. Friction coefficient at 2 m/s and various contact pressures, tested at room temperature (RT).

Concerning the coefficient of friction (µ) under various testing conditions, all the composites possessed, as expected, a running-in and a steady state phase over the testing time. In the latter, the coefficient of friction remained nearly on a constant level. Figure 23 illustrates, that (at a given sliding velocity v) the transition was reached earlier the higher the contact pressure was. At the same time, typical for PTFE materials, the value of µ was reduced with p.

4. Tribology of Selected PEEK- and PTFE-based Composites in Cryogenic Environments

4.1. *Room Temperature Air vs. 77K Liquid Nitrogen*

When it comes to a comparison between the friction and wear test results obtained at room temperature (in air environment) and at a temperature of 77K (in liquid nitrogen), both the coefficients of friction as well as the specific wear rates are found to be lower under the presence of the low temperature medium. The corresponding mechanisms were discussed in detail by Theiler.[3] Briefly, when considering that the coefficient of friction μ is composed of an adhesive (μ_a) and a deformation (μ_d) component, i.e.

$$\mu = \mu_a + \mu_d,$$

the higher modulus (E) of the material at 77K leads to less deformation (i.e. μ_d is reduced, since $\mu_d \sim 1/E^{1/3}$), and to a smaller real contact area (A_r) between the mating surfaces , so that also μ_a is reduced (according to $\mu_a \sim T_s\, A_r \,/\, F_N$, where T_s is the materials shear strength and F_N is the normal contact load). Besides, there may also be some kind of lubricating effect due to the liquid environment. It should be mentioned here, that, according to an early work by Tabor and Ludema on PTFE[42], μ is also strongly dependent on the damping factor (tan δ) of the material (μ increases with increasing tan δ). Although PTFE is exhibiting several transitions at low temperature, the damping factors at 77K and at temperatures below are clearly lower than in the room temperature range[34], so that also due to this fact a lower coefficient of friction can be expected.

Figure 24 illustrates this effect for three optimized material compositions, consisting of a PTFE matrix, 15% of carbon fibers and either ca. 10% of a harder polymer filler (Ekonol® or PEEK) or 30 wt% bronze balls. It is obvious that no dramatic effect was found with regard to the type of the harder polymer particles on the frictional properties (especially in the low temperature range). This is not surprising, since PEEK and PAR (Ekonol®) have a similar molecular structure (see Fig. 8), and the PTFE matrix is also much stiffer under the low temperature conditions.

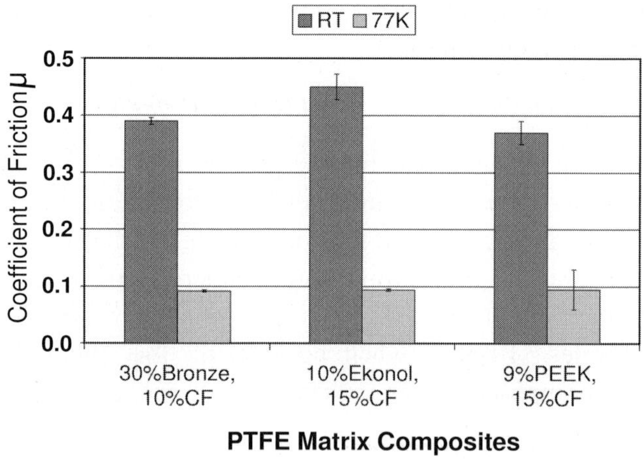

Fig. 24. Coefficient of friction of three different PTFE composites at RT and 77 K.

Similar conclusions can be drawn for the specific wear rates, as shown in Fig. 25, except that here the bronze filled composites exhibits a much better wear resistance than the two other materials , both at room temperature and at 77 K. This result seems to be due to the fact that a major factor which can affect the wear behaviour of PTFE materials is the thermal conductivity of the composite. This can be improved by incorporating fibres or metal particles. As shown in Table 12, the thermal conductivity of the bronze filled PTFE composite is much higher than in the case of Ekonol® filled PTFE. This leads to a better dissipation of the frictional heat at the friction contact and can contribute to the improvement of the wear resistance.

Comparing the wear results of the PTFE composites in Fig. 25, the wear of bronze filled composite does not decrease at low temperatures. This may be due to the fact that the wear value at RT is already quite small.

But the thermal conductivity of bronze filled PTFE is not the only factor influencing the wear resistance. Surface analyses of the discs give some other indications. Indeed, after the experiments at RT, the transfer film developed by the bronze composite is more homogenous compared to the ones developed by other composites. It is reasonable to suggest

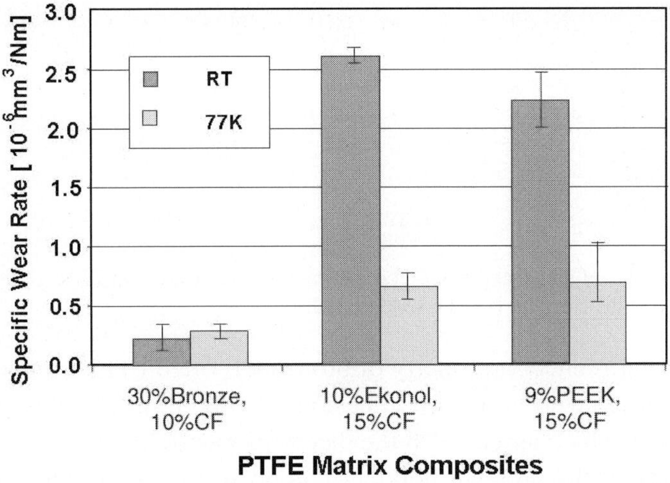

Fig. 25. Specific wear rate of three different PTFE composites at RT and 77 K.

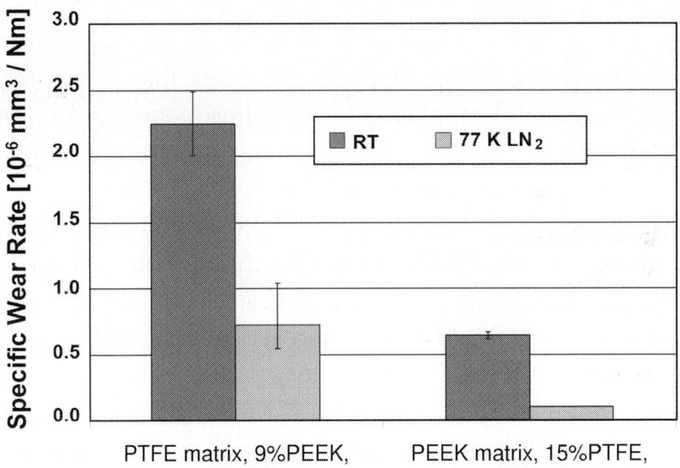

Fig. 26. Effect of matrix type on specific wear rate at room temperature and at 77 K.

Table 12. Thermal conductivity of PTFE composites.[43]

Material	Thermal conductivity [W/mK]
PTFE + 30% bronze + 10% CF	0.46
PTFE + 10% Ekonol® + 15% CF	0.27

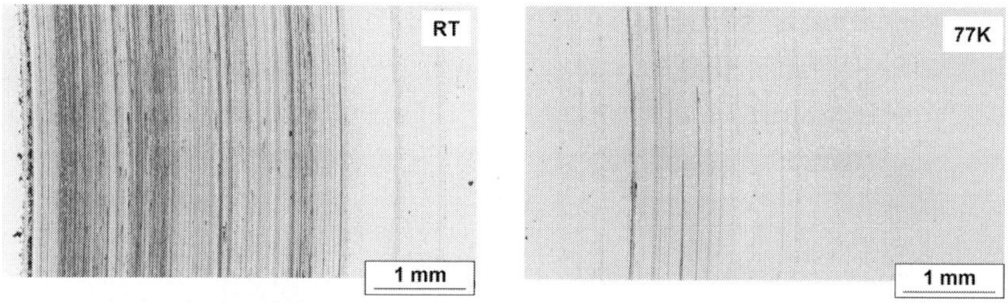

Fig. 27. Surfaces of the steel discs after experiments at RT (left) and 77 K (right) against PEEK filled with 15 wt% CF and 15 wt% PTFE.

that the wear reduction property of bronze/CF filled PTFE composite can be associated with a better adhesion of the transfer film onto the counterface, which agrees with the theory proposed in Ref[44].

Looking at the polymer/polymer composites only, an interesting fact is that a change in the type of matrix from PTFE to PEEK, and for the filler particles vice versa, the harder PEEK matrix composites exhibit a better wear resistance at both temperature levels than the PTFE ones (Fig. 26). The differences can also be detected on the surfaces of the metallic counterparts, which show in all cases that a transfer of thin polymer films had taken place. These films were thicker and more consistent in case of the PTFE based composites, and in both cases the amount of transferred material was clearly reduced when the experiments took place in liquid nitrogen. Figure 27 illustrates this fact in a very pronounced way for the PEEK composite, filled with 15% CF and 15% PTFE particles. This means, at low temperatures the transfer of material onto the disc does not build a homogenous film, as it does at RT. Only scattered particles get attached to the surface of the steel disc. The destruction of the PTFE band structure is hindered by the frozen state of the molecules, which prevents also the composite material from wearing out. The coefficient of friction depends no more on the formation and adhesion of the transfer film, but more on the real surface in contact between pin and disc, as described above. In other words, in LN_2 the tribological behaviour of the polymer composites does not depend so much anymore on the composition because of similar properties of thermoplastic materials at low temperatures.

On the worn surfaces of the composite pins, an opposite tendency was visible. When the tests were carried out at room temperature, the worn surfaces were characterized by clean fragments of carbon fibres, surrounded by compacted patches of wear debris and some grooves in between. This was especially visible on the PTFE based composites (Fig. 28) which is an evidence that this matrix wears away easier than the PEEK. Looking at corresponding surfaces after tests in LN_2, the fibres were all covered with a thick layer of compacted polymer wear debris (i.e. the fibres were almost not visible), indicating that under this condition the creation of fibre fragments was hindered and therefore a reduced amount of wear of the composites occurred.

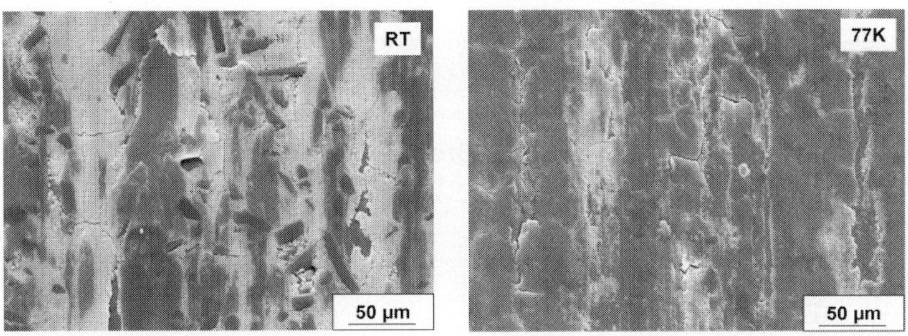

Fig. 28. SEM images of a PTFE filled with 9 wt% PEEK and 15 wt% CF after testing at RT (left) and 77K (right).

4.2. *Influence of the Cryogenic Medium*

Figures 29 and 30 represent the friction coefficients of a bronze and CF reinforced PTFE- composite at the same (77 K) and at different temperatures in various cryogenic media for two sliding speeds, i.e. v = 1 m/s and v = 0.2 m/s. The importance of the cryogenic medium is clearly visible. The friction coefficient at T = 77 K is lower in LN_2 than in He gas for the two sliding speeds. Comparing the friction behaviour of the bronze –CF filled PTFE in the three cryogenic liquids, a higher amount of friction is detected for the liquid helium, especially at the higher speed. At lower speeds, on the other hand, the coefficient of friction remains on a similar level for the three cryogens.

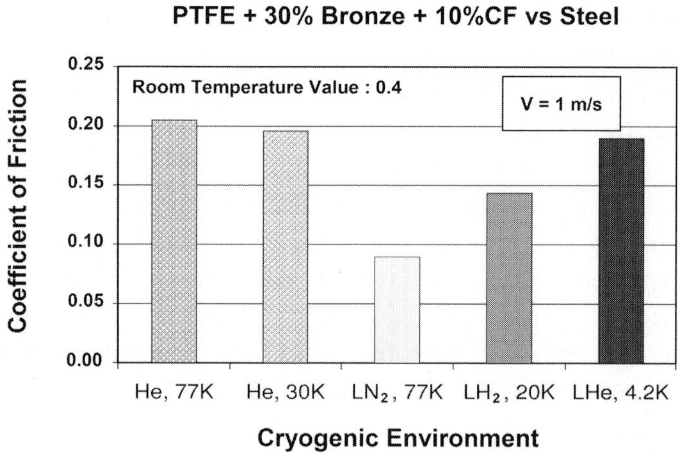

Fig. 29. Coefficient of friction under various low temperature testing environments (v = 1 m/s).

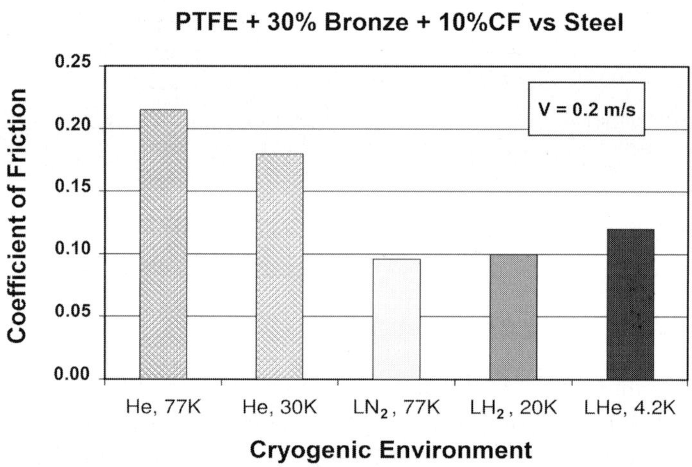

Fig. 30. Coefficient of friction under various low temperature testing environments (v = 0.2 m/s).

While the influence of the speed is obvious in He environment, it is not the case for LN_2, in which the friction coefficient remains on a constant, low level for both sliding speeds. This was further verified for a very low speed of 0.06 m/s and suggests that an optimum cooling is reached in LN_2 under all three conditions, whereas this is not the case for the helium environment, and also not for the liquid hydrogen, which ranks between the two extremes.

This means that for a better understanding of these differences, the temperature at the friction contact must be more carefully considered. Since it was not possible in this study to determine this temperature, special attention was given to the physical, especially thermal properties of the cryogenic media, which affects the temperature at the friction contact. A good example is given by the friction results at T = 77 K comparing He-gas and LN_2 environments (Figs. 29 and 30). Although both experiments are carried out at T = 77 K, it was shown that the liquid LN_2 has a better cooling efficiency than He-gas.

Moreover, the decrease of friction with temperature of the cryogen is not observed down to extreme low temperatures. Indeed, a higher coefficient of friction is measured at T = 4.2 K compared to the one at T = 77 K in LN_2, especially for the experiments carried out at a high sliding speed (Fig. 29).

This behaviour cannot be explained by the bulk structure of PTFE, whose lowest transition temperature is at about T = 170 K, neither by the temperature dependence of its mechanical characteristics, because at T = 4.2 K the hardness and mechanical properties should only be slightly higher than at T = 77 K according to Ref [10].

The influence of the diffusion of helium into materials has not been investigated here. However, this should not have a significant effect on the behaviour of the friction couple since at v = 0.2 m/s, similar friction results are obtained in LN_2, LH_2 and LHe.

The thermal properties of the cooling medium can be responsible in this case for a higher friction coefficient in LHe compared to LN_2. Since the frictional heat rises the temperature at the contact area, especially for polymers with low heat conductivity, the heat transfer to the environmental medium is a significant factor to consider.[45,46] As soon as a relative motion occurs and generates frictional heat, the liquid in the contact area begins to evaporate, and bubbles appear. Thereby, by thermal conduction, convection and evaporation, the heat is extracted from the contact. This leads to an efficient cooling effect. During this bubble formation phase (nucleate boiling), the temperature difference between the contact area and environmental medium is relatively small, because the evaporation requires a large amount of heat. With increasing friction, the size and number of bubbles increase largely until finally reaching a vapour film at the contact area (film boiling) (Fig. 31). Since the heat transfer to a gaseous environment is much less efficient than to a liquid, the transition from nucleation to film boiling is accompanied by a steep temperature rise which can amount several hundred K. For free radiating heaters, the temperature increase as function of the heat flux density can be taken from boiling curves.[47] The data from such curves cannot be transferred directly to the conditions in a frictional contact, which represents a confined geometry and forced-flow conditions, but they may serve as a rough estimation of the temperature conditions in the contact zone.

A direct measurement as well as a calculation of the actual temperature distribution in the contact zone are extremely complex and have not been carried out in this study. However, the heat flux density occurring at the sliding contact of area (A) can be calculated as followed:

$$q = \mu \, v \, F_N / A$$

Table 13 gives the heat flux density corresponding to the friction measurements carried out at v = 1 m/s with the bronze/CF filled PTFE composite (from Figs. 29 and 30).

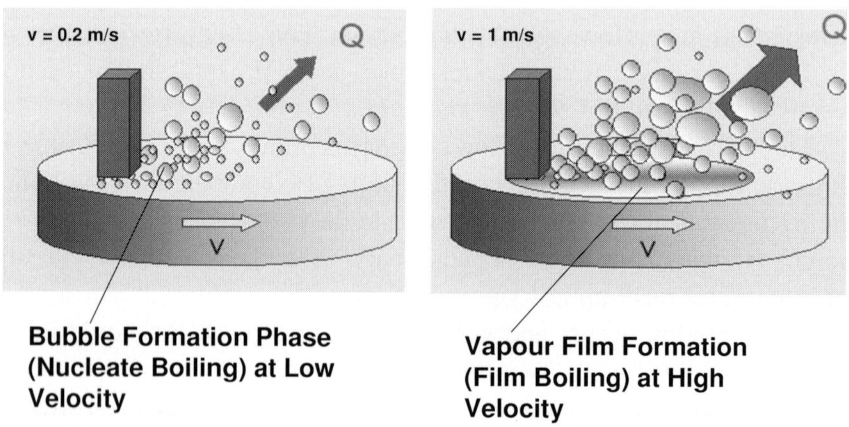

Fig. 31. Schematic illustration of the vapour film formation ("film boiling") at two different velocity levels.

Table 13. Friction coefficient and heat flux density as a function of sliding speed in different cryogenic media.

Cryogenic Medium	T (K)	v = 1 m/s μ	v = 1 m/s q [W/cm^2]	v = 0.2 m/s μ	v = 0.2 m/s q [W/cm^2]
He Gas	77	0.21	62.5	0.22	13.4
He Gas	30	0.19	61.2	0.18	11.2
LN_2	77	0.09	30.0	0.09	6.0
LN_2	21	0.14	44.0	0.10	6.0
LHe	4	0.19	60.0	0.12	7.5

In He-gas and LHe, no differences of the coefficient of friction appear at high speed (v = 1 m/s). The calculated heat flux density (q_r) in both cases is 60 W/cm^2 for the bronze filled composite. Since for LHe the heat flux density at the onset of film boiling, i.e. the critical heat flux density (q_c), is only 1 W/cm^2 [47], the contact area at 60 W/cm^2 is far in the film boiling regime with a corresponding ΔT of approximately 500 K for static free radiating heaters.[47] In this case the entire contact is certainly surrounded by gas, which separates the hot contact zone from the surrounding cryogenic liquid. That means that the heat transfer at the friction contact is similar for each test in He-environment (at T = 77 K and T = 4.2 K), and thus the same temperature leads naturally to the same coefficient of friction.

In LN_2, a heat flux density of 30 W/cm^2 was calculated, which is just above the transition to film boiling for static free radiating heaters in LN_2.[47] Because the rotating disc continuously transports liquid and provides a cold surface area in the contact zone, it can be assumed that this transition is shifted to higher heat fluxes and ΔT is still within a range of 10 K. The low coefficient of friction and only small material transfer confirm this assumption.

In LH_2, the critical heat flux density (q_c) is 20 W/cm², which is below the calculated q_r (46 W/cm²).

At v = 0.2 m/s, the heat flux density for LN_2, LH_2 and LHe reach only 6 to 7.5 W/cm², which is respectively below or near the critical values q_c. The similar friction behaviour in the three cryogenic liquids can be explained by comparing the critical heat flux density.

Friction measurements with other PTFE composites indicated similar friction behaviour as with the bronze/CF filled PTFE at low sliding speed in LN_2 and LH_2. However, the friction coefficient of Ekonol® filled PTFE is higher in LHe than in LN_2. The lower heat conductivity of this composite could be responsible for the bad evacuation of the frictional heat developed in LHe, leading to the formation of gas bubbles at the contact area.

Hence, the tribological behaviour of the composites investigated in different environments can be partly explained by the thermal properties of the cryogenic media. As seen above, LN_2 is an efficient coolant since friction stays constant whatever the sliding velocity. The hydrodynamic effect has not been considered, since the viscosity of the cryogens are too low to establish a hydrodynamic film of sufficient thickness to separate the two surfaces.[48] However, a possible influence of the cryogenic fluid, in particular on the displacement of the wear particles from the sliding track can be considered. Indeed, powder-like wear debris were found on the surfaces of the pin and disc after testing in LHe and LH_2. Surface analyses after the experiments in LN_2 at $v = 0.2$ m/s, using AFM, SEM, XPS and EDX[3], show however less wear debris on the pin. This suggests that LN_2 has an influence in removing wear particles from the surface, probably due to its relatively higher viscosity compared to LHe and LH_2.[3]

Regarding the specific wear rates of the composites in the different cryogenic environments, the trends can be shown best for the PTFE- and PEEK- based composites filled with CF and the corresponding polymer particles. Figures 32 and 33 compare the wear rates measured at room temperature in air with those determined in LN_2, LH_2, and LHe. The results indicate that the composites wear much less in LN_2 or LH_2, whereas in LHe the wear rates reach higher values, close to those measured at room temperature in air.

The same tendency could also be seen for the PTFE-composite containing bronze balls and CF, but here the differences were not as pronounced, due to the very low wear rates of this material in general (cf. Fig. 25). Nevertheless, observations of the transferred material formed on the surface of the steel discs after the test against bronze-CF filled PTFE,

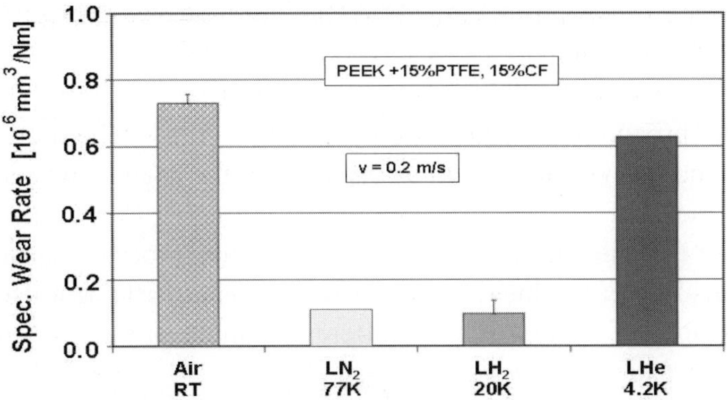

Fig. 32. Specific wear rate of a PEEK composite under various low temperature testing environments (v = 0.2 m/s).

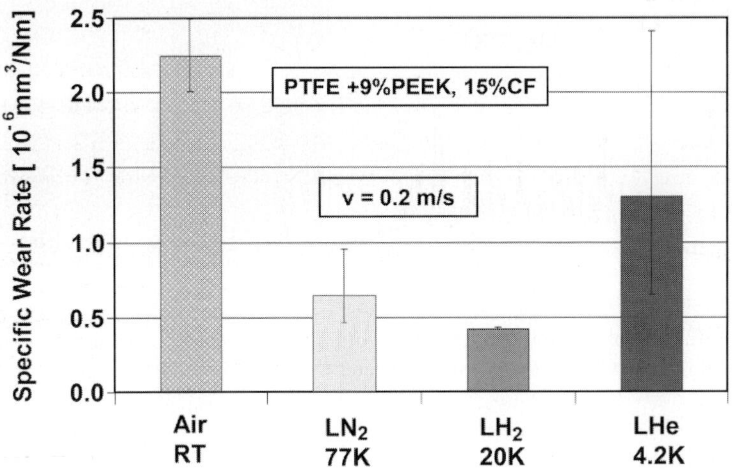

Fig. 33. Specific wear rate of a PTFE composite under various low temperature testing environments (v = 0.2 m/s).

which were investigated by means of optical microscopy and profilometry, confirm the differences in the wear behaviour of this material, when tested in LN_2 compared to LHe. At v = 1 m/s, profile measurements and optical microscopy images show the important influence of the temperature as well as the environment on the topography (Fig. 34). The experiments carried out in LN_2 produced a much thinner transfer film (Fig. 34a) that consists of scattered polymer particles. The average thickness is 100 nm. The transfer film built up in LHe is thicker compared to the one formed in LN_2 (Fig. 34b). The film consists of broad tracks, which are up to 1.5 μm thick. The transfer film is brittle and shows many small wear debris at the edge as well as at the centre of the film.

For the experiments carried out at v = 0.2 m/s, profile measurements indicate also a clear influence of temperature and environment. However, the topography of the disc is quite different from the one at v = 1 m/s.

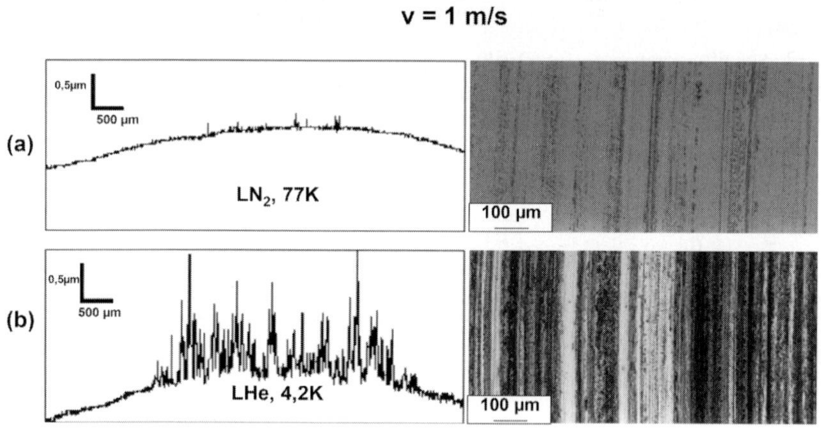

Fig. 34. Influence of the low temperature medium on the transfer of material to the steel counterpart: liquid helium allows a much greater transfer than liquid nitrogen.

5. Conclusions

The PTFE- and PEEK-based composites tested in this chapter with regard to their friction and wear behaviour in a temperature range between liquid helium and room temperature were all suitable for this application. Nevertheless, there were differences among them as a result of the type of polymer matrix and the type of filler combination used. PEEK matrix composites performed in most of the cases slightly better than corresponding PTFE matrix composites, except those PTFE – compounds that contained bronze particles in addition to the other tribo-fillers. But since metallic fillers are not always acceptable, these composites can only be recommended for special cases. It should also be mentioned that the type of cryogen (i.e. He vs. H_2 vs. N_2) has a remarkable influence on the friction coefficient and the specific wear rate. The lower the thermal conductivity of this medium is, the higher is the friction coefficient and the specific wear rate of all the composites tested.

References

1. E. Spinadel, F. Gamallo, G. Nunez, F. Spinadel and M. Cervino, *Hydrogen Energy Progress XII* in *Proceeding. 12th World Hydrogen Energy Conference*. Vol. 1 (Buenos Aires, June 1998), p. 369.
2. The German Hydrogen Association (2004). http://www.hyweb.de/h2cars
3. G. Theiler, *PTFE- and PEEK- Matrix Composites for Tribological Applications at Cryogenic Temperatures*. Dissertation, Technische Universität Berlin, BAM-Dissertationsreihe, Band 14, (Berlin, Germany, 2005), ISBN 3-86509-378-7.
4. W. Hübner, T. Gradt, T. Schneider and H. Börner, *Wear*, 216 (1998).
5. G. Hartwig and R. Hübner, *Adv. Cryog. Eng.*, 42 (1997).
6. K. Ahlborn, *Mechanische Eigenschaften von kohlenstoff-faserverstärkten Thermoplasten für die Anwendung in der Tieftemperaturtechnologie*. Dissertation Uni. Karlsruhe (TH) (1989).
7. K. Ahlborn, *Cryogenics*, 31 (1991).
8. R. P. Reed, in *Materials at Low Temperatures*, ASM, Ed. A. F. Clark (Metals Park, 1983).
9. W. Hübner, G. Mulet and O. Berndes, *Investigations in Materials Behaviour in Cryogennic Tribosystems for Formulation of Safety Regulations* in *Proceeding. 4th International Symposium on Hydrogen Power*, Vol. 2 (2001), p. 397.
10. G. Hartwig, *Polymer Properties at Room and Cryogenic Temperatures*, (Plenum Press, 1994).
11. K. Sanada, Y. Shindo and K. Horiguchi, *Non Linear Fracture Behaviour of G-10 Woven Glass-Epoxy Laminates at Liquid Nitrogen Temperature* in *Proceeding. 16th International Cryogenic Conference,* (Kitakyushu, Japan 1996), P3, (Elsevier, 1997), p. 1981.
12. T. Suzuki, S. Usami, H. Tsukamoto and M. Otsuka, *Effects of Friction on Shear/Compressive Strength of High-Density GFRP at Low Temperature* in *Proceeding. 16th International Cryogenic Conference,* (Kitakyushu, Japan 1996), P3, (Elsevier, 1997), p. 1985.
13. T. Takao, T. Nemoto, H. Konda, D. T. Kashima and A. Yamanaka, *Frictional Characteristics of High Strength Polyethylene Fiber Reinforced Plastics in*

Proceeding.16th International Cryogenic Conference, (Kitakyushu, Japan 1996), P3, (Elsevier, 1997), p. 1993.
14. G. Hartwig, *Properties of Fibre Composites (F2)* in *Handbook of Applied Superconductivity*, Ed. B. Seeber (Univerisity Geneva, 1998), p. 1007.
15. G. Heinicke, *Tribochemistry,* (Akademie-Verlag, Berlin, 1984), Chapter 4.5.
16. W. A. Glaeser, J. W. Kissel and D. K. Snediker, *Polym. Sci. Technol.*, **5B** (1974).
17. R. L. Fusaro, *Tribol. Int.*, 23 (1990).
18. G. Hartwig and R. Hübner, *Overview of the Fatigue Behavour of Fibre Composites at Low Temperatures in Proceeding.16th International Cryogenic Conference,* (Kitakyushu, Japan 1996), P3, (Elsevier, 1997), p. 1977.
19. D. Wissander, *ASLE Trans.*, 2 (1) (1959).
20. D. E. Brewe, H. W. Scibbe and W. J. Anderson, *NASA Technol. Note,* D-3730 (1966).
21. M. N. Gardos, in *Friction and Wear of Polymer Composites*, Ed. K. Friedrich (Elsevier, Amsterdam, The Netherlands, 1986), p. 397.
22. P. C. Michael, E. Rabinovicz and Y. Iwasa, *Cryogenics,* 31 (1991).
23. P. C. Michael and Y. Iwasa and E. Rabinowicz, *Wear*, 174 (1994).
24. J. L. Bozet, *Wear*, 162-164 (1993).
25. K. Tanaka, in *Friction and Wear of Polymer Composites*, Ed. K. Friedrich (Elsevier, Amsterdam, The Netherlands, 1986).
26. J. Brydson, in *Polymer Science*, Ed. A. Jenkins, Vol. 1 (1972), p. 193.
27. F. N. Cogswell, *Thermoplastic Aromatic Polymer Composites.* (Butterworth-Heinemann Ltd., Oxford, UK, 1992).
28. K. Friedrich, J. Karger-Kocsis, T. Sugioka and M. Yoshida, *Wear*, 158 (1992).
29. J. Flöck, K. Friedrich, Q. Yuan and K. Takano, *Wear*, 225-229 (1999).
30. Database on Material Properties (2003). http://www.matweb.com
31. Saint-Gobain Product Informations (2003). http://www.ekonol.com
32. B. J. Briscoe, L. H. Yao and T. A Stolarski, *Wear*, 108 (1986).
33. K. Friedrich, Z. P. Lu and A. M. Häger, *Wear*, 190 (1995).
34. P. Klein, *Tribologisches Eigenschaftsprofilkurzfaserverstärkter Polytetrafluorethylen/ Polyetheretherketon- Verbundwerkstoffe*, Dissertation, University of Kaiserslautern, IVW–Schriftenreihe, Bd. 54, (Kaiserslautern, Germany, 2005), ISBN 3-934930-50-6.
35. H. Ziebland and J. T. A. Burton, *Br. J. Appl. Phys.*, 9 (2) (1958).
36. H. M. Roder and D. E. Diller, *J. Chem. Phys.*, 52 (11) (1970).
37. R. Bowers, *Proc. Phys. Soc.*, 65 (1952).
38. K. Friedrich, Z. Lu and A. M. Häger, *J. Theor. Appl. Fract. Mech.*, 19 (1993).
39. Z. Lu and K. Friedrich, *Materialwiss. Werkstofftech.*, 28 (1997).
40. K. Friedrich, in *Friction and Wear of Polymer Composites*, Ed. K. Friedrich, (Elsevier, Amsterdam, The Netherlands, 1986), p. 233.
41. K. Friedrich, Z. Zhang and P. Klein, in *Wear – Materials, Mechanisms*, Vol. II, Ed. G. W. Stachowiak (Practice, John Wiley & Sons, Ltd., Chichester, England, 2005), p. 269.

42. K. C. Ludema and D. Tabor, *Wear*, 9 (1966).
43. P. Klein, K. Friedrich, G. Mulet and W. Hübner, *Untersuchungen zum Verschleißverhalten ausgewählter PTFE-Werkstoffe bei tiefen Temperaturen* in: *Gleitverschleiß polymerer Verbundwerkstoffe zwischen Raumtemperatur und 4,2 K.* BAM-Sonderheft 3, (2001).
44. S. Bahadur and D. Gong, *Wear*, 158 (1992).
45. T. Gradt and H. Börner, *Einführung in die Tieftemperatur-Tribologie* in *Gleitverschleiß polymerer Verbundwerkstoffe zwischen Raumtemperatur und 4,2 K.* BAM-Sonderheft 3, (2001).
46. G. Theiler, W. Hübner, T. Gradt, P. Klein and K. Friedrich, *Tribol. Int.*, 35 (2002).
47. E. G. Brentari and R.V. Smith, *Adv. Cryog. Eng.*, 10 (1965).
48. J. L. Bozet, *Wear*, 162-164 (1993).

CHAPTER 12

MECHANICAL AND TRIBOLOGICAL BEHAVIOR OF POLYMERS FILLED WITH INORGANIC PARTICULATE FILLERS

Shyam Bahadur[1] and Cris Schwartz[2]

[1]*University Professor Emeritus, Mechanical Engineering Department*
Iowa State University, Ames, Iowa 50011, USA
[2]*Assistant Professor, Mechanical Engineering Department*
Texas A&M University, College Station, TX 77843, USA

Polymers are filled with particulate materials to enhance the tribological properties and often to change the physical and mechanical properties as well. The composites made with such fillers are amenable to common molding processes unlike fiber-reinforced composites which makes them very attractive economically. As such a large number of studies on the tribological behavior of filled polymers have been made. The representative results from such studies are presented in this chapter. In this context, the friction and wear behavior of these composites is examined. The results show that some particulate fillers improve the wear resistance of polymers while others affect it adversely. The mechanism of the improvement in wear behavior is discussed by referring to the transfer film characteristics such as texture, uniformity and thickness as well as bonding to the counterface surface. In order to support these observations, the micrographs of transfer films along with the experimental results from transfer film-counterface bonding are presented. In addition the results from energy dispersive X-ray analysis are presented to support the hypothesis of bonding. The similarities and differences with respect to the tribological behavior of the composites made with micro and nano fillers are also discussed.

1. Introduction

Polymer composite materials are employed in a number of applications because they offer properties that cannot be obtained with polymers alone. Whereas continuous fiber-reinforced polymer composites are the best for enhancement of mechanical as well as tribological properties, they suffer from the disadvantage of processing because the traditional production processing techniques such as injection molding and extrusion etc. cannot be used with such composites. This makes these composites expensive for use in common applications. When instead of the continuous fibers, chopped fibers are used for reinforcement which makes these amenable to the above production processes, the improvement in mechanical properties is lost and the tribological properties also suffer considerably. As an alternative, filled polymer composites have been found very attractive because these can be processed easily and inexpensively by the processes used for unfilled polymers and at the same time offer very superior tribological properties comparable to those of the fiber-reinforced polymers.

Filled polymer composites (or simply filled polymers) are made using fillers in the polymer matrix. These fillers are in micro and nano meter sizes. Many kinds of inorganic materials have been used as fillers in polymers such as metal powders, minerals, oxides, mica, carbon black, and solid lubricants. Metal powders are more effective in improving the thermal and electrical properties than the tribological properties. Mineral fillers are commonly used to modify properties such as stiffness, hardness, heat resistance, thermal expansion, electrical behavior, and density but some fillers have also been found to be effective in improving the tribological properties. Many inorganic compounds in both micro and nano sizes have been found to be very effective in improving the tribological behavior *which is the thrust of this chapter.* Solid lubricant fillers such as graphite, MoS_2 and PTFE have shown effectiveness in reducing the coefficient of friction and also increasing the wear resistance in many cases.[1]

2. Effect of Fillers on Mechanical Properties

Fillers are used in polymers to achieve varying goals. While in some instances fillers are used merely as a bulk filling agent to reduce material cost, their more targeted applications typically involve modifying the properties of the polymer such as stiffness modulus, fracture strength, and toughness. A challenge to the product developer has been to fully understand and predict the property changes brought about by the addition of particulate fillers. It has been shown that some fillers provide significant beneficial effects while others are detrimental to the properties of the polymer. The complication arises because the enhancement in one property often comes at the cost of degrading other properties. Experimental observations are thus necessary to determine the effects of the filler on the matrix. Some models have been developed that have sown limited success in predicting properties. These models involve filler parameters such as particle composition, particle size and shape, orientation, distribution in the matrix, and volume or weight fraction of the filler. In this section, relevant research related to the effects of fillers on various properties is discussed.

2.1. *Stiffness Modulus*

The most common application of filler particles in polymers is to enhance the stiffness of the polymer, as measured by the elastic (storage) modulus. In their neat form, many polymers lack the stiffness required for engineering applications. Fillers have been used to create engineering composites from such polymers. There are a great variety of fillers that have been used for stiffness enhancement. Models that have been proposed to explain the properties of filled polymers have had varied levels of successful application based on the assumptions made and the materials being used. A comprehensive review of the development of models to describe modulus enhancement is given by Ahmed and Jones.[2] The earliest model for the shear modulus of filled elastomers had been adapted from the viscosity relationship of a fluid filled with rigid spheres and as developed by Einstein[3] was

$$\eta_c = \eta_m (1 + K_E V_p) \tag{1}$$

where η refers to the viscosity of the suspension (c) and matrix (m), K_E is the Einstein coefficient (equal to 2.5 for spherical filler particles), and V_p is the volume fraction of filler particles. While this relationship shows agreement with test data at very low weight fractions of filler, it has not been useful largely because it does not take into account filler particle size and does not address the effects of interaction among filler particles within the polymer matrix.

Einstein's model has been used as a basis for refinement and evolution of stiffness models. Mooney[4] modified the above relation to address matrices filled with closely packed spheres, while Brodnyan[5] extended it to account for non-spherical fillers by introducing a term that incorporates the aspect ratio of the filler particle. Further refinement of the models came about when stress field interactions around filler particles were considered. Guth[6] produced such a model and he was one of the first to show that the shear modulus models based on viscosity could be used to describe elastic modulus behavior by the following relationship

$$E_c = E_m(1 + K_E V_p + 14.1V_p^2) \tag{2}$$

where E_c and E_m are the elastic moduli of the composite and the matrix, respectively. This model was not intended for filler volume fractions in excess of 10 percent.

More complex models address the interactions among filler particles and their effect on the composite modulus. One of the most detailed models for spherical filler particles, as proposed by Kerner,[7] was

$$G_c = G_m\left(1 + \frac{V_p}{V_m}\frac{15(1-\nu_m)}{(8-10\nu_m)}\right) \tag{3}$$

where G_c and G_m are the shear moduli of the composite and the matrix materials, respectively, V_m is the volume fraction of the matrix material, and ν_m is the Poisson ratio of the matrix material. The Kerner model is important because it was used as a building block to produce the Halpin-Tsai model, which is widely known by composites researchers. Nielsen generalized the Halpin-Tsai model to make it dependent on the theoretical maximum filler packing fraction (based on filler size and geometry).[8] The modified Halpin-Tsai relation[9] is

$$\frac{M_c}{M_m} = \frac{1 + AB\phi_f}{1 - B\psi\phi_f} \tag{4}$$

where M_c and M_m are the moduli (elastic or shear) of the composite and matrix materials, respectively, A is a constant related to filler geometry and matrix Poisson ratio, ϕ_f is the filler volume fraction, ψ is based on the maximum packing fraction of the filler, and B is determined from the following equation

$$B = \frac{(M_f / M_m) - 1}{(M_f / M_m) + A} \tag{5}$$

Currently, the modified Halpin-Tsai model enjoys widespread use due to the fact that it encompasses many of the relevant parameters that have been found to be important in the interaction between filler and matrix. Somewhat surprisingly, none of the above models incorporates the quality of the bond between the filler and the matrix. In fact, Sato and Furukawa[10] argued that interfacial bonding does not play a large role in the modulus of a composite. While bonding does not appear to play a large role in the modulus of particulate-filled composites, it is responsible for many of the effects on composite strength as described below.

2.2. *Strength*

There has not been the same level of progress in producing models that describe filler effects on composite strength. The reason for this is that it requires bond strength between the filler particles and the matrix, which has been challenging to model. Separate phase filler particles may act as internal stress raisers within the polymer, thus lowering strength, in opposition to their load supporting role. Thus there are two competing phenomena that are at play in the context of the effect of fillers on strength. To complicate matters further, many polymers are susceptible to water absorption that can detrimentally affect the interfacial bonding between the filler and the matrix. Therefore, strength models may entail great complexity if they attempt to address all of the pertinent parameters.

The fundamental mechanism of strength enhancement by filler particles is by the transfer of shear stress from the matrix to the filler. Obviously, this transfer is highly dependent on the bond quality between the filler and the matrix. Typically, the particulate filler materials (inorganic compounds, clays, glass beads) do not bond well to polymeric materials. Therefore, coupling agents are often used to enhance the filler-matrix bond. Many of these agents are molecules that have specialized functionalities at each end, such that one end bonds to the polymer matrix while the other end bonds to the filler material. Coupling agents include materials such as silanes, methacrylates, and titanates.[9]

Among the models that have been proposed to predict composite strength is that developed by Schrager[11] and is as follows

$$\sigma_u = \sigma_m e^{-r\phi_f} \tag{6}$$

where σ_u and σ_m refer to the ultimate strength of the composite and unfilled matrix, respectively, while r is an interfacial factor based on the volume of filler particles and the matrix volume that is affected by the filler particles. This factor was experimentally determined to be equal to 2.66 for spherical particles. The limitation of the Schrager model is that it considers a somewhat idealized composite that has well-bonded filler particles; however, it shows good agreement with experimental results up to volume fractions of approximately 50%.

It has also been noted that particle size plays a significant role in strength enhancement as it does with modulus. Hojo *et al.*[12] proposed the following model to describe this size effect

$$\sigma_u = \sigma_m + kd^{-1/2} \tag{7}$$

where k is an empirical constant and d is the mean particle diameter. Particle size has been shown experimentally to be a very important parameter for the properties of composites. In the stiffness models above, smaller particles lead to higher moduli because smaller particles allow a greater maximum packing factor than do larger particles. Considering that smaller particles are beneficial to stiffness and strength, the case becomes clear for reducing filler particle size. This size effect is one of the reasons behind the growing use of nanoparticle fillers in polymer matrices.

Some work has been done to examine the strength of nanoparticle-filled polymers. A direct method for determining the failure strength of a rigid composite is the use of three-point bend flexure testing to failure. Schwartz and Bahadur[13] investigated the flexure strength of PPS filled with alumina nanoparticles as shown in Fig. 1. No attempts were made to bond the nano-Al_2O_3 to the PPS matrix, and the results show the strength decreases at very low weight fractions of filler. Interestingly however, flexure strength begins to recover somewhat as weight fraction is increased, though the composite never quite regains the flexure strength of the unfilled polymer. Such an increase in strength after a minimum follows the general predictions of the Schrager model with increasing filler fraction.

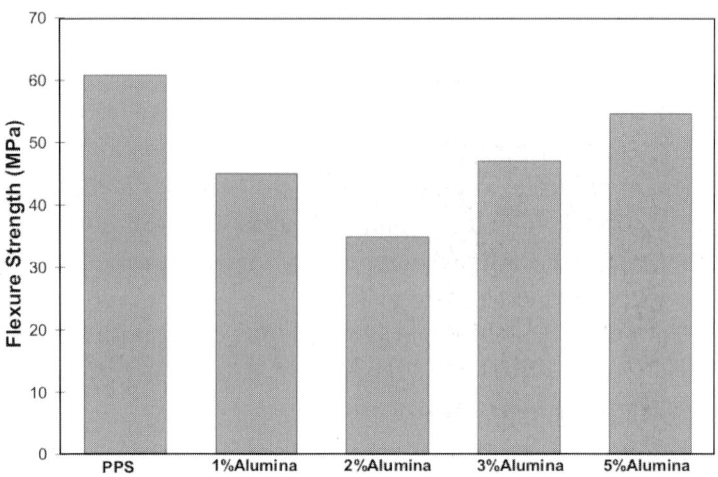

Fig. 1. Three-point flexure strength of PPS filled with varying amounts of nano-Al2O3.[13]

2.3. *Toughness*

Nielsen proposed a simple illustrative model that says that poorly-bonded particulate fillers typically reduce the impact toughness of the polymer.[14] This is due to the fact that tensile stress can lead to separation at the interface between filler and matrix. This separation may, in turn, lead to localized stress concentration and cracking in the matrix in the interfacial boundary with the filler. Such matrix damage can severely degrade the performance of the composite in impact loading. Such a situation is illustrated in Fig. 2. On the other hand, systems using well-bonded elastomeric fillers can actually increase the impact toughness of the composite. An analogy to this situation is the use of elastomeric polybutadiene particles to improve the toughness of polystyrene to produce high-impact polystyrene (HIPS).[9] It should be noted that many filler materials used for modulus reinforcement or wear reduction tend not to be well-bonded to the matrix.

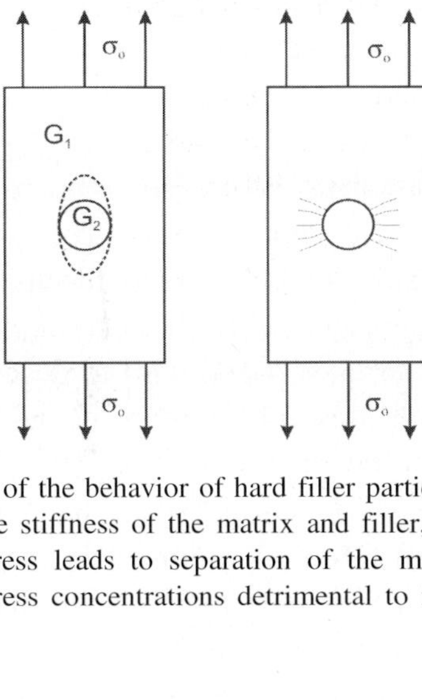

Fig. 2. Conceptual model of the behavior of hard filler particles in a less rigid polymer matrix (G_1 and G_2 are the stiffness of the matrix and filler, respectively, and $G_1<G_2$). Application of tensile stress leads to separation of the matrix from filler, and thus produces cracking and stress concentrations detrimental to impact toughness. Adapted from ref. 14.

3. Effect of Fillers on Friction and Wear

The particulate fillers are added to polymers in order to improve tribological properties such as wear resistance and friction coefficient. Tribology has relied more on empirical investigations than modeling primarily because of the complexity of the wear process. For this reason, predictive models of the effects of filler material, size, and shape have not been successfully validated by experiment and thus the materials designer often must focus on the particular characteristics of the polymer-filler system in order to propose methods of improvement.

One of the most common tribological applications of filled polymers is that of polymer-on-metal sliding, where a polymeric material slides against a metallic counterface. Such applications are common in engineering practice. This system has shown itself to be useful for two primary reasons: under proper conditions, the polymer becomes self-lubricating so that wear and friction coefficient are low; and the metallic counterface has high thermal conductivity thus effectively avoiding a large buildup of heat at the wear interface that would otherwise degrade the performance of the polymer. The addition of particulate fillers to polymers has resulted in the increase in wear resistance with some fillers and the degradation in wear resistance with others. Therefore, it is useful for the materials designer to have some idea *a priori* which filler materials are useful for increasing wear resistance in a particular polymer.

The inorganic particulate fillers are commonly characterized by chemical composition, size and hardness. This section will report the findings from investigators based on their results from both micro-size and nano-size particulates and the further divisions among hard and soft fillers. The use of the terms soft and hard to describe filler materials is open to debate, however a useful reference is chosen to be the hardness of the counterface material. In this section, abrasive materials that can scratch the counterface will be considered as hard fillers, while the others will be described as soft fillers. A rudimentary method of distinguishing between hard and soft fillers is found by using the Mohs scale used to assign relative hardness rankings to minerals. On this scale, hardened tool steel, used as a counterface in much of the research, has a value of 7 to 8. The small number of filler materials that exceed this value are thus classified as hard fillers.

3.1. *Microparticle Fillers*

The vast majority of tribological research work done with particle-filled composites has used fillers that have length dimensions greater than 1 μm. These are termed microparticles in practice. There are significant differences in the tribological behavior of the composites with specific fillers and polymers that have been investigated. A few microparticle copper-based inorganic compounds have shown benefits in nearly all of their uses in tribological applications. Such a study showing the effectiveness of copper compound fillers such as CuO, CuS and CuF_2 was reported by Bahadur and Gong.[15] The proportion of the fillers used in this study was 35 vol.% which was based on an earlier reported analysis of critical packing by these workers.[16] The friction and wear experiments were performed in pin-on disk configuration with the polymer pin sliding against a hardened steel disk finished with 320-grit SiC paper to a surface roughness of 0.11 μm Ra. Figure 3 shows the wear volume of the polymer and its composites as a function of sliding distance. It is obvious that all the fillers reduced wear and the ability of the fillers in reducing wear was in the following order: CuF_2, CuS and CuO. The steady state wear rates of the last two fillers were almost equal. The coefficient of friction in steady state sliding by and large increased with the addition of fillers.

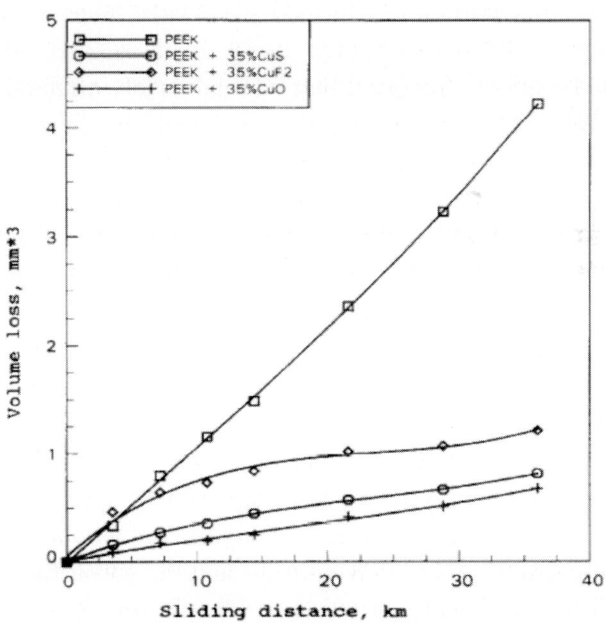

Fig. 3. Wear loss vs. sliding distance for PEEK and its filled composites sliding against a hardened steel disk. Test conditions: sliding velocity 1.0 m/s, contact pressure 0.654 MPa, counterface roughness 0.11 μm Ra.[19]

Similar observations on the effectiveness of copper compounds as fillers in increasing the wear resistance of many kinds of polymers have been made. For example, Yu and Bahadur reported that the wear resistance of poly(phenylene sulfide) (PPS) increased when filled with CuO particles.[17] Bahadur and coworkers showed the benefits of CuO filler on wear resistance in high-density polyethylene (HDPE),[18] nylon 11,[19] and thermosetting polyester,[20] and Vande Voort and Bahadur showed its wear benefits in poly(ether etherketone) (PEEK).[21] Another copper compound, CuS, has been shown by Bahadur and coworkers to increase the wear resistance of polytetrafluoroethylene (PTFE),[22] PEEK,[21] and PPS.[23,24] In addition to CuO and CuS, the filler CuF_2 has also been reported to reduce the wear of nylon 11.[25]

Some soft filler materials have shown mixed results in terms of the tribological performance. For example, Schwartz and Bahadur[24] showed that Ag_2S and CuS reduced the wear of PPS, while ZnF_2 and SnS led to higher wear rates. They hypothesized that this behavior was due to whether the filler particles deformed plastically under compression as Ag_2S and CuS did, or crumbled under load as ZnF_2 and SnS did. Bahadur and Kapoor[26] reported that ZnS adversely affected the wear of nylon 11. Working with thermosetting polyester as the matrix material, Zhang[27] found that both ZnO and CaF_2 decreased the wear resistance of the polymer.

Zhao and Bahadur[28,29] reported that Ag_2S and NiS enhanced the wear resistance of PPS while PbSe and PbTe degraded its wear resistance. In order to illustrate the difference in wear behavior, the wear results for PPS filled with NiS and PbSe fillers are shown in Figs. 4 and 5. Figure 4 shows the wear loss vs. sliding distance diagram for unfilled PPS and that filled with three proportions of NiS. For all the compositions, the transient and steady state wear behaviors are observed. The effect of 10 vol.% NiS on steady state wear rate is negligibly small but with 30 vol.% NiS wear rate decreased by a significant amount. However, the steady state wear rates for PPS with 20 and 30 vol.% filler were about the same. The coefficient of friction of PPS decreased with the addition of filler, and the higher the filler proportion the smaller was the coefficient of friction.

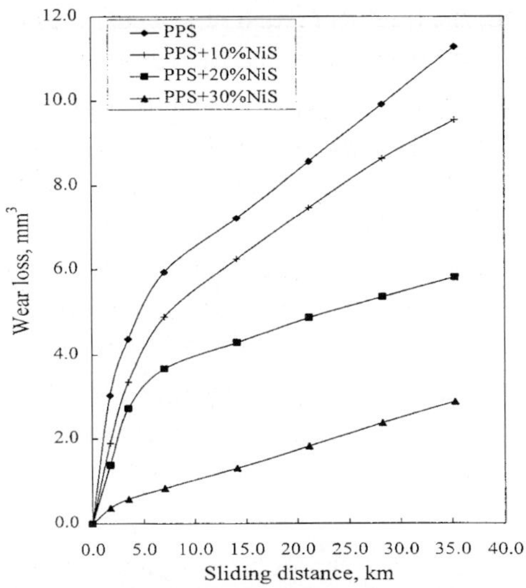

Fig. 4. Wear loss vs. sliding distance for PPS filled with NiS. Test conditions: sliding speed 1.00 m/s, nominal pressure 0.65 MPa, counterface roughness 0.10 μm Ra.[28]

Figure 5 shows the wear behavior with PbSe filler in PPS. Contrary to the behavior of NiS, the wear rate of PPS increased with PbSe as the filler. The wear plots show that the higher the proportion of PbSe, the higher was the steady state wear rate. With 30 vol.% PbSe the steady state wear of PPS increased by a factor of about 8. Similar to the case of NiS filler, the coefficient of friction of PbSe-filled PPS decreased with the addition of filler and the higher the filler proportion, the lower was the coefficient of friction. Because of their abrasive nature, very little work has been done with micro-sized hard particles of alumina, silicon carbide, or silicon nitride. In these relatively large sizes, the typical result is a combination of counterface scratching and aggregation of the filler particles at the wear interface that abrades the polymer and leads to extremely high wear rate. For instance, Shao and Xue[30] studied microparticle SiO_2 filler in poly(phthalazine ether sulfone ketone) (PPESK) and found considerable degradation of wear resistance due to abrasion.

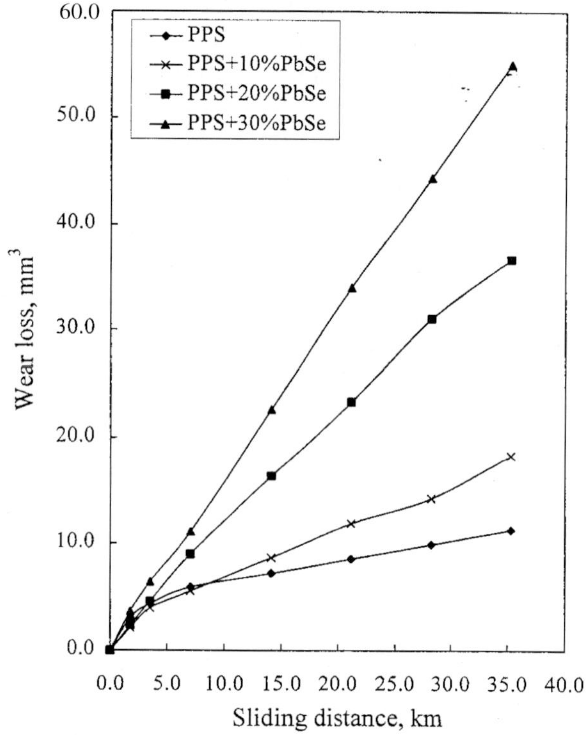

Fig. 5. Wear loss vs. sliding distance for PPS filled with PbSe. Test conditions: sliding speed 1.00 m/s, nominal pressure 0.65 MPa, counterface roughness 0.10 μm Ra.[28]

3.2. *Nanoparticle Fillers*

Examination of both the modulus and strength models suggests that smaller filler particles contribute more to strengthening than do larger particles of the same compound. The size effect has been found to be important for tribological behavior as well because nanoparticles are effective in increasing wear resistance in much smaller volume percentages than the microparticle fillers. For instance, micro-size fillers show maximum tribological benefit at volume fractions of 25 to 35 vol.% whereas nanoparticles often produce optimal results from 1 to 3 vol.%. This behavior makes the nanoparticles more attractive than the microparticles as the fillers in polymers for tribological applications.

Cho and Bahadur[31] found that nano-sized CuO increased the wear resistance of PPS, as did the micro-sized fillers of the same compound, with an optimal fraction of 2 vol.% nanoparticles versus 35 vol.% microparticles.[17] Bahadur and Sunkara[32] showed that in addition to CuO, TiO_2 nanoparticles also increased the wear resistance of PPS. In the same study, they reported that nano-size ZnO and SiC fillers degraded the wear resistance of PPS. On the other hand, Li *et al.*[33] reported that nano-ZnO filler enhanced the wear resistance of PTFE. This shows that the same filler material can produce different results based on the polymer matrix. The nanoparticle TiO_2 filler was shown by Chang *et al.*[34] to increase the wear resistance of poly(etherimide). Wang *et al.* reported similar tribological benefits with nano-ZrO_2 filler in PEEK.[35]

The wear behaviors of PPS filled with CuO, TiO_2, ZnO and SiC in 30-50 nm sizes, as reported by Bahadur and Sunkara,[32] are given in Fig. 6. The steady state wear rate is the lowest for PPS filled with 1-2 vol% CuO filler proportion. With 3 vol.% CuO filler, the wear rate increased to about twice that of 1 vol.% filler and with 10 vol.% filler it became even higher. In all the CuO filler proportions tried, the wear rate was lower than that of unfilled PPS. With nano size TiO_2 filler, the maximum reduction was obtained with 2 vol.% filler proportion. With 3 vol.% TiO_2 filler, the wear rate was higher than that of unfilled PPS and with 5 vol.% filler it was even higher. Contrary to these two cases, nano size ZnO and SiC fillers increased the wear rate of PPS. The coefficient of friction for any filled composition was lower than that of the unfilled PPS.

The wear results for hard nanoparticle fillers in polymers are intriguing due to the fact that they often provide benefits without the gross abrasive behavior expected of larger particles. The most common hard compound used as a nanoparticulate filler is Al_2O_3.

It has been used in a variety of polymer matrices with good results, so long as its volume fraction is held below 5 vol.% approximately. For example, Schwartz and Bahadur[13] demonstrated the increase in wear resistance of PPS with the addition of nano-Al_2O_3. They attributed this wear benefit to enhanced bonding of the transfer film to the counterface due to the interlocking of nanoparticles with asperities on the counterface surface. Bhimaraj *et al.*[36] found that Al_2O_3 nanoparticles in poly

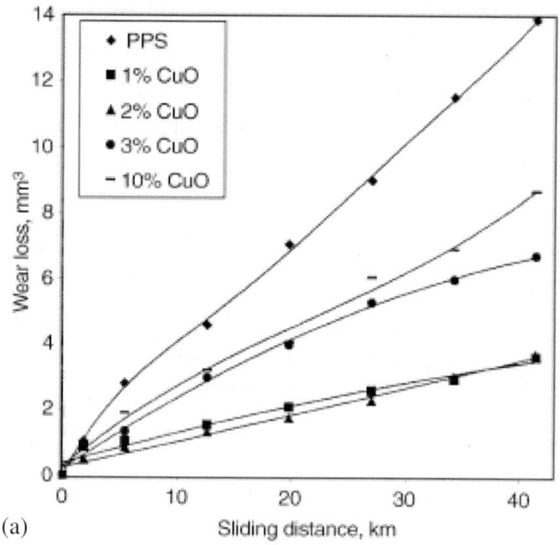

(a)

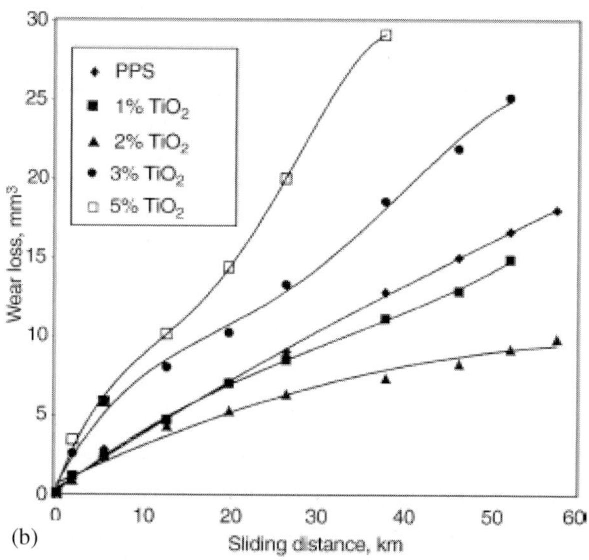

(b)

Fig. 6. Wear loss vs. sliding distance plots for PPS filled with nano particles of (a) CuO, (b) TiO_2, (c) ZnO, and (d) SiC. Sliding conditions: 1 m/s sliding speed, 0.65 MPa pressure, 0.10 μm counterface roughness.[32]

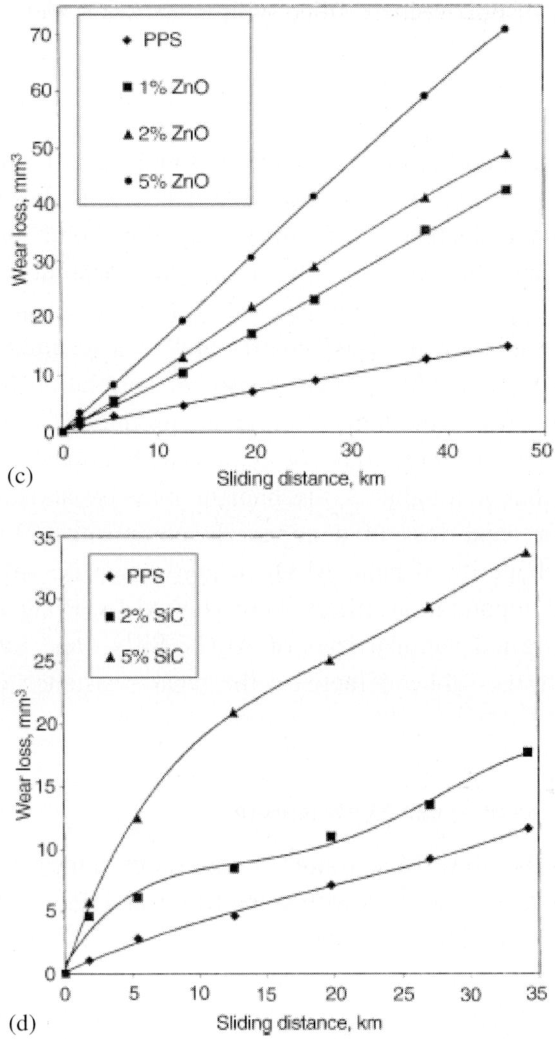

Fig. 6. (*Continued*)

(ethylene terephthalate) (PET) increased the wear resistance up to a point, but the filler volume fractions over 5 vol.% led to agglomeration of the nanoparticles at the wear interface and abrasive wear similar to what is witnessed with microparticles. Recently, Zhao *et al.*[37] studied the wear behavior of polyamide 6 filled with nano Al_2O_3 and found that the filler contributed to increased wear resistance. Shao and Xue[30] reported the increase in wear resistance of PPESK when filled with nano SiO_2 particles.

In sections 2.2 and 2.3 on strength and toughness, the importance of interfacial bonding between the filler and matrix was discussed. In this light, some investigators tried enhancing the bond between hard nanoparticles and the polymer. Shi *et al.*[38] demonstrated that using a silane coupling agent or grafting nano-Al_2O_3 to polyacrylamide (PAAM) led to a strong bond with epoxy matrix and thus resulted in enhanced wear resistance. Interestingly, they report no correlation between wear resistance and stiffness modulus of polymers produced from their grafted fillers, but did find substantial correlation with impact strength. This may suggest that a toughness mechanism may be at least somewhat responsible for wear loss in nanoparticle composites. Wetzel *et al.*[39] found similar benefits of nano-Al_2O_3 in epoxy matrix. Additional hard nanoparticle compounds as fillers were studied by Rong *et al.*[40] They used PAAM-grafted nanoparticles of Al_2O_3, Si_3N_4, and SiC and found that each filler was able to increase the wear resistance of the epoxy matrix.

4. Mechanisms of Wear Modification

As seen from the above discussion, the wear resistance of polymers is increased when they are filled with some fillers and decreased with some others. This raises the question why some fillers are effective in increasing the wear resistance while others are not. This has been the subject of investigation by many researchers but there is no single answer because of the complexity of the processes involved. As would be expected, some have looked at the mechanical aspects of the fillers while others have paid attention to the modification of the physical processes involved in sliding by fillers. The following are some of the main ideas behind the reduction in wear in filled polymer composites:

1. In case of hard fillers, the bulk of the load is shared by the hard fillers so the soft polymer is subjected to lower load. This effectively reduces the damage to the polymer sliding surface.
2. The fillers help in the formation of transfer film on the metallic counterface which promotes sliding of the polymer surface against its own transfer film.
3. The effective fillers increase bonding of transfer film to the counterface so that the loss of transfer film from the counterface is reduced during sliding which effectively reduces wear.

4.1. *Sharing of Load by Hard Particles*

The wear reduction produced by the use of filler particles has been theorized to be due to the ability of hard fillers to partially support normal loads during sliding. In simple terms, the use of a filler material that is stiffer than the polymer matrix may lead to a significant amount of the normal load being carried by the filler particles as opposed to the polymeric material. On the other hand, hard filler particles can accumulate at the wear interface and thus increase wear in some instances. This dual role of filler materials has been described by investigators. Schwartz *et al.* showed that the wear of ultra-high molecular weight polyethylene (UHMWPE) was significantly reduced by using platinum-zirconium quasicrystal filler particles.[41] They attributed this wear reduction to the high stiffness of the filler particles and their disproportionate support of the normal load at the wear interface compared to the UHMWPE matrix. Blanchet found that the use of hard filler particles in PTFE reduced the wear of the polymer, but not due to load sharing by the fillers but rather because of the specific wear resistance of the filler material.[42] Han and Blanchet further studied the issue and found that an inverse rule-of-mixtures model described the wear of PTFE composites. This model focused more on the accumulation of filler particles at the wear interface as opposed to their role in load sharing.[43] Because of the challenge of directly observing mechanical behavior at the wear interface, the exact mechanisms of filler reinforcement require further investigation.

4.2. *Transfer Film Topography*

This is well known that when polymer slides against a metal counterface, the transfer film of polymer is formed on the counterface. The role of transfer film is very important for wear because in the initial transient state the wear rate is very high because of the absence of transfer film. The wear rate is reduced considerably in the changeover from transient state to steady wear state which occurs due to the deposition of polymer transfer film on the metallic counterface.[44] It would thus imply that the role of filler in affecting the formation of transfer film is important for wear.

In section 3.1 the wear results for PEEK filled with CuS, CuO and CuF_2 were presented. It was found that all the three fillers reduced the steady state wear rate of PEEK considerably. In order to show how transfer film affects the wear behavior, the transfer films of PEEK and its filled compositions are given here.[19] Figure 7 shows the surface of an unrubbed steel disk with abrasion marks left over from the finishing operation. From Fig. 8 it is seen that some patchy transfer of unfilled PEEK has occurred but the bulk of the counterface surface is uncovered as the finishing marks in Fig. 7 can be seen in this case as well.

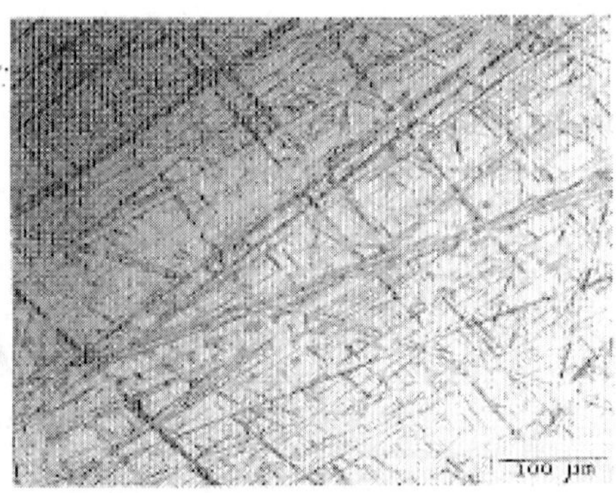

Fig. 7. Optical micrograph of unrubbed tool steel counterface.[19]

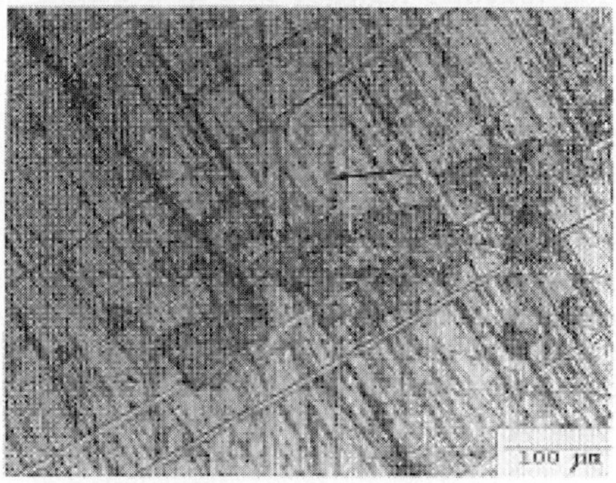

Fig. 8. Optical micrograph of unfilled PEEK on steel counterface. Sliding conditions same as in Fig. 2. Arrow shows sliding direction.[19]

The wear rate of unfilled PEEK as seen in Fig. 3 was thus high and the role of transfer film was basically absent because the wear rate did not moderate throughout the entire sliding period. Figures 9 through 11 show transfer films of the filled composites covering the counterface uniformly. In all of these cases, there is a thin and uniform (as opposed to patchy and lumpy) deposition of the filled polymer which makes the counterface surface smooth. Thus the polymer is not subjected to the abrading action as would be the case if the counterface had exposed sharp asperities. The formation of transfer film in case of the filled polymers is very efficient because in a short duration of 0.5 h a thin transfer film has covered the entire rubbing surface, as shown in Fig. 9 (a) for CuS-PEEK composite. Because of this, the transient wear state lasts only a very short time and is followed by the steady wear state, as seen in Fig. 3. The same was observed in cases of the other filled polymers as well.

In order to emphasize the role of fillers in the formation of transfer film, the transfer films of unfilled PPS and that filled with NiS and PbSe are presented in Figs. 12 and 13.[28] Unlike PEEK (Fig. 8), PPS formed a good transfer film on the abraded steel surface as shown in Fig. 12. The transfer film of PPS is thick and grainy and is covering the entire sliding track.

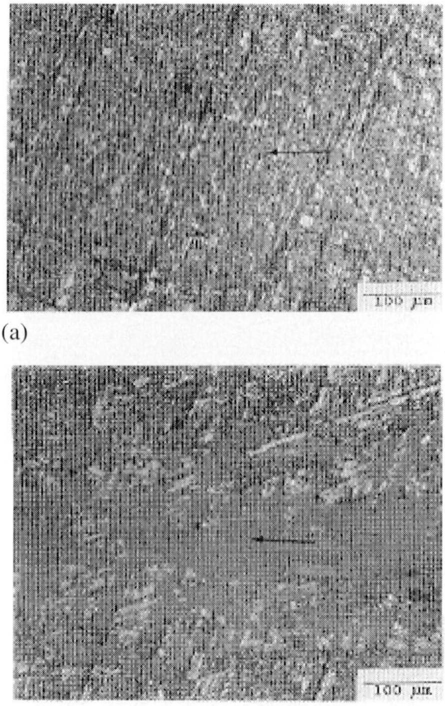

(a)

(b)

Fig. 9. Optical micrographs of transfer film of 35% CuS-PEEK composite on steel counterface after rubbing for (a) 0.5 h and (b) 10 h. Arrow indicates sliding direction. Sliding conditions same as in Fig. 3. Arrow shows sliding direction.[19]

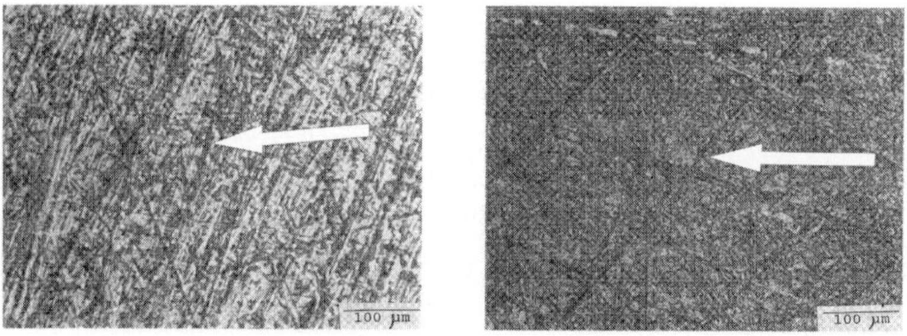

(a)

(b)

Fig. 10. Optical micrograph of 35% CuO-PEEK composite sliding against a steel counterface: (a) transfer film after rubbing for 0.5 h; (b) transfer film after rubbing for 10 h. Sliding conditions same as in Fig. 3. Arrow shows sliding direction.[19]

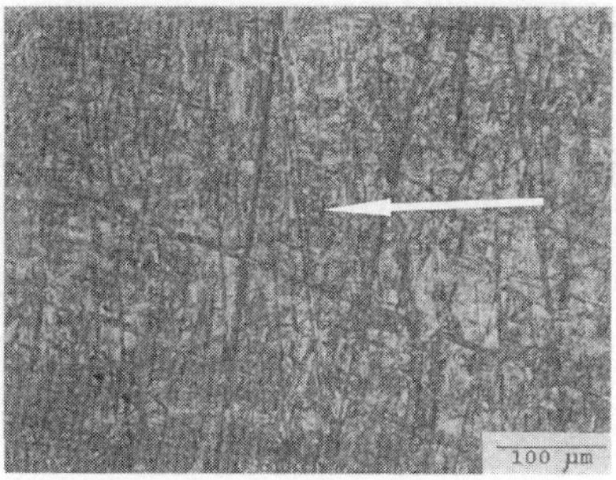

Fig. 11. Optical micrograph of the transfer film of 35% CuF_2-PEEK composite sliding for 10 h against a steel counterface. Sliding conditions same as in Fig. 3. Arrow shows sliding direction.[19]

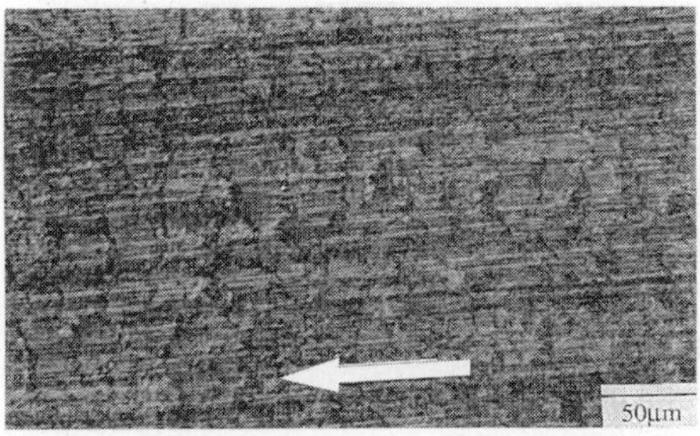

Fig. 12. Transfer film of unfilled PPS. Sliding conditions same as in Fig. 4. Arrow shows sliding direction.[28]

The transfer films formed on the counterface of PPS with NiS and PbSe fillers are shown in Fig. 13. With 10 vol.% NiS the topographical features of the transfer film changed slightly because the thickness was reduced and the texture became finer but the film was still grainy. With the increase in NiS proportion to 30 vol.%, the transfer film became much thinner and more uniform. This led to the decrease in wear rate of PPS with NiS filler. The transfer film in the case of PPS filled with PbSe is different. There is a meagre transfer film on the counterface only in discrete patches, as seen in Figs. 13c and d for 10 and 30 vol.% filler proportions, respectively. Most of the sliding track in these cases is not covered by the transfer film and so wear occurred mostly by abrasion. This led to the increase in wear rate with the addition of PbSe filler in PPS.

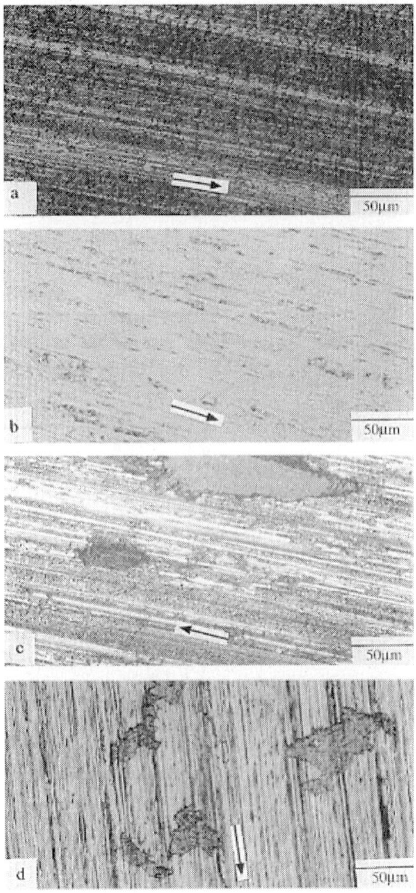

Fig. 13. Transfer films of PPS filled with NiS and PbSe fillers: (a) PPS + 10 vol.% NiS, (b) PPS + 30 vol.% NiS, (c) PPS + 10 vol.% PbSe, and (d) PPS + 30 vol.% PbSe. Sliding conditions same as in Figs. 4 and 5. Arrow shows sliding direction.[28]

Figure 14 shows the optical micrographs of the transfer films formed during sliding of nano CuO-filled PPS composites. The transfer film of unfilled PPS is thick and grainy and it exhibited small pores in discrete locations at higher magnification. The transfer film of PPS filled with 2 vol.% CuO filler is thinner and more uniform than above. It should be noted that this composition gave the lowest wear rate. With 10 vol.% CuO filler, the transfer film became non-uniform and presumably for that reason the wear rate increased.

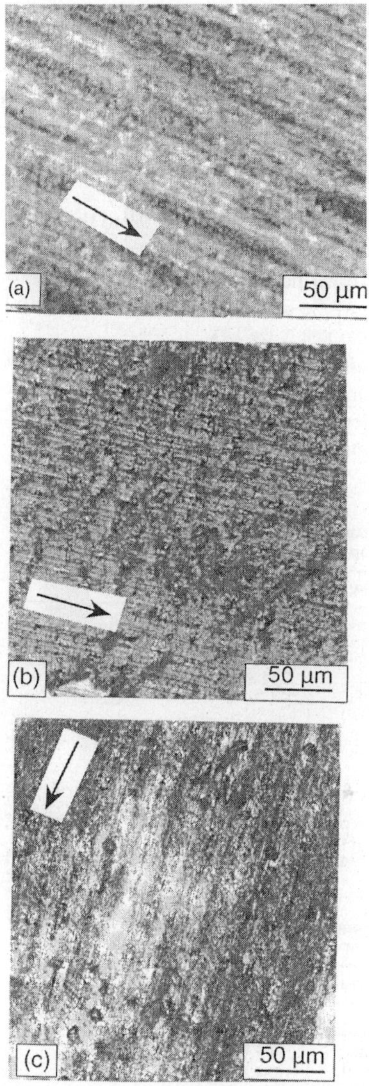

Fig. 14. Optical micrographs of transfer films formed during sliding on steel counterface for (a) unfilled PPS, (b) PPS filled with 2 vol.% nano CuO, and (c) 10 vol.% nano CuO. Sliding conditions same as in Fig. 6. Arrow shows sliding direction.[32]

Figure 15 shows the transfer films of PPS filled with 2 and 5 vol.% TiO_2. For 2 vol.% TiO_2 filler, the transfer film is thin and uniform and this gave the lowest wear rate of the unfilled and filled compositions with

1-5 vol.% TiO_2 filler. The transfer film of PPS with 5 vol.% TiO_2 filler is thick and lumpy and it also does not cover the counterface completely. The wear rate in this case was higher than that of any TiO_2-filled or unfilled compositions. As also reported earlier, the wear rate of PPS increased when filled with ZnO or SiC. The transfer films for these two cases are shown Fig. 16. The transfer film in Fig. 16a is similar to that of unfilled PPS and in Fig. 16b is thick and lacks in coverage of the counterface. Thus again a correlation between transfer film features and wear rate is seen.

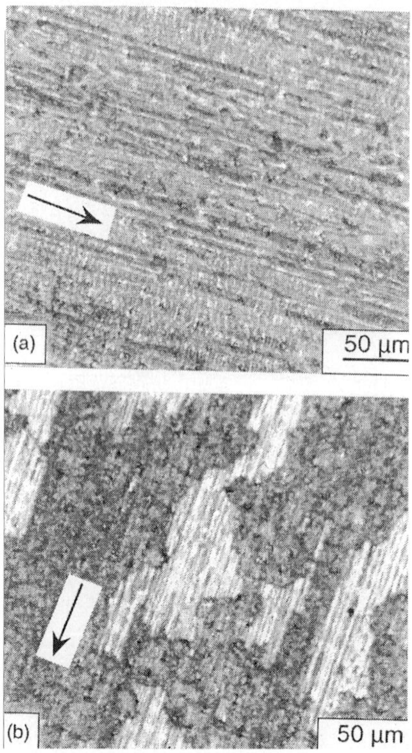

Fig. 15. Optical micrographs of transfer films formed during sliding on steel counterface for PPS filled with (a) 2 vol.% nano TiO_2, and (b) 5 vol.% nano TiO_2. Sliding conditions same as in Fig. 6. Arrow shows sliding direction.[32]

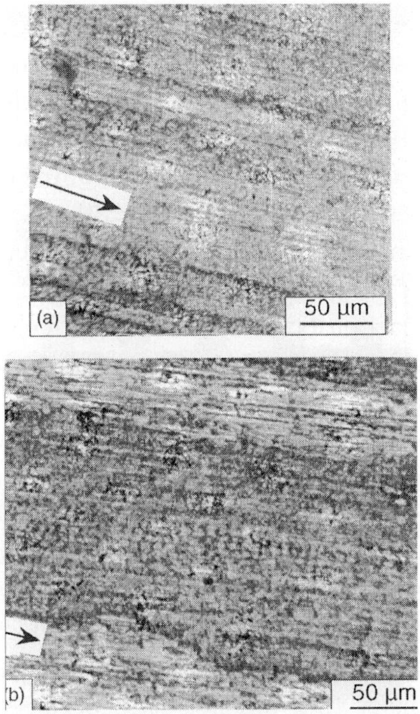

Fig. 16. Optical micrographs of transfer films formed during sliding on steel counterface for PPS filled with (a) 2 vol.% nano ZnO, and (b) 5 vol.% nano SiC. Sliding conditions same as in Fig. 6. Arrow shows sliding direction.[32]

4.3. *Transfer Film Bonding*

As indicated above, the formation of transfer film and its stability are important in sliding for wear reduction. A part of the contribution to wear also arises by the loss of transfer film from the counterface during repetitive sliding. In order to reduce this loss and to maintain its stability on the counterface, the transfer film needs bonding to the counterface. The experimental findings have revealed that the stronger the bonding of the transfer film to the counterface, the smaller is the wear loss.

The bonding of transfer film has been an important topic of discussion in polymer-metal sliding. The obvious form of bonding is mechanical interlocking of the transfer film or parts thereof into the crevices of asperities on the counterface surface. This is further augmented by van der Waal's forces of attraction between the contacting surfaces. It is also believed that tribochemical reactions play an important role in the adhesion of transfer film to the counterface. This has been documented by XPS analysis results of the transfer films. Zhao and Bahadur[28] have reported the XPS spectra of the transfer films formed on steel counterfaces filled with microsize 30 vol.% NiS and 30 vol.% PbSe fillers. From these spectra, the species detected in the transfer film have been determined and are given in Tables 1 and 2.[28] In Table 1 Fe_2O_3 is the product of reaction between counterface Fe and atmospheric O_2, and FeS and $FeSO_4$ are the products of reaction between Fe and NiS or PPS. In addition the element Ni is also generated in the process. Since Ni has high affinity with Fe, it provides a mechanism of adhesion between the transfer film and counterface. In Table 2 for PPS filled with 30 vol.% PbSe, there is no elemental Pb and so the decomposition of PbSe is not indicated. Similar results for other fillers have also been reported all indicating that the fillers that reduce wear are those that decompose and generate an element that has high affinity with Fe in the counterface, and vice versa. In the absence of tribological reaction for PbSe-PPS composite, the transfer film did not form because the mechanism of adhesion as indicated above was missing.

For the composites of PPS made with nano fillers of CuO, TiO_2, ZnO and SiC, whose wear behavior was given in Fig. 6 and discussed in section 3.2, the transfer film bond strengths were measured. This provided a measure of the tangential shear stress needed to strip the transfer film from the counterface. The measured values are shown in Fig. 17.[32] There appears to be a direct correlation with wear resistance and transfer film bond strength. For example, 1-2 vol.% CuO-PPS composites show the highest bond strengths and so was the case with their wear resistance. With higher CuO proportions, the bond strengths decreased and so did the wear resistance. The bond strength of 2 vol.% TiO_2 -filled PPS was the highest and so was also its wear resistance. The bond strengths of PPS filled with 3 and 5 vol.% TiO_2 were lower than that of the unfilled PPS and so was the case with wear resistance as well. For PPS filled with ZnO and SiC, the bond strengths were lower than that of the unfilled PPS and the wear resistance of filled composites was also lower.

Table 1. Identified species in the XPS spectra of the transfer film of PPS + 30 vol. % NiS.

Binding energies of peaks (eV)					
Compositions	C(1s)	O(1s)	S(2p)	Fe(2p)	Ni(2p)
Contaminated C	284.8				
C in PPS	284.8				
S in PPS			163.7		
Fe_2O_3		530.2		710.9	
$FeSO_4$		532.4	168.8	712.1	
FeS			161.6	712.2	
Fe				707.0	
NiS					855.1
Ni					852.7

Table 2. Identified species in the XPS spectra of the transfer film of PPS + 30 vol.% PbSe.

Binding energies of peaks (eV)						
Compositions	C(1s)	O(1s)	S(2p)	Fe(2p)	Pb(4f)	Se(3d)
Contaminated C	284.8					
C in PPS	284.8					
S in PPS			163.7			
Fe_2O_3		530.2		710.9		
Fe				707.0		
PbSe					137.4	53.4

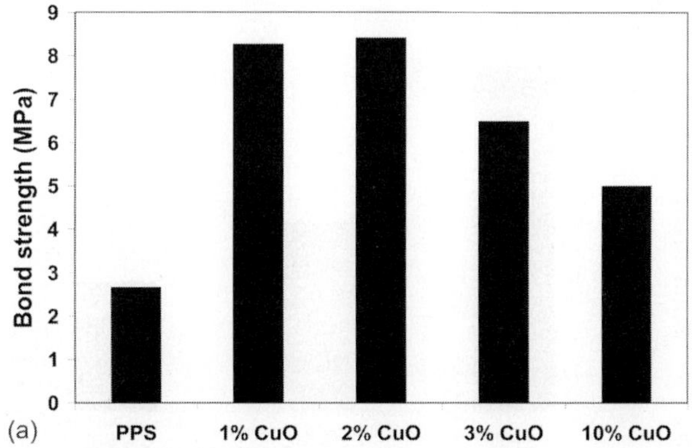

Fig. 17. Transfer film bond strengths of PPS composites filled with varying proportions of (a) CuO, (b) TiO_2, (c) ZnO and (d) SiC.[32]

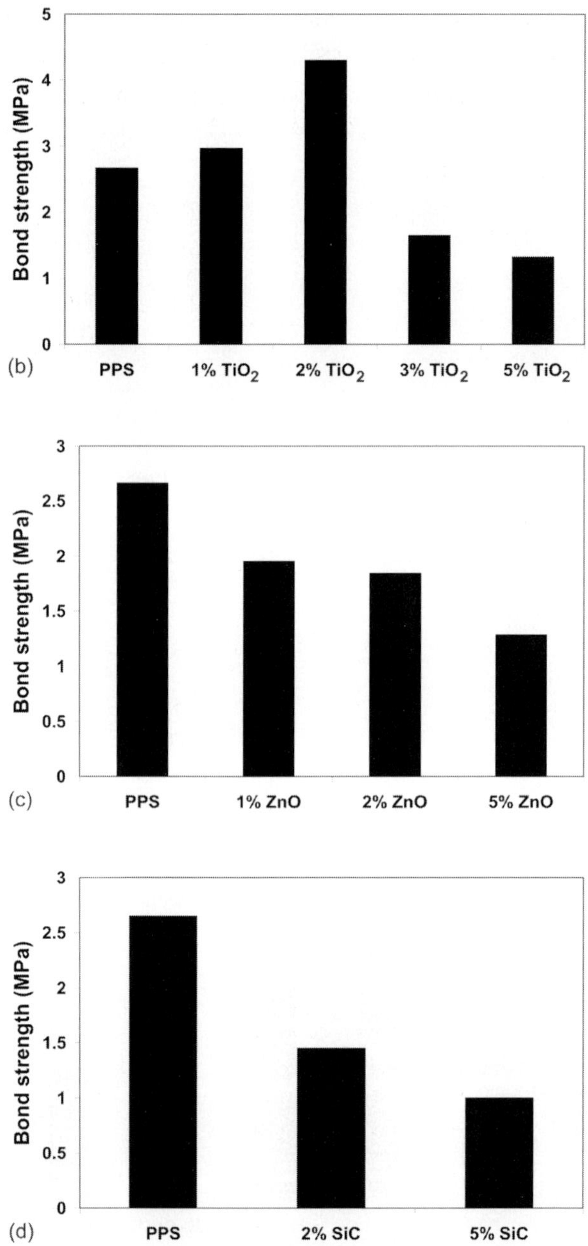

Fig. 17. (*Continued*)

In order to get insight into the bonding of above fillers, XPS analyses of the transfer films of PPS filled with CuO, TiO_2, and ZnO fillers in nano sizes were performed and the results are given in Tables 3-5.[32] In Table 3 for nano CuO-filled PPS, Fe_2O_3 is the product of reaction between counterface metal Fe and atmospheric oxygen. $FeSO_4$ is the product of reaction between the counterface element Fe and PPS. There is presence of elemental Cu which was presumably generated as a result of the tribochemical reaction between CuO and Fe. The XPS results in Table 4 for TiO_2-filled PPS show similar results further indicating the generation of elemental Ti. The high values of transfer film bond strength for these two fillers are accounted in terms of the strong attraction between the nascent Cu or Ti element and the steel surface. In the case of PPS composite filled with ZnO (Table 5), no reduced Zn was

Table 3. Identified species in the XPS spectra of the transfer film of 2 vol.% CuO-filled PPS.

Binding energies of peaks (eV)					
Compositions	C(1s)	O(1s)	S(2p)	Fe(2p)	Cu(2p)
Contaminated C	284.8				
C in PPS	284.8				
S in PPS			163.7		
Fe				707.0	
$FeSO_4$		532.4	168.8	712.1	
Fe_2SO_3		530.2		710.9	
Cu, Cu_2O		529.8			929.8
CuO		526.8			932.3
$Cu(OH)_2$		528.6			933.3

Table 4. Identified species in the XPS spectra of the transfer film of 2 vol.% TiO_2-filled PPS.

Binding energies of peaks (eV)					
Compositions	C(1s)	O(1s)	S(2p)	Fe(2p)	Ti(2p)
Contaminated C	284.8				
C in PPS	284.8				
S in PPS			163.7		
Fe_2O_3		530.2		710.9	
$FeSO_4$		532.4	168.8	712.1	
Ti					454.1

Table 5. Identified species in the XPS spectra of the transfer film of 2 vol.% ZnO- filled PPS.

Binding energies of peaks (eV)					
Compositions	C(1s)	O(1s)	S(2p)	Fe(2p)	Cu(2p)
Contaminated C	284.8				
C in PPS	284.8				
S in PPS			163.7		
Fe				707.0	
Fe_2O_3		530.2		710.9	
ZnO		530.4			1021.8

detected and so the tribochemical reaction did not promote adhesion. In SiC too no such mechanism of adhesion between the two metal elements can be expected.

5. Concluding Remarks

The use of fillers in polymers for modification of the tribological properties is relatively new. In the early days, polymers were filled with fillers such as carbon, MoS_2 and PTFE and also pure metals. The significant benefit to the tribological properties of polymers has just been realized in the last 2-3 decades by the incorporation of inorganic particulate fillers in polymers alone as well as with PTFE. The latter contributes to the reduction in the coefficient of friction and often improves the wear resistance of the filled polymer composites.

In this chapter, the discussion has been limited to the inorganic fillers only. It has been shown that some filler materials drastically improve the wear resistance of polymers sliding against metals. In this situation, the filler material helps in the formation of a smooth, uniform and thin transfer film on the metal counterface. The counterface roughness is also important and all of the studies reported in this chapter pertain to the optimum surface roughness of 0.10-0.11 μm Ra. These benefits are due both to the mechanical and tribochemical action of the filler particles during the sliding process. This involves the ability of the filler particles to aid in interlocking of polymer in the metallic surface asperities, and the tribochemical reaction provides improved adhesion between the transfer film and counterface.

Nanoparticle fillers behave similar to the microparticle fillers in respect to the modification of the tribological behavior of the polymer. However, whereas the optimum benefits in terms of the enhanced wear resistance are obtained with 30-35 vol.% micro particle filler proportion, the corresponding proportion for nano fillers is 1-3 vol.%. This is a great advantage considering that most of these fillers are expensive. However, the filler in nano size is more expensive than the filler in micro size because of the increased processing cost. Since some fillers improve the wear resistance while others do not, it would be of immense benefit if one could predict *a priori* which filler to use with a polymer for enhanced wear resistance. Some guidance on this aspect was provided in one study.[28] However further work is needed for optimum prediction. It is believed that the principles discussed in this chapter will go a long way in tackling this possibility.

References

1. S. Bahadur and D. Gong, *Wear*, **158**, 41 (1992).
2. S. Ahmed and F.R. Jones, *J. Mat. Sci.*, **25**, 4933 (1990).
3. A. Einstein, *Investigations on the theory of the Brownian movement*, [New York]: Dover Publications. 119 (1956).
4. M. Mooney, *J. Coll. Sci*, **6**, 162 (1951).
5. J.G. Brodnyan, *J. Rheol.*, **3**, 61 (1959).
6. E. Guth, *J. Appl. Phys.*, **16**, 20 (1945).
7. E.H. Kerner, *Proc. Phys. Soc. Sec B*, **69**, 808 (1956).
8. L.E. Nielsen, *J. Appl. Phys.*, **41**, 4626 (1970).
9. J.R. Fried, *Polymer science and technology*. Englewood Cliffs, N.J.: Prentice Hall PTR. xiii, 509 (1995).
10. Y. Sato and J. Furukawa, *Rubber Chem. Tech.*, **36**, 1081 (1963).
11. M. Schrager, *J. Appl. Poly. Sci.*, **22**, 2379 (1978).
12. H. Hojo, W. Toyoshima, M. Tamura, and N. Kawamura, *Poly. Engg. Sci.*, **14**, 604 (1974).
13. C.J. Schwartz and S. Bahadur,. *Wear*, **237**, 261 (2000).
14. L.E. Nielsen and R.F. Landel, *Mechanical properties of polymers and composites*. 2nd ed. New York: M. Dekker. xiii, 557 (1994).
15. S. Bahadur and D. Gong, *Wear*, **154**, 151 (1992).

16. S. Bahadur and D. Gong, *Wear*, **157**, 229 (1992).
17. L. Yu and S. Bahadur, *Wear*, **214**, 245 (1998).
18. S. Bahadur and D. Tabor, *Polymer Wear and its Control*, ACS Symposium, Ser. 287, L.H. Lee, Editor. American Chemical Society: Washington, D.C. 253 (1984).
19. S. Bahadur, D. Gong, and J.W. Anderegg, *Wear*, **154**, 207 (1992).
20. S. Bahadur, L. Zhang, and J.W. Anderegg, *Wear*, **203**, 464 (1997).
21. J. Vande Voort and S. Bahadur, *Wear*, **181**, 212 (1995).
22. S. Bahadur and D. Tabor, *Wear*, **98**, 1 (1984).
23. L. Yu, S. Bahadur, and Q. Xue, *Wear*, **214**, 54 (1998).
24. C.J. Schwartz and S. Bahadur, *Wear*, **251**, 1532 (2001).
25. S. Bahadur, D. Gong, and J.W. Anderegg, *Wear*, **160**, 131 (1993).
26. S. Bahadur and A. Kapoor, *Wear*, **155**, 49 (1992).
27. L. Zhang, *The role of particulate inorganic fillers on the tribological behavior of polyester*, M.S. Thesis, Iowa State University, 1995.
28. Q. Zhao and S. Bahadur, *Wear*, **225-229**, 660 (1999).
29. Q. Zhao and S. Bahadur, *Wear*, **217**, 62 (1998).
30. X. Shao and Q.-J. Xue, *Mater. Mech. Eng. (China)*, **28**, 39 (2004).
31. M.H. Cho and S. Bahadur, *Wear*, **258**, 835 (2005).
32. S. Bahadur and C. Sunkara, *Wear*, **258**, 1411 (2005).
33. F. Li, K.A. Hu, J.L. Li, and B.Y. Zhao, *Wear*, **249**, 877 (2001).
34. L. Chang, Z. Zhang, H. Zhang, and K. Friedrich, *Trib. Int.*, **38**, 966 (2005).
35. Q. Wang, Q. Xue, H. Liu, W. Shen, and J. Xu, *Wear*, **198**, 216 (1996).
36. P. Bhimaraj, D.L. Burris, J. Action, W.G. Sawyer, C.G. Toney, R.W. Siegel, and L.S. Schadler, *Wear*, **258**, 1437 (2005).
37. L.X. Zhao, L.Y. Zheng, and S.G. Zhao, *Mat. Lett.*, **60**, 2590 (2006).
38. G. Shi, M.Q. Zhang, M.Z. Rong, B. Wetzel, and K. Friedrich, *Wear*, **256**, 1072 (2004).
39. B. Wetzel, F. Haupert, and M.Q. Zhang, *Comp. Sci. Tech.*, **63**, 2055 (2003).
40. M.Z. Rong, M.Q. Zhang, G. Shi, Q.L. Ji, B. Wetzel, and K. Friedrich, *Trib. Inter.*, **36**, 697 (2003).
41. C.J. Schwartz, S. Bahadur, and S.K. Mallapragada, *Wear*, **263**, 1072 (2007).
42. T.A. Blanchet, *Trib. Trans.*, **38**, 821 (1995).
43. S.W. Han and T.A. Blanchet, *J. Trib. –Trans. ASME.*, **119**, 694 (1997).
44. L. Chang, Z. Zhang, H. Zhang, and K. Friedrich, *Trib. Inter.*, **38**, 966 (2005).

CHAPTER 13

THE SLIDING WEAR OF POLYPROPYLENE AND ITS BLENDS

Navin Chand* and S. A. R. Hashmi

Advanced Materials and Processes Research Institute (AMPRI)
(Formerly - Regional Research Laboratory)
Council of Scientific and Industrial Research (CSIR)
Hoshangabad Road, Bhopal 462026, India

Thermoplastic polymers are light in weight and easily process-able materials hence they are extensively used in auto parts and in several other industries. The materials with the unique combination of good friction and wear properties with the easy processing-ability are in demand. Polypropylene and its blends exhibit excellent mechanical and thermal properties coupled with excellent processability. The sliding wear performance of polypropylene or its blends have been improved significantly by introducing UHMWPE. This chapter gives a review on the sliding wear properties of PP and its blends with polyethylene terephthalate (PET) and recycled PET. Mechanism of wear under the sliding wear condition has been explained with the help of wear rates, mechanical properties and worn surfaces. The improved wear performance on loading a small amount of UHMWPE is an interesting feature of these materials. The sliding wear of UHMWPE modified polypropylene and its blends depend primarily upon the composition, sliding conditions, wear mechanism, the coefficient of friction and rise in temperature of interface.

1. Introduction

Polymers are used in automobiles and other industries as gears, liners, fenders, trim guard blades, grills, filters, bearings etc. In most of the

*Corresponding author

examples single virgin polymers have been replaced with the blends or special polymeric compounds that exhibit the required functional performance. In the case of sliding wear applications several modifiers are being used to make polymers wear resistant. The choice of basic raw materials is limited due to economic considerations and therefore commercially available polymers such as polypropylene has been investigated recently for modifications that has led to a blend/composite of high wear resistant properties. Graphite, molybdenum sulphide etc have been incorporated to improve wear resistance of PP, on the other hand the blending with the wear resistant polymers such as ultra high molecular weight polyethylene (UHMWPE) and polytetra-fluoro-ethylene (PTFE) is also being employed. UHMWPE is preferred in several applications, particularly in joint replacement of human beings due to the non-halogen composition of this material.

2. Polypropylene

Polypropylene (PP) is a versatile polymeric material has high crystallinity, which imparts high tensile strength to weight ratio. The high melting point of PP allows moulded parts to be sterilized and retaining high tensile strength at elevated temperatures. The crystallisability of isotactic polypropylene makes it with properties of commercial interest. Isotactic PP is essentially a linear, highly crystalline polymer. PP also has excellent electrical properties, chemical inertness and resistance to moisture.[1,2] It can be easily processed on conventional processing machines. It has low processing temperature; low cost but has poor toughness and wear properties. Polypropylene has been modified using methods such as blending with different thermoplastics,[3,4] reinforcing with fibres[5] and incorporation of different fillers.[6]

3. Ultra High Molecular Weight Polyethylene

UHMWPE is a linear polyethylene with a molecular weight more than 3 million. It is an outstanding polymer with the best self lubrication, impact absorbing properties and the highest resistance to abrasion, impact, compared to any thermoplastic polymer. UHMWPE's water absorption is less than 0.01%, nearly 1% that of nylon66. In water environment, it is very stable, resists the attachment of algae and fungus. It displays self-lubricating properties and the bearings of UHMWPE work better than steel or bronze bearing with grease. The coefficient of friction of UHMWPE is very low as compared to other polymers as well as metals.[7,8] The additional characterstics of UHMWPE include the high notched impact strength, energy absorption capacity at high loading rates, resistance to stress cracking, and the extremely low embrittlement temperature.

UHMWPE is made up of coarse particles (in some cases an average diameter is 300 μm) and these coarse particles are in turn made up of smaller particles (average diameters 10 μm) as shown in Fig. 1.[9] The smaller particles are aggregates of yet smaller, sub-micron particles. Some fibrillar bridges exist between these submicron particles in the structure of UHMWPE. The low mobility of the molecular chain and the gaps that exist between the particles retard the transmission of heat within the material during processing.

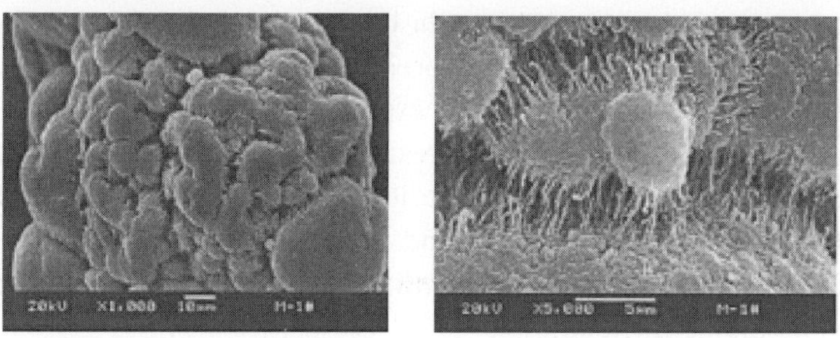

Fig. 1. SEM micrograph of commercial UHMWPE powder at magnifications x1000 and x5000 [Source ref: 9].

UHMWPE exhibits a very low wear coefficient or a moderate coefficient of friction when it is sliding against steel or other materials.[8,10-14] Wear of UHMWPE proceeds by two mechanisms: an adhesive process, which occurs immediately after sliding begins, and a fatigue process, which appears after a long period of sliding.[13,14]

The high molecular weight of UHMWPE results in a very high viscosity and pronounced viscoelastic characteristics in the melt that makes it unsuitable for conventional processing operations, except the compression molding and ram extrusion.[15,16] Solution or gel-processing methods have been employed to reduce the viscosity of UHMWPE.[17-19] Another way to reduce the melt viscosity is to dilute the UHMWPE with polyolefins, those having lower average molecular weights.[20-23] The polypropylene and its blends exhibit excellent mechanical and thermal properties coupled with excellent process ability. The sliding wear performance of polypropylene or its blends can be improved significantly by introducing UHMWPE in these materials. The friction and wear behaviours of polymer blend are fairly complex because the individual components have their unique response towards the friction and wear.

4. Polyethylene Terephthalate /Polypropylene (PET/PP) Blends

Polyethylene terephthalate (PET) is a semi crystalline polymer and is an important engineering polymer which is generally used to pack beverages, electronic parts food and other liquids and for making synthetic fibres. It has applications in injection stretched blow moulding and biaxial films. PET is used as thermally sprayed coating for surface protection against corrosion. PET is especially suitable for making gears and bearings, because of high wear resistance, general toughness, fatigue resistance and availability of high purity recycled resin.[24] The sensitivity to brittle fracture due to notch and stress concentration restricts its applicability. The considerable distortion, shrinkage and clouding of PET make its moulding difficult.

PET provides higher temperature and fatigue resistance because of the stiffening action of the aromatic phenylene group. Sensitivity to brittle fracture due to notch and stress concentrations restricts its applicability.[24] As compared to PET, PP has processing advantages and can be processed easily on all conventional processing equipments whereas PET has good wear resistance and therefore properties of the PP/PET blends would be superior. The usefulness of PET in PP under dry sliding conditions was recently demonstrated.[25]

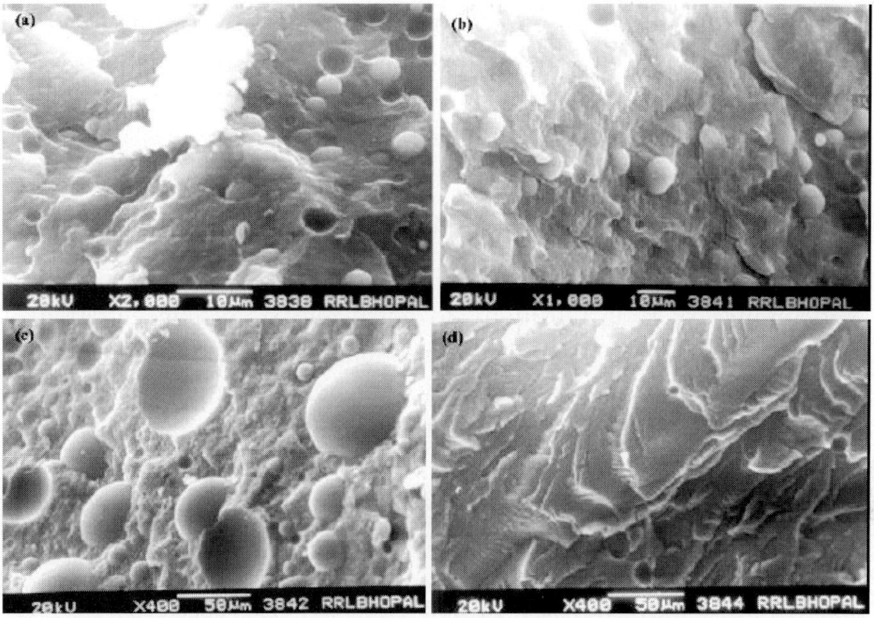

Fig. 2. The cryogenic fractured surfaces of PP/PET blends of (a) 10 wt.% (b) 15 wt.% and (c) 30 wt.% of PET in PP and (d) pure PET [Source ref: 25].

The blends of PP and PET of various compositions were reported to have different morphologies.[25] The cryogenic fractured surfaces of PP/PET blends of 10, 15 and 30 wt.% of PET in PP and pure PET are shown in Fig. 2 (a, b, c, d respectively). PET spherical balls were observed distributed in PP matrix. The melting point of PET is 265°C and that of PP is 165°C. During solidification, PET solidifies first, whereas PP remains in the molten state due to the large difference in the melting points of two polymers. Therefore, PET is found in discrete form and PP forms the matrix. The cavities were observed in the fractured surface. These cavities indicate the removal of PET balls. Smooth surface of PET balls confirmed that PP and PET form the distinct phases and do not adhere with each other. The diameters of PET spherical balls were different in different blends and increased with the increased weight percent of PET in PP. The diameters of PET spherical balls were less than 5 µm in a blend having 10 wt.% PET. A few balls were observed up to 8 µm in a blend having 15 wt.% of PET and 50 µm in a blend having

30 wt.% of PET. The balls of different sizes ranging from 3 μm to 50 μm were observed in PP/PET blend having 30 wt.% of PET. The increase in size of few PET domains was attributed to the coalition of small PET balls on the reduction of inter-domain distances.

5. Recycled Polyethylene Terephthalate

Recycled polyethylene terephtalate (RPET) has all the benefits and shortcomings such as sensitivity to notch and stress concentration of PET. The recycled PET has also been a subject of active research due to the excellent properties retained by this polymer after recycling. In another study[26] PP/PET and PP/RPET blends were prepared with and without maleic anhydride (MAH), in an elastic melt extruder.[27] The extruded material was quenched in water at 12°C and was immediately cut by a pelletizer to a length of approximately 5 mm.

Table 1. Some mechanical and physical properties of PP, PP/PET(90/10) and PP/RPET(90/10) blends [Source ref: 26].

Material	Density g/cm^3	Shore A Hardness	Tensile MPa	% Elongation	Impact St. kJ/m^2
PP	0.904	310	26.109	38.9	2.667
PP/PET(90/10)	0.9395	340	28.3	11.5	6.66
PP/RPET(90/10)	0.935	346	27.5	9.2	6.08

In PP/PET blends the PET concentration was 5, 10, 20 and 50 wt.% in PP. The MAH addition was 1% of PP by weight for MAH modified PP/PET blends. Table 1 summarizes some mechanical and physical properties of PP, PP/PET and PP/RPET blends.[26]

6. Wear Properties of PP/PET Blends

The wear rate (W) is generally defined as the wear volume per unit sliding distance as given below.

$$W = V/D \tag{1}$$

where V is volume loss and D is the sliding distance. The volume loss $V= m/\rho$, where ρ is the density of the material, m is the mass loss. In some cases specific wear rate is also estimated and in that case the wear rate is normalized by applied load on the specimen. Figure 3 is a plot of sliding wear-rate versus applied pressure for PP, PET and their blends of different compositions.[25]

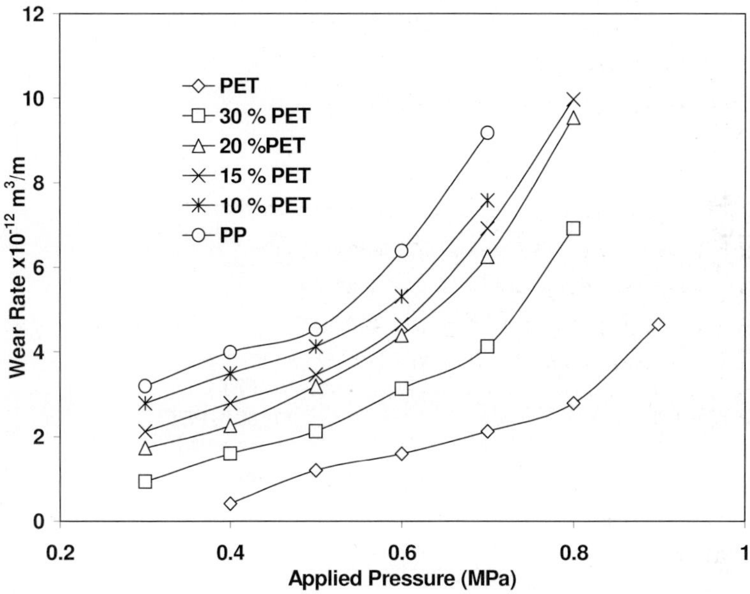

Fig. 3. Plots between the wear rate and applied pressure for PP, PET and PP/PET blends [Source ref: 25].

An increased applied pressure on the specimens increased the wear rate. PP showed maximum wear loss as compared to PET and PP/PET blends whereas PET showed minimum wear loss. The wear rate was observed in two stages, viz. moderate wear and high wear. At the low applied pressure a moderate increase of wear was observed, whereas at higher applied pressure, the wear rate increased sharply. With the increase of PET content in the blend the high-wear stage shifts to higher applied pressure. PP showed moderate-wear rate up to 0.5 MPa of applied pressure beyond which the wear rate increased sharply with the applied pressure. PET showed moderate wear up to 0.8 MPa of applied pressure. The moderate wear of PP/PET lies in between 0.5 and 0.8 MPa pressure at constant sliding speed of 1.33 m/s.[25]

The wear resistance of PP was improved on addition of PET. At a fixed sliding-speed, an increase in applied pressure on specimen increased the wear rate of PP, PET and PP/PET blends as can be observed in Fig. 3.

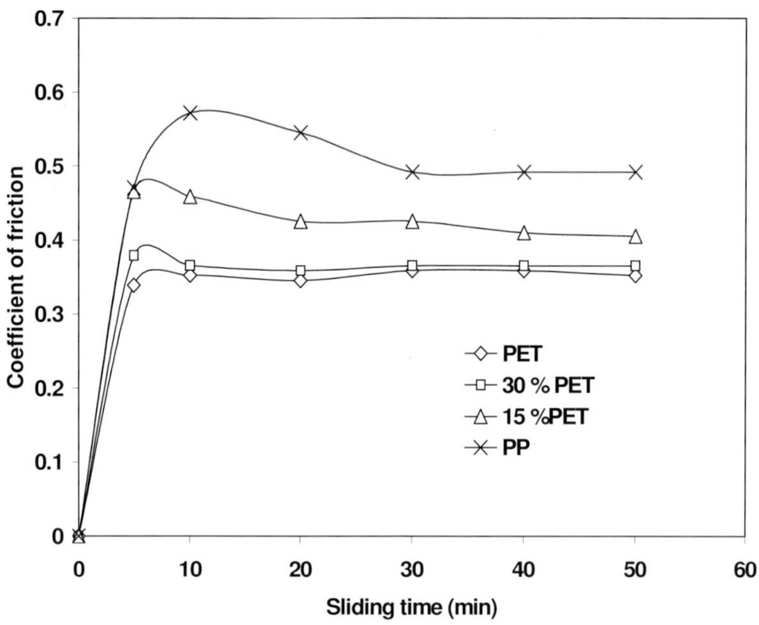

Fig. 4. Plot between the coefficient of friction and time for PP, PET and PP/PET blends at 0.4 MPa applied pressure [Source ref: 25].

Figure 4 shows the variation of coefficient of friction with time for PP, PET and PP/PET blends at 0.4 MPa pressure and 1.33 m/s sliding speed. The coefficients of friction increase rapidly and reached a maximum value and thereafter reduced to a stable lower value. The coefficient of friction of PET was less as compared to virgin PP. The blends of PP/PET show the values of coefficient of friction in between the values of PET and PET.

6.1. *Influence of Addition of RPET on Wear of PP*

The mechanical properties of PP/PET blends having virgin as well as recycled PET (RPET) is given in Table 1. Figure 5 shows the effect of load on the sliding wear rate of PP, PP/PET (90/10) and PP/RPET (90/10) blends after 4.806 km sliding distance.[26]

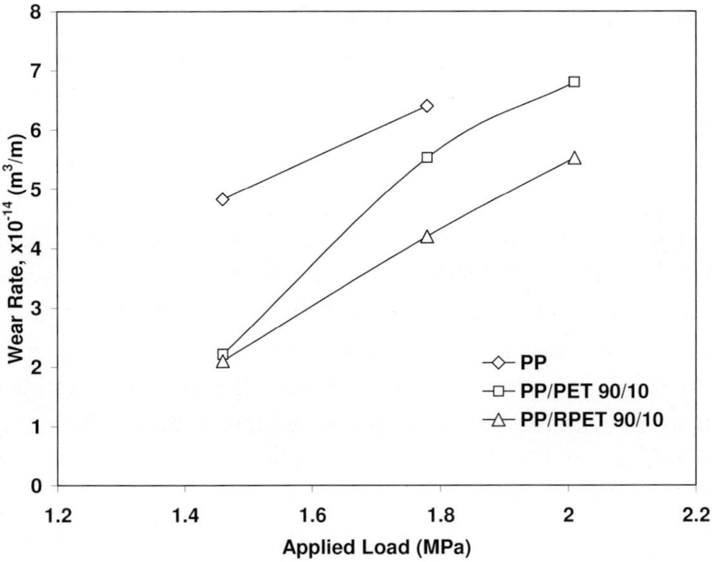

Fig. 5. Effect of load on sliding wear rate of PP, PP/PET (90/10) and PP/RPET (90/10) blends after 4.8 km sliding distance [Source ref: 26].

This graph shows that pure PP has maximum wear rate followed by PP/PET (90/10). Addition of RPET to PP has further reduced the wear rate at all the loads as compared to PP/PET (90/10). It was argued that PET is more amorphous compared to PP, which softened during sliding and form a film on the counter surface, which provided the wear resistance to PP. Another reason of improved sliding wear resistance on incorporating the PET is that the slipping of PET chains restricted the cracking of PP during shearing under load. But, on increase of load PET film would have cracked and immiscible structure of PP/PET (90/10) would have separated in different phases, which would have resulted in debris formation. At higher load of 60 N shearing would have easily removed more number of debris and increased the wear rate. Recycled PET has shorter chains as compared to virgin PET present between the phases of PP and PET. Shorter chains would have filled the gap and did not allow easy fragmentation or debris formation. This would have further reduced the sliding wear rate as compared to PP/PET (90/10).

Sliding wear rates of PP at 40 and 50 N loads are 4.83×10^{-14} and 6.21×10^{-14} m^3/m respectively. PP/PET (90/10) blend show the wear rate values at 40, 50 and 60 N load of 2.21×10^{-14}, 5.53×10^{-14} and 6.86×10^{-14} m^3/m respectively. PP/RPET (90/10) blend shows wear rate values of 2.1×10^{-14}, 4.2×10^{-14}, and 5.53×10^{-14} at 40, 50 and 60 N load respectively. Virgin PP failed at 50 N load, while PP/PET (90/10) and PP/RPET (90/10) blends survived up to 60 N load. At 40 N load wear rate of PP/PET (90/10) and PP/RPET (90/10) is nearly same. Increase of load from 40 to 60 N creates a large difference in the wear rate values of these two blends. PP/PET (90/10) gives very high wear rate at higher loads. This is because at high load the shearing action breaks the film and separates the two phases, which removes the debris and material behaves as PP alone. RPET has higher hardness as compared to PET, because of the usage during its virgin stage and reprocessing. Therefore, PP/RPET (90/10) blend is harder than PP/PET (90/10). That is why, PP/RPET (90/10) is more sliding wear resistant than PP/PET (90/10).

Figure 6 shows the sliding wear worn surface of pure PP at 1.46 MPa pressure after 4.8 km sliding distance. This shows that there is a film formation and the removal of the film has resulted in sliding wear loss. The sheared layer of material near the crack, tries to overlap the other side of the crack and due to which the material fragmented and removed.

Figure 7 illustrates the worn surface of PP/PET (90/10) and smoother surface was observed as compared to Fig. 6 of virgin PP which shows the reduction in sliding wear loss. In the case of PP/RPET (90/10), Fig. 8 even better smooth surface was formed and no fracture of film was observed.

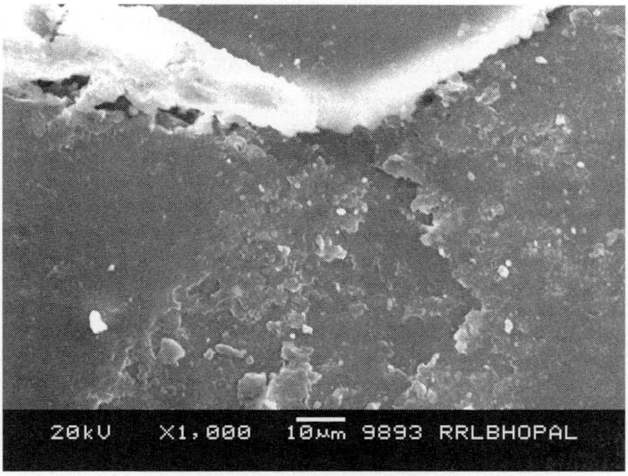

Fig. 6. Sliding wear worn surface of pure PP at 40 N load and 4806m sliding distance [Source ref: 26].

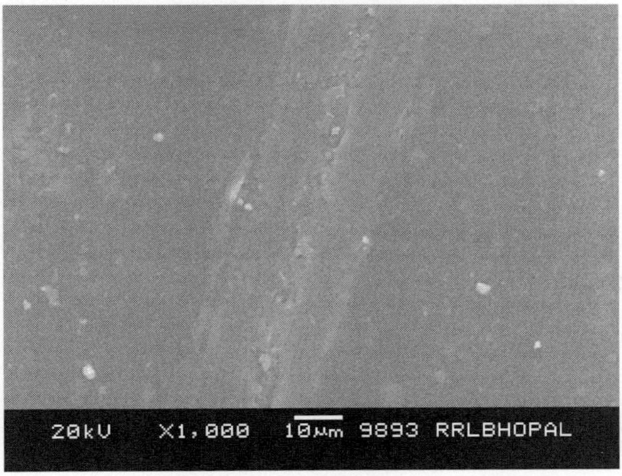

Fig. 7. PP/PET (90/10) sliding wear worn surface at 40 N load and 4806m sliding distance [Source ref: 26].

Tensile strength of PP/PET (90/10) blend shows that incorporation of PET and RPET gives 28.3 and 27.5 MPa strength values at 20 mm/min cross head speed. Sliding wear rates of PP at 40 and 50 N load are 4.83×10^{-14} and 6.21×10^{-14} m^3/m respectively. PP/PET (90/10) blend shows the wear rate values at 40, 50 and 60 N load 2.21×10^{-14}, 5.53×10^{-14}

and 6.86x10^{-14} m^3/m respectively. PP/RPET (90/10) blend shows wear rate values of 2.1x10^{-14}, 4.2x10^{-14}, and 5.53x10^{-14} at 40, 50 and 60 N load respectively. These values show that pure PP ceases at 50 N load, while PP/PET (90/10) and PP/RPET (90/10) blends survives up to 60 N load. Minimum wear rate is found for sliding wear in the case of PP/RPET (90/10). Increase of load from 40 to 60 N creates a big difference in wear rate values of these two blends. PP/PET (90/10) gives very high wear rate at higher loads. This is due to shearing action, which breaks the film and separates the two phases and removes the debris.

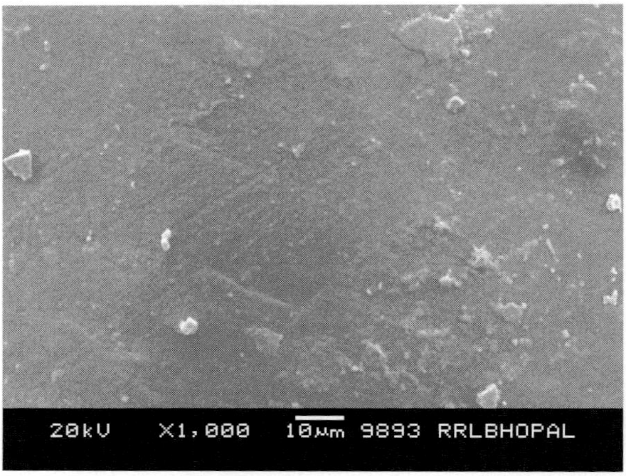

Fig. 8. PP/RPET (90/10) sliding wear worn surface at 40 N load and 4806m sliding distance [Source ref: 26].

6.2. *Effect of UHMWPE Addition on Wear Rate of PP and Its Blends*

The PP has been blended with polyolefins such as polyethylene.[27] The mechanical properties of the PP/UHMWPE blends were reported superior to those of PP/HDPE blends of the same composition.[28] The compatibility of PP/UHMWPE blends was better than that of PP/HDPE blends due to the enhancement of molecular chain entanglement of UHMWPE with PP in the amorphous phase because of the chain flexibility and low crystallinity of UHMWPE.[29] Therefore, low content of UHMWPE might be able to enhance PP even in the absence of compatibilizing agents under the conditions that the two components are mixed homogeneously.[30] The very high viscosity of UHMWPE in the molten state requires special processing methods.[30]

In the case of PP/UHMWPE blend, the molten PP does not penetrate into the UHMWPE domains because of the poor miscibility of UHMWPE and PP. The molten PP prefers to flow into the gaps between particles and the skin layer, which increases the stickiness of the melt to the roller surface and the adhesive bonding between the UHMWPE particles. Therefore, PP/UHMWPE blends are plastic with a twin roller. With increasing plasticating time, greater amounts of PP penetrate into the gaps, and the PP content in the skin decreases. On comparing the processing of PP/UHMWPE with that of UHMWPE/HDPE, it was found that the processing of UHMWPE/HDPE by a twin roller was very difficult under the same condition. It has been reported that the UHMWPE/HDPE (80:20) blend was plastic during the first few minutes, and the melt sheet could be peeled off from the roller but on increasing the plasticating time, the melt sheet fragmented. The phenomenon has been attributed to the good miscibility of UHMWPE and HDPE. The molten HDPE easily penetrated into the UHMWPE domains within the first few minutes during processing. With the increased plasticating time, an insufficient HDPE melt exists on the skin of the specimens and the gaps between the particles, thus hindered the plasticating of UHMWPE/HDPE blends.[9]

A blend containing 10 wt.% PP in the PP/UHMWPE blend exhibits significantly low apparent melt viscosity than that of UHMWPE. The melt viscosity of PP/UHMWPE blends decreased with the increase of PP content in the blends. PP prefers to locate in the amorphous or low crystallinity zones of UHMWPE matrix. The addition of PE did not effectively reduce the chain entanglement density of UHMWPE. The improvement of processability of UHMWPE by the addition of PE is rather limited as compared to that of PP.[31]

PP/UHMWPE blends could be effectively processed with a twin-roller and general single-screw extruder. In the extrusion of PP/UHMWPE blends, PP enriched at the surface of the blend adjacent to the barrel wall, thus increased the frictional force on the wall. Also the

PP penetrated into the gaps between the particles, acting as a heat-transfer agent and adhesive, thus enhancing the heat-transfer ability of the material.[32]

Friction and wear of PP/UHMWPE tested against EN-24 steel showed that the wear volume of PP, UHMWPE and their blends at 1.09 m/s sliding speed increased with the applied pressure as shown in Fig. 9.[23] Maximum wear was observed in PP samples. UHMWPE

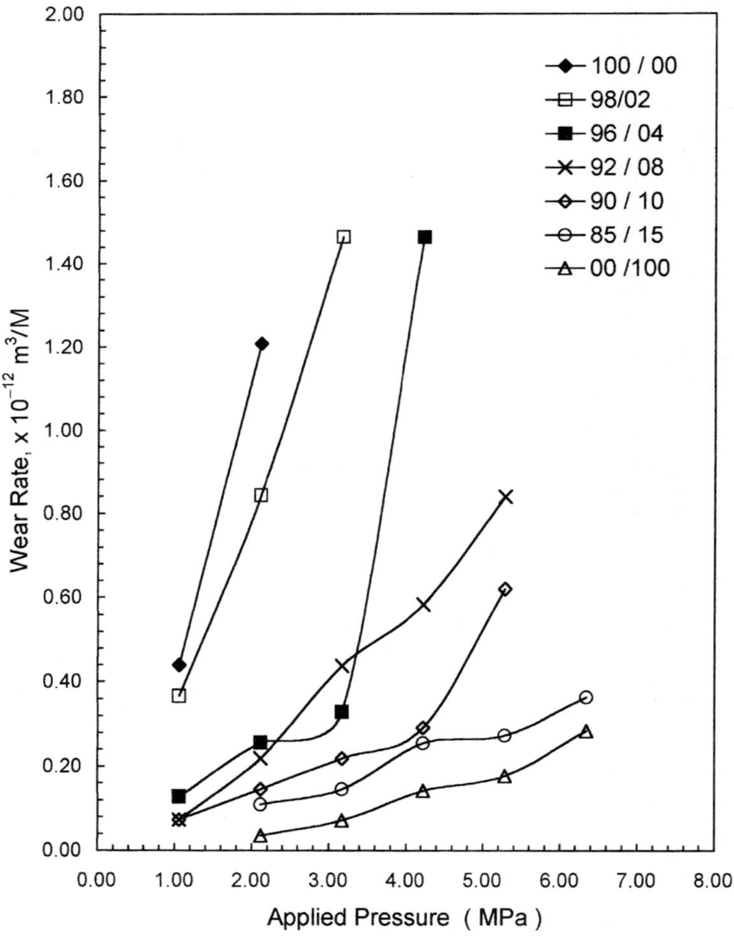

Fig. 9. Plots between wear rate and applied pressure for PP, UHMWPE and PP/UHMWPE system of composition at constant pressure [Source ref: 23].

showed the minimum wear loss at various pressures. PP could not be tested beyond 2.11 MPa pressure because the specimen deformed. It was reported that PP was highly sensitive to the pressure as compared to UHMWPE. On addition of UHMWPE to PP the test could be conducted up to 3.17 MPa, 4.23 MPa and 5.28 MPa pressures on the samples containing 2, 4 and 8 wt.% UHMWPE in PP.

Figure 10 shows the variation of wear volume as a function of sliding speed. Large variations in wear volume data were observed for samples A, B, C, D, E, and F wherein UHMWPE was 0, 2, 4, 6, 8, 10, 15, and 20 wt.% in PP respectively. Pure PP (sample A) could not be tested beyond 1.09 m/s sliding speed. The variations in wear volume decreased with the increase of weight fraction of UHMWPE.

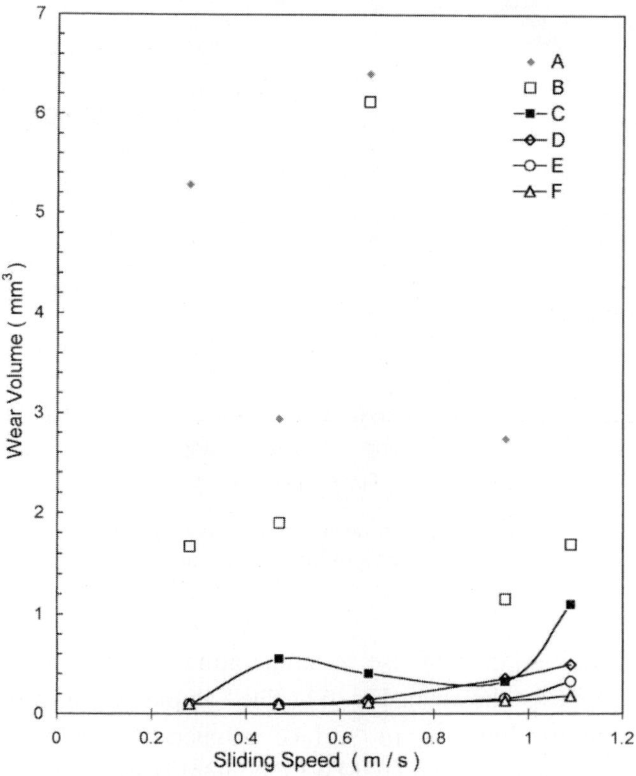

Fig. 10. 0, 2, 4, 6, 8, 10, 15, 20 % of UHMWPE in PP as shown A, B, C, D, E, and F respectively after 300m sliding distance and at 2.11MPa pressure. [Source ref: 23].

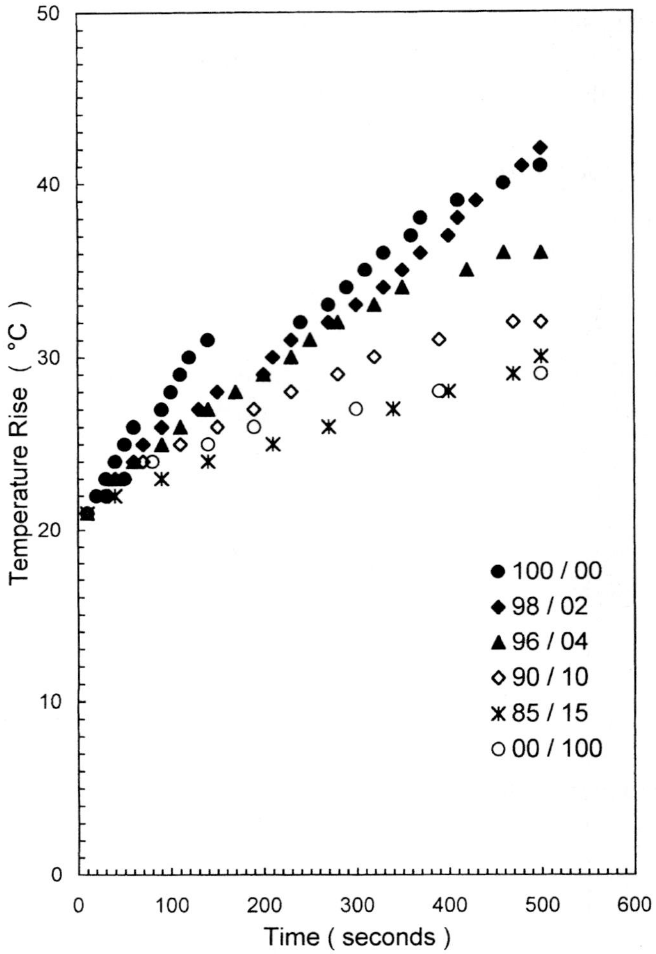

Fig. 11. Plots for rise in temperature against increasing time for PP, UHMWPE and PP/UHMWPE blend systems at sliding speed 1.03 m/s and pressure 2.11 MPa [Source ref: 23].

Plot for the variation of rise in temperature near the contact surface and sliding time is shown in Fig. 11. The sample A that represents PP shows maximum increase in surface temperature, which gradually decreased with the increased UHMWPE content in PP. The coefficient of friction of PP and UHMWPE are reported as 0.67 and 0.35, respectively in the literature.[8,20] Consequently the amount of frictional heat generated by PP during sliding was more as compared to UHMWPE. Therefore the temperature rise decreased gradually with the increased UHMWPE content. As the amount of UHMWPE increased in the blend, the area covered by UHMWPE at the contact surface also increased, which helped in reducing frictional heat and the temperature of contact surface reduced. PP, on the other hand, softened and deformed while sliding and lost its structural integrity.

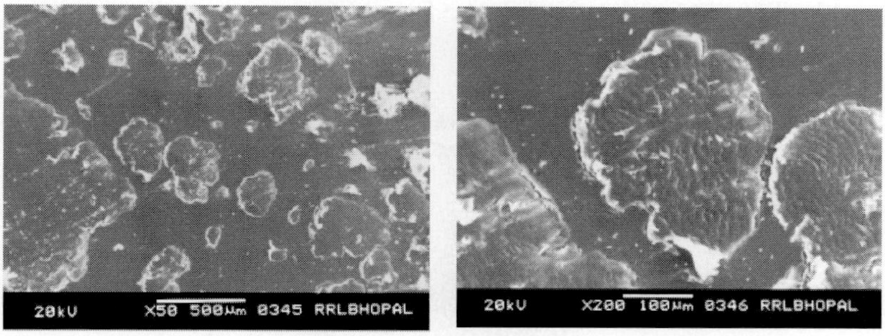

(a) (b)

Fig. 12. Micrographs of worn surfaces of PP/UHMWPE blend at different magnifications [Source ref: 23].

The softened portion of test specimen transferred to the counter face and a new surface having low surface temperature came in contact with the counter face. The cycle of material transfer repeated through out the test causing increased wear.[18]

A 15 wt.% of UHMWPE in PP demonstrated almost similar wear behaviour of pure UHMWPE. This improvement was attributed to the morphology of the blend, as shown in Figs. 12a and b. In this figure PP was in continuous phase whereas UHMWPE was in discrete phase. Once sliding starts, PP softens due to frictional heat and deforms easily. UHMWPE did not flow due to highly entangled molecular chains. While a part of PP was removed from the surface, more number of UHMWPE particles came in contact with the counter surface and reduced the frictional heat and consequently reduced wear losses.

The morphology of the specimen revealed interesting details. The two different regions of the worn surface of specimen showed different morphology as shown in Fig. 13 by the schematic diagram. UHMWPE and PP/UHMWPE both worn differently as illustrated in this figure.

The left region shows UHMWPE particles adhering with each other. This region belongs to the leading edge of test specimen. The featureless right region belongs to the trailing edge of the specimen. This indicates that softening is initiated from the trailing edge. The microstructures of worn surfaces of UHMWPE and PP/ UHMWPE (96/04) are also shown in Figs. 14a and b, respectively.[23] Two different morphological regions of UHMWPE can be observed in the Fig. 14a.

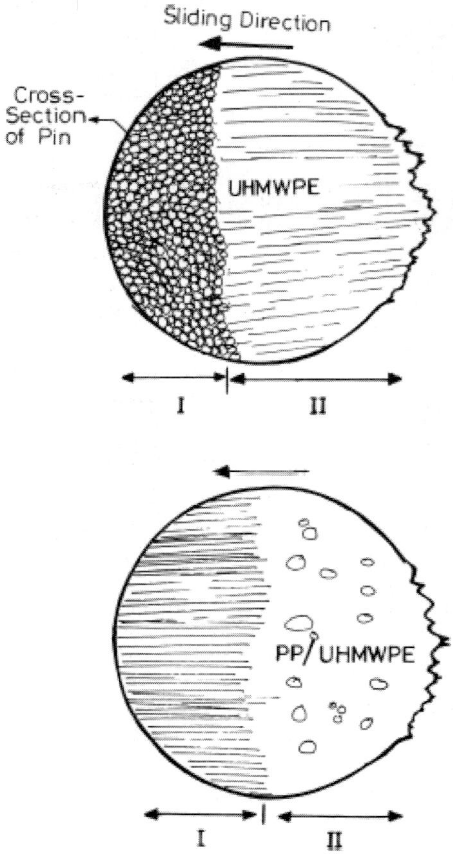

Fig. 13. Schematic diagram of worn surface of UHMWPE and PP/UHMWPE blend [Source ref: 23].

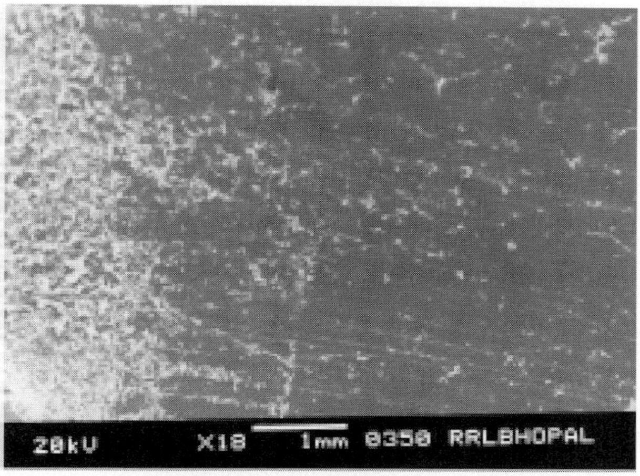

(a)

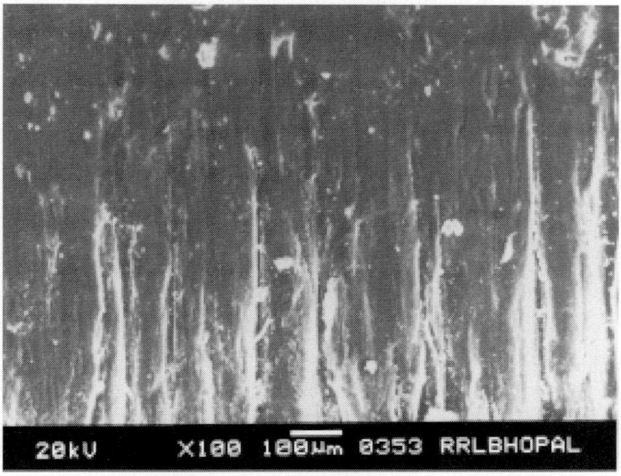

(b)

Fig. 14. Micrographs of worn surface of (a) UHMWPE and (b) PP/UHMWPE blend [Source ref: 23].

Figure 14b shows specimen containing 96 wt.% of PP. In this micrograph, bottom region shows the trailing edge of the pin. The severe damages in the form of deep grooves starting from the edge of sample were observed.

A combination of UHMWPE and PP provides a material, which is suitable for the wear application with the added advantage of ease in processing. As far as PP/PET blends are concerned, these materials also exhibit two-phase morphology in which PET spherical balls disperse in the PP matrix. The size of PET balls would depend upon the volume fraction of PET in PP. The increased content of PET improves the wear resistance of PP/PET blends by reducing wear losses.

References

1. L. E. Nielsen and R. F. Landel, *Mechanical Properties of Polymers and Composites* (Marcel Dekker, New York, 1994), Ch. 8.
2. P. K. Mallick, *Fiber Reinforced Composites* (Marcel Dekker, New York, 1988).
3. L. M. Zhang and D. S. Wang, *J. Appl. Polym. Sci.,* 77(13) (2000).
4. L. Incarnato, P. Scarfato, O. Motta and D. Acierno, *Polym. Compos.,* 21(3) (2000).
5. Y. J. Maeng, B. S. Yoon, M. H. Suh, B. Wan and S. H. Lee, *Polym. Compos.,* 21(1) (2000).
6. D. M. Yu, J. S. Wu, L. M. Zhou, D. R. Xie and S. Z. Wu, *Compos. Sci. Techol.*, 60(4) 2000.
7. http://www.dunone.com/ENGLISH/product.asp?prono=6.
8. K. Marcus and C. Allen, *Wear*, 178 (1994).
9. G. Liu, Y. Chen and H. Li, *J. Appl. Polym. Sci.*, 94 (2004).
10. T. S. Barrett, G. W. Stachowiak, and A. W. Batchelor, *Wear,* 153 (1992).
11. R. M. Rose, H. U. Goldfarb, E. Ellis, and A. M. Crugnola, *Wear*, 92 (1983).
12. A. Abdallah and D. Treheux, *Wear*, 142 (1991).
13. J. R. Atkinson, K. J. Brown, and D. Dowson, *J. Lubr.Technol.*, 100 (1978).
14. J. R. Atkinson, K. J. Brown, and D. Dowson, *J. Lubr.* 104 (1982).
15. C. Whitehouse, M. L. Liu and P. Gao, *Polym. Eng. Sci.,* 39 (1999).
16. A. E. Zacbarriades, T. Kanamoto, *AIChE Annu. Meet. Symp.* (1984), p. 25.
17. P. Smith and P. J. Lemstra, *J. Mater. Sci.,* 15 (1980).
18. M. Mastuo and R. Manley, *Macromol.* 15 (1982).
19. C. Anton, M. R. Mackiey and S. B. Solbal, *Polym. Bull.,* 17 (1987).
20. V. Parimal and K. Thein, *Polym. Eng. Sci.,* 27 (1987).
21. H. A. Scheetz, and R. C. Gilles, *U.S. Pat.,* 4,281,070 (1981).
22. Y. Bin, M. Lin, and R. Alachi, *Polym.* 42 (2001).
23. S. A. R. Hashmi, S. Neogi and N. Chand, *Wear*, 247 (2001).
24. N. G. McCrum, C. P. Buckley and C. B. Bucknall, *Principles of Polymer Engineering,* (Oxford Science Publication, Oxford, 1988).
25. S. Neogi, S. A. R. Hashmi and N. Chand, Role of RPET in improving wear properties of PP in dry sliding condition, Bull. Mater. Sci., 26 (2003), p. 579.

26. A. Naik, PhD Thesis submitted to Rajeev Gandhi Technical University, Bhopal (2007).
27. S. A. R. Hashmi, T. Kitano and N. Chand, *Polym. Compos.*, 24(1) (2003).
28. C. M. Tai and N. C. Lirky, *Polym. Test.* 19(2) (2000).
29. W. Zhu, Z. Q. Zhang, Z. L. Feng and B. T. Huang, *J. Macromol. Sci. Part B Phys.*, 35(5) (1996).
30. T. Tincer and M. Coskun, *Polym. Eng. Sci.*, 33(19) (1993).
31. G. Liu, Y. Chen and H. Li, *J. Appl. Polym. Sci.*, 92 (2004).
32. G. Liu, Y. Chen and H. Li. *J. Appl. Polym. Sci.*, 94 (2004).
33. G. Liu, Y. Chen and H. Li, *Wear*, 256 (2004).

CHAPTER 14

ENGINEERING POLYMERS AND COMPOSITES FOR MACHINE ELEMENTS

K. D. Dearn[1], S. N. Kukureka[2] and D. Walton[1]

[1]*Power Transmission Laboratory*
Department of Mechanical Engineering, University of Birmingham
Edgbaston, Birmingham, B15 2TT, UK
[2]*Department of Metallurgy and Materials, University of Birmingham*
Edgbaston, Birmingham, B15 2TT, UK

This chapter reviews the use of engineering polymers and short-fibre reinforced thermoplastics for machine elements such as gears, cams and pulleys where the polymers are in non-conformal contact. Both twin-disc testing and gear tests are examined for a variety of engineering polymers as well as the effects of temperature, fibre-reinforcement and geometry on performance. Gear efficiency, failure mechanisms and wear of polymer gears are also covered in detail. Finally polymer gears for high-performance applications are considered together with the effects of external lubrication.

1. Introduction and Overview

Injection-moulded polymers and polymer composites are increasingly being used in engineering applications as non-conformal engineering components such as gears, cams and rolling-element bearings. The reasons are the low cost of polymers, the ease with which complex components can be produced and critically, their ability to run unlubricated. Most of the published work on the tribology of non-conformal polymer pairs relates to the performance of gears. For a pair of gears the dominant operating parameters such as slip ratio, sliding velocity and load and the geometric parameters such as module and curvature of the contacting surfaces vary with the contact position on the tooth profile. The alternative method to studying wear by running gears at an applied load and speed is to simulate the contact conditions experienced by gears, but applied to a much simpler geometry. An example of such a simulation is the use of two circular discs loaded against one another in edge-to-edge contact and rotating at different speeds. By varying the relative speeds of the discs and the applied load, normal load, slip ratio and sliding speed then the conditions experienced by gear teeth in contact may be approximated. An introduction to this technique is given below followed by experimental results, interpretation and wear mechanisms for a variety of polymers including acetal (polyoxymethylene), nylon 66 and polymers reinforced with glass, carbon and aramid fibres.

The chapter then moves on to consider experimental and modelling work on gears which is subsequently compared with the twin-disc work. The wear rates of different polymers and polymer composites are compared, including polymers with internal lubricants. Other forms of failure such as root and pitch-line fracture are covered. Polymer gear performance is dominated by running temperatures and methods of predicting the bulk and surface temperatures are explored. The concept of a transition torque, at which the wear rapidly increases due to local melting, is described.

The measured mechanical efficiencies of polymer spur gears are shown as a function of load and speed for a range of materials. These results show the interesting feature that as the load is increased the efficiency also increases. Experimental data for polymer/polymer and polymer/steel running combinations are shown as are the effects of running polymer gears with grease lubrication. Under dry running conditions the coefficients of friction can be very high, much higher than commercially published data obtained from pin-on-disc and twin-disc tests. The high friction has repercussions in terms of contact stresses and generated temperatures.

Finally, the performance of polymer gears running in oil is presented. The conditions for elastohydrodynamic lubrication are considered in terms of oil-film thicknesses, and are compared with the surface roughness to determine the operating regimes. It can be shown that polymer gears mainly operate under mixed-film lubrication; except when heavily loaded the evidence suggests that full-film conditions prevail. Tooth-flank pitting, which does not occur on dry-running gears, is seen as a feature for heavily-loaded gears when running in oil. High pressure-angle gears, not normally employed in steel-gear applications, are shown to be beneficial for polymer gears. The performance of a high temperature polymer, high pressure-angle gear, is shown in terms of durability and demonstrates the current state-of-art for polymer gears.

2. Rolling-Sliding Tests by the Twin-Disc Technique

2.1. *Principles*

Gear geometry brings special problems for polymer wear testing. Gear contact is concentrated and non-conformal throughout, except for internal gears. Gears have an involute profile and a constant velocity ratio; the sliding velocity, contact radius, and contact load all vary over the tooth flank. Thus the motion is a combination of rolling and sliding.

An alternative is twin-disc (or two-roller) tests – an example of a thermally self-regulating, cyclic-energy system. Two discs with parallel axes are loaded and run together. Two discs of equal size (in this case face-width 10 mm and diameter 30 mm) are run one above the other with the slip ratio being adjusted by changing gears in the drive train. The lower disc is mounted in a rigid block. The upper disc is supported in a pivoted block. The load is applied by using a lever arm and a sprung, dead-weight load. Sliding friction is measured by a force transducer recording the horizontal force on the lower disc. Continuous measurement of the movement of the upper block by a displacement transducer allows wear to be monitored throughout the tests[1,2].

In this way we can separate rolling from sliding. If the discs rotate at the same velocity then we have pure rolling; a velocity difference allows controlled rolling and sliding[2] as shown in Fig. 1. Slip ratio is defined here as the ratio of sliding velocity to rolling velocity. For two discs running at velocities v_1 and v_2 the sliding velocity is $(v_1 - v_2)$ and the rolling velocity is $(v_1 + v_2)/2$. As an example, since the disc speeds are controlled by gears in the drive chain: a pair of gears with 50 and 52 teeth in the drive chain produce a slip ratio of 2/51 or 3.92%. The twin-disc tests are not meant to be a substitute for gear tests but to complement them. The slip ratios used in disc testing are lower than the maximum ones in gear testing in order to investigate the development of wear and to study materials behaviour.

Samples can either be injection-moulded (often over a steel, knurled insert for the hub), or machined from extruded bar. In both cases a final light machining followed by a running-in period are required to ensure sample uniformity with uniform curved surfaces and face-width parallel to the shaft axis and normal to the line of contact. The running-in period typically removes about 5 μm from each disc.

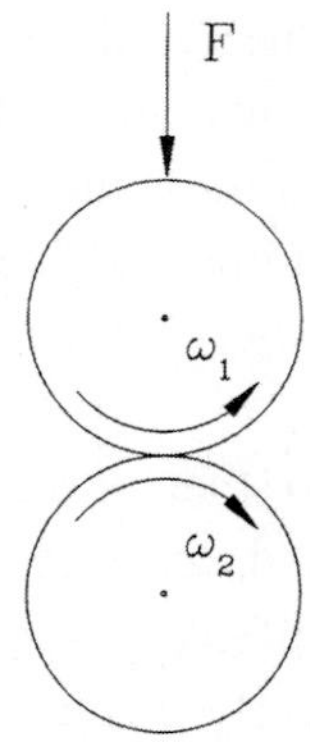

Fig. 1. Twin-disc configuration[2].

Discs are run at varying normal loads, speeds and slip ratios. Wear and coefficient of friction are measured continuously. Wear is measured by on-line displacement monitoring and by weighing samples[2]. Because the continuous-monitoring system takes displacement as a measure of wear it was necessary to allow for the fact that thermal expansion, creep and moisture absorption may occur simultaneously. This was done by drying the samples (typically at 70°C for five hours) after the running-in period then weighing them before allowing them to stand for several days until equilibrium moisture content was restored before testing began. At the end of the test the discs were dried and weighed again to provide an alternative measure of wear. In all the following work with the twin-disc rig, wear rate is defined as the average depth of material removed from each disc per rolling cycle. This is a more appropriate measure than, say, the Archard wear coefficient[19] which assumes that wear is proportional to load and sliding distance. There is no evidence that this is the case in rolling-sliding tests under non-conformal contact[2,8].

During the tests a considerable amount of heat is generated by sliding friction and viscoelastic hysteresis. Estimates show that sliding friction strongly dominates and this was used to calculate heat generation. Heat dissipation is by conduction through the discs to the shaft, by radiation from the disc surfaces and by convection with air moving round with the discs. It has been estimated to be conduction (40%), convection (40%) and radiation (20%) with superimposed local flash temperatures[2] (see 2.2 below). These flash temperatures are generated at the contact which moves round the disc. The final maximum temperature at any position is the sum of the mean temperature and the flash temperature. It is not generally possible to measure the local temperature during running with any accuracy.

Finally it should be noted that surface morphology of polymers, especially for injection-moulded specimens which show a strong skin-core effect, is often markedly different from that in the bulk[9]. In all cases there is a surface skin with clearly different spherulite size and a proper analysis of surface effects and their magnitude in relation to wear damage and loading is important.

2.2. *Unreinforced Thermoplastics — Polyoxymethylene and Polyamide 66*

A typical engineering polymer which is widely used in tribological applications is polyoxymethylene (POM) or acetal. It is a semi-crystalline thermoplastic with a relatively high crystallinity and a melting point of about 175°C. A variety of rolling-sliding tests at either different applied loads (e.g. 100 to 500 N) or different slip ratios (e.g. 2% to 8%) was performed[1]. Figure 2a shows the coefficient of friction as a function of slip ratio and it can be seen that after an initial period it increases steadily from about 0.30 at a slip ratio of 2% to about 0.35 at 8.5%.

When wear rate (in mm^3/cycle) is plotted as a function of slip ratio for a particular load and speed (here 200 N and 1000 revs/min) as shown in Fig. 2b, it can be seen that the wear rate remains fairly constant in a 'mild wear' regime below a slip ratio of about 4%. Above that the wear rate increases rapidly ('severe wear') and failure ensues beyond about 8% slip. As slip ratio increases it is clear that there is a greater degree of frictional heating and the severe wear appears to be associated with substantial temperature rises, approaching the melting point of the material.

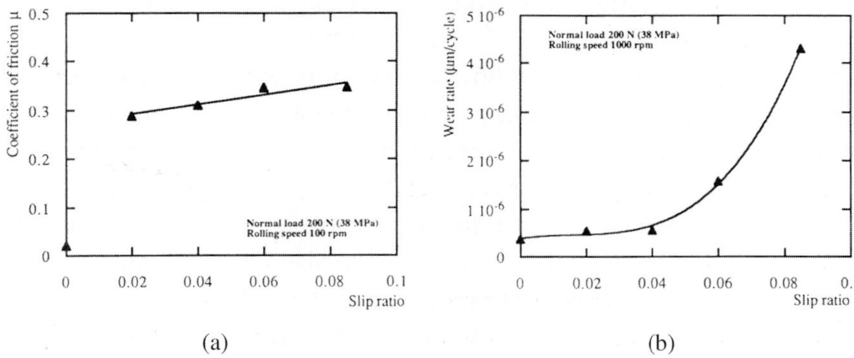

Fig. 2. (a) Friction and (b) wear rate of acetal discs[1].

Estimates of the heat dissipation from the discs were made as follows. Heat conduction to the shaft was calculated and heat convection to the surrounding air was also estimated[2]. The convection was estimated by assuming a laminar thermal boundary layer around the disc. Equating the frictional power input to the sum of these components and the estimated effects of radiation enabled the average surface temperature of the discs to be determined. Calculation of the flash temperature generated by the effects of high local Hertzian contact stresses showed it to be typically only around 10°C and so the difference between maximum and average surface temperatures was slight.

Although differences in slip ratio and speed have some influence on wear rate, the primary factors controlling it are contact stress and surface temperature[2]. At low surface temperatures (below say 80°C) the wear rate is low (below about 10^{-6} μm/cycle); for higher surface temperatures the wear rate remains below about 10^{-5} μm/cycle up to a surface temperature of around 150°C beyond which wear increases rapidly to failure. Wear patterns were found to be a combination of pitting and lateral cracking. Similar dependence of the wear of acetal on temperature was found in pin-on-disc tests by Lancaster, and similar severe wear behaviour was found by Lawrence and Stolarski in a four-ball test[1].

For the design of non-conformal components in acetal it is thus recommended to keep the maximum contact stress below about 50 MPa and the component temperature to 80°C. This will limit the wear rate to about 1 mm every 10^6 cycles and give a typical component life of 10^8 or 10^9 cycles.

In contrast, polyamide 66 (nylon 66) exhibited considerable differences[2,6]. The coefficient of friction was higher than for polyoxymethylene and was typically around 0.4 though peaking at a value of 0.6. Wear appeared uniform but an apparently higher creep rate meant that the wear rate as measured by displacement was considerably higher than that by weight loss. There were also deep transverse cracks from the surface at higher slip ratios though they appeared to have little effect on disc performance. In comparison with acetal it appears that unreinforced nylon 66 is likely to be inferior as its higher coefficient of friction (and hence heat generation) will have a greater effect than its higher melting point of 256°C compared with that of 175°C for acetal. For instance, at 1000 rpm and 200 N load the wear rate of nylon 66 is always higher than that of acetal for equal values of slip.

2.3. *Fibre-Reinforced Thermoplastics — Polyamides*

In order to attempt to improve the tribological properties of engineering polymers, they are often reinforced with short fibres of glass, carbon or aramid, usually in loadings of 30 or 40% by weight. Such reinforcement typically improves both the stiffness and strength of polymers considerably – by a factor of one to four[3,4,6,7].

For polyamides reinforced with either glass or carbon, it is found that wear rate increases with load and the wear rates are increased by about a factor of 10 over those with unreinforced material. Carbon fibres and glass fibres significantly reduce the coefficient of friction (e.g. nylon with glass fibre has a coefficient of friction around 0.1 to 0.2 compared with 0.4 to 0.6 for unreinforced nylon). This means that the material can withstand higher loads and slip ratios before thermal softening (and contrasts with pin-on-disc behaviour against steel where fibre-reinforcement reduces wear).

The wear process for reinforced polyamides involves the formation and removal of a thin surface layer of matrix-rich material – i.e. nylon. After a running-in period the nylon layer builds on the surface and protects the abrasive fibres from meeting and interacting with each other. The surface layer can be regarded as effectively a 'self-lubricating' layer. The wear rates are determined by the ease with which this surface layer is removed. When the contact surfaces in glass-fibre reinforced nylon are separated by a thin film, the wear rate is rather low at approximately 10^{-6} μm/cycle and the friction coefficient is around 0.1. Once this layer has been removed the wear rate increases sharply and the friction coefficient rises too. The rapid wear stage depends strongly on load and wear increases tenfold as the load is doubled.[2] The transition to severe wear begins when approximately 50 μm of the surface has been worn, or when the surface temperature exceeds approximately 200°C. Under a typical operating load of 300 N (corresponding to 80 MPa surface stress) the wear rate is around 10^{-4} μm/cycle. Thus this material is suitable for high-duty, short-life applications up to a maximum life of approximately 10^7 cycles[2].

A comparison of wear rates with load for different fibre reinforcements is given in Fig 3. NGF(F) is the final wear-rate for nylon glass fibre; NGF(I) is the initial wear-rate for nylon glass fibre; NCF and NAF refer to nylon carbon-fibre and nylon aramid fibre respectively whilst N represents unreinforced nylon.

It has also been found that after severe wear at high slip ratios on glass-fibre reinforced nylons the fibres become oriented[4] – presumably because softening of the nylon matrix enables them to rotate (Figs. 4a & b). Figure 4a shows random fibre arrangements beneath the surface at fairly low slip ratios (4%). After considerable running at a high slip ratio (here 29%) it can be seen from Fig. 4b that the fibres are almost entirely oriented in the direction of friction (top to bottom in the figure).

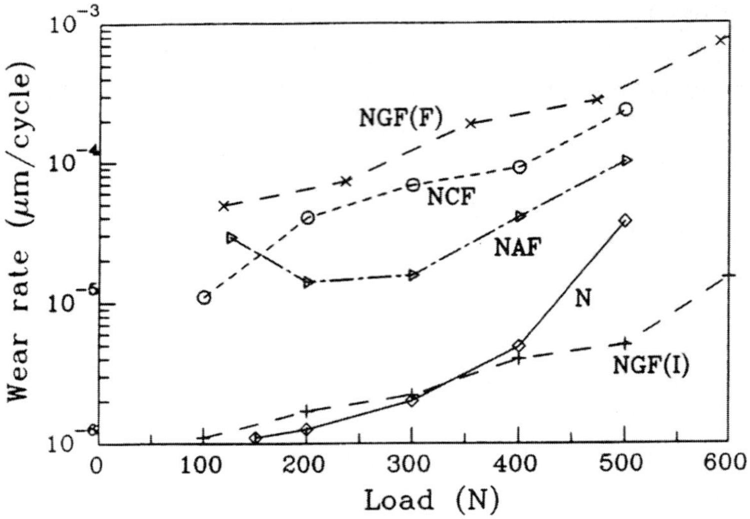

Fig. 3. Wear rates for thermoplastic composites[6].

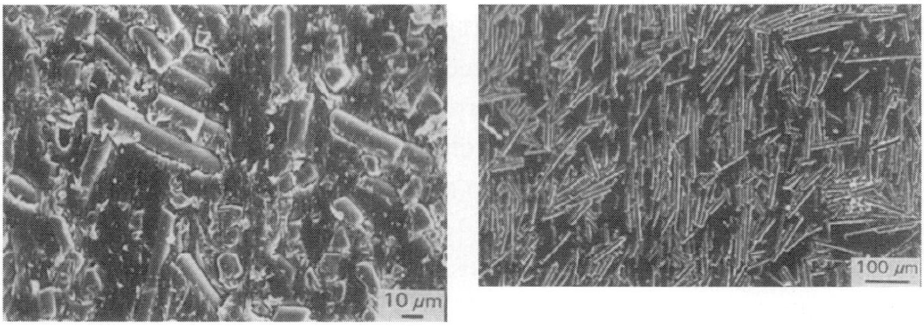

(a) (b)

Fig. 4. Orientation of fibres after severe wear[4].

2.4. *Internal Lubricants and Additives — PTFE*

The effects of additives such as polytetrafluoroethylene (PTFE typically to about 15 wt%) are to enhance performance for both POM and nylon[5] (Figs. 5a & b). Typical commercial formulations might contain both glass-fibre reinforcement and PTFE as an 'internal lubricant'. The addition of PTFE reduces the coefficient of friction of acetal from around 0.3 to 0.1 and that for polyamide 66 from around 0.45 to 0.11. This appears to be due to the formation of a thin film of PTFE on the contacting surfaces. The effect on wear rate is indirect and temporary; because the surface layer decreases friction and hence temperature, it decreases wear rate by effectively shifting the curves of wear rate against load to higher loads as shown in Figs. 5a & b. Overall, the performance with PTFE is superior to that with no internal lubricant.

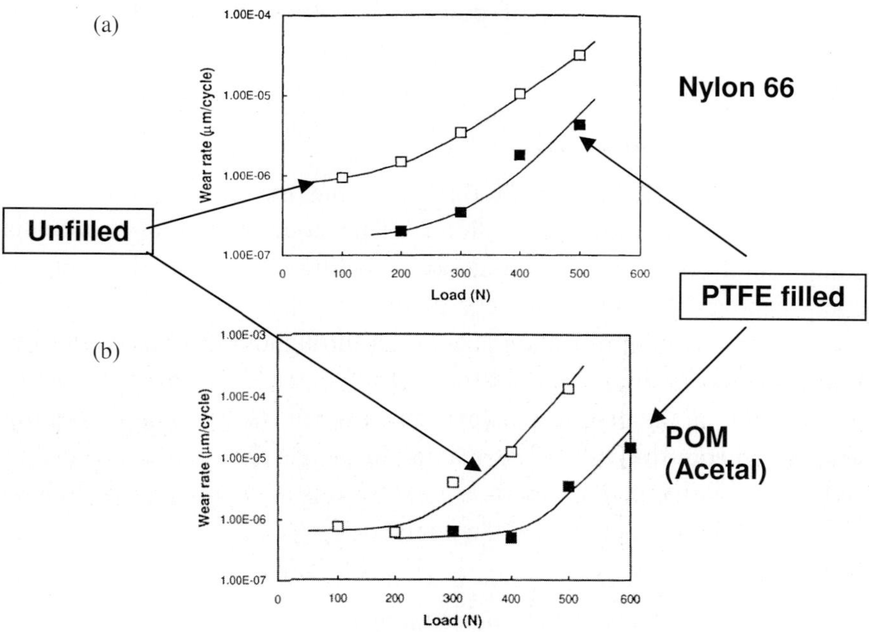

Fig. 5. Wear rates of (a) PA66 and (b) POM with PTFE fillers[5].

2.5. *Comparison of Results and Validity of Data*

Some work has compared the usefulness of pin-on-disc or roll-slide work for simulating gear behaviour[8]. It was found that twin-disc roll-slide tests were a much more successful model than pin-on-disc testing. Comparisons with gears can never be truly representative since changes in sliding, tooth flexibility and differences in the thermal conditions make contact conditions very different. Typically the slip ratios found in many published polymer gear tests are higher than those in the twin-disc work described above and in fact a slip ratio above 28% is found in most of the gear meshing cycle. Correlation between twin-disc and gear results is within an order of magnitude at best.

Nevertheless twin-disc tests provide fundamental information about materials behaviour in rolling-sliding under non-conformal contact and this is their strength. They should be seen as complementing the applied information from direct gear testing and as contributing to our more fundamental understanding of polymer tribology under rolling-sliding conditions.

3. Gears and Gear Testing

The motion of involute tooth flanks, during the meshing cycle is a mixture of both rolling and sliding. This combination of motion is complicated given that the direction of sliding changes as the point of contact moves through each stage of motion: approach, pitch point and recess. The direction of rolling is constant throughout the meshing cycle. Thus, on the driving gear, the frictional force acts away from the pitch-point, while on the driven it always acts towards the pitch point. Sliding velocity, theoretically, will be maximum at the first and last point of contact (when the path of approach/recession is a maximum) and will be zero when the tooth mesh is at the pitch point.

4. Thermal Aspects of Polymeric Gearing

Temperature dominates all aspects of polymer gear performance, and due to the inherent weakness of polymers to heat, it means they can only operate within a limited temperature range. Temperature exacerbates all of the main modes of plastics gear failure and affects load capacity as mechanical properties are reduced. Thus, the importance of accurate running-temperature estimations of polymer gears for a given application is emphasised. However, in practice, accurate estimates of operating temperature are difficult to determine because the exact conditions and the properties of the material needed to determine these temperatures are not known to any great accuracy. There are however, models available to estimate temperatures and these are discussed below.

4.1. *Heat Generation*

The mechanism of heat generation in polymer gears is a complex combination of frictional heating, viscoelastic hysteresis losses, heat flow (conduction) through shafts and ambient radiation. The conduction and radiation, which, in most cases, represent a negligible contribution to overall heat, can be minimised through careful design.

Gauvin *et al.*[16] suggest that the majority of heat generated at the tooth-contact interface is the result of frictional heating. The frictional heat-flux is caused by relative sliding which generates a frictional force that resists tooth motion. The frictional heat (or the power loss) generated during one meshing cycle will be a product of the frictional force and the velocity of sliding. Geometrically, this means that the frictional force must always be perpendicular to the line of action and in the direction of the velocity of sliding. It follows then, that frictional heat will be greatest at the first and last points of contact and will vary along the path of contact. The power loss, across the entire active tooth profile, is given by:

$$\Delta P_{MAX} = \mu W_n \Delta v$$

$$\Delta P_{MAX} = \mu \Delta v \frac{T}{r_p \cos\varphi} \quad \text{(per tooth)}$$

$$\Delta P_{MAX} = \mu \Delta v \frac{T}{r_p \cos\varphi}\left[1+\frac{C_R}{2}\right] \quad \text{(per meshing cycle)} \qquad (1)$$

where μ is defined as the coefficient of friction, W_n the normal load, Δv the sliding velocity, T the applied torque, r_p, the pitch radius, φ the pressure angle and C_R denotes the theoretical contact ratio.

Viscoelastic hysteretic losses are caused by principal strains in the bulk material and are comprised of three components that are generated as a result of contact, bending and shear stresses. Hysteresis resulting from contact is at a maximum at the tooth surface and decreases to zero at the neutral axis of the tooth. Bending hysteretic losses increase from zero at the tooth tip to a maximum at the root fillet, rapidly returning to zero through the gear body[25]. It is useful to note that the tooth is at maximum tension on the active profile, zero at the neutral axis and maximum compression on the non-active flank. Finally, there are the viscoelastic shear losses which are likely to be similar in nature to those in bending.

Cropper[15] sought to quantify the magnitude of viscoelastic losses in a range of common engineering polymers employed under moderate power transmission conditions. He concluded that the combined viscoelastic losses in polymer gear teeth are not a major factor in heat generation when compared with those generated by adhesive friction at the contact interface. Typically, viscoelastic losses account for approximately 0.5% of the total losses. Figure 6 shows the effect of temperature on the combined viscoelastic losses of different materials.

4.2. *Operating Temperatures of Polymeric Gears*

The in-service operating temperature of a polymeric gear is a combination of three separate heat sources. These are the ambient temperature, the bulk temperature, and the flash temperature. The ambient temperature is the temperature of the fluid medium in which the gear operates and in most cases may be considered as constant. The bulk temperature refers to the temperature of the entire gear body when run continuously and after steady state conditions have been achieved. The flash temperature is caused by adhesive friction at the instantaneous contact interface.

For simplicity, the bulk temperature may be considered as the temperature of the gear body above ambient and equally the flash temperature is that above the bulk temperature. At the contact interface, temperature estimations will be an accumulation of all three factors, whereas at a large distance from the surface the temperature will be a combination of bulk and ambient only. Bulk and flash temperatures are dependent on materials properties, the operating conditions and environmental factors.

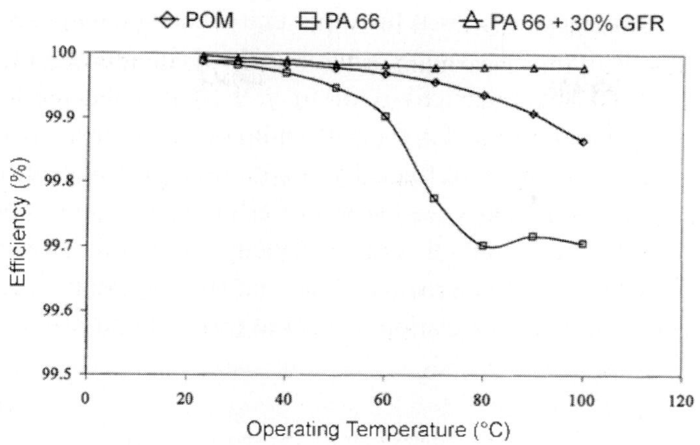

Fig. 6. Viscoelastic losses of engineering polymers at 7 Nm and 1000 rpm with temperature (after Cropper[15]).

4.2.1. *Flash Temperature*

At the point of tooth-mesh the sliding velocities create a localised, instantaneous, high temperature - the flash temperature - first proposed by Blok in 1963[11] (following earlier work on surface temperatures of oil lubricated steel gears[12]). This was developed by Hooke *et al.*[20] for polymer gears with more accurate predictions of the build-up of the flash temperature. They assumed an unsteady process where the intensity distribution and velocity of the heat source changed as meshing proceeded. The flash temperature affects the coefficients of friction between rubbing surfaces and can, in extreme cases, lead to local melting and subsequent rapid wear. This is often exacerbated due to the fact that polymer gears are often run unlubricated where friction forces can be high. Temperature rises also lead to thermal expansion and this is important because plastics have coefficients of thermal expansion between four to ten times higher than steel. Blok's modified equation for gearing applications is given as:

$$\Delta\theta_{flash} = \frac{1.11\mu P\left(V_1^{\,0.5} - V_2^{\,0.5}\right)}{\left(2k\rho C_p a\right)^{0.5}} \tag{2}$$

where P is defined as the normal force per unit face width, V_1 and V_2 are the rolling velocities of specific pinion and wheel interacting tooth flanks, k is the thermal conductivity of air, ρ the specific density of air, $C_{p,}$ the heat capacity of air and a ,the half-width of the contact patch.

This model was further refined by Mao[24] using a finite difference model to predict the dynamic variation in flash temperature (accounting for changes in load distribution, sliding velocity and radius of curvature of the tooth profile). A comparison of Mao and Blok's results is given in Fig. 7 and show a good correlation, except at the first and last points of contact.

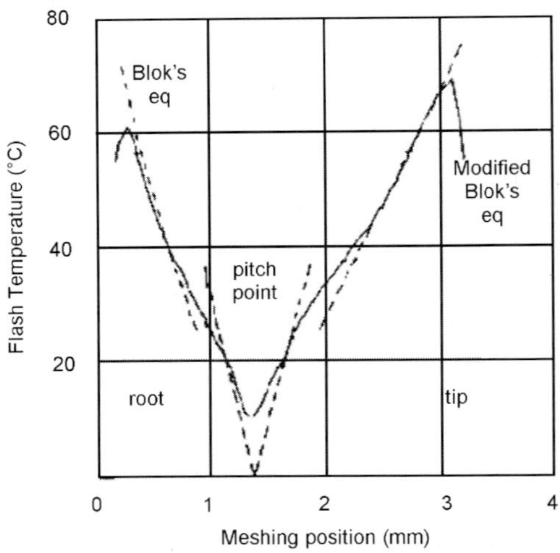

Fig. 7. Calculated flash temperature along the tooth profile (unbroken line) compared with temperature distribution calculated according to Blok (after Mao[24]).

4.2.2. *Bulk Temperature*

The bulk temperature of a polymer gear affects the mechanical properties of the selected material (such as the modulus of elasticity, Poisson's ratio, tensile and shear strengths). There are several models available for calculating bulk temperatures for plastic gears and most of the major standards related to the design and rating of polymer gears are based on these. Three major models are discussed below.

4.3. *Bulk Temperature Estimation*

Hachmann and Strickle[18] investigated the average body and surface temperatures for polymer gear teeth. They assumed that heat is generated only at the surface due to the sliding contact and hysteresis losses were ignored during the investigation. This is the model employed in the German plastic gear rating standard, VDI 2545 (1981)[31].

The British Standard, BS6168 (1987)[14] equation states that bulk temperature of a non-metallic gear consists of three heat generation or transfer components, i.e. heat due to sliding friction, heat loss due to forced convection from the gears to the inside of the gearbox casing and conduction of heat from the gearbox to the surroundings. It is clearly based on the work of Hachmann and Strickle and takes a very similar form. Both the Hachmann and Strickle and the BS equation predict that temperature is a weak function of velocity and that temperature is proportional to torque.

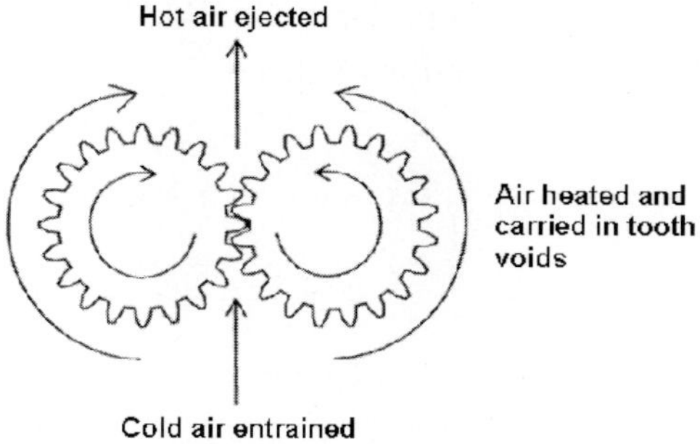

Fig. 8. Schematic of air flow between dry running meshing gears (after Mao[25]).

According to the Birmingham heat dissipation model, the gears behave as a gear pump, as shown schematically in Fig. 8[25]. They pump air from one side of the contact zone to the other. As the gears rotate, pockets of air are trapped between the gear teeth and are carried round and heated by the surface temperature of the gear flank. The warm air is expelled as the gears mesh and is replaced by fresh, cold air as the teeth disengage. It was assumed that the air within each tooth pocket would reach a temperature close to that of the gear surfaces and that none would be lost by convection from the open ends. The temperature rise can be expressed as:

$$\Delta\theta = \frac{1.25\pi\mu T}{C_p \rho N b \left(r_0^{\ 2} - r_p^{\ 2}\right)} \tag{3}$$

where $\Delta\theta$ is the increase in gear temperature (i.e. bulk temperature), T is the transmitted torque, N is the number of gear teeth, b is the gear face width and r_0 denotes the addendum circle radius (remainder are as defined above). This model predicts a temperature rise that is directly proportional to the applied torque and the coefficient of friction and is independent of the running speed.

Figure 9 shows the bulk temperatures above ambient predicted by each of the models over a range of transmitted powers for dry running POM gears. The Birmingham model predicts the highest temperature rise followed by BS6168 and then the Hachmann and Strickle equation. All models predict that temperature rise is directly proportional to load, while the BS and Hachmann and Strickle models also display temperature rise as a weak function of speed.

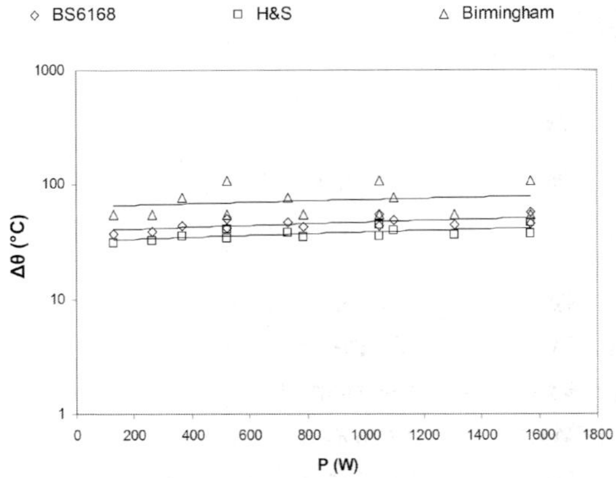

Fig. 9. Comparison of the predicted temperature rise above ambient (23°) by Hachmann and Strickle, BS 6168 and the Birmingham Heat Dissipation models.

5. Failure Modes of Polymer Gears

The failure of polymer gears is governed both by the limitations of the material properties and the operational idiosyncrasies imposed by upon them as geared machine elements. Many of the failure modes observed in metallic gearing are seen when plastics are employed but with additional failures attributed to differences in the properties of the respective materials.

Hachmann and Strickle[18] observed, during extensive testing of a range of polymers gears, four main groups of failure. These were root bending fatigue, pitch line fracture, excessive wear and pitting damage (in lubricated experiments). As mentioned above, subsequent studies have also highlighted the importance of temperature estimation in the rating and performance prediction in polymer gears: eg Yelle and Poupard[38], Gauvin *et al.*[16] and Walton *et al.*[32-34]. Each of these failure modes are briefly introduced below with the exception of wear. Wear is not included in the discussion below, as extensive testing has shown it to be the most common form of failure in polymer gears; the topic is therefore given its own section.

5.1. *Thermal Failure*

The importance of temperature and heat generation in polymer gears has already been discussed. There is, however, a distinct form of failure that can be attributed to thermal overheating and is separate from other failures, such as root fracture, which are exacerbated by high temperatures. It occurs in one of two forms related to the melting temperature of the polymer. The first occurs as large and permanent tooth deflection (such that conjugate motion is no longer maintained) when the operating temperature reaches the melting temperature of the polymers. The second type of failure, which occurs when the melting point of the polymer is exceeded by the operating temperature, is characterised by rapid wear followed speedily by tooth failure. Both are the result of plastic flow of the material caused by overloading. Other observed modes which may be attributed to thermal factors include expansion causing extreme wear and a change in material properties inducing other forms of failure. Figure 10 shows the local melting of a PA 6/6 matrix composite gear containing 30% glass fibre and 15% PTFE gear. This benchmark gear[a] was run dry at 15 Nm and 1500 rev/min.

Fig. 10. Nylon 6/6 composite gear (30% Glass fibre and 15% PTFE) that has thermally failed resulting in gross wear (after Cropper[15]).

The effect of lubrication on thermal failure is dependent on the type of lubrication employed. By definition, all forms of lubrication will reduce the coefficient of friction and subsequently the flash temperatures during the meshing cycle. If full external lubrication is selected (for example oil or water) the lubricant will also act as a coolant, reducing operating temperatures and the risk of hard asperity contact. A partial external lubricant, such as grease will reduce the level of friction in the mesh but will have no effect on operating temperatures[36]. The beneficial effects of running polymer gears under lubricated conditions are discussed further below.

5.2. *Fatigue*

Fatigue failures are commonly observed in plastic gears both under dry running and lubricated conditions. It is manifest, usually, in two forms that are described below and caused by overloading and can occur

[a]The Birmingham standard or benchmark gear has the following specification 2 mm module of standard proportions, 17.4 mm facewidth, 30 teeth, 20° nominal pressure angle and is employed in the majority of gearing experiments conducted in the Power Transmission Laboratory at the University of Birmingham.

without discernable wear. As with all polymer gearing failures, it is highly influenced by temperature.

5.2.1. *Root Failure*

Fatigue failure occurs as root fracture caused by bending stresses concentrated in the root fillet of the gear teeth. Fractures either remove a single tooth or, if the gear rim is not correctly designed, can propagate radially, removing a section of the gear rim together with several teeth. Figure 11 shows root and rim failure of a benchmark gear made from unreinforced PEEK that was run under oil lubrication at 17 Nm and 4500 rev/min.

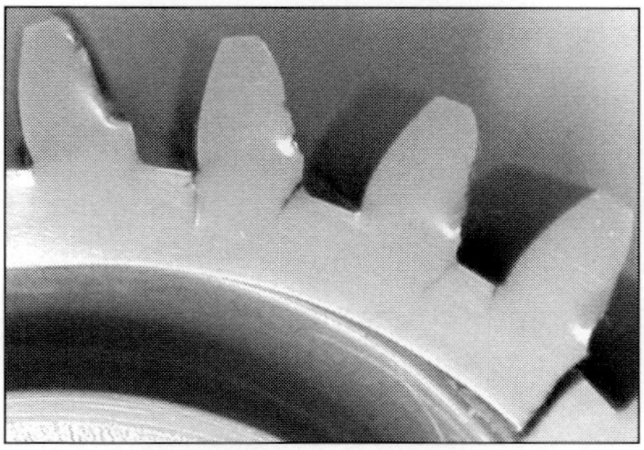

Fig. 11. Unreinforced PEEK gear displaying both root and rim fracture (after Kono[23]).

5.2.2. *Flank Failures*

Excessive contact stresses are the cause of the other fatigue failures. The most prominent of which is pitch line fracture (PLF). This form of failure was first reported by Hachmann and Strickle[18] and many since, however it has proved to be a complex phenomenon that is still not fully understood. Fracture in the region of the pitch point is a phenomenon that is almost exclusively witnessed in plastic gears. A re-occurring theory attributes pitch line fracture (PLF) to highly reversed shear stresses induced by compressive contact stresses produced by contact between tooth flanks (Breeds *et al.*[13]). The position of the maximum shear stress is highly dependent on the friction coefficient at the contact interface. A coefficient greater than 0.3 (von Mises criterion) or 0.275 (Tresca) will transpose the maximum shear from beneath the contact interface to the tooth surface. It is likely then, that when the maximum shear stress is at the surface a crack will form and propagate, first, under the influence of contact stress leading to combined tension as well as bending stresses.

Observations made suggest that failure is a complex combination of factors. Cropper[15] observed that PLF occurred on the active flank between the first point of single tooth contact and the pitch circle. He noted a thumbnail crack was initiated at the centre of the face width and then worked across the tooth. The failure mechanism, therefore, is initially influenced only by contact however as meshing progresses the influence of contact stresses is diminished to be replaced by a prevalent bending stress which ultimately destroys the operating ability of the tooth. PLF has also been known to occur in the absence of serious wear suggesting that wear is not a contributory factor. Figure 12 shows a typical PLF failure of a carbon-fibre reinforced nylon benchmark gear. Notice how the crack has propagated across the entire tooth flank. Of cardinal importance is that PLF only occurs on the driven gear where reversed sliding occurs and the friction forces act so as to open any small surface cracks.

Fig. 12. Carbon-fibre reinforced nylon gear with pitch line fracture (after Cropper[15]).

Lubricating plastic gears appears to lower the chances of PLF. However, fatigue experiments conducted with PEEK gears, run lubricated at high temperatures, gave a limited number of PLF failures. More common was the occurrence of pitting, which in this case is likely to be caused by cracks (themselves initiated by sub-surface shear stresses) originating below the surface in the vicinity of the pitch circle (where contact stresses are greatest for a non-integer contact-ratio gear set). Pits tend to occur at the edge of the face width (likely to be a result of the manufacturing process) but are also seen at the tooth tip caused by premature contact between tip and flank of mating teeth (Kono[23]). Figure 13 shows pitting damage on an unreinforced PEEK, benchmark gear that was run under oil lubrication at 21 Nm and 4500 rev/min. Note that under these conditions the gears were transmitting 15 kW which gives an indication of the levels of power that polymer gears are capable of carrying.

Fig. 13. Unreinforced PEEK gear showing pitting and wear damage (after Kono[23]).

5.3. *Other Forms of Failure*

The above has dealt mainly with tooth failure modes. However, failure is by no means limited to those discussed above: other forms of indirect failure occur such as noise, transmission error, static overload, creep, weld-line fracture (when the gears are moulded using pin-gating) and hub failure (dependent on the selected method of power transmission coupling). Thus, polymer gears may fail in a wide variety of ways.

6. Wear of Polymer Gears

In dry-running applications, wear has consistently been shown to be the most prevalent mode of failure. It is also one of the most difficult forms of failure to predict. The rate at which wear occurs in a given polymer gear set is dependent on many different factors which include:

- Operational factors – running speed, applied loads and ambient temperature

- Materials factors – similar material meshes, dissimilar materials, fillers and the addition of internal lubricants
- Gear geometry – pressure angle, module and facewidth.

6.1. *Commercial against Experimental Wear Data*

Consequently, the prediction of wear for a particular gear set is a convoluted, ill-defined process. Commercially available design data are given in the form of wear coefficients but these must be used with caution, as data are usually correlated experimentally from thrust washer or pin-on-disc tests that do not recreate the complex kinematics (loading, deformations and simultaneous rolling and sliding motions which change direction during a cycle) experienced by the gear teeth during a typical meshing cycle.

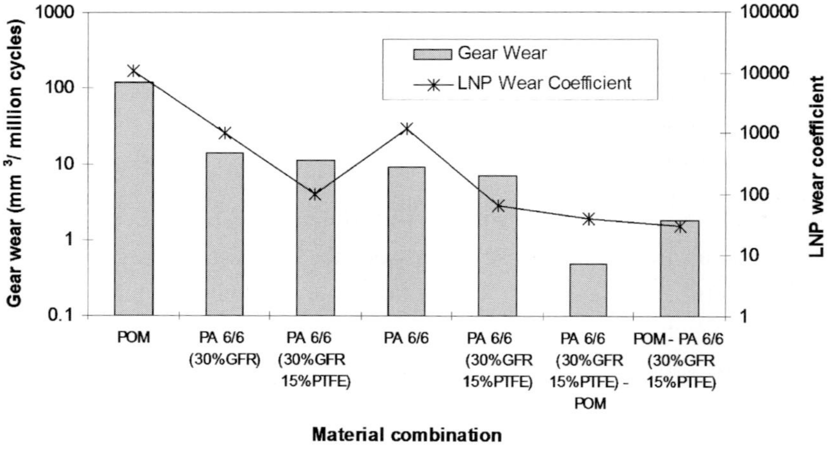

Fig. 14. Correlation between LNP catalogue wear coefficients and polymer gear test data (after Breeds *et al.*[13]).

Figure 14 shows a comparison between catalogue wear coefficients and measured gear wear. There is a weak correlation between the sets of data indicating that catalogue data is suitable for comparative means between the various polymer compounds and should not be used as the principal means of selecting a material for a gearing application.

Wear occurs in the form of gross material removal such that conjugate motion between teeth is lost or backlash (defined as the amount of clearance between a tooth and that of the thickness of the engaging tooth on the pitch circle) is deemed to be unacceptable. Cropper[15] found that adequate backlash allowances are vital to wear resistance (if too little backlash is provided, rapid unsustainable wear occurs after a short test duration). The result of such tooth degradation often leads to a reduction in tooth strength which in turn can lead to secondary failure caused by bending fatigue (root failure) or overloading (deformation).

6.2. *Wear Manifestation*

Typical surface topography of a severely worn polymer gear tooth is shown in Fig. 15. The raised pitch-line is clearly visible across the entire flank and indicates that this was the driven gear of the test pair. Above this point wear patterns may be observed where the roll/slide is in the same direction. Below the pitch-line roll/slide are in opposition and show typical scalloped wear damage. In addition to the expected patterns, a zone of premature contact is also visible at the tooth tip. The importance and effects of the combined roll/slide are thus underlined. Cropper[15] during his extensive test programme found that some polymers were far more susceptible to the transition between the phases of approach and recess, such that significant differences in wear resistance between driver and driven were observed.

Fig. 15. Severe wear damage to a polymer gear and effects of roll slide motion (Breeds *et al.*[13]).

Despite the difficulties faced in predicting wear rates for different materials, the manifestation of wear in polymer gears displays similar characteristics despite a plethora of material variations. Figure 16 exhibits three of the most common results of typical gear wear experiments.

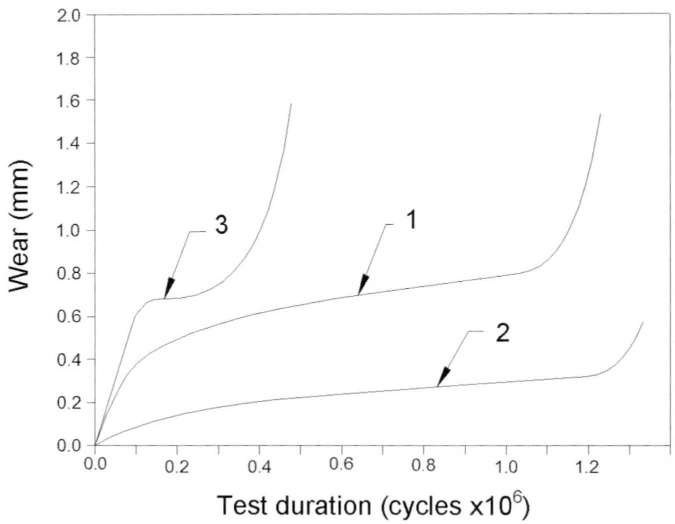

Fig. 16. Typical wear signatures which tend to be independent of selected material.

The most common wear history signature is that shown by profile 1, where three distinct regions may be observed. The gear first experiences a 'running-in' period where the tooth profile adjusts to optimum geometry (this is usually the period where manufacturing errors are removed). There is then a period of steady-state wear followed by a period of rapid wear and failure. After an extensive test programme[13,20,24] it was concluded that the period of steady-state wear be used to assess the tribological performance of a gear material, as the periods of rapid wear at the beginning and end of the cycle are also affected by gear geometry and manufacturing accuracy. This was the recommendation adopted by all researchers at the University of Birmingham. The second data set is similar to the first with the running-in period removed; this response is indicative of an accurate gear set designed for a specific application. Finally, data set three shows typical wear characteristics of an ill-design system, where material or gear geometry is unsuitable for the operating conditions.

The sudden onset of rapid wear is most likely to be associated with the surface melting of the active tooth flanks as the flash temperature approaches the material melting point. Hooke *et al.*[20] defined this operational limit as the transition torque and indicated that it was a weak function of operational speed and a much stronger function of applied load.

6.3. *Wear Rates of Various Polymers*

White[36] compared the relative wear rate of gears for a variety of commercially available polymers. His results showed a large range in wear resistance between materials. Figure 17 shows the wear rates of nine different materials, differentiating between driver and driven gears. The gears were tested at 1000 rev/min and 10 Nm. Observing the trends of the graph, the highest wear occurs in POM gear set, followed by a glass-bead reinforced PA; the remainder being composites of a POM or PA matrix with either glass or carbon fibres. The greatest wear resistance occurred in the composites also containing a PTFE internal lubricant. When these materials were run in similar materials meshes, there was little difference in wear between driver and driven. However, wear increased significantly under certain dissimilar material meshes.

7. Mechanical Efficiencies of Polymer Gears

It is the mechanical inefficiency (losses) of gears that gives rise to heating and therefore knowledge of the efficiency of gears is useful, not least in understanding how losses might be minimized. Walton *et al.*[32,33] carried out a large test programme to investigate the efficiency of polymer gears; this was done in two parts, the first looking at the effect of different polymer materials, the second to examine the effects of changes in gear tooth geometry.

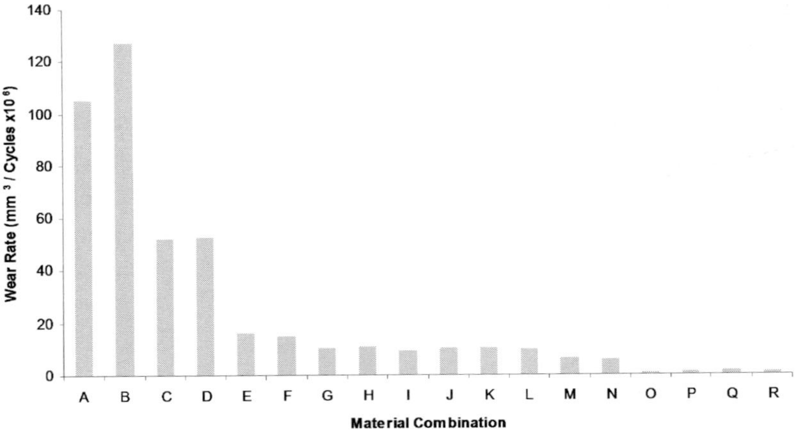

Fig. 17. Relative wear rates of polymer and polymer composite gears. Note: A meshes with B, C with D… etc, with \A, C…as driver and B, D … as driven. (after White[36]).

7.1. *Theoretical Efficiencies*

The theoretical efficiency of a pair of spur gears can be calculated in two ways, the main difference emanating from whether the coefficient of friction is taken as a constant throughout the meshing cycle (i.e. an average value is employed) or a different coefficient of friction is used during the phases of access or recess. The efficiency is solely a function of tooth geometry, gear ratio and the coefficient of friction (assuming an average coefficient of friction)[32].

7.2. *Experimental Efficiencies and Coefficients of Frictions*

POM/POM efficiencies displayed strong speed dependence up to speeds of 500 rev/min but not beyond this speed; equally they showed a dependence on load with efficiency increasing with torque. PA/PA meshes displayed a different frictional profile showing a strong dependence on speed while being almost unaffected by load. However, when the materials are run against each other (POM/PA) the frictional characteristics lie somewhere between the POM/POM and PA 6/6/PA6/6 behavior, with the general trend being similar to the all-POM results. Efficiencies were shown to further increase in all cases with the application of grease. An example of the measured efficiencies of a POM/PA mesh for a variety of loads and speeds is given in Fig. 18.

In the composite material tests, both the PA glass and carbon reinforced composites showed similar characteristics and a slight efficiency improvement compared with the unreinforced PA meshes. A stronger load dependence was introduced however when a solid phase internal lubricant (PTFE) was added. Meshing a PA composite with a steel driver showed improved efficiencies at high speeds but a reduction at lower speeds compared with the PA composite (without internal lubricant). The measured efficiencies for PA composite/PA composite gears containing 30% glass fibres for a range of loads and speeds are shown in Fig. 19.

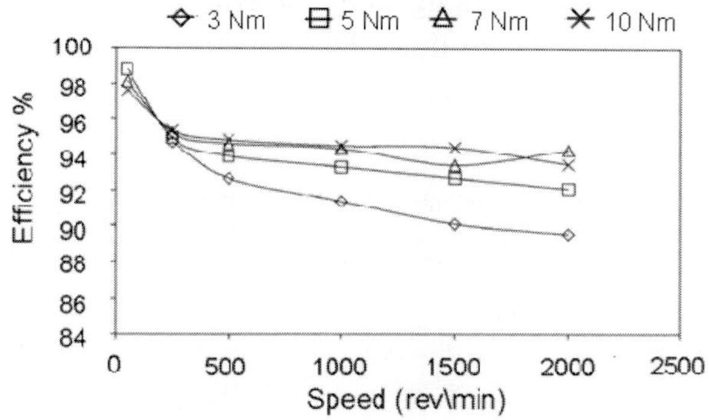

Fig. 18. Efficiencies for Nylon 6.6/POM gears over a range of loads and speeds (Walton *et al.*[32]).

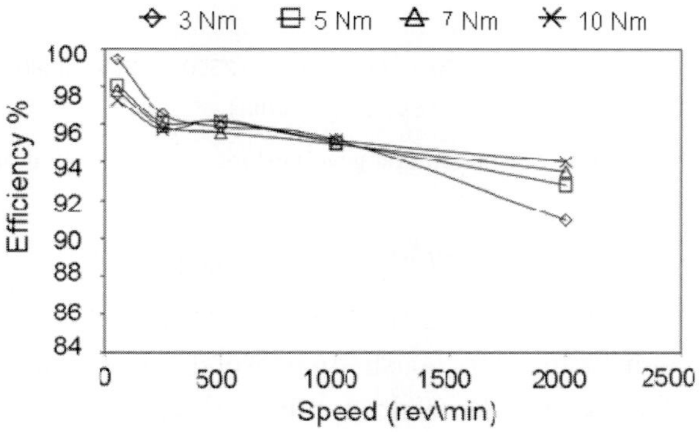

Fig. 19. Efficiencies for PA composite/PA composite gears containing 30 per cent glass fibres for a range of loads and speeds (Walton *et al.*[32]).

Geometric changes to a gear set had a variety of positive and negative effects on the efficiency of spur gears. Centre distance extension showed little effect but a reduction in centre distance lowered efficiencies by as much as 14%. Tooth geometry modification yielded mixed results. A reduction in tooth module from the benchmark to 1 mm resulted in an improvement in efficiency whereas employing a novel tooth profile designed to ANSI/AGMA 1106-A97 (1997) proportions[10] showed little benefit. Similarly helical gears, employing an 8° helix angle, displayed comparable frictional characteristics to the benchmark gears with only a slight reduction in efficiency. Improved performance was observed when gear ratios greater than one were employed, with the greatest improvement in reduction ratios followed by speed increasers. All measured efficiencies (assessing both material and geometric effects) were shown to decrease with time. The measured efficiencies for 1mm module and benchmark gears carrying 5 and 7 Nm at a range of speeds are shown in Fig. 20.

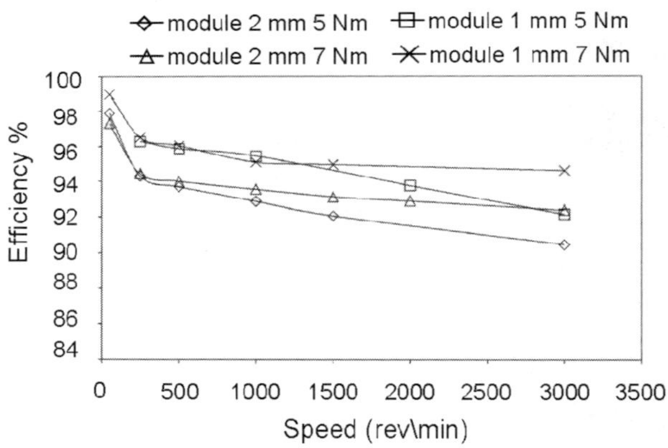

Fig. 20. Effect of tooth size (module) on gear meshing efficiency (Walton *et al.*[3]).

8. Polymer Gears in High Performance Applications

The majority of polymer gears are employed in low load and speed applications, such as printers, cash dispensers, car window and door latching mechanisms and domestic appliances. However, the push for cheaper and lighter gears has resulted in a trend towards polymer gears capable of running at higher torques and speeds and also higher temperatures. These requirements have made it necessary to run the gears with external lubrication.

8.1. *Power Density Comparisons*

If the benchmark gears were made of typical gear steel, their torque capacity would be around 60 Nm with a power rating of 32kW for a life of at least 5000 hours running in typical gearbox oil. If the gears were made from a high performance polymer, say PEEK, the gears could transmit 22 Nm with a power rating of 10 kW running, in automotive engine oil for an observed life of at least 100 hours before the test was stopped. This is the transmissible power for a polymer gear designed to conventional standards. The same size gear designed to a standard which is non-conventional (employing a pressure angle of 30° instead of 20°) is capable of carrying around 15 kW. Figure 21 shows a comparison of the benchmark gear with 30° and 40° pressure angle PEEK gears. The 40° pressure angle gear represents a geometric limit to increasing the pressure angle in order to improve power density.

Fig. 21. Left: the benchmark gear, centre: 30° pressure angle gear, and right: a 40° pressure angle gear; all made from PEEK (Kono[23]).

Running polymer gears in oil has been shown to increase power density but also introduces new forms of failure. Now instead of wear, the gears are subject to extreme pitting as seen in Fig. 13. In an attempt to simulate under-bonnet conditions PEEK gears were tested at elevated temperatures including tests with oil heated to 150°C. An interesting phenomenon was that as the temperature of the oil was raised, so too was the gear fatigue life increased, as shown in Fig. 22. Running polymer gears in oil lubricated conditions necessitated a study of the film conditions at the contact interface.

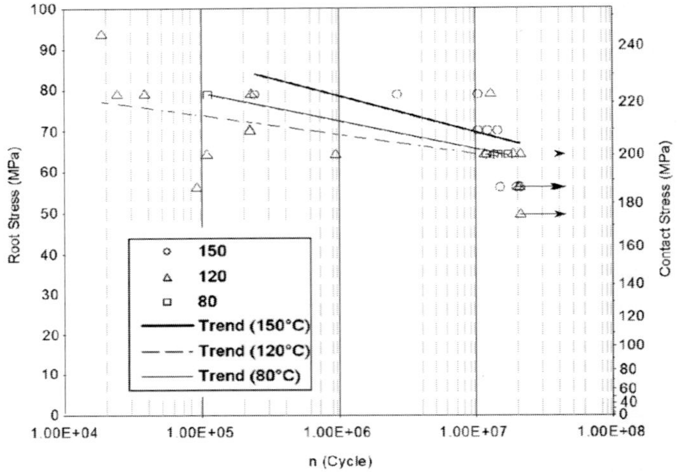

(a)

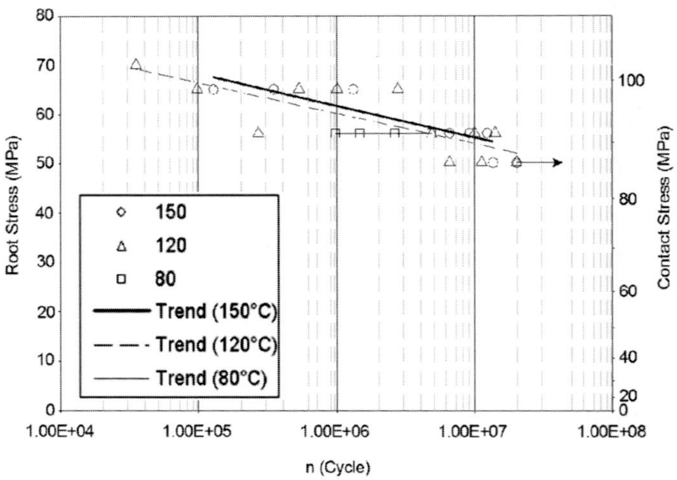

(b)

Fig. 22. Life of (a) PEEK with 30% carbon fibre and (b) unreinforced PEEK against contact and bending stresses for a variety of temperatures.

8.2. *Elastohydrodynamic Aspects of Polymer Gears*

Little has been published on the behaviour of polymer gears operating under lubricated conditions. Song *et al.*[28] compared the lubricating oil film-thickness and pressure in a polymer/steel mesh with those found in lubricated steel gears. They reported higher film thicknesses and lower film pressures in the dissimilar material mesh, neglecting viscoelasticity and temperature increases dues to friction.

Theoretical studies based on the work of Johnson[21], Myers *et al.*[26] and Gu[17] indicated the lubrication regime under which polymer gears operate, lubrication film thicknesses through the theoretical path of contact, and the effects that these may have on performance of polymer/polymer and steel/polymer meshes.

Lubricating film thicknesses were shown to increase with operating speed and are a much weaker function of load (decreasing as load was increased). Geometrically, film thicknesses increased when pressure angle was increased from 20° to 30° (as a result of the increased entrainment velocities); however the reduced contact ratio (1.32 instead of 1.65) meant that the magnitude of curvature change during single tooth contact was greater. The effects of gear material combinations showed that film thicknesses were lower for dissimilar meshes (steel/polymer) compared with polymer/polymer meshes. The minimum film thicknesses across the line of contact for polymer/polymer and polymer/steel 30° pressure angle gears are shown in Fig. 23.

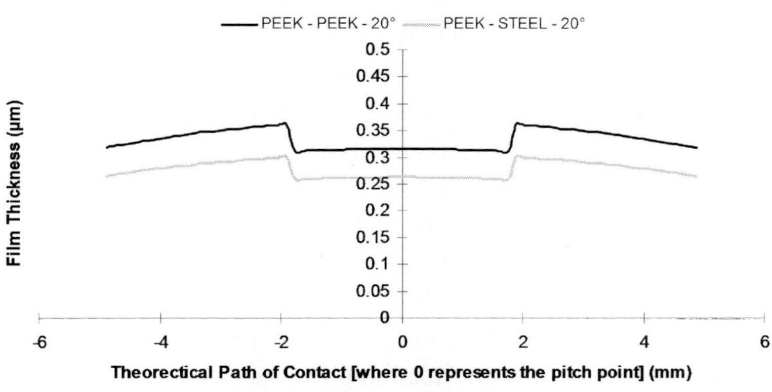

Fig. 23. Minimum film thicknesses across the line of contact for PEEK/PEEK and PEEK/steel 20° pressure angle gears.

Mechanical efficiencies as discussed above were measured for gears run under motion control conditions. Efficiency measurements displayed a load dependence for all tests similar to those seen by Walton *et al.*[32,33] in their dry-running experiments. This suggested that the asperity elastic smoothing effect suggested by Hutchings[19] still existed under external lubrication for the given test conditions. Efficiencies were shown to improve for the higher pressure angle gears but were lower when polymer gears were run against steel pinions (probably due to the large tooth deformations). The measured efficiency of PEEK/steel gear meshes having 30° pressure angles is shown in Fig. 24.

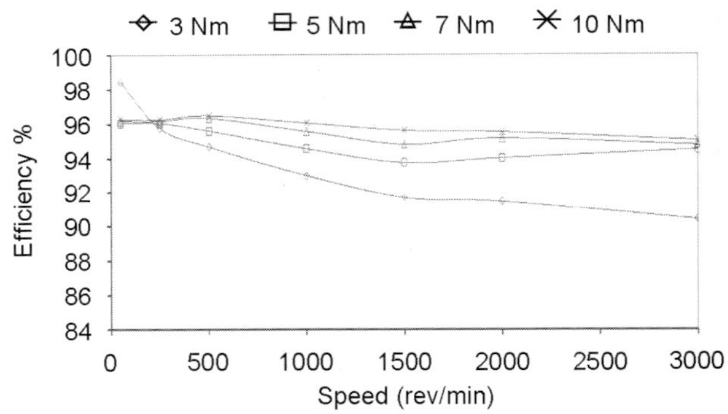

Fig. 24. Measured efficiency for 30° pressure angle PEEK/steel gear combinations.

Lubrication effectiveness under more demanding experimental conditions (4500 rpm and 27 Nm) was assessed using surface topography measurements. An examination of the gears running up to 20 x 10^6 cycles, showed varying degrees of wear with very little observed in the 30° pressure angle gears while the 20° pressure angle gears exhibited extreme wear and gross pitting as shown in Fig. 13.

Polymer gears were shown to operate within the transitional region of Johnson's lubrication regime map, validating the use of Myers *et al.*'s solution and under mixed/boundary lubrication conditions for operational parameters up to 10 Nm and 3000 rev/min. Lubrication quality was shown to be a strong function of speed and a lesser function of load. Results suggested that for low load applications, the use of an external lubricant offers little performance benefit with full-film lubrication regimes unable to develop (this is not to say that lubrication will not have a positive effect on running temperatures).

Efficiencies were not as high as expected and bore similarities with dry running polymer gears. Derived coefficients of friction showed much higher values than for steel gears and were likely to be a contributive factor in causing severe pitting damage. The use of the Walton *et al.*[32] efficiency equation is limited given that gear load distribution, local rolling and sliding velocities, local radii of curvature, surface roughness as well as lubricant parameters, which will all affect the coefficient of friction, are neglected in their formula. However, surface topography measurements taken from the power transmitting gears displayed similar trends to the theoretical predictions of full-film thickness and lubrication efficiency.

The application of the smooth surface film thicknesses formulae of Johnson and Myers *et al.* were limited when pitting damage of the magnitude shown in Fig. 13 is observed. Under such conditions, when surfaces accumulate wear to the extent shown, the equations used to calculate lubrication parameters would not be valid. However, the simple theoretical predictions showed similar trends to those observed in the experimental programme and so, for the given conditions present a first approximation of lubrication parameters. Given that the 30° pressure angle gears run at higher loads showed no wear or pitting, this suggests that full film lubrication occurred as predicted by the Myers *et al.* solution.

At loads typically associated with polymer gears the operating regime is shown to be mixed film or boundary lubrication. When high pressure angle gears were tested at high loads the operating regime became full film lubrication and relatively little tooth flank damage occurred. From these results the beneficial effects of oil lubrication and high tooth pressure angles in polymeric gearing are demonstrated.

8.3. *Novel Forms of Lubrication*

Finally, several workers have suggested that water could be employed as an external lubricant. Tsukamoto *et al.*[30] showed significant increases in durability and wear performance as a result of lubricating unreinforced polymer gears with water. Water has several advantages such as a smaller viscosity-temperature change rate, lower cost, is easier to obtain, has lower energy consumption and is safe for use because it is fire resistant (Wang *et al.*[35]). There are clear environmental benefits for using water instead of oil.

References

1. S. N. Kukureka, Y. K. Chen, C. J. Hooke and P. Liao, *Wear*, **185**, 1 (1995).
2. C. J. Hooke, S. N. Kukureka, P. Liao, M. Rao and Y. K. Chen, *Wear*, **200**, 83 (1996).
3. C. J. Hooke, S. N. Kukureka, P. Liao, M. Rao and Y. K. Chen, *Wear*, **197**, 115 (1996).
4. Y. K. Chen, S. N. Kukureka and C. J. Hooke, *Journal of Materials Science*, **31**, 5643 (1996).
5. M. Rao, C. J. Hooke, S. N. Kukureka, P. Liao and Y. K. Chen, *Polymer Engineering and Science*, **38**, 1946 (1998).
6. S. N. Kukureka, C. J. Hooke, M. Rao, P. Liao and Y. K. Chen, *Tribology International*, **32**, 107 (1999).
7. Y. K. Chen, S. N. Kukureka, C. J. Hooke and M. Rao, *Journal of Materials Science*, **35**, 1269 (2000).
8. N. A. Wright and S. N. Kukureka, *Wear*, **250**, 1567 (2001).
9. S. Apichartpattanasiri, J. N. Hay and S. N. Kukureka, *Wear*, **251**, 1557 (2001).
10. American National Standard, *Tooth proportions for plastic gears*, ANSI/AGMA 1106-A97 (1997).
11. H. Blok, *Wear*, **6**, 483 (1963).
12. H. Blok, Proceedings of the general discussion on lubrication and lubricants, Institution of Mechanical Engineers, **2**, 222 (1937).
13. A. R. Breeds, S. N. Kukureka, K. Mao, D. Walton and C. J. Hooke, *Wear*, **166**, 85 (1992).
14. British Standards Institution, BS6168, *Specification for non-metallic spur gears*, (BSI, London, 1987).
15. A. B. Cropper, *PhD dissertation*, University of Birmingham (2003).
16. R. Gauvin, P. Girad and H. Yelle, *Trans ASME*, 80-C2/DET-108 (1980).
17. A. Gu, *Soviet Atomic Energy* 72 PTG-34 8-12 (1972).
18. H. Hachmann and E. Strickle, Konstruktion im Maschinen Apparate und Gratebau, **18**, 81 (1966).
19. I. M. Hutchings, *Tribology*, (Arnold, London, 1992).
20. C. J. Hooke, K. Mao, D. Walton, A. R. Breeds and S. N. Kukureka, *Trans. ASME Journal of Tribology*, **115**, 119 (1993).
21. K. L. Johnson, *Journal of Mechanical Engineering Science*, **12**, 9 (1970).

22. K. L. Johnson, *Contact Mechanics*, (Cambridge University Press, Cambridge, 1989).
23. S. Kono, *PhD dissertation*, University of Birmingham (2002).
24. K. Mao, *Wear*, **262**, 1321 (2007).
25. K. Mao, *PhD dissertation*, University of Birmingham (1993).
26. T. G. Myers, R. W. Hall, P. H. Savage and P. H. Gaskell, *Proc. Roy. Soc.: Mathematical and Physical Sciences*, **432**, 467 (1991).
27. J. E. Shigley and J. J. Uicker, *Theory of Machines and Mechanisms* (McGraw Hill, New York, 1995).
28. K. Song, Y. Wang, B. Huang and M. Wang, *Lubrication Engineering,* **1**, 50 (2005).
29. T. A. Stolarski, *Tribology in Machine Design*, (Butterworth-Heinemann, London, 1990).
30. N. Tsukamoto, H. Maruyama, H. Mimura and Y. Ebata, *Japan Society of Mechanical Engineers International Journal,* **C 36**, 241 (1993).
31. Verein Deutscher Ingenieure, VDI 2545, *Zahnrader aus thermoplastichen kunstoffen* (Gear wheels made of plastic), (VDI, 1981).
32. D. Walton, A. B. Cropper, D. J. Weale and P. Klein Meulemann, *Proc. I. Mech. E. Part J: Engineering Tribology,* **216**, 75 (2002).
33. D. Walton, A. B. Cropper, D. J. Weale and P. Klein Meulemann, *Proc. I. Mech. E. Part J: Engineering Tribology,* **216**, 93 (2002).
34. D. Walton, A. B. Cropper, D. J. Weale and S. Kono, *Proceedings of the International Tribology Conference*, (Nagasaki, Japan, 2000).
35. J. Wang, T. Li, J. Peng and Z. Chen, *Proceedings of the International Conference on Gearing, Transmissions and Mechanical Systems*, 761 (Nottingham, 2000).
36. J. White, *PhD dissertation*, University of Birmingham (1999).
37. A. R. Breeds, *MPhil dissertation*, University of Birmingham (1993).
38. H. Yelle and M. Poupard, *Polymer Engineering and Science*, **15**, 90 (1975).

CHAPTER 15

BRAKE FRICTION MATERIALS

Ho Jang[1] and Seong Jin Kim[2]

[1]*Department of Materials Science and Engineering, Korea University*
Seoul, South Korea, 136-713
E-mail: hojang@korea.ac.kr
[2]*Materials Research Team, Advanced Technology Center*
Hyundai Motor Company, Gyeonggi, South Korea, 445-706
E-mail: seongjin@hyundai-motor.com

Tribological properties of brake friction materials are presented considering compositional effects, processing parameters, and microstructural aspects of sliding interfaces during brake applications. In the first part, the role of ingredients on the brake performance is summarized by grouping the constituents into binders, reinforcements, and property modifiers. The manufacturing parameters that affect compressibility, hardness, and porosity are also examined for optimization of brake performance. Secondly, the sliding interfaces between brake pads and a disk are studied since the friction films on the disk surface and the plateaus on the pad surface are closely related to the critical brake performance including noise propensity, fade phenomena, etc. In the last section, the environmental issues of the brake friction material and the current development efforts toward next generation friction materials are discussed.

1. Introduction

The brake friction material in the automotive brake system is one of the most complicated polymer matrix composites since it contains more than 10 disparate ingredients comprising polymeric, metallic, and ceramic constituents. The optimization of its tribological properties is, therefore, difficult to obtain in a systematic manner because of the large number of parameters. A great deal of effort has been given to the development of the multiphase composite for a brake friction material since a monolithic material or the composite with a few ingredients has never been successful to meet numerous performance related demands.[1-6] The development effort also includes possible applications of new ingredients and the improvement of the manufacturing process. Therefore, the control of the tribological properties of the brake friction material has been relied on the engineering experience rather than the scientific database.

In general, the brake friction materials have been classified as asbestos-based organic materials, non-asbestos organic materials (NAO), semi-metallic materials, sintered materials, and carbon composites in terms of the ingredients and their friction characteristics. Currently, majority of commercial automobiles uses NAO type brake friction materials and they are subdivided into two different types; known as non-steel type and low-steel type materials according to the steel content.

The ingredients in the friction materials are normally classified as binders, fibrous reinforcements, non-fibrous reinforcements, and property modifiers[1] and more than 200 raw materials are currently used to produce commercial friction materials. The ingredients in the friction material are classified in terms of the expected role as binders, reinforcing fibers, lubricants, abrasives, wear resistant space fillers, and miscellaneous friction modifiers,[7] although the ingredients often play multifaceted roles in many cases. However, the information regarding the ingredients is seldom found in the literature due to proprietary reason. In particular, the composition of the commercial friction materials and technical information about the manufacturing processes are not released.

For fundamental understanding of various tribological phenomena observed during brake applications, microscopic examinations of the sliding surfaces of the friction material and the gray iron counter disk have been performed.[8-12] This is because the sliding surface is an important source to understand various brake-induced phenomena. The sliding interface experiences the transfer of atoms during sliding and the amount of atomic flux at the interface is determined by bonding characteristics of the molecules at contact junctions, chemical and electrostatic states of mating surfaces, and local stress distribution at the junctions. The material transfer during sliding continuously build up and destroy transfer films by forming transient plateaus on the sliding interface and the transient plateaus play important roles in determining the wear rate, noise propensity, and other friction characteristics.[8,13]

Recently, environmental issues of the ingredients attract public attention again after decades of struggle to replace carcinogenic asbestos in the friction material since new regulations concerning the use of heavy metals and acicular fibers are being established.[14-16] Research published in Sweden has shown that the friction materials are among the major offender for copper pollution and EU has taken steps to ban or severely limit the use of heavy metals in the vehicle components.[17] Several ceramic fibers are also considered as potential health-hazardous materials and a serious concern has been raised to remove them in the friction material. The replacement of the heavy metals and whisker fibers by environmentally acceptable alternatives is a serious challenge for the friction material industry since they play crucial roles for brake performance.

The performance of the brake friction material has been improved much during the last two decades to keep pace with the fast improvement of the vehicle quality. The demand from consumers for effective and comfortable braking is also increased and extensive development efforts have been given to meet the high standards from vehicle manufacturers and high expectation from consumers. However, it is well known that the commercial friction material with full satisfaction is not easy to accomplish and shortcomings still exist. This is because the brake performance is determined by the ingredients in the formulation showing unique tribological properties. Examples are insufficient high temperature wear resistance of non-steel friction material and the aggressive nature of low-steel friction materials, among many others.[18] Recent research efforts, therefore, have been given to the development of the "hybrid" or "global" type friction material by adopting advantages from each friction material type while eliminating the disadvantages.

In this chapter, tribological properties of resin bonded NAO type brake friction materials are presented in terms of ingredients, manufacturing processes, and microscopic interpretation of sliding interface. Environmental issues of the commercial friction material and efforts to develop prospective "hybrid" or "fusion" type friction materials are also introduced.

2. Ingredients of Brake Friction Materials

Typical ingredients used in the commercial brake friction materials are listed in Table 1. Each ingredient in the table changes its tribological properties as the size, chemical composition, and morphology are changed.

Table 1. Typical ingredients for brake friction materials.

Ingredients	Type	Materials
Binder	Straight	Phenolic resin, Polyimide, Resol
	Modified	Si-modified, P-modified, B-modified, Acryl-modified, Aralkyl-modified, Epoxy-modified, Rubber-modified
Reinforcements	Fibrous	Aramid pulp, Steel wool, Bronze fiber, Cu fiber, Brass fiber, Stainless fiber, Rock wool, Glass fiber, Ceramic fiber, Potassium titanate (whisker), Cellulose fiber, Carbon fiber, Acrylic fiber
	Non-fibrous	Mica, Vermiculite, Potassium titanate (plate), Woolastonite
Property Modifiers	Abrasives	Al_2O_3, SiO_2, MgO, $ZrSiO_4$, ZrO_2, Zeolite, Fe_3O_4, SiC, ZnO, Chromite
	Lubricants	Graphite, MoS_2, CaF_2, ZnS, Sb_2S_3, Sb_2O_5, WS_2, CuS, Cokes
	Space Filler	$BaSO_4$, Clay, Diatomaceous earth, $CaCO_3$
	Miscellaneous modifiers	$Ca(OH)_2$, Carbon black, Cork, Cashew, Rubber powder, Al powder, Cu powder, Bronze powder, Cut copper, Zn powder, Sn powder

Detailed information about the source of the ingredient is, therefore, normally considered as proprietary information among friction material manufacturers. In the following sections, the effects of the ingredients on the tribological properties of brake friction materials are presented.

2.1. *Binder Resins*

The binder resin used in the brake friction material is a thermosetting phenolic resin and various products are available in the form of liquid or solid powder. In general, novolac type phenolic resin is used and several modified resins have been developed to improve heat resistance and to optimize other tribological properties at elevated temperatures[19] since the fade and wear resistance of the binder resin is crucial for high temperature performance of the brake friction material. Figure 1 shows the wear amounts of phenolic resins measured by a Krauss type tester using two pieces of friction materials (total contact area 8 cm^2) while the disk temperature was maintained at 300°C.

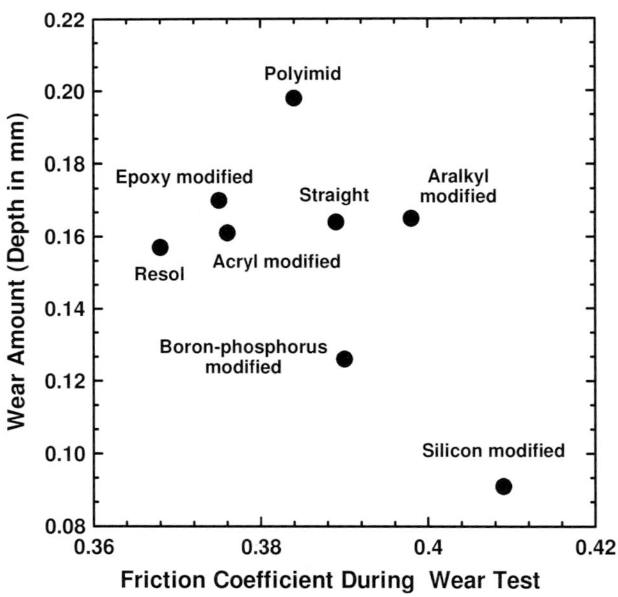

Fig. 1. Friction and wear of binder resins tested at 300°C. Wear amount was measured using a Krauss type tester[23] using a formulation with 16 ingredients.

The figure indicates that Si-modified resin has the best wear resistance in spite of the high coefficient of friction. The improved high temperature properties of the Si-modified resin are attributed to the Si-O bonds produced during the reaction of siloxanes with phenolic compounds during modification.[19] Heat resistance of two different phenolic resins for non-steel friction materials; straight resin and aralkyl modified resin, were also compared using a simple formulation.[20] Figure 2 shows the change of friction coefficient and temperature during constant interval tests. It shows that the aralkyl-modified resin has better thermal stability than the straight resin. However, the aralkyl modified resin exhibits extended heat affected zone on the sliding surface and the wear amount was approximately two times greater than that of the straight resin.[20] Bijwe *et al.*[21,22] also examined the influence of modified resins on the fade and recovery of friction materials. They found improvements in the some tribological properties using modified resins but no clear correlations between overall friction characteristics including wear resistance at elevated temperatures and mechanical properties were obtained.

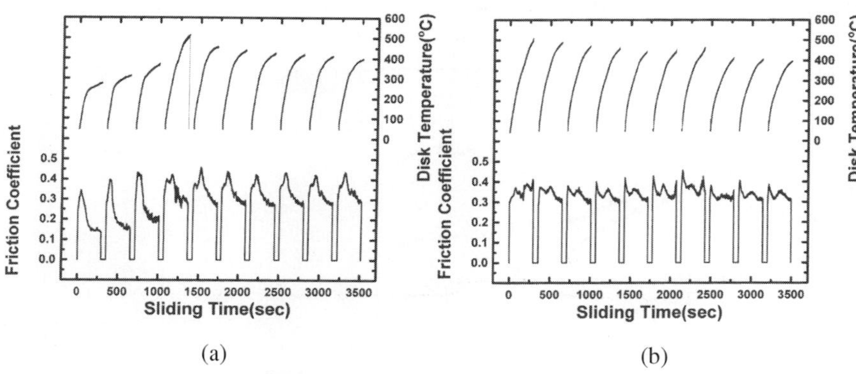

Fig. 2. The change of the friction coefficient and disk temperature during constant initial temperature tests. The test was carried out on a Krauss type tester using friction materials containing straight phenolic resin (a) and aralkyl modified phenolic resin (b).[20] (Reprint from Tribology International, copyright 2000, with permission from Elsevier Science).

2.2. *Reinforcements*

The reinforcing fibers commonly used for brake friction materials are aramid pulp,[24-26] metallic fibers,[27] ceramic fibers,[28,29] glass fibers,[30] acrylic fibers,[31,32] cellulose fibers,[33] carbon fibers,[33,34] and other natural fibers.[35] In general, NAO commercial friction materials contain 5~25 vol. % of fibrous ingredients and the type and the relative amount of the reinforcing fibers affect various aspects of brake performance and wear rate.[6,36,37] The study of metal fibers on the friction and wear of sliding couples showed that copper and steel fibers were unfavorable in terms of noise and vibration, while they improved fade resistance.[18,27] Synergistic effects from aramid pulp and whisker fibers were also reported and the improved friction stability and wear resistance by mixing the two disparate ingredients were closely related to the durability of friction films at the rubbing interface.[23] Figure 3 shows that the friction coefficient at elevated temperatures is retained well when the friction material is produced with aramid pulp and potassium titanate whiskers at a certain ratio. Non-fibrous reinforcements such as mica, vermiculite, woolastonite, and potassium titanate (plate type) are often used in the friction material to reduce the raw material cost.

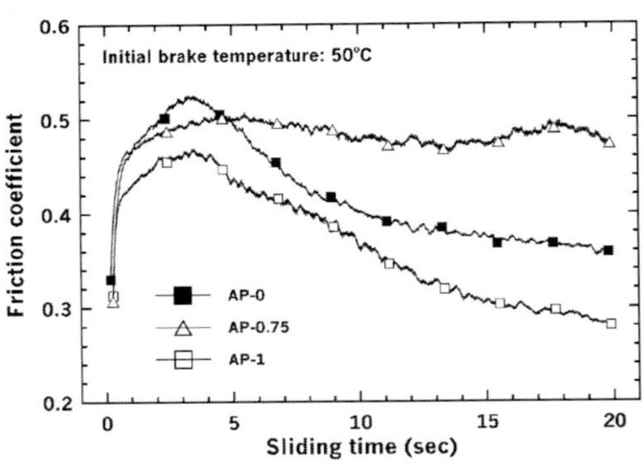

(a)

(b)

Fig. 3. The change of the friction coefficient as a function of sliding time using friction materials containing different ratios of aramid pulp and potassium titanate whiskers (a) and an SEM micrograph showing well dispersed whiskers in fibrillated aramid pulp (b).[23] (Reprint from Wear, copyright 2001, with permission from Elsevier Science.)

2.3. *Property Modifiers*

Property modifiers are added not only to adjust basic tribological properties such as friction level, fade resistance, and wear rate but also to fine-tune specific friction characteristics such as noise propensity, pedal feel, and aggressiveness against a counter surface, which are often deficient in the base formulation. Various modifiers have been used to tailor brake performances at a wide range of conditions. The property modifiers are divided into non-abrasive and abrasive materials, and the non-abrasive materials are sub-classified as high or low friction modifiers.[1] Friction dusts from cashew nutshell liquid and ground rubber are typical non-abrasive modifiers. The friction dusts are added to obtain friction effectiveness at low and medium pressure and also to reduce noise propensity.[4,38] Rubber particles are used to increase friction at low to moderate temperature and also to increase damping capacity of the friction material.[4,7] However, the cashew dusts and rubber particles are prone to increase the brake fade (reduction of stopping power) at elevated temperatures.[7,39]

Solid lubricants are used to improve wear resistance and to reduce noise propensity. Various materials such as carbonaceous and chalcogenides are used as solid lubricants.[40,41] Different types of solid lubricants are often used to maintain lubricity at different temperature ranges since each solid lubricant has its own decomposition temperatures beyond which the lubricant does not play a role as a lubricant.[42] Therefore, complementary effects of solid lubricants are utilized by compounding different solid lubricants with different thermal decomposition ranges.[43,44] Figure 4 shows the temperature ranges of typical solid lubricants used in the commercial brake friction materials; natural graphite, MoS_2, and Sb_2S_3.

The DSC (differential scanning calorimetry) curves in the figure clearly indicate that the decomposition takes place at different temperature ranges and MoS_2, and Sb_2S_3 show multiple reactions at different temperatures. The figure also indicates that the decomposition temperature of graphite is higher than the other two solid lubricants.[43] On the contrary, the decrease of the friction coefficient at elevated temperatures is pronounced when graphite is used in the friction material

(Fig. 5), suggesting that the graphite looses its lubricating capability at the temperature lower than the thermal decomposition temperature.[45] The figure also shows that the mixture of two different solid lubricants improves fade resistance by complimentary effects.

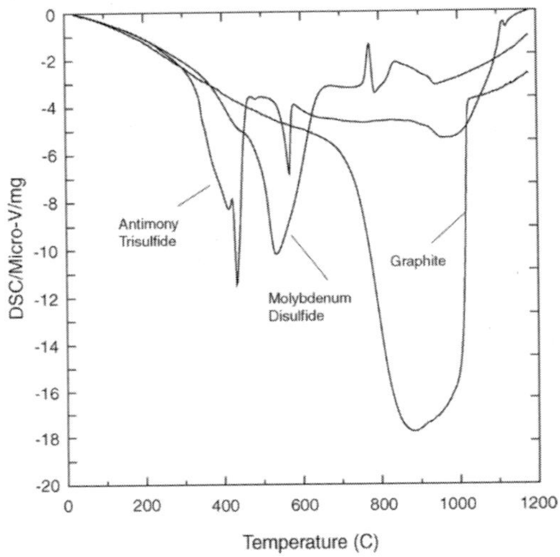

Fig. 4. DSC (differential scanning calorimetry) curves showing heat flow due to decomposition of solid lubricants (graphite, MoS_2, and Sb_2S_3) as a function of temperature.[43] (Reprint from Wear, copyright 2005, with permission from Elsevier Science.)

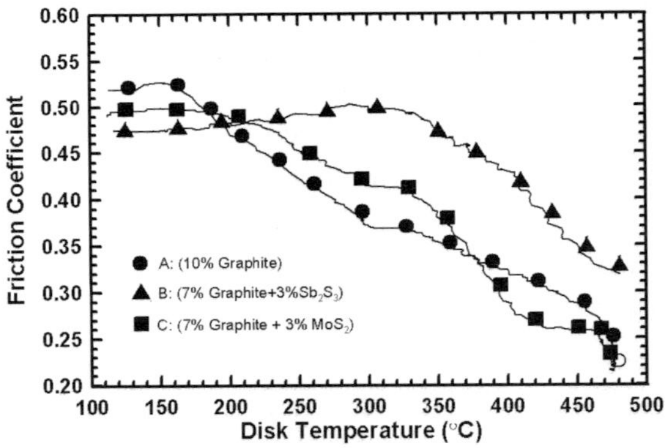

Fig. 5. The change of the friction coefficient as a function of disk temperature during the Krauss type drag tests.[43] (Reprint from Wear, copyright 2005, with permission from Elsevier Science.)

Abrasives in the friction material are used to eliminate pyrolized friction film on the disk surface and to control friction level.[1,4] The particle size of the abrasives is an important factor to achieve optimized performance of the friction material.[46,47] In general, fine particles are employed when hard materials such as silicon carbide (SiC), alumina (Al_2O_3), and kyanite (Al_2SiO_5) are used. The effect of the relative amount of a solid lubricant (Sb_2S_3) and an abrasive ($ZrSiO_4$; average size < 70 μm) on friction characteristics was studied using a full scale inertial dynamometer.[46] It showed that the relative amount of the lubricant and abrasive significantly affected the friction material wear (Fig. 6) and torque variation during brake applications (Fig. 7). The size effect of the abrasives on the tribological properties of the friction material was also studied considering friction level and wear of friction materials and gray iron disks.[47] It showed that the thermal stability of the friction material was improved and the amplitude of the friction coefficient was decreased when large abrasive particles were used (Fig. 8). The friction material with fine abrasives was less aggressive to the counter surface while the friction material wear was relatively large compared to those with large size particles (Fig. 9).

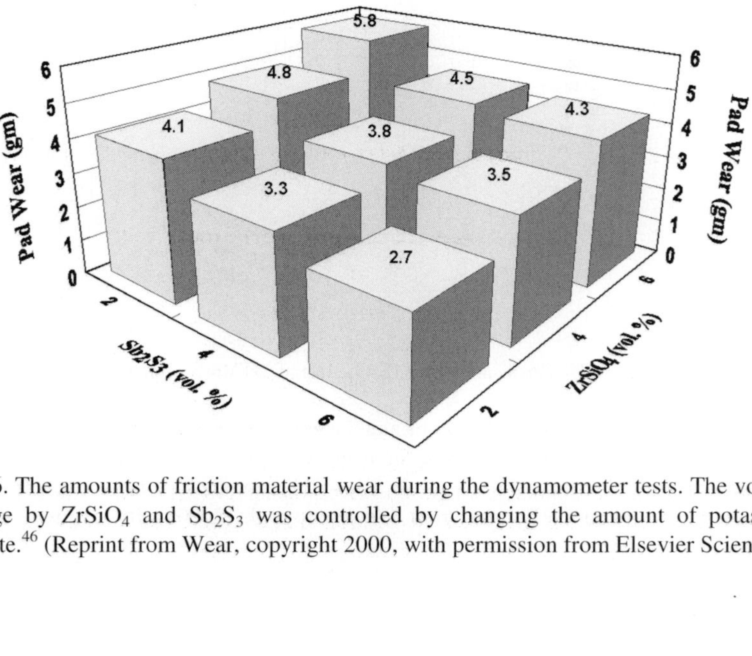

Fig. 6. The amounts of friction material wear during the dynamometer tests. The volume change by $ZrSiO_4$ and Sb_2S_3 was controlled by changing the amount of potassium titanate.[46] (Reprint from Wear, copyright 2000, with permission from Elsevier Science.)

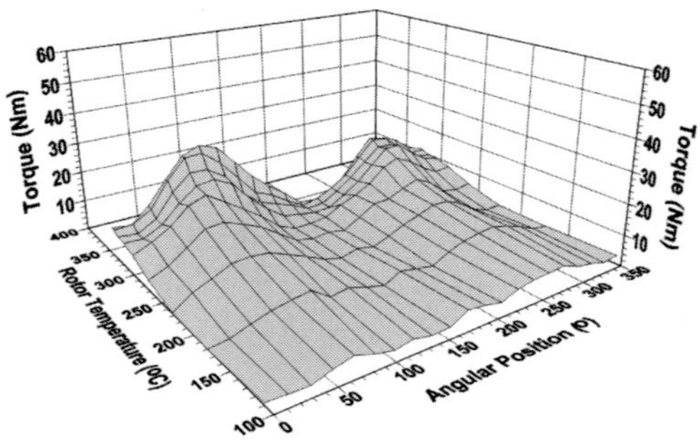

(a) 6% Sb_2S_3 and 2% $ZrSiO_4$, Initial DTV= 4.16 μm

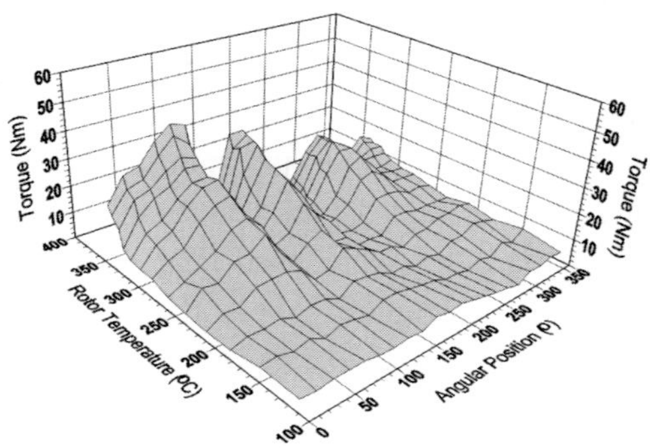

(b) 4% Sb_2S_3 and 4% $ZrSiO_4$, Initial DTV= 3.91 μm

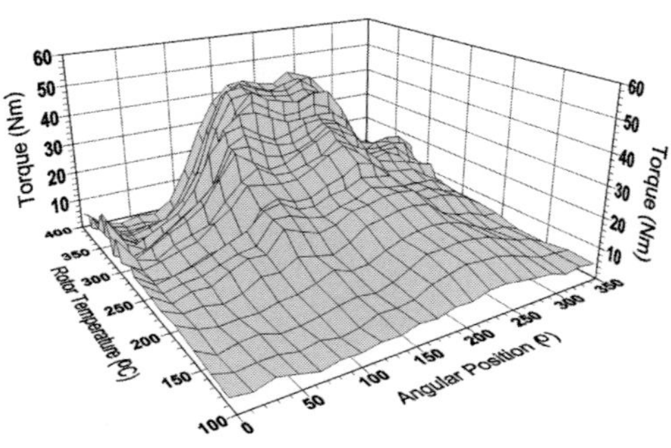

(c) 2% Sb_2S_3 and 6% $ZrSiO_4$, Initial DTV= 3.85 μm

Fig. 7. Torque variation as a function of angular position at different disk temperatures. Torque intensity was high in the case of using the most aggressive friction material containing 6 vol. % zircon and 2 vol. % Sb_2S_3. The amounts of initial disk thickness variation were similar.[46] (Reprint from Wear, copyright 2000, with permission from Elsevier Science.)

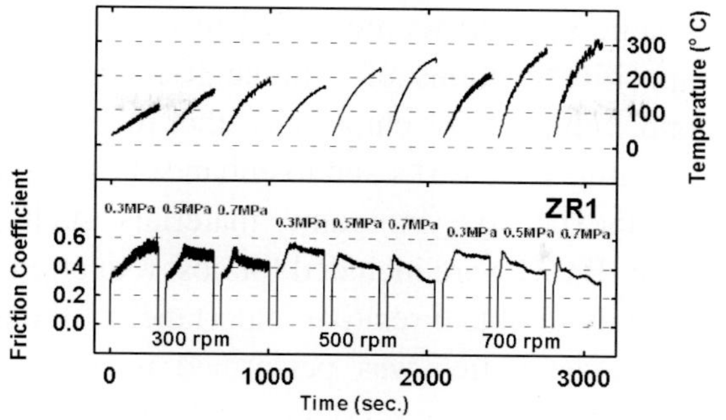

(a)

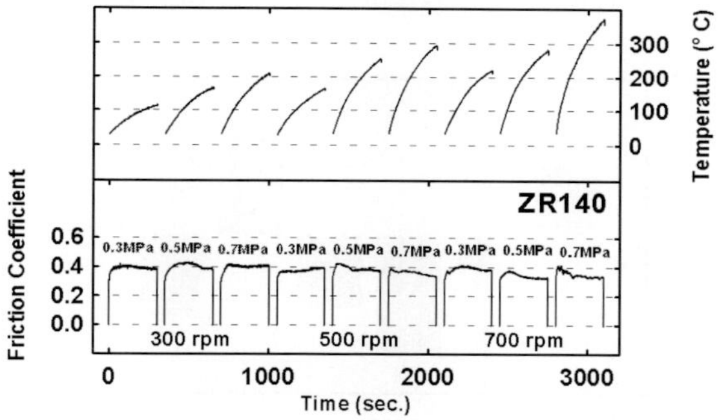

(b)

Fig. 8. The friction coefficient and temperature as a function of applied pressure and speed during Krauss type drag tests using the friction materials containing zircon powders with the average sizes of 1 μm (a) and 140 μm (b). 100 rpm is equivalent to 0.476m/sec.[47] (Reprint from Wear, copyright 2007, with permission from Elsevier Science.)

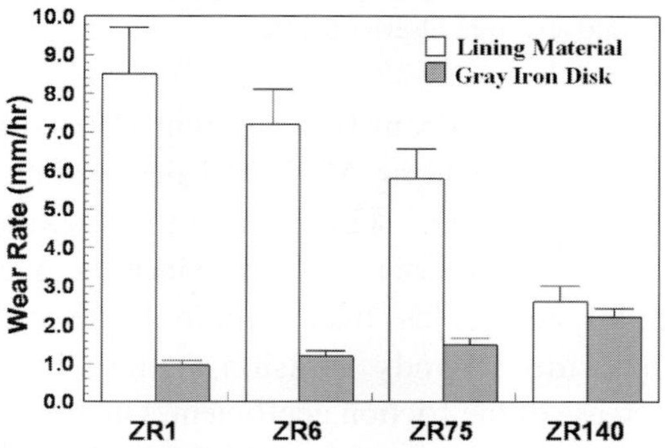

Fig. 9. Wear rate of friction materials and the gray iron disks tested with the friction materials containing different sizes of zircon.[47] (Reprint from Wear, copyright 2007, with permission from Elsevier Science.)

Various inorganic ingredients are added to improve wear resistance or to provide specific functions of a brake friction material. Metallic chips such as copper chips, brass chips, and other metal powders are used to remove surface films from disks and to enhance heat dissipation on the rubbing surface. Inexpensive inorganic materials such as limestone ($CaCO_3$) and barite ($BaSO_4$) are included for cost reduction.

A systematic study to obtain relative importance of the ingredients on various friction characteristics was performed using an experimental formulation, which is close to a commercial product.[7] The effect of ingredients on the friction coefficient is shown in Fig. 10.

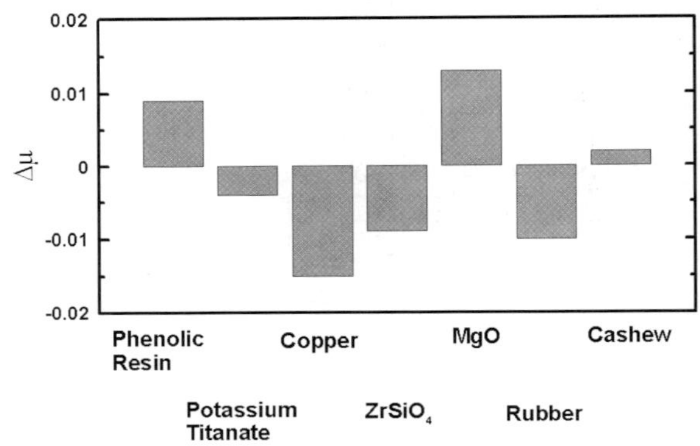

Fig. 10. Relative influence of the ingredient on the coefficient of friction. The influence was obtained by normalizing the difference in a property from ±50% modified formulations by the volume change of an ingredient. Negative values indicate that the friction coefficient is decreased with addition of an ingredient.[7] (Reprint from Wear, copyright 2004, with permission from Elsevier Science.)

It shows a contrary effect from two different abrasives; increasing the friction coefficient by adding MgO and decreasing the friction coefficient by adding $ZrSiO_4$. This is a typical example of the morphological effect of the abrasive particles since the average size of the zircon used in this study is less than 1 μm in diameter and the small zircon particles provide 3rd body abrasion at the sliding interface, resulting in the decrease of the friction coefficient. On the other hand, the large (100μm in diameter) MgO particles effectively increase the friction coefficient although the hardness of MgO is lower than zircon. Figure 11 shows the relative influence of the ingredients on fade, wear rate, and

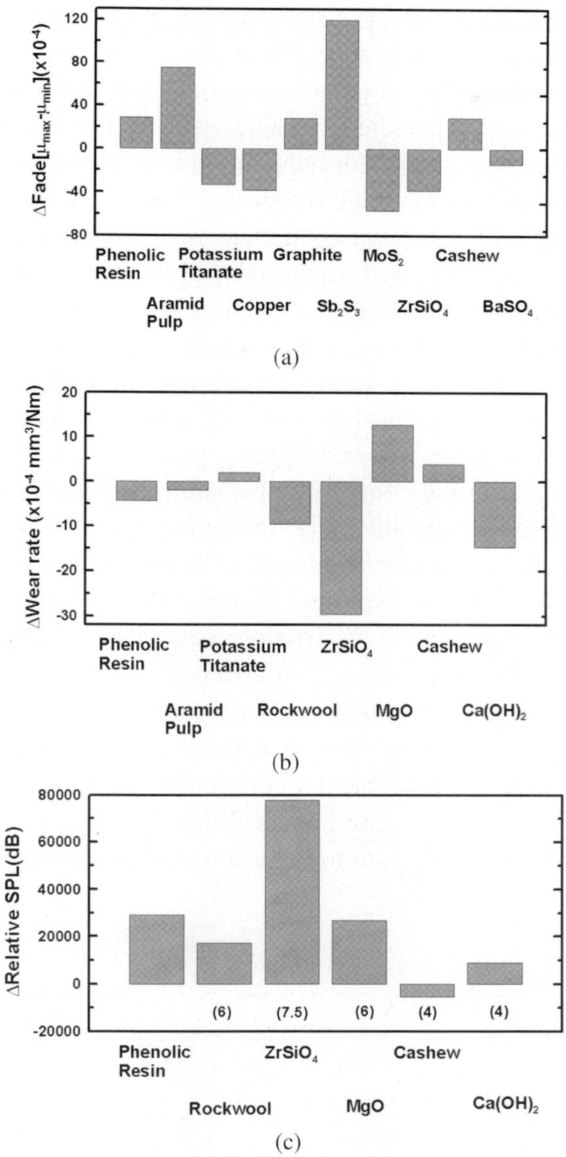

Fig. 11. Relative influence of the ingredient on the fade (a), wear rate (b), and noise propensity (c). The influence was obtained by normalizing the difference in the property from ±50% modified formulation by the volume change of an ingredient.[7] (Reprint from Wear, copyright 2004, with permission from Elsevier Science.)

noise propensity based on the data obtained from a Krauss type tester with apparent sliding contact area of $8cm^2$.

Tribological properties of composites containing barite and potassium titanate whiskers were studied by Kim *et al.*[48] The composites were produced based on a simple formulation and the tribological properties were measured using a lab scale tribotester. The effect of extremely low sliding velocity on the friction coefficient was investigated in the range of 1.57-47.1 mm/s since the friction-induced vibration can be evaluated by analyzing the μ-v (coefficient of friction versus sliding velocity) and it is often related to creep groan (self-excited brake vibration in the range of 200-500 Hz) phenomena during brake applications. Results showed a strong negative μ-v relationship from the composite containing barite while potassium titanate showed slightly positive slope (Fig. 12). The potassium titanate in the composite also showed higher stiffness due to effective reinforcing capability (Fig. 13) and the friction force oscillation was smaller in the case of the composite with potassium titanate (Fig. 14), suggesting the whisker type potassium titanate can improve the tribological properties of brake friction materials. A study to improve creep groan of a friction material was also performed by Jang *et al.* They measured the static and dynamic coefficients of friction before and after the composition modification using a tribometers with full size brake assembly.[49] Figure 15 shows the μ-v curves obtained from the two friction material specimens. The figure indicated that the slope of the μ-v curve was decreased after modification, suggesting low stick-slip propensity.

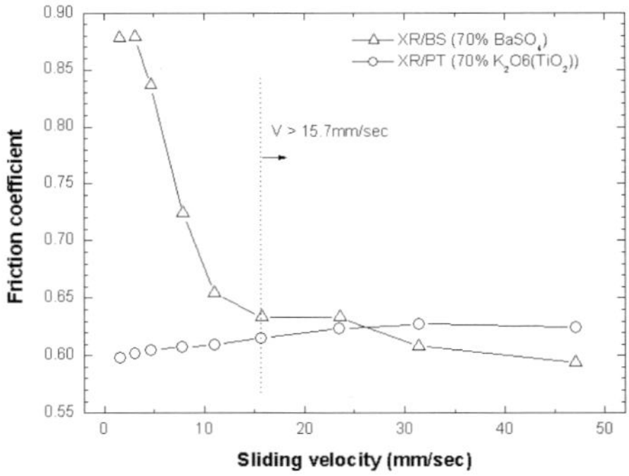

Fig. 12. The change of the friction coefficient as a function of sliding velocity (normal force = 15N).[48] (Tribology Letters, ©2004, with kind permission from Springer Science and Business Media.)

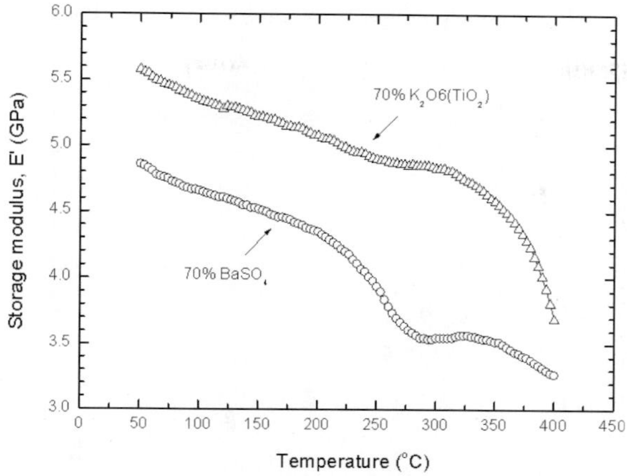

Fig. 13. The storage modulus as a function of temperature.[48] (Tribology Letters, ©2004, with kind permission from Springer Science and Business Media.)

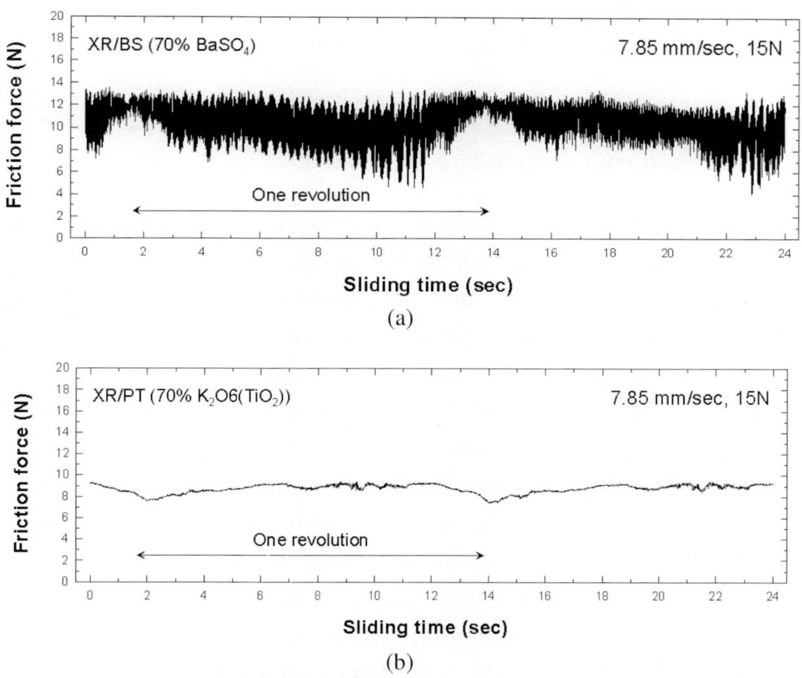

Fig. 14. The change of friction force at 7.85 mm/sec of sliding speed obtained from composite containing barite (a) or potassium titanate (b).[48] (Tribology Letters, ©2004, with kind permission from Springer Science and Business Media.)

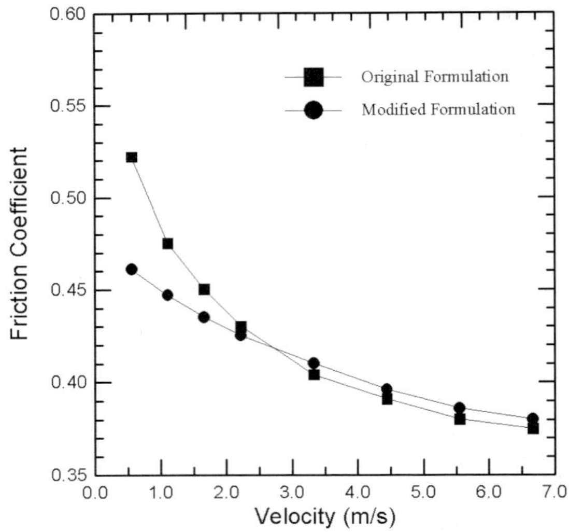

Fig. 15. The coefficient of friction as a function of sliding velocity at 0.687 MPa. The negative slope of the μ-v curve was reduced by the modification of the original composition.[49] (Reprint from Wear, copyright 2001, 1477-1483, with permission from Elsevier Science.)

2.4. *Manufacturing Parameters*

Commercial brake friction materials are produced by hot molding of a mix at high pressure followed by post heat treatments.[4] The manufacturing parameters are important factors for tribological performances of the brake friction material since they determine the intrinsic physical properties and control the quality assurance of a product. The mixing is one of the critical steps for the tribological properties since it determines the distribution of the numerous ingredients. The sequence of charging the ingredients into a mixer and the type of the blades in the mixer are often classified as proprietary information. In general, a chopper is used to open preloaded organic reinforcing fibers such as aramid pulp and subsequent mixing sequence is determined by the specific gravity, size, morphology, and hardness to prevent segregation of specific ingredients without severely changing the aspect ratio or size of the ingredients. Hot molding condition is also very important because the temperature setting around the mold and the gas release cycle significantly affects the friction and wear of the product.[50] During hot molding, the temperature setting is determined by the thermal properties of the resin such as flow distance, B-transformation time, and gel time.[4] Post heat treatment is carried out with no pressure or clamping to ensure uniform curing and to reduce thermal expansion during service. In order to find optimum manufacturing parameters for a friction material, a large number of experiments are required. To reduce number of experiments, several experimental design techniques have been suggested.[50]

3. Sliding Interface

The sliding interface between the friction material and a disk surface has attracted much attention since the friction force and related phenomena are determined by the interface. However, the study of the sliding interface has been considered a difficult task since the interfacial structure is transient during sliding and is difficult to assess. The investigation of sliding interface for brake system has been carried out by several research groups by focusing on the change of plateaus on the friction material,[10,13,51-53] measuring the thickness of the friction films on the gray iron disk surfaces,[9,54] and by analyzing the chemical composition of the friction film.[11,12,55,56]

A pioneering work of the interface analysis was performed by Eriksson *et al.*[52] by examining the formation of contact plateaus on the commercial brake pad. They studied the size and distribution of plateaus and found that the plateaus were closely related to the change of friction coefficient and squeal noise generation. The formation of contact plateaus during sliding was also studied using a glass plate slid against a block of the friction material[51] and found that the sliding interface was supported by hard primary plateaus and the soft secondary plateaus which were produced by compacted wear debris. Cho *et al.*[9] also reported that the formation of primary plateaus by large abrasive particles reduced the oscillation amplitude of the friction coefficient.[49] They found that the plateaus on the pad surface with small abrasive particles were compacted wear debris while large abrasive particles served as primary contact plateaus (Fig. 16). They suggested that the oscillation of friction force during sliding was closely related to the rolling action of the small zircon particles since the rolling hard particles destructed the friction film and generated transient contacts.

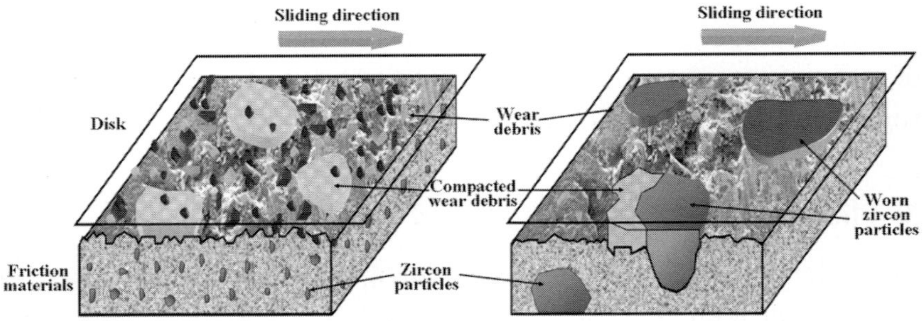

Fig. 16. Schematic drawings of sliding interface containing abrasives of different sizes.

The friction film produced on the disk surface and its influence on the friction characteristics have been studied by Jacko *et al.*[54] They found that a stable friction and low wear rates were obtained when the friction film, often called as "glaze", was presented on the disk surface. They suggested that the cohesive strength of the friction film and the adhesive bond strength between the friction film and counter surface determined the formation of the friction film. Cho *et al.*[9] examined the thickness of the friction film on the disk to study the role of film thickness on the friction characteristics (Fig. 17). They found that the friction film was mainly composed of the ingredients of the friction material transferred during sliding and the solid lubricants and iron increased the film thickness, while abrasive ingredients removed the film during sliding (Fig. 18). The thickness of the film changed as a function of temperature but the film thickness showed no direct correlation with the coefficient of friction and wear rate. Wirth *et al.*[55] also study the friction film on the gray iron disk using EDX (energy dispersive x-ray spectroscopy) and XPS (X-ray photoelectron spectroscopy). They suggested that the friction performance is independent of transfer film thickness or topography but affected by transfer film composition. Filip *et al.*[11]

Fig. 17. An SEM micrograph showing a transfer film on the disk surface (cross section). The bottom gray area is a gray iron substrate and the uneven darker layer indicates transfer films. The white layer on top of the transfer layer is a silver coat.

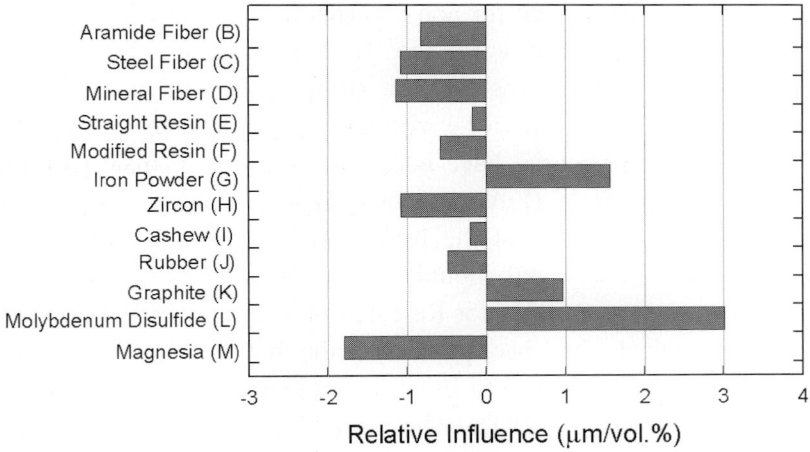

Fig. 18. Relative influence of ingredients on the transfer layer formation on the disk surface. The relative influence was obtained by normalizing the difference of transfer layer thicknesses from base and modified formulations divided by vol.% of ingredients.[9] (Tribology Letters, ©2005, with kind permission from Springer Science and Business Media.)

studied the friction layer using GAXRD (glancing angle x-ray diffraction) and EDX techniques to investigate the composition of the friction films on the friction material surface. They suggested that no simple relationship was found between composition of the friction layer and bulk material formulation. The detailed analysis of friction films was carried out by Österle and Urban[12] using the FIB (focused ion beam) technique. They identified the friction film as three different layers; nanocrystalline oxides with sulfur containing amorphous layers, nanocrystalline compacted wear debris, and severely deformed metal particles. However, they did not attempt to analyze the correlation between composition of the friction layer and tribological performance.

4. Friction Materials and Environmental Issues

Environmental concerns of the ingredients used in the brake friction material began in 1980 when the health issues of asbestos in the work place were addressed.[57] The asbestos ban in the friction material significantly affected the formulation and various organic and inorganic substitutes were developed for non-asbestos friction materials. After the health issues of asbestos in the friction material, most vehicle manufacturers do not accept vehicle components containing asbestos fibers. Recently, new issues for environmentally hazardous ingredients in the brake friction material have been addressed again from EEC (EU regulation) and EPA (Environmental Protection Agency).[14-17] The alternative raw materials in the brake friction material will be more environmental friendly and should not be classified as carcinogens by IARC (International Agency for Research on Cancer). The current focus on the ingredients has been given to the refractory ceramic fibers containing less than or equal to 18 wt.% of alkaline oxide and alkali earth oxide (Na_2O + K_2O + CaO + MgO + BaO) [Directive 2001-41-EC], heavy-metals (cadmium, hexavalent chromium, lead & lead compounds, and mercury & mercury compounds), and other heath hazard inorganic/organic compounds containing antimony, carbon black, and respirable fibers.

Copper content in the brake friction material has been blamed for the contamination of storm water and bay near metropolitan areas.[17,58,59] The lead compounds and antimony compounds also attract attentions because they are classified as carcinogenic substances.[14] In particular, a growing concern has been given to the antimony trisulfide that has not been classified as a carcinogen. This is because the antimony trisulfide changes into toxic antimony oxides when it is exposed to the high temperatures over 300°C[14,43,46] during brake applications.

Fibrous ceramic materials and glass fibers have also been suspected as human carcinogens. This is because the wear debris of these fibers can be changed into acicular form or airborne in the course of sliding against counter disks even if the original aspect ratio of the fiber is rather large. Particularly, respirable acicular ceramic fibers such as potassium titanate whiskers are prone to be imposed by an immediate regulation. Recent studies to replace the whisker type potassium titanate in the friction material by a platy form (aspect ratio < 1)[60] showed that the platy type potassium titanate improved wear resistance with slight deterioration of high temperature properties. However, the synergistic reinforcing capability with fibrillated fibers was not found.

The environmental concern of the ingredients in the brake friction material recently extended from the work environment in the manufacturing plant to the public safety regarding air and water contamination by wear debris produced during brake applications. The recent issues concerning the public health of the brake friction material, therefore, may radically alter the formulation and change the market trend significantly in the near future.

5. Friction Evaluation Tests

Performance evaluation of a brake friction material has long been considered as a difficult task. This is partly because of the poor data repeatability from different tests and insufficient correlation between bench test results and actual brake performance in a vehicle. Much effort has been given to the development of the test machines and test procedures to achieve the same friction characteristics as the real vehicle. Various testers are currently used to analyze tribological properties of the friction materials and they are classified as laboratory scale tribometers, drag mode tribotesters, small-scale inertial brake dynamometers, and full-scale inertial brake dynamometers. Each tester has its own development history to access brake performances of the friction materials emphasizing certain tribological properties.

Laboratory scale tribometers are mainly used to measure detailed tribological properties of a friction material and widely used in the universities and research labs. They are very sensitive to test parameters such as sliding speed, pressure, and temperature and are known as a good tool to investigate the role of ingredients on the friction performance.[48] The lab scale tester has also been used to examine in-situ interfacial contact variation during sliding, which is closely related to the microscopic understanding of the friction properties.[61] However, the lab scale tribometers normally use a small size specimen so that the test results often exhibit serious deviations due to segregation of ingredients in the test samples. At high temperatures, in particular, the small size specimen often shows dissimilar test results compared with full scale dynamometer or in-vehicle test results since the heat accumulated in the friction material is different compared to the full size specimen.

Drag mode tribotesters, often called as a Krauss type machine, use larger specimens compared to the lab scale tribometer to avoid errors from local ingredient segregation in the friction material. Several commercial testers are available and have been used to obtain friction properties in an early stage of the formulation development and for screening the commercial products. The first drag mode tester is FAST (friction assessment screening test) machine that was developed by Ford Motor Company for quality assurance of the friction material. This tester has been used to measure wear rate and fade phenomena at elevated temperatures in a constant drag mode. However, the correlation of the test data with a full scale dynamometer is poor and the evaluation of the NVH related brake performances is not assessable due to its own setting for constant drag mode tests. Another type of drag mode tester is the Chase Machine, which has been used to determine the edge code of the commercial friction materials (SAE J661).[62] Contrary to the FAST machine, which uses disk type gray iron sliding surface, the counter surface is provided by a rotating drum. However, the chase machine also has shown discrepancies in the test results when the data have been compared with full-scale inertial dynamometers.[63]

Recently, dynamometer manufacturers have developed inertial type scale dynamometers to overcome inherent drawbacks of the drag mode tribotester. Two different types are commercially available in terms of the design of the brake module. One of them uses a small size caliper[64] and the other type employs a pad-on-disk type set up using two pieces of specimen.[44,65] The scale ratio is approximately 1/5 of the full size pads and the attached inertial moment is also provided with the same scale ratio. The performance of the scale dynamometer has not been completely confirmed yet but it is believed that they can provide reliable data after careful calibration. In particular, the scale dynamometer has an advantage of isolating the tribological characteristics of a friction material without serious influence from the caliper assembly since they normally adopt a rigid caliper design. A few test procedures such as JASO 406 are available to obtain various performance of the friction material.

The full scale inertial dynamometer has long been accepted as a standard tester for the development and verification of the commercial friction materials when it is properly instrumented and operated. It uses a real fixture from a vehicle and equipped with an inertia that can simulate real situation drags and stops at a wide range of speeds. Compared to other testers, well-developed test procedures are available[62] and repeatability of the test results is better than the above-mentioned smaller testers. Currently full-scale inertial dynamometers are well accepted as standard tools to simulate in-vehicle tests, whereas the test procedures and the dynamometer design concept have been continuously improved to duplicate the brake performance of a vehicle.

6. Concluding Remarks

The brake friction material has been considered as one of the most difficult components in the vehicle to obtain first-class performance. A part of the reason for the unsatisfactory brake performance is attributed to the diverse braking situations and the inherent shortcomings that are caused by the ingredients used in the traditional friction material. Currently, a large amount of effort is given to develop "hybrid" or "fusion" type friction materials that are designed to overcome the limitation from a specific type of the friction material, namely, non-steel type or low-steel type materials. The new friction material also targets the higher friction level, less fade, and less NVH (noise, vibration, and harshness) with a minimal wear rate. In addition, it has to exclude and replace the toxic ingredients with environmentally friendly substitutes.

Fundamental understanding of braking phenomena seems achievable when the behavior of the friction interface, in particular, the friction film is well explored in terms of chemistry, topography, and adherence. Current research efforts to understand tribological properties of brake friction interface are noteworthy and the results from the researchers play important roles in the development of advanced commercial friction materials. In order to develop a superior brake friction material by overcoming the above mentioned technological difficulties, combining efforts from industries, research institutes, and universities are needed.

Acknowledgments

This work was partly supported by the Korea Science and Engineering Foundation (KOSEF) through the National Research Lab. Program funded by the Ministry of Science and Technology (No. ROA-2007-000-10011-0). Thanks are also given to Dr. M. H. Cho and Mr. K. H. Cho for their help during this study and manuscript preparation.

References

1. M. G. Jacko and S. K. Rhee, *Kirk-Othmer Encyclopedia of Chemical Technology*, 4th ed. Vol. 4, 523 (1992).
2. A. E. Anderson, *ASM Handbook, in: Friction, Lubrication, and Wear Technology*, Vol. 18, 569 (ASM International, Ohio, 1992).
3. P. J. Blau, *ORNL/TM-2001/64* (2001).
4. G. Nicholson, *Facts about Friction* (Gedoran America, Winchester, 1995).
5. D. Chan and G. W. Stachowiak, *Proc. Instn Mech. Engrs,* **218**, Part D: *J. Automotive Engineering*, 953 (2004).
6. G. Crosa and I. J. R. Baumvol, *Ch. 16. Tribology of Polymer Composites used as Friction Materials in Advanced in Composite Tribology*, edited by K. Friedrich (Elsevier, 1993).
7. M. H. Cho, S. J. Kim, D. H. Kim and H. Jang, *Wear*, **258**, 1682 (2005).
8. S. K. Rhee, M. G. Jacko and P. S. Tsang, *SAE Technical Paper* 900004 (1990).
9. M. H. Cho, K. H. Cho, S. J. Kim, D. H. Kim and H. Jang, *Tribology Letters*, **20**, 101 (2005).
10. M. Eriksson, F. Bergman and S. Jacobson, *Wear*, **252**, 26 (2002).

11. P. Filip, Z. Weiss and D. Rafaja, *Wear*, **252**, 189 (2002).
12. W. Österle and I. Urban, *Wear*, **257**, 215 (2004).
13. M. Eriksson, F. Bergman and S. Jacobson, *Wear*, **232**, 163 (1999).
14. EPA: California Proposition 65 (2007).
15. DFG: List of MAK and Values (1996).
16. ILO: Occupational Safety and Health Series No. 64 (1989).
17. K. Midander, I. O. Wallinder and C. Leyfraf, *Environmental Pollution*, **145**, 51 (2007).
18. H. Jang and M. J. Chung, 15th SAE Brake Colloquium and Engineering Display, Tempe, Arizona (1997).
19. A. Knop and W. Scheib, *Chemistry and Application of Phenolic Resins* (Springer-Berlag, 1977).
20. S. J. Kim and H. Jang, *Tribology International*, **33**, 477 (2000).
21. J. Bijwe, Nidhi, N. Mjumdar and B. K. Satapathy, *Wear*, **259**, 1068 (2005).
22. P. V. Gurunath and J. Bijwe, *SAE Technical Papers* 2007-01-3945 (2007); *Wear*, **263**, 1212 (2007).
23. S. J. Kim, M. H. Cho, D. S. Lim and H. Jang, *Wear* **251**, 1484 (2001).
24. B. Briscoe, I. Ramirez and P. J. Tweedale, *IMechE* **C452/88**, 15 (1988).
25. T. Kato and A. Magario, *Tribology Transactions*, **37**, 559 (1994).
26. P. Gopal, L. R. Dharani and F. D. Blum, *Wear*, **193**, 199 (1996).
27. H. Jang, K. Ko, S. J. Kim, R. H. Basch and J. W. Fash, *Wear*, **256**, 406 (2004).
28. J. W. Robinson, G. E. Mogensen, K. D. Packard and J. Herman, *Automotive Engineering*, **98**, 47 (1990).
29. F. Dong, F. D. Blum and L. R. Dharani, *Polymers and Polymer Composites*, **4**, 155 (1996).
30. P. Gopal, L. R. Dharani and F. D. Blum, *Wear*, **174**, 119 (1994).
31. R. E. Evans and J. A. Trainor, 15th SAE Brake Colloquium and Engineering Display, Tempe, Arizona (1997).
32. U.S Patent 5106887.
33. B. K. Satapathy and J. Bijwe, *Wear*, **257**, 573 (2004).
34. J. P. Giltrow, *ASLE Transactions*, **16**, 83 (1972).
35. C. Blanco, J. Bermejo, H. Marsh and R. Menendez, *Wear*, **213**, 1 (1997).
36. M. Sloan, L. Savage, K. Evans and B. Hooper, *SAE Technical Paper* 2006-01-3187 (2006).
37. A. Anderson, *Proceedings of Fibers in Friction Materials Symposium*, The Asbestos Institute, 2 (1987).
38. U.S. Patent 5004497.
39. Y. C. Kim, M. H. Cho, S. J. Kim and H. Jang, *Wear*, **264**, 204 (2008).
40. L. Gudmand-Høyer, A. Bach, G. T. Nielsen and P. Morgan, *Wear*, **232**, 168 (1999).
41. B. Melcher and P. Faullant, *SAE Technical Papers* 2000-01-2757 (2000).
42. M. E. Campbell, J. B. Loser and E. Sneegas, *Solid Lubricants*, NASA SP-5059 (1966).

43. M. H. Cho, J. J. Lee, S. J. Kim and H. Jang, Wear, **260**, 855 (2006).
44. S. J. Kim, M. H. Cho, K. H. Cho and H. Jang, *Tribology International*, **40**, 15 (2007).
45. F. P. Bowden and D. Tabor, *The Friction and Lubrication of Solids* (Oxford, 1964).
46. H. Jang and S. J. Kim, *Wear*, **239**, 229 (2000).
47. K. H. Cho, H. Jang, Y. S. Hong, S. J. Kim, R. H. Basch and J. W. Fash, *Wear*, **264**, 291 (2008).
48. S. J. Kim. M. H. Cho, R. H. Basch, J. W. Fash and H. Jang, *Tribology Letters*, **17**, 655 (2004).
49. H. Jang, J. S. Lee and J. W. Fash, *Wear*, **251**, 1477 (2001).
50. S. J. Kim, K. S. Kim and H. Jang, *J. of Materials Processing Technology*, **136**, 202, (2003).
51. M. Eriksson, J. Lord and S. Jacobson, *Wear*, **249**, 272 (2001).
52. M. Eriksson and S. Jacobson, *Proc. Instn. Mech. Engrs*, **215**, 1245 (2001).
53. M. Eriksson and S. Jacobson, *Tribology International*, **33**, 817 (2000).
54. M. G. Jacko, P. H. S. Tsang and S. K. Rhee, *Wear*, **133**, 23 (1989).
55. A. Wirth, D. Eggleston and R. Whitaker, *Wear*, **179**, 75 (1994).
56. W. Österle, M. Griepentrog, Th. Gross and I. Urban, *Wear*, **251**, 1469 (2001).
57. R. P. Nolan and A. M. Langer, *Proceedings of Fibers in Friction Materials Symposium*, The Asbestos Institute, 75 (1987).
58. C. C. Engberg, The Regulation and manufacture of Brake Pads: The Feasibility of reformation to Reduce the Copper Load to the San Francisco Bay, Stanford University (1995).
59. Copper Action Plan Report, City of Palo Alto Regional Water Quality Control Plant (2007).
60. M. H. Cho, K. H. Cho and H. Jang, *The Morphological Effects of Potassium Titanate on the Brake Performance of Friction Materials*, in preparation (2007).
61. S. Jacobson, *Proceedings of International Workshop on Advances in Asbestos Free Friction Composites*, edited by J. Bijwe, IIT, 90 (2006).
62. SAE International, *SAE Surface Vehicle brake Systems Standard Manual H-24* (1998).
63. P. H. S. Tsang, M.G. Jacko, and S. K. Rhee, *Wear*, **103**, 217 (1985).
64. T. K. Sohn, *Proceedings of KSAE Spring Meeting*, KSAE06-F0137, 881 (2006)
65. P.G. Sanders, T.M. Dalka, and R.H. Basch, *Tribology International*, **34**, 609 (2001).

CHAPTER 16

STUDY OF TRIBOLOGICAL AND MECHANICAL PROPERTIES OF MOLD-IN-COLOR POLYPROPYLENE USED IN AUTOMOBILE INDUSTRY

Weidian Shen,[1] Eva Montalvo,[1] Rene Crombez[1] and Rose Ryntz[2]

[1]*Surface Science and Nano-Tribology Laboratory*
Eastern Michigan University
Ypsilanti, MI 48197, USA
[2]*International Automotive Components, North America*
Dearborn, MI 48126, USA
E-mail: wshen@emich.edu

Molded polypropylene is increasingly used as the interior and exterior materials in the automobile industry to replace metals and other materials, due to their much desired properties. To optimize and further expand the applications, correct characterization of their tribological and mechanical properties is important. A single-probe test technique is applied to test the tribological properties of four polypropylene samples, products of Advanced Composites Inc., in the present study, using a Nano-Indenter for controlled damaging and using a scanning probe microscope (SPM) to examine the damages, as well as test their mechanical properties. The study shows the technique is capable of distinguishing the samples' performances. A new method to quantitatively and comprehensively characterize the surface damage is also introduced in the study, *i.e.* the measurement of surface area increase.

1. Introduction

Polypropylene (PP) is the most commonly used thermoplastic olefins (TPO) in commercial applications. PP is a semi-crystalline polymer, and the properties of it can be significantly enhanced by the addition of fillers and reinforcements. Due to its low cost, light weight, easy processing, excellent chemical resistance and good recyclability, injection molded PP are widely used as housing parts for domestic appliances, kitchen utensils, storage boxes, toys, food containers, etc. Since the late 1980s, PP has been increasingly used in the automobile industry as the interior and exterior materials to replace metals and other materials. It started with the molded PP parts coated with polymeric color paints. To simplify the procedure and further reduce the costs, the mold-in-color PP is replacing the painted counterparts, thus eliminating the use of the paints and making the products even more environmentally friendly and less costly. To optimize and further expand the applications of mold-in-color polypropylenes, correct characterizing, clear understanding, and further improvement of their tribological and mechanical properties are critically important.

With the development of the nanotechnology, nano instruments and nano techniques are increasingly used to characterize the materials' properties. Single-probe testing techniques were developed in the middle of 1990s, and are widely used now in studying tribological and mechanical properties of polymeric materials and coatings. Researchers used atomic force microscope (AFM),[1-5] commercial nano indenters/scratchers that became available in the late 1990s,[6-13] home-made devices and other instruments[14-22] to scratch the surfaces of polymeric materials and coatings under well-controlled conditions, and then used AFM, scanning electron microscope (SEM), optical instruments and others to examine the damaged surfaces. Mar/scratch resistance behavior, one of the most important tribological properties of materials and coatings in many applications, were studied from a variety of approaches. Some measured the dimensions of the mars and scratches, and quantified materials' responses to scratching stress, *i.e.*, elastic recovery, plastic deformation and abrasive wear (mass loss);[2,4,10] Some classified the scratch damages into two categories, plastic flow and fracture, and used the depth of plastic flow under a fixed normal load, *e.g.*, 5 mN, and the critical force, at which the plastic flow transits to fracture to characterized the materials' and coatings' resistance against light and severe damages, respectively;[7] Some used two critical forces, instead of one, to characterize the first fracture occurrence and severe fractures occurrence;[23] Some classified scratch damage to more different modes and used a series of critical forces to characterize the resistance to each damage mode;[24] Some used critical depth, instead of critical force, to characterize the fracture occurrence;[13] Some thought the critical strain was a better measure of the fracture occurrence;[11] Some established a map to determine the damage modes, depending on the applied normal load and the nominal contact strain;[14,15,25]. They all met with success in some extent for the studied materials and coatings.

Unprocessed polypropylene (PP) is a brittle and fragile polymeric material. The tribological properties of PP and enhancement of it with additives have been studied by quite a few groups. They applied forces, from milli-newtons up to several dozen newtons, to the tips with various sharpness to scratch the surfaces of polypropylene samples, and used SEM, light reflection, grey level analysis, etc. to examine the damages.[16,18,19,26-30] Effort was made to search the correlation between tribological properties and mechanical properties of the material. Due to the complexity of the material, no consistent and conclusive finding by all the researchers has yet been reached.

Based on the experience gained in the above-mentioned work, a Nano-Indenter XP, made by MTS, equipped with a conical-shaped diamond tip was used to scratch the surfaces of mold-in-color polypropylene samples under progressive loads as well as constant loads, and a scanning probe microscope (SPM) was used to examine the morphologies of the damages, thus studying the mar/scratch resistance behavior of the materials in the present work. The study showed the presented technique is capable of identifying the performances of different polypropylene samples, even distinguishing their slight differences. To better characterize the damages quantitatively and comprehensively, and make the characterization consistent to visual inspection, a new method, measurement of surface area increase, was proposed.

The same Indenter, equipped with a diamond Berkovich tip, was also used to carry out the indentation tests with the increasing multiple maximum loads on the sample surfaces, measuring the elastic modulus and hardness under different loads, *i.e.*, at different indentation depths.

The correlation between tribological and mechanical properties was carefully examined.

2. Tested Samples

The tested four mold-in-color polypropylene samples are products of Advanced Composites, Inc. for the automotive industry. The first one is coded as ATX832N in the company, and is described as talc-filled and impact modified polypropylene. Its ingredients include natural rubber, talc, scratch additive, and others, and colored at press. The second one is coded as ATX880N-1 in the company, and is described as talc-filled polypropylene. Its ingredients are similar to ATX832N's, but it has no natural rubber added. The third one and the fourth one are the same material, but molded at different times. They are coded as ADX-5017 in the company, and are described as talc-filled and impact modified polypropylene. Their ingredients include natural rubber, talc, scratch additive, and others, and they were colored at press. For simplification, they are labeled as AC1, AC2, AC3, and AC4 in the present study, respectively, as shown below.

AC1: ATX832N, Talc-filled, Impact Modified PP
(PP, Natural Rubber, Talc, Scratch Additive, Colored at Press)

AC2: ATN880N-1, Talc-filled PP
(PP, Talc, Scratch Additive, Colored at Press)

AC3: ADX-5017, Talc-filled, Impact Modified PP
(PP, Natural Rubber, Talc, Scratch Additive, Colored at Press)

AC4: Same as AC3, but molded at different time

Compared with the polymeric clearcoats and some other polymer materials, the surfaces of mold-in-color polypropylene are pretty rough due to the incompatibility of the fillers with the semi-crystal structure of PP. Figure 1 is 100 μm by 100 μm topographic images of the four samples with a z brightness scale of 1.5 μm, showing their rough surfaces.

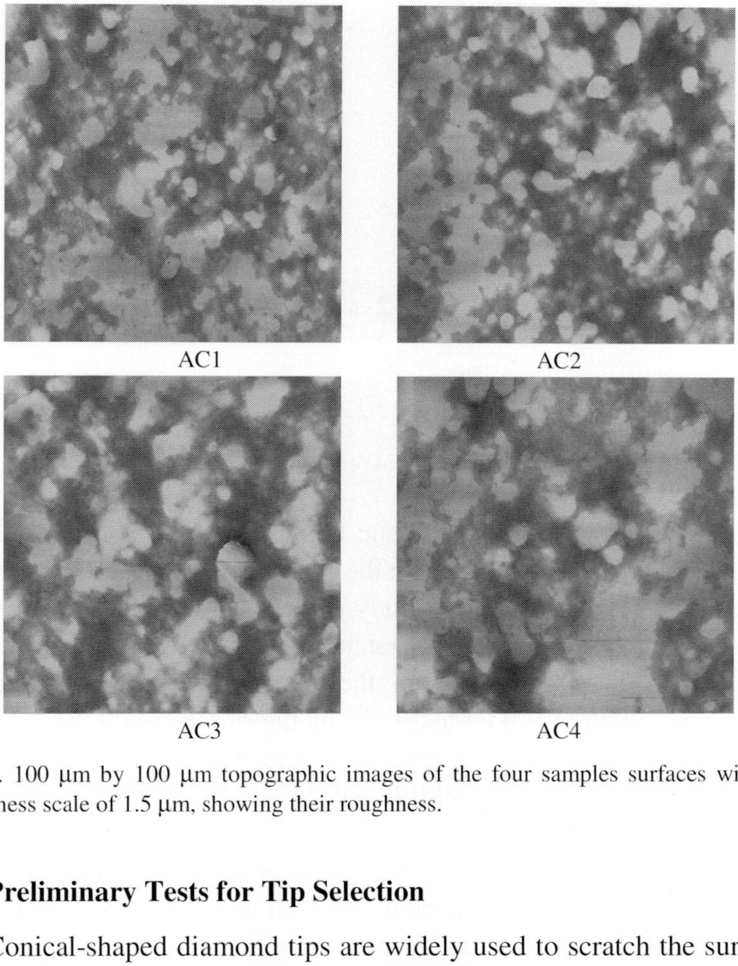

Fig. 1. 100 μm by 100 μm topographic images of the four samples surfaces with a z brightness scale of 1.5 μm, showing their roughness.

3. Preliminary Tests for Tip Selection

90^0 Conical-shaped diamond tips are widely used to scratch the surfaces in the tribological studies due to their symmetrical shape. The practical shape of the tips is not exactly converged to a geometric point, but rounded with a curvature at its apex, as shown in Fig. 2. Using geometry, the height h can be found to be $(1 - (\sqrt{2}/2)R)$, *i.e.*, 0.293R, where R is the radius of the apex. When the scratch depth D is smaller than h, only the spherical part of the tip is involved in the scratching. As D exceeds h, the conical part begins to be involved, too. The tip with a radius of one micron or so is often used in the micro characterization of tribological properties of materials. However, such a tip cracks the surfaces of polypropylene as soon as it indents the surface, even before

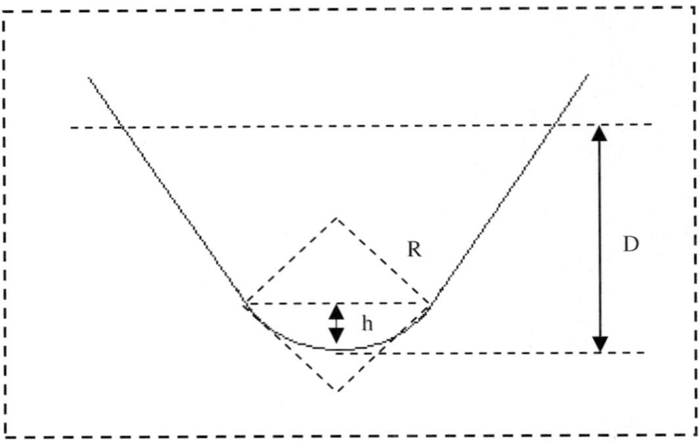

Fig. 2. Practical 90^0 conical-shaped tip with a rounded apex of radius R.

scratching, due to the brittleness and fragileness of polypropylene, thus making it impossible to distinguish the behaviors of different samples.

To select a proper tip for the study, the tips with a radius at its apex of 2, 5, and 20 μm were used to scratch the surface of samples under an increasing load. As an illustration, the scratches made by the three tips under an increasing force up to 30 mN on the surface of the AC2 sample are shown in Fig. 3. As predicted by the map established by Briscoe and his coworkers in studying PMMA, PTFE and UHMWPE, the larger loads and higher nominal strain, *i.e.*, shaper tip will result in more severe damages, such as fracture and cutting. The principle is applicable to PP, too. As shown in Fig. 3, a scratch appeared and cracks occurred at the very beginning in the 2 μm tip case. The scratch in the 5 μm case appeared a little bit later, and the cracks occurred immediately with the scratch. To better distinguish the behaviors of different polypropylene samples, a 20 μm tip was selected and used in the study.

4. Progressive Load Scratch Tests

Progressive load scratch tests were carried out on the four samples. Each run consisted of three steps. First, the tip under a light load of 10 μN scanned the surface to be tested over a distance of 420 μm to collect the

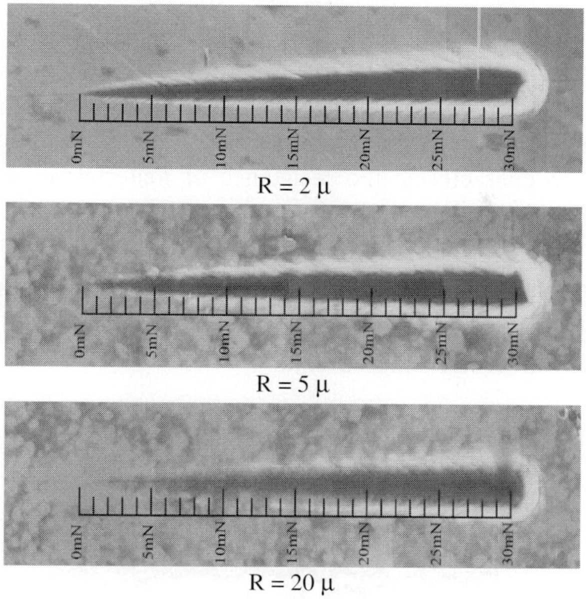

Fig. 3. The scratches, from top to bottom, made by three tips of a radius 2, 5, and 20 μm at its apex, respectively, under an increasing load up to 30 mN on the surface of AC.

topographic information of the surface, which would be used to calculate the real-time scratch depth during the scratching and residual depth of the ditch after the scratching. 10 μN was chosen for obtaining a contour of the original surface without damaging it. Then, a progressive load, from 0 to 30 mN, was used to scratch the surface over a distance of 300 μ in the middle of the pre-scan path of 420 μm at a speed of 10 μm/s. It was followed by a post-scan under a load of 10 μN again to obtain the residual depth of the ditch. The difference between the real-time scratching depth and the residual depth of the ditch indicates the immediate recovery of the scratched surface. During the scratching, the frictional force encountered by the scratching tip was sensed and recorded. The fluctuation in the frictional force was mainly caused by the rough features on the surface. Figure 4 is an illustration of the progressive load test. Three runs were performed on the surface of each sample. Scanning probe microscope (SPM) was used to examine the morphologies of the scratches and take images along the scratches. The measurements were well repeatable and reproducible. Figure 5 shows the results of the tests on the four samples.

All the scratches appeared pretty early, but the cracks occurred under the different critical loads. Onset, under which the cracks begin to occur, can be used to characterize the resistance of material against cracking. They were 4, 3, 11, and 11 mN, for AC1, AC2, AC3, and AC4, respectively, and are shown in Fig. 6. Group of AC3 and AC4 possesses higher onset values than group of AC1 and AC2, while the onset of AC1 is a little higher than AC2's, which may be attributable to the existence of the natural rubber in its ingredients.

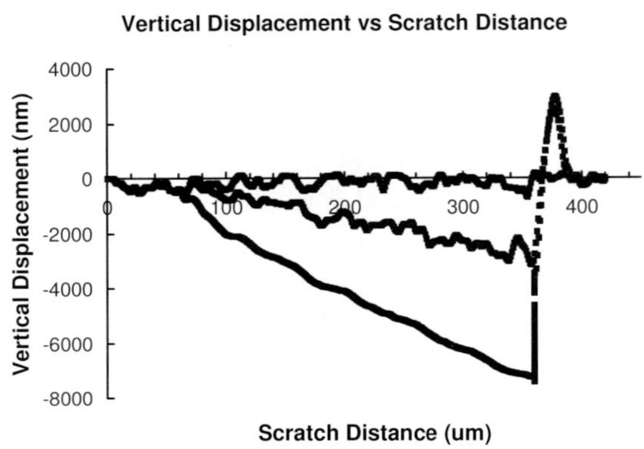

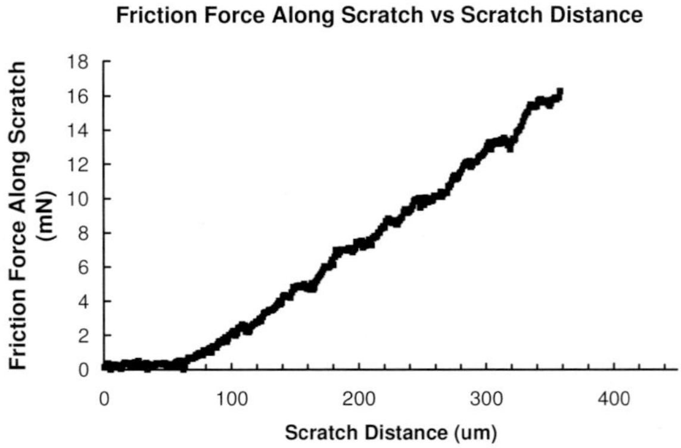

Fig. 4. A typical progressive load scratching test. The three curves in the upper figure are, from top to bottom, pre-scan, post scan, and real-time scratching depth under the progressive load, respectively. The lower figure shows the friction force encountered in the scratching.

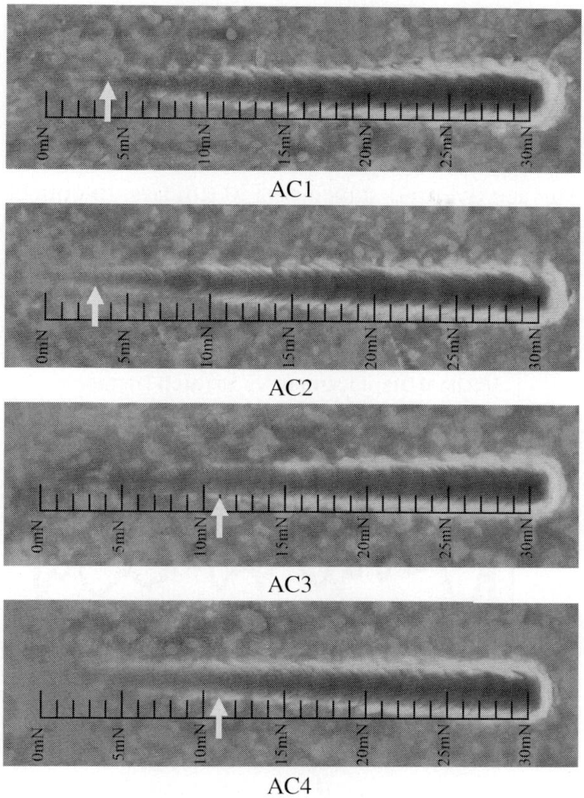

Fig. 5. The four samples possessed different onset values, under which the cracking began.

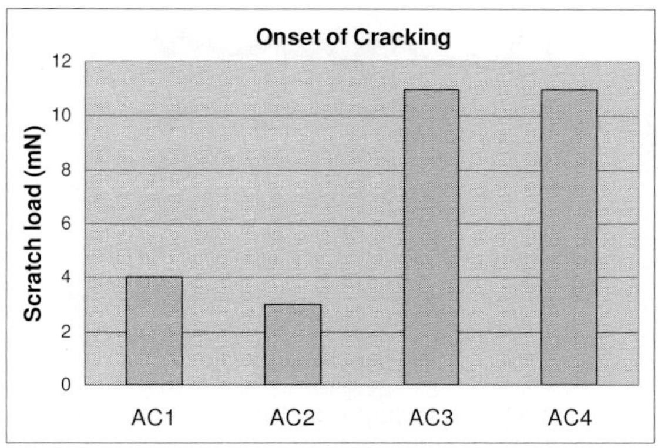

Fig. 6. The measured onset values of cracking of AC1, AC2, AC3, and AC4, were 4, 3, 11, and 11 mN, respectively.

5. Constant Load Scratch Tests

Constant load scratch tests, under selected loads of 8, 14, 20 and 26 mN, were carried out on the four samples. Each run consisted of three steps, similar to the progressive load tests. The tip under a light load of 10 μN scanned the surface over a distance of 330 μm first to collect topographic information of the original surface. Then, a constant load was used to scratch the surface over a distance of 300 μm in the middle of the pre-scan path 330 μm at a speed of 10 μm/s. It was followed by a post-scan

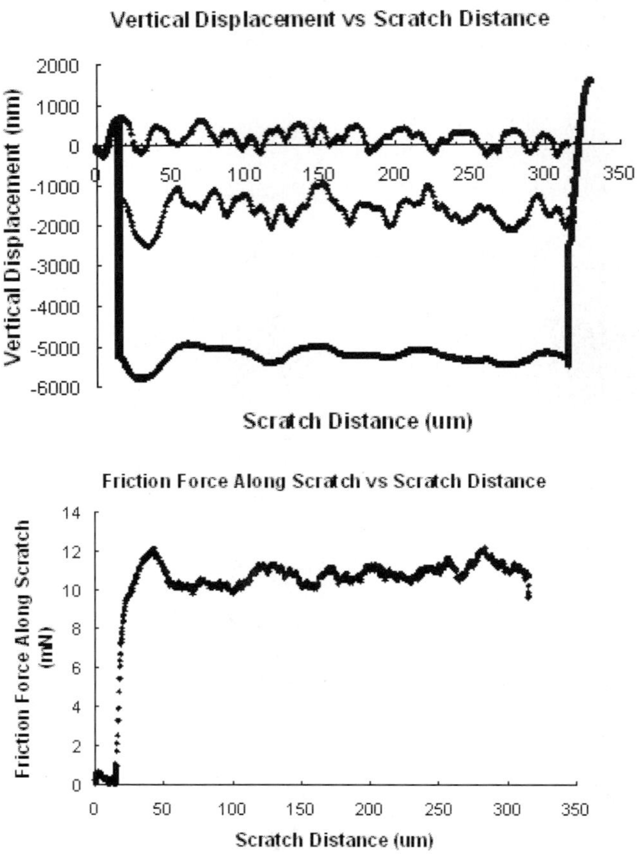

Fig. 7. A typical constant load scratching test. The three curves in the upper figure are, from top to bottom, pre-scan, post scan, and real-time scratching depth under the constant load, respectively. The lower figure shows the friction force encountered in the scratching.

under a load of 10 μN to obtain the residual depth of the ditch. The frictional force encountered by the scratching tip was sensed and recorded. Figure 7 is an illustration of the constant load test. Three runs under each load were performed on the surface of each sample. Scanning probe microscope (SPM) was used to take images of the scratches, and measure the dimensions of them. The measurements were well repeatable and reproducible. Figure 8 are the scratches made on the surface of AC2, as an example, showing the scratches with cracks on both sides which became deeper and wider under the incremental loads.

The number of cracks along the scratches made under the loads of 20 mN and 26 mN on four samples were counted. The average numbers of cracks per 100 μm on the surface of AC1, AC2, AC3, and AC4 were 25, 26, 31, and 32, respectively, see Fig. 9. Although the onset values of group AC3 and AC4 are higher than group AC1 and AC2, their crack densities are higher than AC1 and AC2 after cracking occurred.

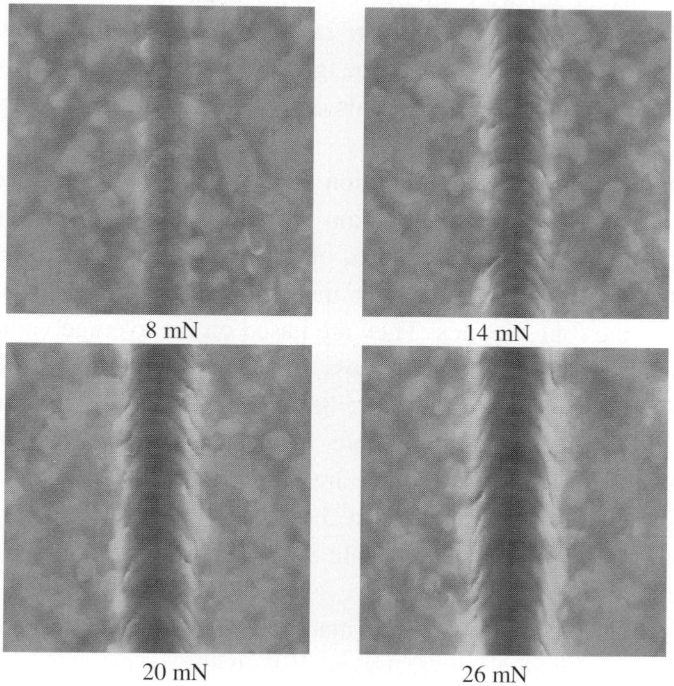

Fig. 8. Scratches made under different constant loads, 8 mN, 14 mN, 20 mN, and 26 mN, respectively, on the surface of AC2.

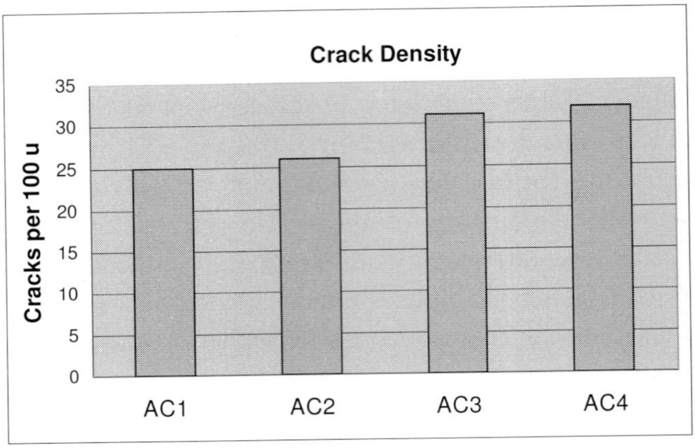

Fig. 9. The average numbers of cracks per 100 μm of scratches on the surfaces of AC1, AC2, AC3, and AC4, made under 20 mN and 26 mN, were 25, 26, 31, and 32, respectively.

The analysis software of SPM can be used to measure the cross-sectional dimensions of the ditch, *i.e.*, the part of scratch below the original surface, and is called the groove in some publications, and shoulders, *i.e.*, the piled-up materials on both sides of ditch, if any, above the original surface.

Figure 10 shows the cross-section area of ditches, represented by the blue bars, and cross- cross-section area of ditches, represented by the blue bars, and cross-section area of the two shoulders, represented by the purple bars, of the scratches made under different loads on the surfaces of the four samples. They are based on the average values along the scratches. Due to the roughness of the surfaces, the cross-section profiles varied from one line to the next, which made the measurements difficult and the results less reliable. Over all, the cross-section area of ditch, as well as the cross-section area of shoulders, increased with the loads. The area of shoulders of AC3 and AC4 were less than group of AC1 and AC2, which might indicate AC3 and AC4 were less plastically deformed.

To have some insight of the materials' response to the scratching stress, the immediate elastic recovery (IER) was estimated, using the data collected by the Indenter during the scratch test. Set D_{rl} is the real-time

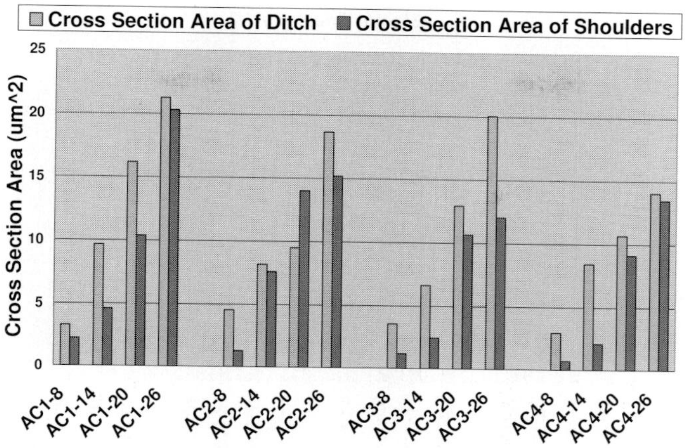

Fig. 10. The cross-section area of ditches, represented by the blue bars, and cross-section area of the two shoulders, represented by the purple bars, of the scratches made under different loads of 8, 14, 20, and 26 mN on the surfaces of the four samples, AC1, AC2, AC3, and AC4.

depth during the scratching and D_{rs} is residual depth of the ditch measured in the post-scan immediately after scratching, and IER can be estimated as

$$\text{IER} = \frac{(D_{rl} - D_{rs})}{D_{rl}} \times 100\%$$

Since the residual depth of the ditch was measured only one or two minutes after the scratching, the difference between the two depths reflects the immediate recovery. After it, the recovery continued via viscocreep, *i.e.*, viscoelastic recovery. The depth of ditch measured after 24 hours or longer by SPM is usually less than the residual depth measured immediately after scratching. The total elastic recovery (TER), including the viscoelastic recovery, was calculated by

$$\text{TER} = \frac{D_{rl} - D_{fl}}{D_{rl}} \times 100\%,$$

where D_{fl} is the depth of the ditch measured by SPM 24 hours or longer after the scratching. The IER and TER, based on the above described one-dimensional measurements and calculations, are shown in Fig. 11.

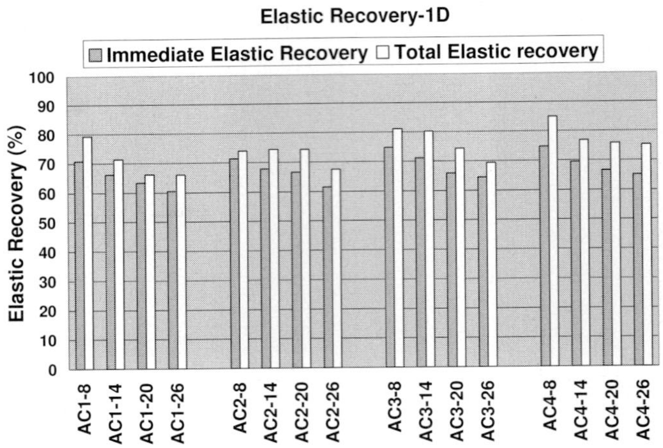

Fig. 11. The immediate elastic recovery, represented by the blue bars, and total elastic recovery, represented by the yellow bars, of four samples, AC1, AC2, AC3, and AC4, under the four constant loads, 8, 14, 20, 26 mN, based on the one-dimensional measurements and calculations.

Although polypropylene is fairly brittle, it recovers a lot elastically after scratching. The recoveries of AC3 and AC4 were a little bit more than AC1 and AC2. Over all, the immediate recoveries, as well as the total recoveries, decreased with the increasing loads.

6. Quantitative Characterization of Scratch Damages

Surface damage is evaluated by the vision of the users. The human eye is an optical device, thus using optical measurements to characterize damage is a common practice, such as light reflection, gross retention, etc. However, the results of these measurements depend on the position of the light source and the position of the detector. The users may view the surface in all directions, thus a specific measurement setup is not comprehensive. To better characterize surface damages, a new method is proposed in this study. It is surface area increase measurement.

As shown above, a scratch usually consists of a ditch and two shoulders, if any, on both sides of the ditch with or without cracks. All the ditches, shoulders, cracks, etc. make their contributions to the visibility of damage in different weights. One thing is common for all of them, i.e. their occurrence increases the surface area. Deeper ditches, higher shoulders, and more severe cracks will increase surface area more, thus making the damage more severe and visible. Using the surface area increase can characterize the damage quantitatively and comprehensively, and it is consistent with the vision.

Figure 12 is a 100 μm by 100 μm topographic image, consisting of 512 by 512 pixels. If the surface was ideally flat and smooth, i.e., all the 262,144 pixels lay on the same plane, the surface area would be 10,000 μ^2. Due to the roughness and damages, all the pixels have different vertical positions, as well as horizontal positions. Based on the positions of 262,144 pixels, analysis software calculates the actual surface area. Compared with the projected area, surface area increase, due to the roughness and damage, is obtained.

A box is picked along the scratch with a width of 110% of the width of the scratch, as shown in the middle of Fig. 12, and the surface area increase of the box is obtained in percentage format, compared with the projected area of the box. It is used to characterize the damage of the scratches made under different loads on the surfaces of four samples quantitatively. The results are shown as blue bars in Fig. 13. The surface area increases of AC3 and AC4 were less than AC1 and AC2. Over all, the area increased more and more as the loads increased and the damages became more severe.

Considering the fact, the same scratch on a rough surface will be less visible than that on a smooth surface. A relative surface area increase (RSAI) can be introduced, which is defined as

$$\mathrm{RSAI} = \frac{SAI_{scr} - SAI_{org}}{SAI_{org}},$$

where $\mathrm{SAI}_{\mathrm{scr}}$ is the surface area increase measured in the box with scratch in it and $\mathrm{SAI}_{\mathrm{org}}$ is the average surface area increase measured in the eight boxes, left and right to the four scratches made under the different loads, as shown in Fig. 12, due to the original surface roughness. The calculated RSAI are plotted in Fig. 14. Since the original surface of AC4 was less rough, compared with the other three samples, its RSAI turned out larger than the others. It is the payoff of the smooth surface; it looks nice and glossy, however, any damage on it is easily noticed.

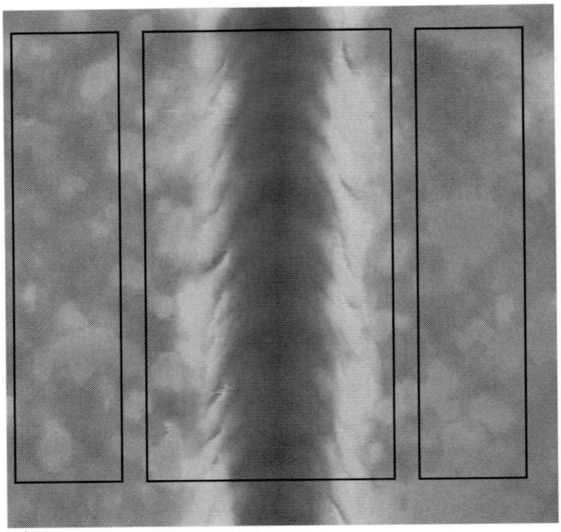

Fig. 12. A 100 μm by 100 μm topographic image, consisted of 262,144 pixels, has a projected area of 10,000 $(\mu m)^2$. The middle box with a width of 110% of the width of the scratch was used to calculate the surface area increase by the scratch, and the left and right boxes were used to calculate the surface area increase by the original roughness.

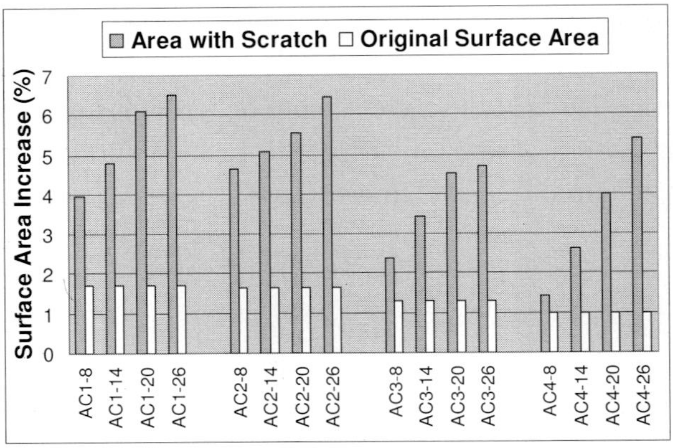

Fig. 13. The measured surface area increases by the scratches made under the four different loads, 8, 14, 20, and 26 mN, on the surfaces of the four samples, AC1, AC2, AC3, and AC4.

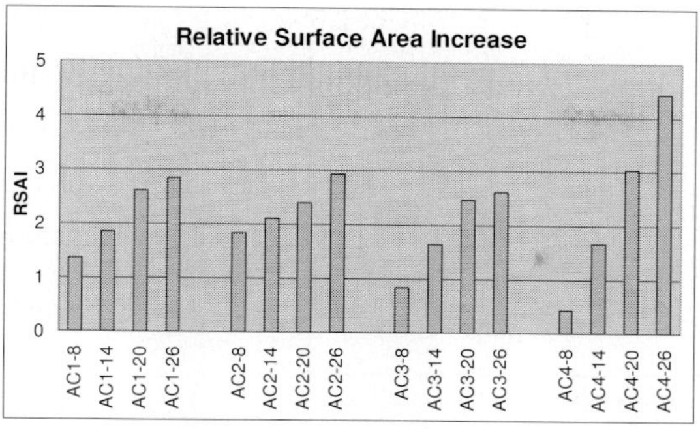

Fig. 14. Relative surface area increases (RSAI) by the scratches made under the four different loads, 8, 14, 20, and 26 mN, on the surfaces of the four samples, AC1, AC2, AC3, and AC4.

7. Indentation Tests

Indentation test is a reliable technique to measure the mechanical properties of materials, such as elastic modulus and hardness, and is used widely. A diamond Berkovich tip was used in the Indenter to perform the tests. A typical run of the test consists of three steps, as shown in the left in Fig. 15. In the loading process, the load on the tip is increased to a pre-determined maximum load in 15 seconds, followed by holding at the maximum load for 30 seconds to check any creep, and then load decreased to zero in 15 seconds. The elastic modulus and hardness were calculated by

$$E = \frac{S\sqrt{\pi}}{2\beta\sqrt{A_c}} \quad \text{and} \quad H = \frac{F_{max}}{A_c},$$

where S is the slope of the tangential line of the unloading curve at the turning point; β is a correction term based on the tip's geometry, and it equals to 1.034 in this case; A_c is the projected contact area on the original surface, determined by the contact depth h_c. Due to the elasticity of the surface, the configuration of contact is usually as shown in the right in Fig. 15. The contact depth h_c is less than the indentation depth under the maximum load h_{max}, and is calculated by

$$h_c = h_{max} - \varepsilon \frac{F_{max}}{S},$$

where $\varepsilon = 0.75$ for Berkovich tip.

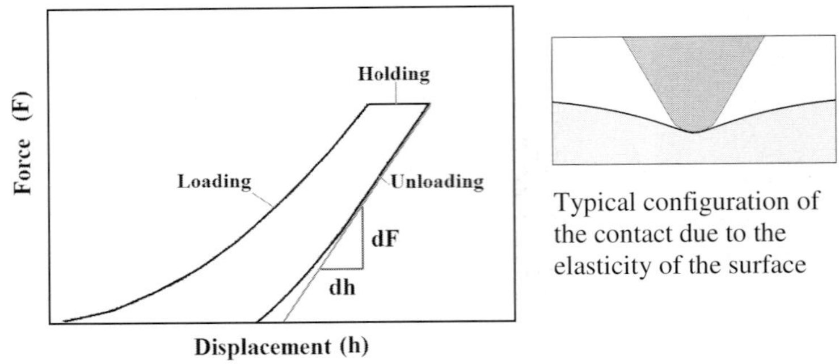

Fig. 15. Typical indentation test (left) and the configuration of the contact of the tip and the samples surface (right).

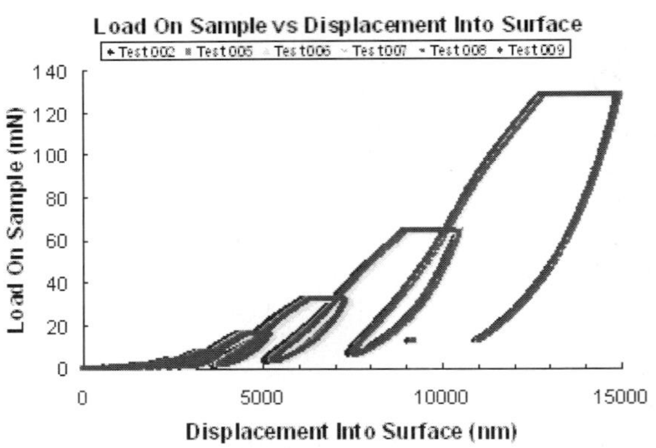

Fig. 16. Indentation test with incremental multiple maximum loads at each site of a total of nine sites with a spacing 1000 μm apart in a 2 mm by 2 mm area on the surface of each sample.

In the preliminary tests, all four samples showed large creep. To secure reliable and repeatable results, the holding time was extended to five minutes in the study. Each sample was indented at nine sites with 1000 μm spacing to measure their modulus and hardness in a 2 mm by 2 mm area. Each site was indented multiple times with an incremental maximum load, starting from 1.0 mN up to 128 mN, doubling the load each time, as shown in Fig. 16. The average modulus and hardness with its standard deviation bars are plotted versus the indentation depth, as shown in Fig. 17. Due to the roughness and inhomogeneity of the surfaces, the deviations of the results, especially under the lighter loads,

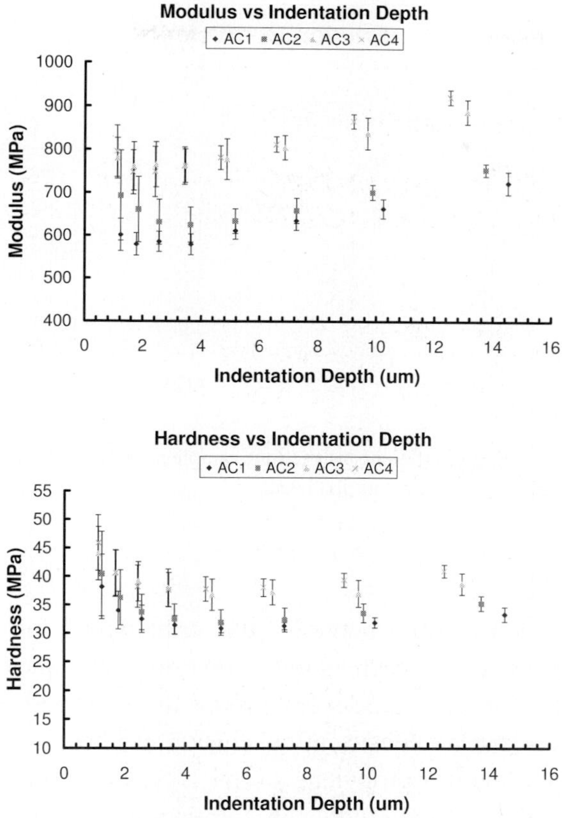

Fig. 17. Measured average modulus (upper) and hardness (lower) of the four samples under the eight different loads.

i.e., shallow surface, were pretty large. AC3 and AC4 possess higher modulus and hardness than AC1 and AC2. AC1's is a little bit lower than AC2's, which may be attributable to its natural rubber ingredient.

As for the friction coefficients of the surfaces of the four samples, measurements in the progressive load tests and measurements in the constant load tests were very consistent. The measured coefficients of the friction of AC1, AC2, AC3, and AC4 were 0.51, 0.53, 0.45, and 0.43, respectively, as shown in Fig. 18. Group AC3 and AC4 showed less friction than group AC1 and AC2.

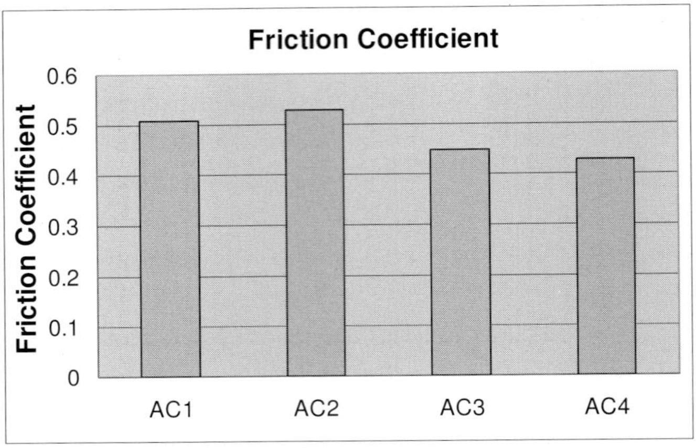

Fig. 18. Measured friction coefficients of the four samples, AC1, AC2, AC3, and AC4, in the progressive load tests and constant load tests.

8. Correlation

Searching the correlation between the tribological properties and mechanical properties has been an interesting research topic for years. As mentioned in the introduction, several groups have pursued the goal from different aspects, but no consistent conclusion has been reached, due to the complexity of the polypropylene's semi-crystal structure with various fillers in it. In the present study, the correlation between the onset of cracking, crack density, elastic modulus, and frictional coefficients, correlation of the ratio of elastic modulus to the hardness, E/H, cross-section area of ditch, cross-section area of shoulders, and surface area increase, *etc.* have been examined carefully, but no clear pattern can be identified at this point. It will be a topic for further studies.

9. Summary

We have applied the Single-Probe technique, using a Nano-Indenter XP and a Scanning Probe Microscope (SPM), to test four mold-in-color polypropylene samples used in the automotive industry. The progressive load tests and constant loads tests were carried out to measure their tribological properties and the indentation tests were carried out to measure their mechanical properties. Although all the samples, products of Advanced Composites, Inc., were very similar, their performances could be distinguished clearly, thus showing the presented characterization technique is a valid one.

1. AC3 and AC4 were better than AC1 and AC2 in the resistance against the cracking occurrence, possessing higher onset values. However, after cracking occurred, their crack densities were a little bit higher than AC1 and AC2.
2. AC3 and AC4 plastically deformed less than AC1 and AC2 in the scratching with less cross-section area of the shoulders, and elastically recovered more than AC1 and AC2 after scratching.
3. A new method to characterize the surface damage quantitatively and comprehensively is introduced, *i.e.*, surface area increase by the scratching. The surface area increases of AC3 and AC4 were less than AC1 and AC2. Due to the relative smoothness of the surface of AC4, its relative surface area increase turned out to be the largest.
4. Elastic modulus E and hardness H of the four samples were measured in the indentation tests. E and H of AC3 and AC4 were higher than AC1 and AC2.
5. AC1 and AC2 are very similar polypropylene material except that AC1 has a natural rubber ingredient. The tests showed AC1 has a slightly higher onset value, a slightly lower crack density, and slightly lower elastic modulus and hardness, due to the natural rubber. AC3 and AC4 had the same composites, but were processed differently. The tests showed their very close but slightly different performances, too.
6. Correlation between the tribological and mechanical properties has not been conclusively identified yet, due to the complexity of polypropylene structure. It will be further studied.

Acknowledgments

Support of this work by the National Science Foundation Industry/University Cooperative Research Center in Coatings at Eastern Michigan University (NSF I/U CRCC at EMU) is greatly appreciated.

References

1. W. Shen, C. Ji, F. N. Jones, M. P. Everson and R. A. Ryntz, *Polym. Mater. Sci. Eng.*, **74**, 346 (1996).
2. F. N. Jones, W. Shen, S. M. Smith, Zh. Huang and R. A. Ryntz, *Progress in Organic Coatings*, **34(1-4)**, 119 (1998).
3. R. A. Ryntz, B. D. Abell, G. M. Pollano, L. H. Nguyen and W. Shen, *J. Coat. Tech.*, **72(904)**, 47 (2000).
4. W. Shen, B. Jiang and F. N. Jones, *J. Coat. Tech.*, **72(907)**, 89 (2000).
5. B. Du, M. R. VanLandingham, Q. Zhang and T. He, *J. Mater. Res.*, **16(5)** 1487 (2001).
6. V. Jardret, H. Zahouani, J. L. Loubet and T. G. Mathia, *Wear*, **218(1)**, 8 (1998).
7. L. Lin, G. S. Blackman and R. R. Matheson, *Prog. Org. Coat.*, **40(1-4)**, 85 (2000).
8. P. Bertrand-Lambotte, J. L. Loubet, C. Verpy and S. Pavan, *Thin Solid Films* **398**, 306 (2001).
9. V. D. Jardret and P. Morel, in *Polymer Interfaces and Thin Films*, Materials Research Society Symposium Proceedings, **710**, 93 (2002).
10. W. Shen, J. Sun, Zh Liu, W. Mao, J. D. Nordstrom, P. D. Ziemer and F. N. Jones, *JCT Research* **1(2)**, 117 (2004).
11. M. R. VanLandingham, L. P. Sung, N. K. Chang, T. Y. Wu, S. H. Chang and V. Jardret, *JCT Research* **1(4)**, 257 (2004).
12. Janet S. S. Wong, H. J. Sue, K. Y. Zeng, Robert K. Y. Li and Y. W. Mai, *Acta Materialia* **52**, 431 (2004).
13. V. Jardret and R. Ryntz, *JCT Research* **2(8)**, 591 (2005).
14. B. J. Briscoe, E. Pelillo and S. K. Sinha, *Poly. Eng. Sci.*, **36(24)**, 2996 (1996).
15. B. J. Briscoe, P. D. Evans, E. Pelillo and S. K. Sinha, *Wear* **200(1-2)**, 137 (1996).
16. R. S. Kody and D. C. Martin, *Poly. Eng. Sci.*, **36(2)**, 298 (1996).
17. B. J. Briscoe, E. Pelillo and S. K. Sinha, *Poly. Internl.*, **43(4)**, 359 (1997).
18. J. Chu, C. Xiang, H. J. Sue and R. D Hollis, *Poly. Eng. Sci.*, **40(4)**, 944 (2000).
19. C. Xiang, H. J. Sue, J. Chu and B. Coleman, *J. Poly. Sci. Part B – Polymer Physics* **39(1)**, 47 (2001).
20. C. Gauthier, S. Lafaye and R. Schirrer, *Trib. Internl*, **34(7)**, 469 (2001).
21. R. A. Ryntz and D. Britz, *J. Coat. Tech.*, **74(925)**, 77 (2002).
22. A. Krupicka, M. Johansson and A. Hult, *Prog. Org. Coat.*, **46(1)**, 32 (2003).
23. J. L. Courter and E. A. Kamenetzky, *presented at the 5th Nurnberg Congress, Nurnberg, Germany*, April 1999.
24. W. Shen, B. Jiang, S. M. Gasworth and H Mukamal, *Trib. Internl.*, **34(2)**, 135 (2001).
25. S. J. Bull, *Trib. Internl.*, **30(7)**, 491 (1997).
26. M. Wong, G. T. Lim, A. Moyse, J. N. Reddy and H. J. Sue, *Wear* **256**, 1214 (2004).
27. M. Wong, A. Moyse, F. Lee and H.J. Sue, *J. Mater. Sci.*, **39**, 3293 (2004).

28. S. Wu, K. Sehanobish, C. Christenson and J. Newton, *Poly. Meter. Sci. Eng.*, **79**, 212 (1998).
29. H. Tang and D. C. Martin, *J. Mater. Sci.,* **38**, 803 (2003).
30. R. D. K. Misra, R. Hadal and S. J. Duncan, *Acta Materialia* **52**, 4363 (2004).

PART III

Polymer Films

CHAPTER 17

MECHANICAL PROPERTIES OF THIN POLYMER FILMS WITHIN CONTACTS

Antoine Chateauminois, Christian Fretigny and Eric Gacoin

Laboratoire de Physico-chimie des Polymères et des Milieux Dispersés (PPMD)
CNRS UMR 7615, UPMC
Ecole Supérieure de Physique et Chimie Industrielles, ESPCI
10, rue Vauquelin, 75231 Paris Cedex 5, France
E-mail: antoine.chateauminois@espci.fr

This chapter addresses the problem of the mechanical properties of thin polymer films geometrically confined within contacts between elastic substrates. Analytical contact mechanics solutions for coated substrates are used to derive, within the limits of confined contacts situations, an approximate oedometric solution for the indentation of a thin film lying on a more rigid substrate. From a discussion of this approximate model, it is shown that lateral contact methods are an interesting alternative to indentation experiments for the measurement of the visco-elastic properties of polymer films in their glass transition range or rubbery state. As an example, the hydrostatic pressure dependence of the visco-elastic properties of confined polymer films is analyzed in the light of lateral contact stiffness measurements. The effects of hydrostatic pressure on the onset of plastic flow within confined polymer coatings are also discussed.

1. Introduction

Thin polymer films are widely used in many tribological applications where their mechanical properties are of primary importance. Within this context, contact experiments such as nanoindentation have emerged as a promising route for the measurement of the visco-elastic and plastic properties of thin polymer films. However, contacts on coated substrates are often characterized by a high level of geometrical confinement. Such a confinement results in very specific mechanical conditions which are characterized by high strain rates and/or elevated hydrostatic pressures within the coating. Polymers being very sensitive to these parameters, the mechanical response of confined coatings is thus likely to be very different from that of bulk polymer substrates. However, the implications of confinement on the measurement of the mechanical properties of thin films remain largely to establish. Such an issue relies on the development of appropriate coated contact models which can provide insights into the deformation modes of confined films. In the context of indentation, the analysis of experiments usually relies on contact mechanics models (such as the popular Oliver and Pharr model[1]) which were initially developed for isotropic and homogeneous solids. In the case of layered solids, these approaches are no longer valid when the deformation of the substrate comes into play, i.e. when the contact radius is of the order of about one tenth of the film thickness. Accordingly, indentation models are often modified by introducing the concept of an equivalent modulus which incorporates the depth dependent contributions of the film and the substrate. In order to extract the film modulus from this measured composite modulus, various empirical and semi-empirical models have been proposed on the basis of experimental data and finite element simulations. Most of them express the composite modulus as a linear or non linear combination of the modulus of the substrate and of the film[2-6]. Alternatively, analytical approaches have also been developed by Gao *et al.*[7] in order to derive approximate expressions for the contact compliance within the limits of moduli ratios not exceeding a factor of two. These approaches have been successfully applied to nanoindentation experiments where the ratio of the contact radius to the film thickness remains close to unity, i.e. for contact conditions which are characterized by a low geometrical confinement. However, when the indentation depth becomes of the order of magnitude of the film thickness, there is some indications that the above mentioned indentation models can provide erroneous[8] or unexpectedly high[9] values of the layer Young's modulus.

In this chapter, we show that within the limits of confined contact geometries, an approximate solution for the elastic indentation of coated substrates can be derived from analytical contact mechanics solutions. This approximate solution will serve as a basis to discuss the adequacy of indentation experiments to measure the visco-elastic properties of confined polymer films. When such measurements are to be done with polymer coatings in their glass transition range or in the rubbery state, i.e. close to incompressibility, it turns out that lateral contact situations offer an interesting alternative to normal indentation. As an example, it will be shown how lateral contact experiments can be used to determine the hydrostatic pressure dependence of the linear visco-elastic properties of confined polymer films in their glass transition range. The effects of hydrostatic pressure will subsequently be analyzed in the context of plastic contact deformations. The onset of plastic flow in confined films will especially be discussed in the light of the pressure dependence of the yield stress in polymers.

2. Contact Mechanics of Confined Polymer Coatings

2.1. *Oedometric Approximation for the Normal Indentation of Confined Layers*

As reviewed by Fretigny and Chateauminois[10], theoretical contact mechanics solutions for layered substrates have long been derived by integral transforms methods. Starting from the basic equations of elasticity and writing the boundary conditions at the interface between the layers, analytical expressions for the Fourier or Hankel transforms of displacements and stresses can be obtained for a prescribed distribution of the surface stresses.

This approach is applied here to the system depicted in Fig. 1, which consists of a coated elastic substrate in contact with a rigid sphere. E_i and ν_i denote the Young's modulus and the Poisson's ratio of the layer (i=1) and of the substrate (i=0). An expression for the Green function of this coated system has been initially derived by Burminster[11] and rediscovered later by others[12-14]. It relates the Hankel transform of the normal displacement, $u_1(r)$, to that of the applied normal stress, $\sigma(r)$, at the surface of the elastic layer:

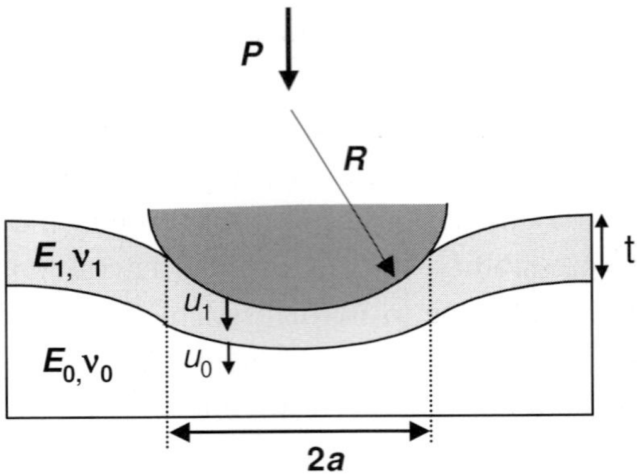

Fig. 1. Schematic description of the elastic indentation of a coated substrate by a rigid sphere. The layer is assumed to be perfectly bound to the substrate and the contact interface is frictionless. E_i and ν_i are the Young's modulus and the Poisson's ratio of the substrate ($i=0$) and the layer ($i=1$).

$$\bar{u}_1(\xi) = \frac{2}{E_1^*}\frac{X(\xi)}{\xi}\bar{\sigma}(\xi) \tag{1}$$

where E_1^* is the reduced modulus of the layer defined as $E_1/(1-\nu_1^2)$ and $X(\xi)$ is defined as:

$$X(\xi) = \frac{1 + 4b\xi t e^{-2\xi t} - abe^{-4\xi t}}{1 - \left(a + b + 4b(\xi t)^2\right)e^{-2\xi t} + abe^{-4\xi t}} \tag{2}$$

$$a = \frac{\alpha\gamma_0 - \gamma_1}{1 + \alpha\gamma_0}, \quad b = \frac{\alpha - 1}{\alpha + \gamma_1}, \quad \alpha = \frac{G_1}{G_0},$$

$$\gamma_1 = 3 - 4\nu_1, \ \gamma_0 = 3 - 4\nu_0$$

where G_0 and G_1 denote the shear moduli of the substrate and the layer, respectively. In equation (1), the quantities in the form of $\bar{q}(\xi)$ correspond to the 0th-order Hankel transform of the function $q(r)$ defined as:

$$\bar{q}(\xi) = \int_0^\infty dr r J_0(\xi r) q(r) \tag{3}$$

with $J_0(x)$ the 0th-order Bessel function of the first kind.

In a recent paper, Perriot and Barthel[15] have developed a semi-analytical methodology which allows the inversion of this Green tensor in the real space. Using this approach, an exact solution is provided to the problem of the determination of the penetration and the contact load. It was also established by Fretigny and Chateaminois[10] that this theoretical model can also be extended to the calculation of the stress and displacement field within the layered substrate. It will be shown below that a simple approximation for the indentation response of a layered substrate can be derived, within the limits of confined contact geometries, from this exact contact mechanics solution.

In the above expressions, the Hankel transform of the fields are equivalent to their two-dimensional Fourier transforms, as they are radially symmetric. Then, as the normal stress is zero out of the contact area, it is expected that its Hankel transform varies at the scale of $1/a$. On another hand, we can deduce from its expression that $X(\xi)$ varies at the scale of $1/t$. Thus, in the case of confined contacts, i.e. $t<<a$, it is presumably sufficient to consider a first order expansion of $X(\xi)$ in order to invert equation (1). Moreover, assuming that the layer modulus is much lower than the substrate one, the expansion can be written:

$$X(\xi) \approx \frac{2}{E_0^*} + \frac{1}{\tilde{E}_1}\xi t \tag{4}$$

where the effective modulus, $\tilde{E}_1$, can be expressed in the following form:

$$\tilde{E}_1 = E_1 \frac{(1-\nu_1)}{[(1-2\nu_1)(1+\nu_1)]} \qquad E_1 << E_0 \tag{5}$$

It can easily be recognized that $\tilde{E}_1$ is in fact the so-called oedometric or longitudinal bulk modulus of the layer, which corresponds to the elastic response of the constrained material under compression or extension, deformation in the lateral dimension being prevented[16,17]. This finding is physically consistent with the expected deformation mode of a layer confined within a contact. Taking into account equation (4), it can be shown that equation (1) reduces to the following expression in the real space[18]:

$$u_1 - u_0 = \frac{t}{\tilde{E}_1}\sigma \tag{6}$$

where u_0 denotes the vertical displacement of the interface. This expression shows that, at each location within the confined contact, the compression of the layer is proportional to the local value of the applied normal stress.

Accordingly, it turns out that the indentation behaviour of the confined contact can be separated into two components (Fig. 2a). The first one corresponds to the œdometric compression of the layer, which acts as a "mattress" of individual springs, each of them having the same compliance $t / \tilde{E}_1$. It can be noted that this picture of the compressive response of the coating is very similar to that used by Johnson[19] in the so-called elastic foundation model for the contact between a sphere and a layer lying on a rigid substrate. Equation (6) can thus be viewed as an extension of this approach to the case of a deformable substrate, where the oedometric nature of the coating's modulus is fully accounted for. The second component of the contact response corresponds to the elastic deformation of the substrate under the action of the contact stress applied over a contact area of radius, *a*.

An indication of the validity of this approximate description of confined contact is shown in Fig. 2b where the normalized indentation load is represented as a function of the ratio, *a*/*t*, which describes the extent of the geometrical confinement of the contact. In addition to the curves corresponding to the above detailed oedometric approximation, calculations using a theoretical exact contact model derived by Perriot and Barthel[15] are also represented in the figure. It can be seen that when *a*/*t* is greater than about ten, the oedometric approximation allows for a very accurate description of the indentation behaviour. On the other hand, a systematic departure from the exact contact mechanics solution is systematically observed for low confinements. Such an insufficiency of the oedometric description in the low confinement regime is expected as the hypothesis of a non expansion of the normal contact stress outside the contact area is obviously no longer valid when the layer thickness is of the order of the contact radius.

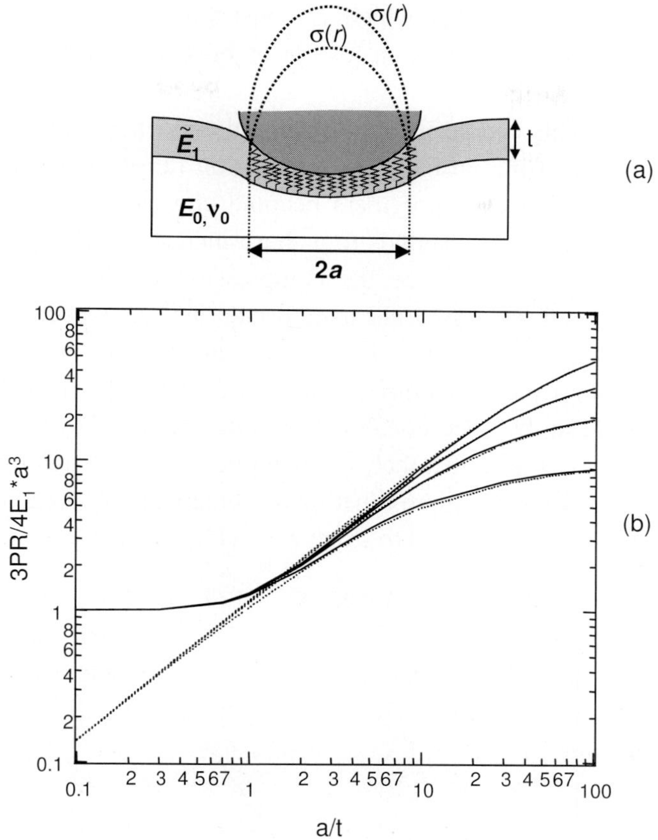

Fig. 2. Oedometric approximation for the indentation of a coated substrate. (a) Approximate description of the indentation behaviour of a confined film lying on a more rigid substrate. $\tilde{E}_1$ is the oedometric modulus of the layer. (b) Non dimensional indentation load, $3PR/4E_1^* a^3$, as a function of the ratio of the contact radius, a, to the layer thickness, t. These curves were calculated for increasing values of the substrate to coating reduced moduli ratio. From bottom to top $E_0^*/E_1^* = 10$; $E_0^*/E_1^* = 25$; $E_0^*/E_1^* = 50$; $E_0^*/E_1^* = 100$; $E_0^*/E_1^* = 200$. The dotted lines correspond to the oedometric approximation schematized in (a), the continuous lines to the exact contact mechanics solution derived by Perriot and Barthel[15] for coated contacts. ($\nu_0 = 0.2$; $\nu_1 = 0.4$). From reference 18, with permissions.

Experimental elastic indentation data using glassy polymer films on glass substrates ($E_0^*/E_1^* \approx 50$, $\nu_1 \approx 0.4$) further support the validity of the oedometric description of confined contacts[18]. For such systems, it comes out that the application of the oedometric model to the data yield a value of the oedometric modulus which is consistent with that determined independently from compression testing of bulk specimens made from the same polymer.

Some comments can be made regarding the indentation behaviour of confined polymer films in their glass transition range or in the rubbery state. In such a situation, polymers become close to incompressibility (i.e. $\nu_1 \rightarrow 0.5$) which corresponds to a sharp increase in their œdometric modulus (equation (5)). Since the compression of the film becomes strongly hampered, increased shear deformations are expected to come into play within the coating during the indentation process. Such deviations from a purely œdometric response of the confined layer are clearly evidenced by theoretical calculations of the $P(a)$ indentation curves. In Fig. 3, it is shown that, when applied to nearly incompressible coatings, the œdometric approximation systematically overestimates the indentation loads as compared to exact contact mechanics calculations.

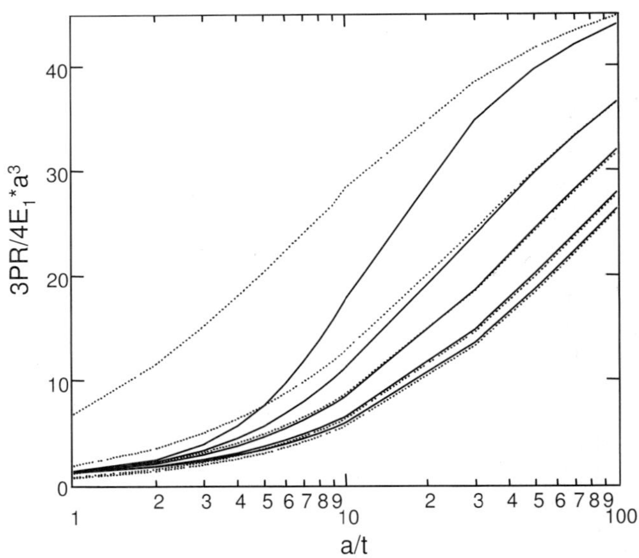

Fig. 3. Non dimensional indentation load, $3PR/4E_1^*a^3$, as a function of the ratio of the contact radius, a, to the layer thickness, t. These curves were obtained for increasing values of the Poisson's ratio of the coating. From bottom to top: ν_1=0.2; ν_1=0.3; ν_1=0.4; ν_1=0.45; ν_1=0.49. The dotted lines correspond to the oedometric approximation, the continuous lines to the exact contact mechanics solution derived by Perriot and Barthel[15] for coated contacts. (ν_0=0.2, E_0^*/E_1^* = 50). From reference[18], with permissions.

This difference may be attributed to the fact that the exact model takes into account shear deformation within the layer, which in turn results in a decreased coating stiffness as compared to a purely œdometric response.

In addition, it is worth noting that the compressibility of the polymer coating also strongly affects the partitioning of the vertical displacement between the layer and the substrate. In the incompressible regime, the compression of the coating is penalized while the shear deformation of the layer is hampered by the confinement. In such a situation, the elastic deformation of the substrate should therefore predominate. An illustration of these effects is provided in Fig. 4 where the vertical displacements of the surface and the film/substrate interface are calculated as a function of the Poisson's ratio of the layer and the confinement using an exact solution to the contact problem[20].

At low penetration ($a/t = 1$) the deformation of the substrate is limited and the response of the compressible and incompressible systems are close. As the contact radii are increased, the substrate starts to deform while the coating is increasingly confined. In this regime, the partitioning of the displacements between the substrate and the coating is strongly dependent on the compressibility of the layer. For an incompressible layer and $a/t = 8$, the surface displacement is almost completely due to substrate deformation; yet a roughly equal contribution from layer and substrate is calculated for a compressible material under the same condition of confinement.

Some practical conclusions may be drawn from these observations regarding the suitability of normal indentation tests to measure elastic properties of thin polymer films in their glass transition or rubbery domains. For common blunt indenters geometries such as Berkovitch tips, high levels of confinement are achieved within thin films even at low indentation depths. As a consequence, most of the measured contact stiffness is provided by substrate deformation and relative errors associated with the determination of the film response will therefore be enhanced. More importantly, Fig. 3 shows that the indentation response becomes increasingly sensitive to small fluctuations in the Poisson's ratio close to incompressibility. Similar conclusions were also drawn from finite elements simulations of the flat punch indentation of thin

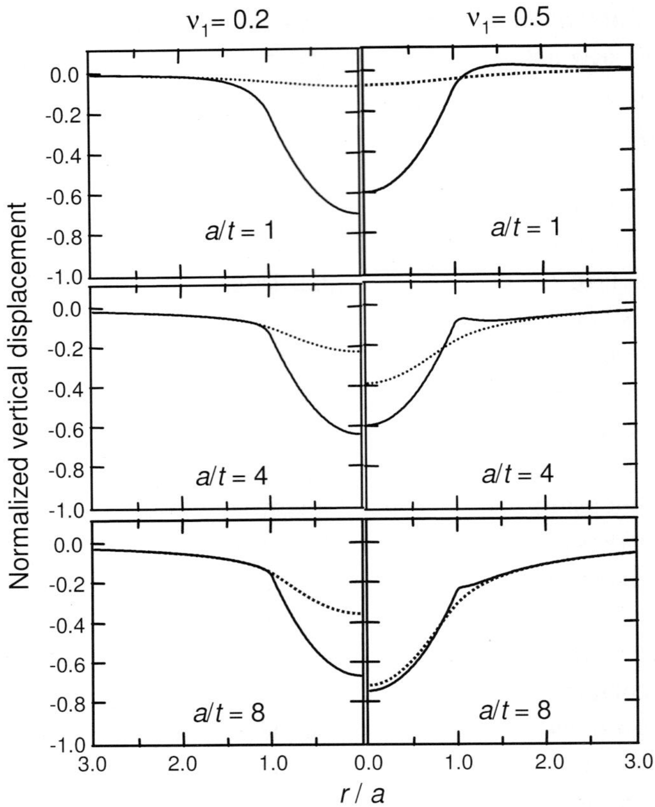

Fig. 4. Computed normal displacement of the surface (plain) and interface (dashed) for the contact of a sphere with a coated substrate. Radii are normalized with respect to contact radius, a, displacements with respect to a^2/R. The coating Poisson's ratio is ν_1 = 0.2 on the left, 0.5 on the right. (ν_0 = 0.2, E_0^*/E_1^* = 50). From reference 20, with permissions.

films on rigid substrate[21]. In this study, the contact stiffness was found to depend strongly on the value of Poisson's ratio up to the 4th decimal close to 0.5.

This sensitivity to the Poisson's ratio close to incompressibility therefore strongly questions the accuracy of Young's modulus measurements of thin films using normal indentation, as ν_1 is generally not known with the required precisions. As it will be detailed below, lateral contact methods do not show such a dependence to the Poisson's ratio of the layer.

They are therefore much more suitable for the mechanical characterization of polymer coatings close to incompressibility which encompasses many systems such as gels, soft adhesives and polymer films in their glass transition or rubbery zone.

2.2. *Lateral Contacts*

The above analysis of the indentation response of a confined coated substrate can readily be extended to a lateral contact loading. We will restrict ourselves to the case where the applied lateral displacement is low enough to prevent any significant micro-slip to occur at the contact interface. In such a situation, the contact response incorporates two components, namely the shear deformation of the film and the lateral contact response of the substrate. Of particular interest is the relationship between the lateral contact stiffness and the shear modulus of the film. As for normal contact configurations, it can be assumed that stresses within the layer do not expand out of the contact zone. The stress applied to the surface is therefore integrally transmitted to the substrate over a constant contact area. Within the limits of confined contact geometries, the layer and the substrate can be considered as two separate stiffnesses in series[22]. Assuming a pure shear response for the coating and neglecting edge effects, the stiffness, k_f, of the film disk enclosed within the contact is simply given by:

$$k_f = \pi G_1 \frac{a^2}{t} \tag{7}$$

where G_1 is the shear modulus of the film.

The second component involved in the lateral response corresponds to the contact deformation of the substrate. The associated stiffness, k_s, is given by the classical Mindlin's theory [23]:

$$k_s = 8G_0^* a \tag{8}$$

where G_0^* is the reduced shear modulus of the substrate defined by:

$$G_0^* = \frac{G_o}{2-\nu_o} \tag{9}$$

The film and the substrate behaving as stiffnesses in series, the overall contact stiffness, k_c, can thus be expressed as follows:

$$\frac{1}{k_c} = \frac{1}{k_f} + \frac{1}{k_s} \tag{10}$$

which can be rewritten as:

$$k_c = \frac{8G_0^* a}{1 + \dfrac{8}{\pi}\dfrac{G_0^*}{G_1}\dfrac{t}{a}} \tag{11}$$

This equation provides the basis to relate the measured contact stiffness, k_c, to the shear modulus, G_1, of the film. As mentioned above, the important point is that, as opposed to the normal indentation approximation which involves a $(1-2\nu_1)^{-1}$ factor, this lateral contact model behaves smoothly with respect to all parameters. Its application to the determination of the linear visco-elastic properties of confined polymer films through their glass transition zone is considered below.

3. Visco-Elastic Properties of Confined Polymer Films in the Glass Transition Range

3.1. *Linearity of the Contact Lateral Response*

From an experimental point of view, the determination of the linear visco-elastic properties of polymer films by lateral contact methods involves the measurement of the contact stiffness under imposed cyclic relative displacements at various frequencies and temperatures. If large displacement amplitudes are considered, sliding friction takes place. In such a situation, the nonlinearities in the contact response make impossible the determination of the linear visco-elastic properties of the layer. However, when the displacement amplitude is reduced below the contact size, a Mindlin's type situation can be encountered where the contact can be divided into two regions: a central zone where no relative displacement of both surface occurs and an outer region where micro-slip takes place. If the lateral displacement amplitude is kept low enough, one can neglect the partial slip within the outer region and describe the cyclic lateral loading as the drag of the sample surface by the slider. Under such condition, the contact response becomes linear since it only involves the visco-elastic response of the film and the elastic deformation of the substrate. Accordingly, the shear visco-elastic modulus can be determined from the lateral contact stiffness using equation (11). In order to check the existence of such a linear regime, experiments can be carried out at increasing displacement amplitudes. An example is shown in Fig. 5 in the case of a glass substrate coated with an acrylate film about 20 μm in thickness in macroscopic contact with a glass sphere.

When the Lissajou curves giving the lateral force as a function of the displacement are normalized with respect to the displacement amplitude, they become nearly superimposed. This demonstrates that, in the considered range of displacement amplitude, the contact response is linear.

3.2. *Pressure Dependence of the Linear Visco-Elastic Modulus*

When considering confined films in their glass transition zone, it emerges that the applied contact pressure can induce major changes in the linear visco-elastic properties as compared to the bulk polymer at ambient pressure. In Fig. 6, the storage modulus, G', and the loss tangent, $\tan\delta$, of confined films is represented as a function of temperature for increasing contact pressures. As expected, the glass transition of the films is associated with a drop in G' and a damping peak. For an unconfined contact with a bulk specimen (dotted line in Fig. 6), the data are perfectly consistent with independent visco-elastic measurements using conventional Dynamic Mechanical Thermal Analysis (DMTA)[24]. On the other hand, a shift of the glass transition to high temperatures is clearly seen when increased contact pressures are applied to confined films. Such a shift can be attributed to hydrostatic pressure dependence of the glass transition of polymers. Within confined contacts, the hampered shear deformation results in the development of stresses of essentially hydrostatic nature when a normal contact loading is applied. As reviewed in references[25,26], it is well known that the visco-elastic relaxation times of polymer glasses are strongly increased under

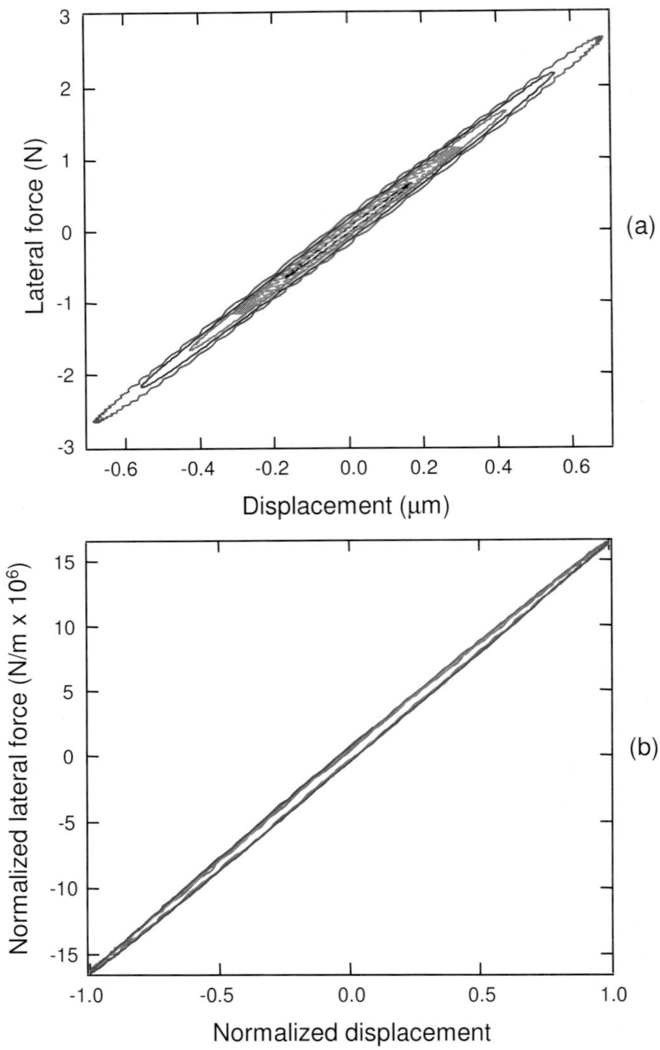

Fig. 5. Lissajou representation of the lateral force as a function of the applied relative displacement for a glassy acrylate film (33 μm in thickness) confined within a macroscopic contact between glass substrates (3.2 Hz, R.T.). (a) Cycles obtained at increasing displacement amplitudes between 0.2 and 0.65 μm. (b) Same data as in (a) after normalization with respect to the amplitude of the applied displacement. The superimposition of the Lissajou cycles in this representation is a proof of the linearity of the contact lateral response.

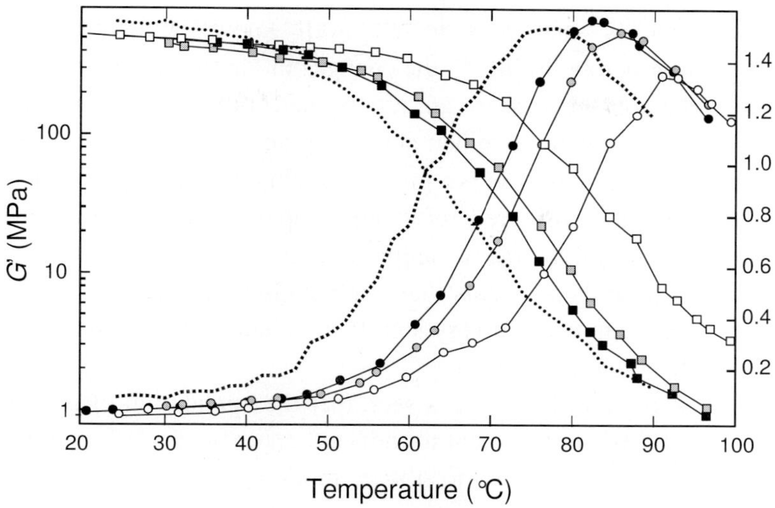

Fig. 6. Storage modulus, G', and tangent of the loss angle, tan δ, of confined polymer films determined from lateral contact stiffness measurements. Circles: tan δ, squares: G'. The tests have been carried out at increasing applied normal loads. Black symbols: P = 40 N; grey symbols: P = 100 N; open symbols: P = 240 N; (frequency = 3.2 Hz, film thickness = 33 μm). The dotted lines correspond to the values of G' and tan δ measured by the same contact method using a bulk acrylate specimen, i.e. under unconfined conditions (from reference[24], with permissions).

static hydrostatic pressure. Qualitatively, this behaviour can be understood in terms of the dependence of the segmental mobility on the free volume since the free volume must decrease with increasing pressure just as it does with decreasing temperature. The occurrence of such effects within confined contacts is supported by the fact that the measured contact pressure dependence of the glass transition of the films (about 0.3 °C/MPa for acrylate coatings) is of the same order of magnitude as that measured using bulk polymers under hydrostatic pressure[18]. It can be noted in passing that hydrostatic pressure effects, although of more limited amplitude, have also been reported by Briscoe *et al.*[27] in the context of sliding friction of polymer films, close to their glass transition.

More generally, the temperature and pressure dependence of the linear visco-elastic properties of confined films in the glass transition zone can be described by means of a temperature-pressure superposition principle which is analogous to the well known time-temperature superposition principle. In Fig. 7, the changes in the storage modulus as a function of the contact pressure are represented for various temperatures in the glass transition zone. When shifted along the pressure axis, these isotherms provide a master curve giving the changes in *G'* as a function of pressure for the considered reference temperature. As shown in the insert box, the shift factor, a_p, is roughly linearly related to temperature.

It emerges from these results that a relatively moderate change in the contact pressure can result in very significant changes in the shear visco-elastic modulus of confined films.

For the example in Fig. 7, it is seen that the application of a contact pressure of 200 MPa is sufficient to induce the vitrification of an initially rubbery film, i.e. a change in the storage modulus of nearly three orders of magnitude.

4. Plastic Properties of Confined Polymer Films

Hydrostatic pressure is also known to affect the yield properties of bulk polymers. Early studies by Rabinowitz and co-workers[28] have revealed a substantial increase in the shear yield stress of glassy polymers such as PMMA up to an hydrostatic pressure of about 300 MPa above which brittle failure occurs. In the context of polymer tribology, this pressure dependence of the yield process has sometimes been invoked to analyse the sliding friction of thin polymer films[29-31]. From the similarities between the contact pressure dependence of the frictional stress and the hydrostatic dependence of polymer yield stress, it was argued that frictional dissipation within confined films results essentially from a plastic flow phenomenon. We will show here that hydrostatic pressure effects can also be involved during plastic indentation of polymer films and affect both the onset of plastic deformation and the yield flow.

4.1. *Plastic Imprints under Normal Indentation*

When a rigid sphere is pressed against a bulk polymer above the elastic limit, plastic flow is known to be associated with the formation of a

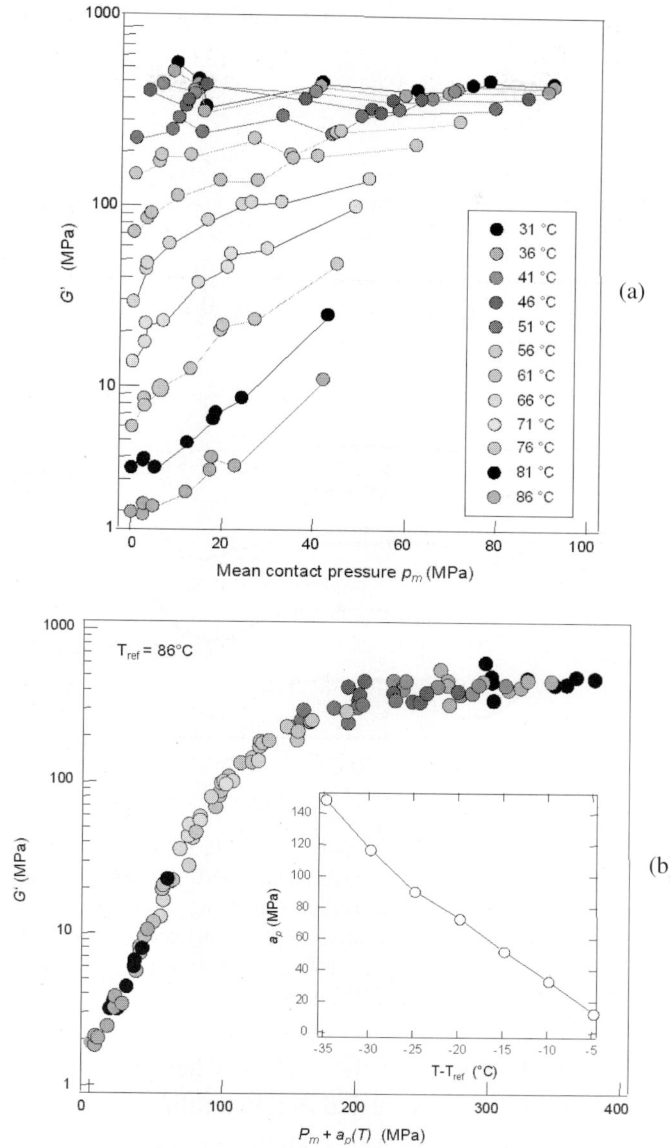

Fig. 7. Pressure-temperature superposition for confined glassy polymer films. (a) Isotherms giving the storage modulus, *G'*, as a function of the mean contact pressure at a frequency of 3.2 Hz. (b) Master curve obtained by shifting the isotherms in (a) along the pressure axis. a_p is the shift factor. As shown in the insert box, it is roughly linearly related to the temperature with $da_p/dp_m \approx 44$ MPa/°C (the glass transition of the film at ambient pressure is 78°C at the considered frequency).

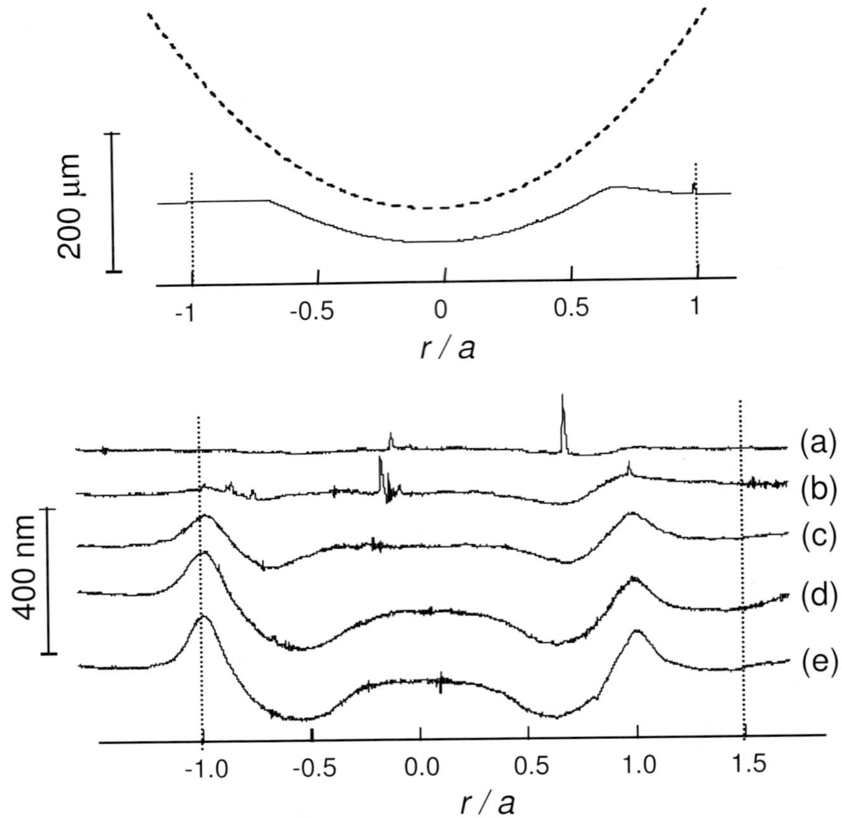

Fig. 8. Plastic imprints left on a polymer surface after sphere indentation. Top: unconfined contact between a glass sphere and a glassy acrylate substrate. Bottom: confined contacts between an acrylate film lying on an elastic glass substrate and a glass sphere. Experiments are carried out at different mean contact pressure p_m. (a) p_m=56 MPa, a/t=6; (b) p_m=78 MPa, a/t=6.4; (c) p_m=131 MPa, a/t=8.6; (d) p_m=143 MPa, a/t=8.9; (e) p_m=171 MPa, a/t=9.6.

permanent spherical imprint on the surface when the load is removed. When the same experiment is carried out on a thin polymer film lying on a more rigid elastic substrate, completely different imprints are surprisingly observed after unloading. Instead of a spherical capped plastic imprint, it is seen that plastic flow is concentrated at the periphery of the contact zone, while no substantial permanent deformation is detected at the centre of the contact (Fig. 8). This behaviour of confined film can be attributed to the essentially hydrostatic nature of the stress distribution at the centre of the contact. Within this region, reduced shear stresses are effectively associated with high values of the hydrostatic pressure which is known to increase the yield limit of polymers.

The occurrence of plastic flow is thus favoured at the periphery of the contact where the reduced confinement results in enhanced shear stresses and in a decreased yield limit as a result of the lowered hydrostatic pressure. This picture is supported by theoretical calculation of the stress field induced within the coating. As a first approach, an elastic calculation can be coupled with some yield criterion in order to predict the location of the initial yield region within the coating.

Such a calculation can be carried out using an exact semi-analytical contact models for coated substrates[10], providing the results shown in Fig. 9. The calculation of a simple von Mises stress from the elastic stress field (Fig. 9(I)) indicates that, for all the confinement levels under consideration, the plastic deformation of the film should start at the center of the contact, as opposed to the experimental observation.

On the other hand, it emerges from the calculation of a modified, hydrostatic pressure-dependent, von Mises stress that yield should initiate close to the border of the contact for confinements, a/t, in the order of a few units (Fig. 9(II)). Such a modified von Mises criterion is simply calculated by subtracting a linear pressure dependent term, αp_H, from the classical von Mises stress[28] (p_H is the hydrostatic pressure and α an empirical coefficient in the order of 0.2 for glassy polymers[28,32,33]). From Fig. 8, it is seen that the first yield deformation is experimentally found to occur within a confinement range (about 6-10) where the maximum modified von Mises stress is located near the periphery of the contact. The pressure dependence of the yield stress of polymer can therefore account for the localization of plastic deformation close to the edge of the contact in the case of confined films.

4.2. *Elastic/Plastic Indentation Limit for Confined Polymer Films*

The pressure dependence of polymer yield stress can further be considered to determine the normal load at the onset of yield within confined coated contacts. A simple dimensional analysis shows that the onset of yield within a coating can be completely defined as a function of

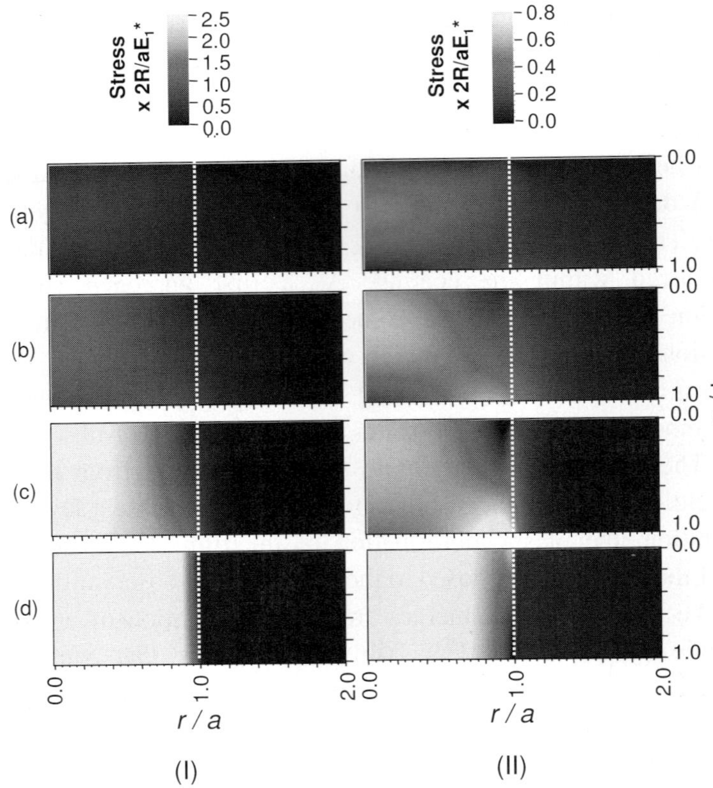

Fig. 9. Calculated non dimensional yield criteria within a coating in contact with a sphere as a function of the confinement level, a/t. (I) von Mises criterion (II) Pressure-dependent modified von Mises criterion, $\alpha = 0.2$. (a) $a/t = 1$; (b) $a/t = 3$; (c) $a/t = 10$; (d) $a/t = 70$. a is the contact radius, t is the thickness of the layer. For symmetry reasons, only one half of the contact zone is shown.

two normalized parameters, t/R and $p_m=P/\pi a^2$. In Fig. 10, the experimental boundary between the elastic and plastic domain has been determined as a function of these two normalized parameters from a set of contact experiments using acrylate films with different thicknesses and spheres with various radii. For each experiment, the occurrence of plastic deformation was deduced from the observation of a residual imprint after indentation. In the same graph, the theoretical boundary between elastic and plastic deformation has also been reported. The later was calculated using a modified pressure dependent von Mises criterion whose parameters where determined from independent constant strain rate compression experiments using bulk acrylate specimens. The strain rate of indentation experiments was ill-defined, but it was assumed to lie in the range 2.10^{-2} s^{-1}-2.10^{-4} s^{-1}. Within these uncertainties in the actual indentation strain rate, it turns out that the application of such a pressure dependent yield stress criterion can adequately describe the boundary between elastic and plastic indentation of confined polymer films. As shown in Fig. 10, this is obviously not the case for a simple von Mises criterion.

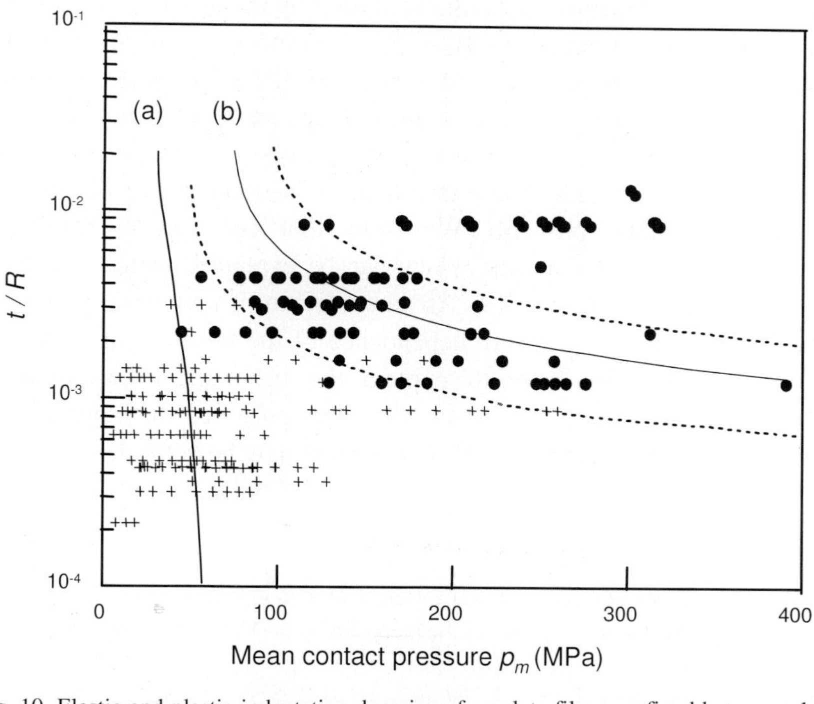

Fig. 10. Elastic and plastic indentation domains of acrylate films confined between glass substrates. (+) Elastic response, no residual indentation imprint (•) plastic response as indicated by a residual contact imprint. Experiments were carried out using different film thicknesses, t (between 18 μm and 110 μm) and sphere radii, R (between 750 μm and 360 mm). (a) Theoretical elastic limit calculated from a von Mises criterion using the bulk compression yield stress of the acrylate polymer at a 2.10^{-3} s^{-1} strain rate; (b) theoretical elastic limit calculated from a pressure dependent von Mises criterion at a strain rate of 2.10^{-3} s^{-1}. The lower and upper dotted lines correspond to the calculated elastic boundaries at strain rates equal to 2.10^{-4} and 2.10^{-2}, respectively.

5. Conclusions

Thin polymer films lying on rigid substrates are encountered in many tribological situations. We have shown that the high values of the hydrostatic pressure which develop in confined geometries – i.e. when the contact radius is larger than the film thickness - may significantly alter the mechanical properties of the polymer layer which in turn affects its tribological behaviour. These effects of confinement can be interpreted in the light of the well-known pressure dependence of the mechanical properties of polymers. In this respect, visco-elastic properties are known to be thermally shifted by the hydrostatic pressure. This shift is experienced under *ordinary* contact conditions and may result in orders of magnitudes of changes in the visco-elastic modulus. Conversely, contact techniques may help analysing the pressure-temperature superposition properties of the polymers, as other methods are technically difficult. Yield-threshold is also known to be affected by the hydrostatic pressure. We show that confinement results in characteristic plastic imprints, which can be explained using a pressure dependent von Mises criterion. As a conclusion, it may be important to take into account the pressure dependence of the mechanical properties of the polymers in order to explain the tribological behaviour of substrates coated with polymers. For that purpose, the approximate oedometric contact model, recalled in the text may be helpful.

References

1. G.M. Pharr, W.C. Oliver and F.R. Brotzen, *J. Mat. Res.*, **7**, 613 (1992).
2. J. Mencik, D. Munz, E. Quandt, E.R. Weppelman and M.V. Swain, *J. Mat. Res.*, **12**, 2475 (1997).
3. M.F. Doerner and W.D. Nix, *J. Mat. Res.*, **1**, 601 (1997).
4. R. King, *Int. J. Sol. Str.*, **23**, 1657 (1987).
5. R. Saha and W.D. Nix, *Acta Mater.*, **50**, 23 (2002).
6. S. Bec, A. Tonck, J.M. Georges, E. Georges and J.L. Loubet, *Phil. Mag.*, **74**, 1061 (1996).
7. H. Gao, C.H. Chiu and J. Lee, *Int. J. Sol. Str.*, **29**, 2471 (1992).
8. A. Strojny, X.Y. Xia, A. Tsou and W.W. Gerberich, *J. Adhesion Sci. Tech.*, **12**, 1299 (1998).

9. S. Roche, S. Pavan, J.L. Loubet, P. Barbeau and B. Magny, *Progr. Organic Coatings*, **47**, 37 (2003).
10. C. Fretigny and A. Chateauminois, *J. Phys. D.*, **40**, 5418 (2007).
11. D. Burmister, *J. Appl. Phys.*, **16**, 89 (1945).
12. T.K. O'Sullivan, RB, *J. Tribology*, **110**, 235 (1988).
13. J.C. Li, TW, *Int. J. Sol. Str.*, **34**, 4463 (1997).
14. E.H. Dimitriadis, F; Maresca, J; Kachar, B & Chadwick, RS, *Biophys. J.*, **82**, 2798 (2002).
15. A. Perriot and E. Barthel, *J. Mat. Res.*, **19**, 600 (2004).
16. Y.Y. Lin, C.-Y. Hui and H.D. Conway, *J. Pol. Sci. B.*, **38**, 2769 (2000).
17. L.D. Landau and E.M. Lifshitz, in *Theory of elasticity.* (Pergamon, 1986).
18. E. Gacoin, A. Chateauminois and C. Fretigny, *Tribology Lett.*, **21**, 245 (2006).
19. K.L. Johnson, in *Contact Mechanics* (Cambridge University Press, Cambridge, 1985).
20. E. Barthel, A. Perriot, A. Chateauminois and C. Fretigny, *Phil. Mag.*, **86**, 5359 (2006).
21. J.F. Ganghoffer and A.N. Gent, *J. Adhesion*, **48**, 75 (1995).
22. A. Tonck, J. Sabot and J.M. Georges, *J. Tribology*, **106**, 35 (1984).
23. R.D. Mindlin, *ASME Trans. J Appl. Mech. E.*, **16**, 327 (1953).
24. E. Gacoin, A. Chateauminois and C. Fretigny, *Polymer*, **45**, 3789 (2004).
25. E.J. Parry and D. Tabor, *J. Mat. Sci.*, **9**, 289 (1974).
26. J.D. Ferry, in *Viscoelastic Properties of Polymers.* Third Edition ed (John Wiley & Sons, New York, 1980).
27. B.J. Briscoe and A.C. Smith, The shear properties of thin organic films., in Reviews on the Deformation Behaviour of Materials III. 1980, Israel Scientific Publications. p. 151.
28. S. Rabinowitz, I.M. Ward and J.S.C. Parry, *J. Mat. Sci.*, **5**, 29 (1970).
29. B.J. Briscoe and D. Tabor, *Wear*, **34**, 29 (1975).
30. B.J. Briscoe and A.C. Smith, *Polymer*, **22**, 158 (1981).
31. B.J. Briscoe and D. Tabor, *J. Adhesion*, **9**, 145 (1978).
32. R. Quinson, J. Perez, M. Rink and A. Pavan, *J. Mat. Sci.*, **32**, 1371 (1997).
33. A.J. Lesser and R.S. Kody, *J. Pol. Sci. B.*, **35**, 16 (1997).

CHAPTER 18

TRIBOLOGICAL BEHAVIOR OF POLYMER BRUSH PREPARED BY THE "GRAFTING-FROM" METHOD

Motoyasu Kobayashi,[§] Zhe Wang,[§] Yasuhiro Matsuda,[†] Masataka Kaido,[#] Atsushi Suzuki[#] and Atsushi Takahara[§*]

[§]*Institute for Materials Chemistry and Engineering, Kyushu University*
744 Motooka, Nishi-ku Fukuoka, 819-0395, Japan
[†]*Faculty of Engineering, Shizuoka University*
3-5-1 Johoku, Naka-ku, Hamamatsu, 432-8561, Japan
[#]*TOYOTA Motor Co., 1-Toyota-cho, Toyota, Aichi, 472-8572, Japan*
**E-mail: takahara@cstf.kyushu-u.ac.jp*

This article describes the tribological properties of well-defined high-density polymer brushes prepared by the "grafting-from" method based on surface-initiated living radical polymerization. The authors have demonstrated here a water-lubrication system using hydrophilic polymer brushes consisting of 2,3-dihydroxypropyl methacrylate and 2-methacryloyloxyethyl phosphorylcholine (MPC). Macro- and microscopic frictional properties were carried out by a ball-on-plate type tribotester and an AFM cantilever attached with a colloidal sphere, respectively. Friction coefficients and adhesion forces of brush surfaces depend on the solvent quality. For example, the friction coefficient of poly(MPC) brush was reduced to be 0.02 under the highly humid atmosphere because water molecules adsorbed into the brush surfaces acted as a lubricant. Furthermore, high-density polymer brushes showed the low friction coefficient even after 100 reciprocating friction test, which clearly indicates an excellent wear resistance of brush film.

1. Introduction

1.1. *Definition of Polymer Brush*

Surface-grafted polymers can be clarified as one of several types based on the graft density, which largely depends on the grafting method. The graft density is the number of chains at the surfaces per unit area. The simplest way to immobilize the polymer chains on the surface is a physisorption approach involving two-component polymer chains, where one part strongly adheres to the surface and the second part extends to generate the polymer layer.[1] The tethering point can be a functional group or a diblock copolymer, both of which are strongly adsorbed at the surface and act as an anchor for the polymer chain. This is a typical "grafting-to" method. Although the polymer layer is thermally stable and is not released into the solvent, the graft density and film thickness are quite low because large polymer chains hardly diffuse to substrate surfaces that are sterically hindered by surrounding bounded chains. The low-density grafted polymer will assume a "mushroom" conformation with a coil dimension similar to that of ungrafted (free) chains.

In contrast, the "grafting-from" technique[2-3] generates the polymer chains by polymerization from a suitable surface immobilized initiator, which are predominantly formed by self-assembled monolayers containing the desired initiating functionality. This surface-initiated polymerization technique, especially combined with a controlled polymerization system, enables us to prepare high-density grafted polymers due to the small molecules of the immobilized initiators and the high initiator efficiency. With increasing graft densities, polymer chains will be stretched away from the surface in order to avoid overlap of each polymer chain, forming a so-called "polymer brush". Figure 1 illustrates the structure of surface-grafted polymers with different graft densities. By definition, polymer brush can be described as polymer chains tethered to a surface or interface with a sufficiently high graft density such that the chains are forced to stretch away from the tethering site.[4]

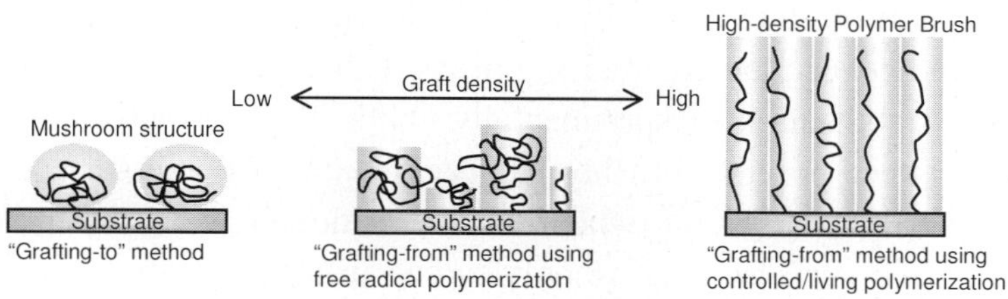

Fig. 1. Chain structure of low- and high-density grafted polymers.

Many papers regarding the frictional property of polymer brush have been published, while very few have described the frictional properties of high-density brush, probably because preparation of well-defined high-density polymer brush is experimentally difficult. The graft chains are expected to exhibit different frictional properties such as wear resistance, by graft density. In this introductory chapter, the authors provide a brief overview of recent works regarding the tribological properties of polymer brush prepared by the "grafting-to" method. In the later section, the macro- and microscopic tribology and mechanical properties of high-density polymer brushes prepared by "grafting-from" are described.

1.2. *Tribology of Polymer Brush Prepared by "Grafting-to" Method*

A number of experimental studies regarding the interaction between solid surfaces bearing polymer brushes have been reported to understand the friction mechanism of brushes.[5-6] Most studies of the shear of polymer brushes to date have involved solvated brushes in a good solvent medium, with graft density limited by the steric repulsion of the self-assembled brush chains, so-called "grafting to". The following early works have investigated the tribological properties of relatively low-density polymer brushes prepared by adsorption of an end-functionalized polymer, by the Langmuir-Blodgett (LB) method using a charged diblock copolymers at the air-water interface, and by the "grafting-from" method using classical free radical polymerization. However, the findings of these previous studies are indispensable in understanding the frictional properties of high-density polymer brush.

When two surfaces with end-grafted chains are brought into contact in a good solvent, they normally repel each other because of the osmotic interaction between the polymer segments. This repulsion is the basis for colloidal stabilization. Experimentally, both the normal[7] and shear forces[8] between polymer brushes has been studied using a surface force apparatus (SFA) by Klein, who found a reduction in frictional forces between solid surfaces bearing polystyrene in a good solvent (toluene) for several molecular weights and graft density. The brushes appear to provide an extremely efficient lubricant for reasonable load, which is 1-2 MPa at the highest compressions. Similar results have also demonstrated that neutral polymer brushes can act as very effective lubricants between sliding surfaces immersed in a good solvent.[9-14] Pelletier *et al.* have also found that adsorbed polystyrene-*block*-polyvinylpyridine on the solid surface shows very low shear stress, even though the opposing solid surfaces are compressed to separation distances of the equilibrium height of grafted chains.[9]

The origin of the efficient lubrication of solvated brush is attributed to the osmotic pressure and weak interpenetration of the graft chains. The osmotic pressure due to excluded volume effects of polymer chains in a good solvent supports the normal load compressing the two surfaces against each other. In addition, weak interpenetration between the well-solvated brushes allows the interface to remain fluid and offers low resistance to the shearing motion of the brush layers sliding past each other. Up to moderate compressions, due to this fluid interface, very good lubrication takes place[8]. At much higher compressions, shear force dramatically rises with compression as a result of the rapid increase in the viscosity of polystyrene at high concentrations (approach to a glassy state).

In the case of a theta solvent, large shear forces are also detected at much milder levels of compression.[15] The changes in the strength of frictional interactions[16] and the interpenetrating depth[17] due to the shrinkage on brush in poor solvent have been offered as reasons, as well as vitrification of the compressed polymer layers.[18-20] Kilbey *et al.* have reported the influence of composition fluctuations on the frictional properties of surface-grafted polystyrene in near-theta cyclohexane at various temperatures.[21]

Interactions between solid surfaces bearing brushes largely depend on not only the solvent quality, but also the chain structure of the brush. Ruths *et al.* have measured the interaction forces between polydispersed polystyrene brushes formed by surface-initiated free radical polymerization from a quartz glass plate,[22] and have pointed out the influence of chain entanglements in a high molecular weight brush on the repulsive force. The shear of telechelic brushes, which have two functional groups at both chain ends, have been also studied, suggesting that telechelic chains interacting with the end group form predominantly bridge-like aggregates connecting opposite surfaces, rather than loops.[23]

A large number of papers focusing on the preparation and characterization of polyelectrolyte brushes have also been published[24-25] since the pioneering theoretical works started by Milavic and Marčelja[26] in the 1980s. The effects of charge density, chain length, graft density,[27-29] ionic strength,[30] and solvent quality[31-32] on the structure of polyelectrolyte brushes have been successively investigated. Experimental studies have also been carried out to measure the repulsive and attractive interactions of polyelectrolyte brushes[7] in order to understand colloidal stability[33] and the electrostatic interactions between proteins and the charged brush.[34] Direct measurements of the forces acting between polyelectrolyte brushes have been carried out by Kurihara[35]. Kelly *et al.* have measured the interactions between polyelectrolyte-covered surface and Si_3N_4 cantilever in salt solution by AFM[36]. The lubrication properties of polyelectrolyte brushes have been investigated by Klein and his coworkers. They have reported that brushes of charged polymer can act as efficient lubricants between mica surfaces in an aqueous medium, even when the graft density is not particularly high.[37] Recently, they prepared an asymmetric case and measured the normal and shear forces between one surface with a polyelectrolyte and another bare charged surface in water.[38] Similarly, Osada and Gong have reported that gel-terminated polyelectrolyte brushes, prepared by controlled radical polymerization using stable nitroxide radical, reduce the friction force between hydrogels and a glass plate across water.[39] They have also described that longer polyelectrolyte molecules exhibit even higher friction than a normal network gel. Sheth *et al.* have measured the normal forces between poly(ethyleneoxide) brush against various surfaces, finding that the brush does not adhere to a neutral lipid bilayer or to bare mica. Spencer and coworkers have studied the lubrication properties of poly(L-lysine)-*graft*-poly(ethylene glycol) adsorbed onto silicon oxide and iron oxide surfaces by using ultra-thin-film interferometry, a mini-traction machine, and pin-on-disk tribometry.[40-41] They reported that the graft polymer formed a stable lubricant layer on the tribological interface in aqueous solution and reduced friction.

However, the great majority of polymer brushes mentioned above are prepared by adsorption of comb-like polyelectrolyte on the solid surface by the "grafting-to" method. Although the various types of well-defined high-density polymer brushes have been synthesized by controlled polymerization system nowadays, the frictional properties of high-density grafted polymer brush have not yet been investigated in detail, with some exceptions.

1.3. *Advantage of Polymer Brush Prepared by the "Grafting-from" Method Combined with Living Polymerization*

Recently, various types of living polymerization techniques, including anionic, cationic, radical, ring-opening, and methathesis polymerization were applied to surface-initiated graft polymerization to prepare a polymer brush. In particular, controlled/living radical polymerization has been the most widely used for preparation of polymer brushes because radical reactions are more tolerant of impurities than the other ionic polymerization systems, and are capable of controlling various monomers. A living polymerization is a chain polymerization that proceeds in the absence of the kinetic steps of termination or chain transfer. As a result, a quantitative initiation reaction takes place to produce a polymer with predictable molecular weight and narrow molecular weight distribution. High initiator efficiency in a living polymerization gives a high graft density of brush. In addition, brush length is directly proportional to the polymer brush molecular weight, which is controlled by the feed ratio of monomer/initiator and the degree of conversion. Classical polymerization, such as free radical polymerization, cannot control brush length and cannot produce high-density brush as a result of the low efficiency of initiator, termination, and chain transfer reaction. Therefore, the useful advantages of living polymerization have improved the design of brush structure, adjusting thickness and graft density as well as the chemical structure of the polymers.

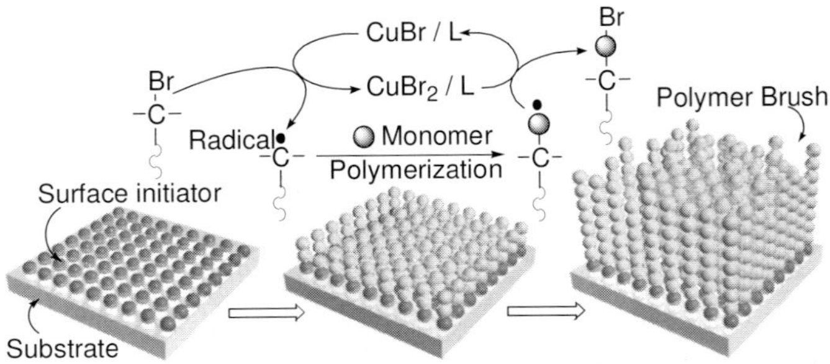

Fig. 2. Illustration of high-density polymer brush prepared by surface-initiated ATRP.

Investigations of the tribological properties of high-density polymer brush prepared by surface-initiated atom transfer radical polymerization (ATRP),[42] which is a typical method of controlled radical polymerization, have begun to be carried out. ATRP is based on the transfer of a halogen atom from the initiator to the monomer and successive transfer to the growing polymer chain catalyzed by a transition metal complex that mediates the propagation.[43] Figure 2 briefly illustrates a chain propagation mechanism. A reversible activation process proceeds continuously starting from cleavage of a carbon-halogen bond to generate a radical species that reacts with monomer until reproduce a new carbon-halogen bond. Tsujii and collaborators have demonstrated force-distance profiles of high-density PMMA brushes immersed in toluene by scanning force microscopy using a micro-silica sphere attached to a cantilever head.[44] At a high graft density, the osmotic pressure extends the grafted polymer chains in the vertical direction, giving rise to extremely strong resistance against compression. Extremely low lateral (frictional) forces have been observed in nano-tribological studies performed with atomic force microscopy (AFM) using a colloidal probe immobilized with a high-density PMMA brush in toluene.[45] The steric repulsion between the polymers and little interpenetration of swollen high-density brushes gives low frictional forces between brush-bearing surfaces. Macroscopic friction tests of high-density PMMA brush have also been carried out using a stainless steel ball as a sliding probe, resulting in findings of a low-friction and good wear-resistant surface under a high normal load.[46] Berger *et al.* have carried out AFM friction experiments on PMMA and polystyrene prepared by the surface-initiated reversible fragmentation chain transfer method.[47] They found that the grafted polymer layer showed lower wear compared with the linear polymer, grafted polymer prepared by the "grafting-to" method, plasma polymerized polymer, and chemically cross-linked polymers.

This review chapter will deal with high-density well-defined polymer brushes produced by a surface-initiated ATRP technique. The next section will briefly describe the preparation procedure for polymer brushes. Section 3 will discuss the tribological properties of high-density polymer brush focusing on water lubrication on a hydrophilic polymer brush.

2. Preparation of High-Density Polymer Brush on Substrate

2.1. *Initiator-Immobilized Silicon Substrate*

Undoubtedly, the most important part of this surface-grafting technique is the immobilization of initiators on the substrate surface. The use of a surface-bond initiator on the surface appears to be promising with regard to obtaining well-defined high-density polymer brushes. Generally, a surface-initiator suitable for ATRP is an alkyl halide with functional groups such as alkoxysilane, chlorosilane, thiol, or phosphoric acid, which can bind through covalent bond with silicon substrate,[48-49] gold,[50] magnetite,[51] and titanic oxides.[52-53] The specially designed alkyl halides are usually immobilized on the substrate by immersion of purified substrate to surface-initiator solution diluted with dry toluene,[54] by the LB method,[42] and by the chemical vapor adsorption (CVA) method.[55] The authors have synthesized (2-bromo-2-methyl)propionyl-oxyhexyltriethoxysilane (BHE)[56] by hydrosilylation of 5-hexenyl 2-bromo- isobutylate treated with triethoxysilane using Karstedt catalyst. The silicon wafers were cleaned by washing with piranha solution (H_2SO_4/H_2O_2 = 7/3, v/v) at 373 K for 1 h and by exposure to vacuum ultraviolet-ray (λ = 172 nm) for 5 min under reduced pressure (30 Pa). BHE was immobilized on silicon wafer by the CVA method to form an stable organosilane monolayer, which was confirmed by AFM and the Fourier transform infrared spectra combined with Brewster angle unit, and X-ray photoelectron spectroscopy. The practical procedure has been described previously.[46]

2.2. *General Procedure for Surface-Initiated ATRP*

Various types of vinyl monomers such as styrene, methacrylate, and their derivatives can be polymerized in a living manner by the ATRP system. In this review, the authors will describe a typical procedure for surface-initiated ATRP of methyl methacrylate (MMA), as follows. A few sheets of the BHE-immobilized silicon wafers, CuBr (catalyst), 4,4’-dihexylbipyridyl (ligand), degassed anisole solution of MMA (monomer), and ethyl 2-bromoisobutylate (free initiator) were introduced into a glass tube with a stopcock. The polymerization solution was degassed by a repeated freeze-and-thaw process to remove the oxygen. The polymerization reaction was conducted at 90 °C for 12 h under argon to simultaneously generate PMMA brush from the substrate and free PMMA from ethyl 2-bromoisobutylate. The reaction was stopped by opening the glass vessel to air, and the reaction mixture was poured into methanol to precipitate the free polymer. The silicon wafers were washed with toluene using a Soxhlet apparatus for 12 h to remove the free polymer adsorbed on their surface, and dried under a reduced pressure at 373 K for 1 h. The free polymer in THF solution was passed through a neutral alumina column to remove the catalyst. Suitable combinations of the chemical structure of alkyl halide, catalyst, ligand, solvent, and temperature for ATRP are dominated by the chemical structure of the monomers. For example, (–)-sparteine and 4,4'-dimethyl-2,2'-bipyridyl were used as a ligand.

Scheme 1 represents the schematic procedure for the preparation of polymer brushes by surface-initiated ATRP of MMA, (2,2-dimethyl-1,3-dioxolan-4-yl)methyl methacrylate (DMM), and 2-methacryloyloxyethyl phosphorylcholine (MPC) from alkylbromide-immobilized silicon wafers was carried out in the presence of EB as a free initiator coupled with copper catalyst to produce polymer brushes. The AFM observation revealed that a homogeneous polymer layer was formed on the substrate, and the surface roughness was 0.8 ~ 1.5 nm in

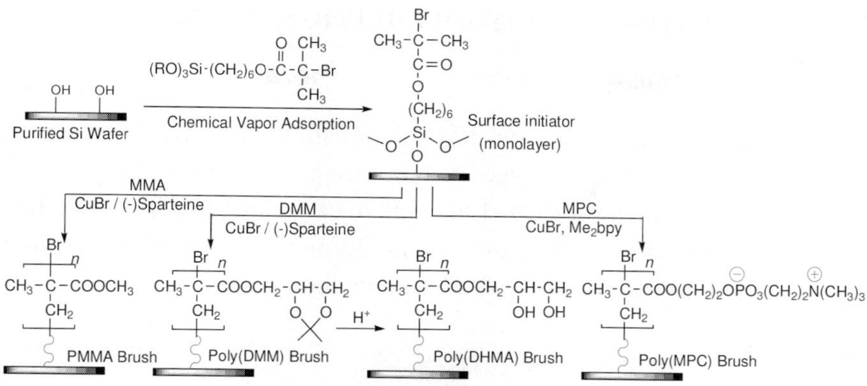

Scheme 1. Preparation of polymer brushes on Si wafer.

the dry state in a 5 × 5 μm^2 scan area. The polydisperisty index of the free polymers obtained was relatively low (M_w/M_n = 1.1~1.5), and the number-average molecular weight (M_n) values estimated by SEC were proportional to the monomer conversion. The M_n of surface-grafted polymer on a silicon wafer was not directly determined here, but it is widely known that a polymer brush should have the same molecular weight as the value for the corresponding free polymer.[51,57-59] The thicknesses of polymer brushes on the silicon substrates were evaluated primarily by imaging ellipsometer equipped with an YAG laser (532.8 nm). The thickness of the polymer brushes increased linearly with molecular weight because the polymer chain forms a fairly extended conformation as a result of the high graft density and the excluded volume effect. Therefore, the graft density σ was estimated by the relationship between the thickness L (nm) and M_n, as follows:

$$\sigma = d \times L \times N_A \times 10^{-21} / M_n$$

where d and N_A are the assumed density of bulk polymer at 293 K and Avogadro's number, respectively. The graft densities of PMMA, Poly(DMM), and Poly(MPC) brushes were calculated to be 0.56, 0.36, and 0.22 chains/nm^2, depending on the cross-sectional area of the monomer. These values are comparatively high, taking into account the volume fraction of the polymer chain.

3. Frictional Property of High-Density Polymer Brush

3.1. *Dependence of Solvent Quality*

Macroscopic friction tests on PMMA brush were carried out on a conventional ball-on-disk type reciprocating tribometer Tribostation Type32 (Shinto Scientific Co. Ltd., Tokyo) by sliding a stainless steel ball on the substrates along a distance of 20 mm at a rate of 90 mm/min in air under a normal load of 0.49 N at room temperature.

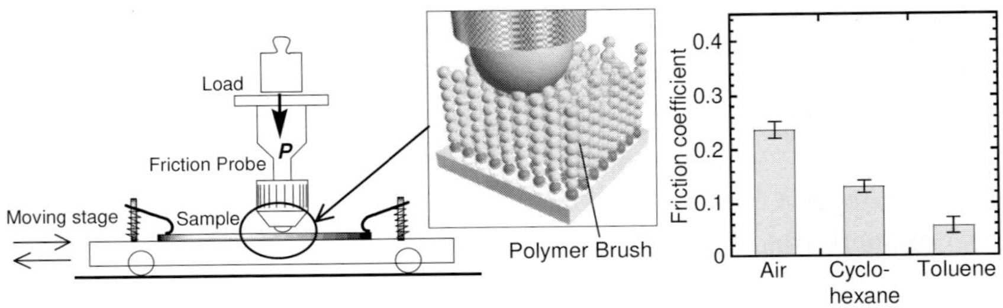

Fig. 3. Friction coefficient of PMMA brush in air, cyclohexane, and toluene by sliding a stainless steel ball over a distance of 20 mm at a sliding velocity of 90 mm/min under a load of 0.49 N at 298 K.

The friction coefficient was determined by a strain gauge attached to the arm of the tester and was recorded as a function of time In the case of a non-modified silicon wafer under a normal load of 50 g (0.49 N), the theoretical contact area between the stainless steel probe and substrate can be calculated as 2.43 x 10^{-9} m^2 by Hertz's contact mechanics theory[60] and the average pressure on the contact area was estimated to be 201 MPa.

Figure 3 shows the dynamic friction coefficient of PMMA brush with 25 nm thick in air, cyclohexane, and toluene. Friction tests in various solvents were conducted using the polymer brush substrates, which were immersed in the corresponding solvents for 24 h prior to the experiments. Compared with the friction coefficient of the polymer brush in the dry state, the value decreased in both organic solvents due to the fluid lubrication effect. The friction coefficient of the PMMA brush in toluene was 0.06, while that in cyclohexane was 0.12. This result indicates that the polymer brush in a good solvent performs well as a lubricant, but that it does not perform well in a poor solvent. The interaction between the brush surface and the stainless steel ball (friction probe) was moderated in toluene to result in low friction. On the other hand, the PMMA brush surface in cyclohexane would strongly interact with the stainless steel probe so as to not be in contact with a poor solvent, thus giving a higher friction coefficient. In addition, thicker brushes tend to show lower friction coefficients, probably because of blocking the direct contact of solid surfaces by brush layer. According to our study, however, a thickness of 20~30 nm is sufficient to reveal a stable friction performance.

3.2. *Water Lubrication by Hydrophilic Polymer Brush*

A water lubrication system is an essential tribology for a living thing; for example, there are excellent biolubrications system with extremely low friction in cartilage, the vitreous body of eyes, and synovial joints such as hip, knee, and finger joints. As mentioned in the Introduction, hydrophilic polymer or polyelectrolyte brushes in aqueous condition artificially afford a water lubrication system. Here we refer to our recent two works on the preparation of hydrophilic brush containing 2,3-dihydroxypropyl methacrylate (DHMA)[61] and MPC[62] respectively. Poly(DHMA) brush was prepared by hydrolysis of poly(DMM) brush with an acidic methanol solution for 6 h at room temperature. As shown in Fig. 4, the friction coefficients of poly(DHMA) brush with a sliding glass probe in air were found to be higher than those in water because hydrophilic poly(DHMA) brush in air interacts with the glass ball to increase adhesion. In the case of friction in water, the friction coefficients of poly(DHMA) brushes were reduced on account of the fluid lubrication effect of water. As water is a good solvent for poly(DHMA), the interaction between poly(DHMA) and the glass probe would be moderated, resulting in a low-friction surface.

MPC is a specially designed methacrylate containing a phosphorylcholine unit in the side chain.[63] Poly(MPC) is well known as a super-hydrophilic and biocompatible polymer;[64] therefore, control of protein adsorption and cell manipulations by well-defined poly(MPC) brush surface have attracted much attention.[65-66] Poly(MPC) brush provides a quite wettable surface. The contact angle of a water droplet on the poly(MPC) brush surface was found to be very low (below 5°). Therefore, water acts as a reasonably good solvent resulting in low friction of poly(MPC) brush. Ho *et al.* have prepared a catheter material consisting of polyurethane coated with poly(MPC) copolymer, which afforded a low-friction surface.[67] They suggested that the water molecules bound to the surface of poly(MPC) due to hydration provide effective and enhanced boundary lubrication. The application to medical implant devices such as artificial joints[68] has also been extended, taking advantage of the low friction property and biocompatibility of poly(MPC) brush.

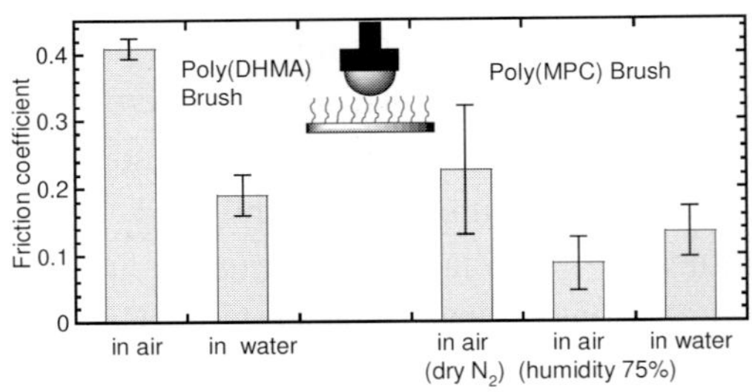

Fig. 4. Friction coefficient of poly(DHMA) (left) and poly(MPC) (right) brushes in dry air, humid air, and in water by sliding a glass ball over a distance of 20 mm at a sliding velocity of 90 mm/min under a load of 0.49 N at 298 K.

The right side bar graph in Fig. 4 shows the friction coefficient of poly(MPC) brush surface measured by sliding a glass ball in air and water at 298 K. Humidity around the sample stage was controlled by blowing dry N_2 gas or by paving a wet paper towel on the stage. Under the dry N_2 gas condition, a high friction coefficient was observed, probably due to the strong adhesion interaction between poly(MPC) brush and the glass probe. In contrast, the brush surface afforded a lower friction coefficient with increasing humidity. It is supposed that water molecules adsorbed on the surface of highly hydrophilic poly(MPC) brush, and worked as a lubricant to reduce the interaction between the brush and probe. Furthermore, water-swollen poly(MPC) brush compressed by the probe sphere would produce a repulsive force against a high normal load due to the osmotic pressure originating from steric repulsion of the high-density brushes.[69] These water-lubrication systems restricted direct contact of the probe with the substrate to reduce the friction force. Contrary to our expectations, the friction coefficient in water was higher than that in a humid air condition. We suppose that swollen and extended poly(MPC) brush chains in water likely have a larger actual contact area between the brush and the probe, resulting in a higher friction coefficient compared with the case of humid air.

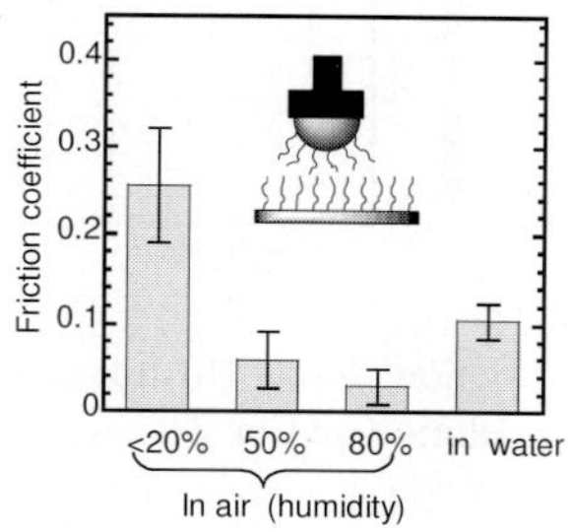

Fig. 5. Friction coefficient of poly(MPC) brush in dry air, humid air, and water by sliding a glass ball immobilized with poly(MPC) brush over a distance of 20 mm at a sliding velocity of 90 mm/min under a load of 0.49 N at 298 K.

When a brush-immobilized glass ball was used as a sliding probe, as shown in Fig. 5, the friction coefficient under the highly humidified air condition was significantly reduced to 0.02, which is lower than that in water. It is supposed that hydrated MPC units effectively reduced the friction force, creating a good water-lubrication system. The excluded volume effects of dense grafting chains might repress a mutual interpenetration of brushes in water to reduce the friction force. The authors do not have any reasonable explanation as to why poly(MPC) brush presented a higher friction coefficient in water than in the humid air condition. This result might be attributed to the interdigitation of brush chains that can occur on the outer surface of the brush layer, in particular, at the thin layer of the brush-brush interface in water because swollen polymer chains were further extended in water than under humidified air condition. The other explanation relies on the formation of a boundary layer at the brush-brush interface under the humid air condition, providing fluid lubrication and low friction, although additional experimental evidence is needed to confirm this hypothesis. Klein and co-workers have pointed out that highly compressed polymer brush layers between surfaces behave in a solid-like manner.[70] For further discussion of the frictional mechanism of poly(MPC) brush the visco-elastic properties of brush and the effects of squeezed hydrodynamic lubrication should be examined.[71]

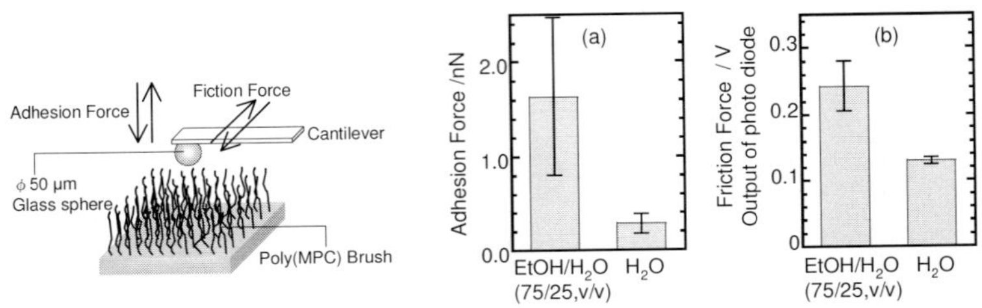

Fig. 6. Adhesion force (a) and friction force (b) measurements of high-density poly(MPC) brush in a ethanol/water solution and in pure water at 298 K: Friction force is represented by output of the photo diode at a scan rate of 10 μm/s under loading of 10 nN.

The nanotribological test for poly(MPC) was carried out by an SPA400 equipped with a 20 mm × 20 mm scanner controlled by an SPI 3800N (SII NanoTechnology Inc., Tokyo) using a colloidal silica bead (ϕ 50 μm) immobilized on cantilever (SI-DR3) by UV-sensitive glue (NOA63, Norland Pro. Inc.). The spring constant of the cantilever was 0.4 N/m. The friction force was measured at a scan rate of 10 μm/s under a load of 10 nN at 298K. The friction stroke was 5.0 μm. Figure 6 shows the adhesive and lateral force for poly(MPC) in aqueous solution. Large adhesion and friction forces were observed in a mixture solution of ethanol and water, which is known to be a poor solvent for poly(MPC).[72-73] These results suggest that the poly(MPC) surface interacts strongly with silica spheres in poor solvent to give a large adhesion force. On the other hand, the interaction between poly(MPC) and colloidal probe decreased in a good solvent such as pure water to give a low adhesion force. A similar tendency was observed in the friction force measurement. Apparently, solvent quality, standing for a good or poor affinity for polymer, affects the interaction between the brush and probe, dominating the frictional properties of polymer brushes.

3.3. *Effect of Graft Density on Wear Resistance*

The most attractive advantage for high-density polymer brush would be an improvement in wear resistance. Since the tethered polymer chain end is strongly bound with the substrate by covalent bonds, the brush layer can hardly be scratched off by a sliding probe. In our previous study, the wear resistance of PMMA brush was evaluated by a continuous friction test involving repeated sliding of a stainless steel ball. In the early stages of the friction test, the friction coefficient increased once, but there was no further increase with friction cycles. The trace scratched by the sliding probe was formed on the substrate, but the remaining PMMA component in the wear track was detected by energy-dispersive X-ray spectra.[46] In contrast, the friction coefficient of the cast film, which was physically adsorbed on a substrate, gradually peeled away to form debris or so-called "wear elements", interfering with the smooth sliding of the probe. Finally, the PMMA cast layer in the wear track had completely peeled off.

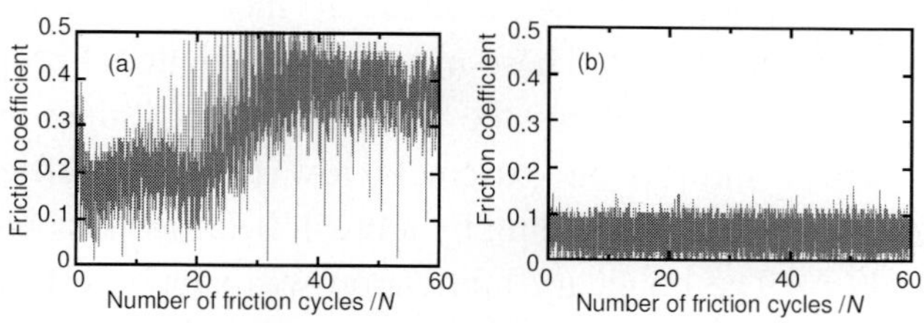

Fig. 7. Evolution of the friction coefficient vs. the number of friction cycles N for the surface of low-density poly(MPC) brush (a) and high-density poly(MPC) brush (b) at a sliding velocity of 90 mm/min under a load of 0.098 N in air: M_n of poly(MPC) = 80800, probe = glass ball, humidity > 75%.

Figure 7(a) shows the evolution of the friction coefficient of low-density poly(MPC) brush film vs. the number of friction cycles by glass ball under a normal load of 0.098 N in humid air (humidity >75%). Low-density brush was prepared by the "grafting-to" method using a poly(MPC) attached with a triethoxysilyl group at the chain end. The friction coefficient of the cast film began to increase in the early stage of the friction test, attaining a magnitude of 0.2 within 20 tracking cycles. This result indicates that the surface polymer was easily scratched away by the sliding glass probe. On the other hand, a low friction coefficient of high-density poly(MPC) brush was continuously observed, as shown in Fig. 7(b), implying a good wear resistance. Actuary, high-density brush kept the low friction coefficient around 0.1 even after 100 friction cycles. As a control experiment, the cast film was prepared by spin casting from methanol solution of poly(MPC). In an air condition over 75% humidity, the cast film was found to adsorb moisture from the air to become a swollen layer, which was then readily swept out by the sliding probe from the silicon wafer. Therefore, densely anchored polymer chains on the substrate could improve wear resistance.

4. Conclusions

This review has described recent studies of the tribological properties of high-density polymer brushes prepared by surface-initiated ATRP, and their dependence on solvent quality and humidity. For instance, the friction coefficient of polymer brush was found to decrease in response to soaking in a good solvent, whereas it increased in response to immersion in a poor solvent. These macroscopic frictional properties are related to adhesion force on the brush surface. Super-hydrophilic poly(MPC) brush showed an extremely low friction coefficient even under a humid air condition sliding by a brush-bearing glass probe under a load of 0.49 N. The boundary lubrication was supposed to have been achieved through the poly(MPC) brush even in humid air, thus significantly decreasing the friction efficient of the surface. High-density poly(MPC) brush under humid conditions demonstrated a stable friction coefficient even after sliding over 100 friction cycles, indicating that the graft density is very effective against wear resistance.

Tribological properties, especially normal and shear forces, of polymer brushes have been well-investigated for the last two decades because of their extensive applications to colloid stability and surface modifications. In contrast, however, experimental analysis relating to the frictional properties of the truly well-defined high-density polymer brushes has just been started. Fortunately, various types of well-defined high-density polymer brushes are now readily available owing to the recent marvelous development of a living polymerization technique. As mentioned above, the "grafting-to" method limits graft densities due to the steric hindrances of the already grafted chains. In contrast, the "grafting-from" method combined with a controlled/living polymerization technique affords the highly "concentrated" brush, which is expected to exhibit different properties from those of low-density brush in consequent of the high-order interaction between graft chains. The repulsive force should increase much more with increasing graft density.[74] Therefore, the authors expect that an excellent lubrication system based on high-density brush can be achieved, and that this work can contribute to the field of tribological study of well-defined and well-designed polymer brushes.

References

1. P. Mansky, Y. Liu, E. Huang, T. P. Russell, and C. Hawker, *Science*, 275, 1458 (1997).
2. R. C. Advincula, W. J. Brittain, K. C. Caster and J. Ruhe, *Polymer Brushes: Synthesis, Characterization, Applications*, Wiley VCH, Weinheim (2004).
3. Y. Tsujii, K. Ohno, S. Yamamoto, A. Goto and T. Fukuda, *Adv. Polym. Sci.*, 197, 1 (2006).
4. B. Zhao and W. J. Brittain, *Prog., Polym. Sci.*, 25, 677 (2000).
5. A. Halperin, M. Tirrell and T. P. Lodge, *Adv. Polym. Sci.*, 100, 31 (1992).
6. J. Klein, *Ann. Rev. Mater. Sci.*, 26, 581 (1996).
7. H. J. Taunton, C. Toprakcioglu, L. J. Fetters and J. Klein, *Nature*, 332, 712 (1988).
8. J. Klein, E. Kumacheva, D. Mahalu, D. Perahia and L. J. Fetters, *Nature,* 370,634 (1994).
9. E. Pelletier, G. F. Belder, G. Hadziioannou and A. Subbotin, *J. Phys II (France)*, 7, 271 (1997).
10. T. Drobek, N. D. Spencer and M. Heuberger, *Macromolecules*, 38, 5254 (2005).
11. J. Klein, E. Kumacheva, D. Perahia and L. J. Fetters, *Acta Polym.*, 49, 617 (1998).
12. R. Tadmor, J. Janik, J. Klein and L. J. Fetters, *Phys. Rev. Lett.*, 91, 115503 (2003).
13. U. Raviv, J. Frey, R. Sak, P. Laurat, R. Tadmor and J. Klein, *Langmuir*, 18, 7482 (2002).
14. A. M. Forster, J. W. Mays and S. M. Kilbey II, *J. Polym. Sci., Part B Polym. Phys.*, 44, 649 (2005).
15. G. S. Grest, *Adv. Polym. Sci.*, 138, 149 (1999).
16. P. A. Schorr, T. C. B. Kwan, S. M. Kilbey II, E. S. G. Shaqfeh and M. Tirrell, *Macromolecules*, 36, 389 (2003).
17. T. Kreer, M. H. Müser, K. Binder and J. Klein, *Langmuir*, 17, 7804 (2001).

18. J. Klein, E. Kumacheva, D. Perahia and L. J. Fetters, *Acta Polym.*, 49, 617 (1998).
19. A. Dhinojwala and S. Granick, *Macromolecules*, 30, 1079 (1997).
20. J. Klein, *Ann. Rev. Mater. Sci.*, 26, 581 (1996).
21. A. M. Forster, J. W. Mays and S. M. Kilbey II, *J. Polym. Sci. Part B Polym. Phys.*, 44, 649 (2006).
22. M. Ruths, D. Johannsmann, J. Rühe and W. Knoll, *Macromolecules*, 33, 3860 (2000).
23. E. Eiser, J. Klein, T. A. Witten and L. J. Fetters, *Phys. Rev. Lett.*, 82, 5076 (1999).
24. J. Ruhe, M. Ballauff, M. Biesalski, P. Dziezok, F. Gröhn, D. Johannsmann, N. Houbenov, N. Hugenberg, R. Konradi, S. Minko, M. Motornov, R. R. Netz, M. Schmidt, C. Seidel, M. Stamm, T. Stephan, D. Usov and H. Zhang, *Adv. Polym. Sci.*, 165-166, 79 (2004).
25. I. Luzinov, S. Minko and V. V. Tsukruk, *Prog. Polym. Sci.*, 29, 635 (2004).
26. S. J. Miklavic and S. Marčelja, *J. Phys. Chem.*, 92, 6718 (1988).
27. R. Israels, F. A. M. Leermakers, G. J. Fleer and E. B. Zhulina, *Macromolecules*, 27, 3249 (1994).
28. Y. V. Lyatskaya, F. A. M. Leermakers, G. J. Fleer, E. B. Zhulina and T. M. Birshtein, *Macromolecules*, 28, 3562 (1995).
29. E. B. Zhulina, J. K. Wolterink and O. V. Borisov, *Macromolecules*, 33, 4945 (2000).
30. P. Pincus, *Macromolecules*, 24, 2912 (1991).
31. R. S. Ross and P. Pincus, *Macromolecules*, 25, 2177 (1992).
32. V. A. Pryamitsyn, F. A. M. Leermakers, G. J. Fleer and E. B. Zhulina, *Macromolecules*, 29, 8260 (1996).
33. S. Minko, *J. Macromol. Sci. Part C Polym. Rev.*, 46, 397 (2006).
34. T. McPherson, A. Kidane, I. Szleifer and K. Park, *Langmuir*, 14, 176 (1998).
35. S. Hayashi, T. Abe, N. Higashi, M. Niwa and K. Kurihara, *Langmuir*, 18, 3932 (2002).
36. T. W. Kelley, P. A. Shorr, D. J. Kristin, M. Tirrell and C. D. Frisbie, *Macromolecules*, 31, 4297 (1998).
37. U. Raviv, S. Giasson, N. Kamph, J. F. Gohy, R. Jérôme and J. Klein, *Nature*, 425, 163 (2003).
38. N. Kamph, J. F. Gohy, R. Jérôme and J. Klein., *J. Polym. Sci. Part B Polym. Phys.*, 43, 193 (2005).
39. Y. Ohsedo, R. Takashina, J. P. Gong and Y. Osada, *Langmuir*; 20, 6549 (2004).
40. S. Lee, M. Müller, M. Ratoi-Salagean, J. Vörös, S. Pasche, S. M. De Paul, H. A. Spikes, M. Textor and N. D. Spencer, *Tribology Lett.* 15, 231 (2003).
41. M. Müller, S. Lee, H. A. Spikes and N. D. Spencer, *Tribology Lett.* 15, 395 (2003).
42. M. Ejaz, S. Yamamoto, K. Ohono, Y. Tsujii and T. Fukuda, *Macromolecules*, 31, 5834 (1998).
43. K. Matyjaszewski, *Advances in Controlled/Living Radical Polymerization, ACS Symp. Ser.* 854, ACS (2003).
44. S. Yamamoto, M. Ejaz, Y. Tsujii and T. Fukuda, *Macromolecules* 33, 5602 (2000).

45. Y. Tsujii, K. Okayasu, K. Ohno, and T. Fukuda, *Polym. Prepr. (Am. Chem. Soc., Div. Polym. Chem.)* 46(2), 85 (2005).
46. H. Sakata, M. Kobayashi, H. Otsuka and A. Takahara, *Polym. J.*, 37, 767 (2005).
47. R. Berger, Y. Cheng, R. Förch, B. Gotsmann, J. S. Gutmann, T. Pakula, U. Rietzler, W. Schärtl, M. Schmidt, A. Strack, J. Windeln and H. -J. Butt, *Langmuir*, 23, 3150 (2007).
48. M. D. K. Ingall, C. H. Honeyman, J. V. Mercure, P. A. Bianconi and R. R. Kunz, *J. Am. Chem. Soc.*, 121, 3607 (1999).
49. B. Zhao and W. J. Brittain, *J. Am. Chem. Soc.*, 121, 3557 (1999).
50. J. -B. Kim, M. L. Bruening and G. L. Baker, *Am. Chem. Soc.*, 122, 7616 (2000).
51. R. Matsuno, K. Yamamoto, H. Otuska and A. Takahara, *Macromolecules*, 37, 2203 (2004).
52. R. Matsuno, H. Otuska and A. Takahara, *Soft Matt.*, 2, 415 (2006).
53. M. Kobayashi, R. Matsuno, H. Otuska and A. Takahara, *Sci. Tech. Adv. Mat.*, 7, 617 (2006).
54. K. Matyjaszewski, P. J. Miller, N. Shukla, B. Immaraporn, A. Gelman, B. B. Luokala, T. M. Siclovan, G. Kickelbick, T. Vallant, H. Hoffmann and T. Pakula, *Macromolecules*, 32, 8716 (1999).
55. H. Tada and H. Nagatama, *Langmuir*, 10, 1472 (1994).
56. K. Ohno, T. Morinaga, K. Koh, Y. Tsujii and T. Fukuda, *Macromolecules*, 38, 2137 (2005).
57. T. von Werne and T. E. Patten, *J. Am. Chem. Soc.*, 121, 7409 (1999).
58. M. Husseman, E. E. Malmstrom, M. McNamura, M. Mate, D. Mecerreyes, D. G. Benoit, J. L. Hedrick, P. Mansky, E. Huang, T. P. Russell and C. J. Hawker, *Macromolecules*, 32, 1424 (1999).
59. J. Pyun, S. Jia, T. Kowalewski, G. D. Patterson and K. Matyjaszewski, *Macromolecules*, 36, 5094 (2003).
60. If a circle with radius a (m) is regarded as the contact area between a stainless steel ball and substrate under a normal load P (0.49 N), Hertz's theory affords the following relationship using Young's modulus of stainless steel and silicon wafer, E_A (1.96×10^{11} Pa), E_B (1.30×10^{11} Pa), and Poisson's ratio υ_A (0.30), υ_B (0.28), respectively;
$2/E = (1 - \upsilon_A^2) / E_A + (1 - \upsilon_B^2) / E_B$
$a = (3/4 \times 2/E \times P \times R_A)3/2$
where R_A is curvature radius (5.00×10^{-3} m) of stainless steel ball. The contact area can be calculated by πa^2.
61. M. Kobayashi and A. Takahara, *Chem. Lett.*, 34, 1582 (2005).
62. M. Kobayashi, Y. Terayama, N. Hosaka, M. Kaido, A. Suzuki, N. Yamada, N. Torikai, K. Ishihara and A. Takahara, *Soft Matter*, 3, 740 (2007).
63. K. Ishihara, T. Ueda and N. Nakabayashi, *Polym. J.*, 22, 355 (1990).
64. K. Ishihara, H. Nomura, T. Mihara, K. Kurita, Y. Iwasaki and N. Nakabayashi, *J. Biomed. Mater. Res.*, 39, 323 (1998).

65. R. Iwata, P. Suk-In, V. P. Hoven, A. Takahara, K. Akiyoshi and Y. Iwasaki, *Biomacromolecules*, 5, 2308 (2004).
66. W. Feng, S. Zhu, K. Ishihara and J. Brash, *Biointerphases*, 1, 50 (2006).
67. S. P. Ho, N. Nakabayashi, Y. Iwasaki, T. Boland and M. LaBerge, *Biomaterials*, 24, 5121 (2003).
68. T. Moro, Y. Takatori, K. Ishihara, T. Konno, Y. Takigawa, T. Matsushita, U. Chung, K. Nakamura and H. Kawaguchi, *Nature Mat.*, 3, 829 (2004).
69. S. Yamamoto, M. Ejaz, Y. Tsujii, M. Matsumoto and T. Fukuda, *Macromolecules*, 33, 5608 (2000).
70. J. Klein, Y. Kamiyama, H. Yoshizawa, J. N. Israelachivili, G. H. Fredrickson, P. Pincus, and L. J. Fetters, *Macromolecules* 26, 5552 (1993).
71. G. H. Fredrickson and P. Pincus, *Langmuir*, 7, 786 (1991).
72. Y. Kiritoshi and K. Ishihara, *J. Biomater. Sci. Polym. Ed.*, 13, 213 (2002).
73. Y. Matsuda, M. Kobayashi, Y. Annaka, K. Ishihara and A. Takahara, *Polym. J.*, 40, 479 (2008).
74. P. Y. Laid and A. Halperin, *Macromolecules*, 24, 4981 (1991).

CHAPTER 19

AFM TESTING OF POLYMERIC RESIST FILMS FOR NANOIMPRINT LITHOGRAPHY

Zygmunt Rymuza

Warsaw University of Technology, Institute of Micromechanics and Photonics
sw.A.Boboli 8, 02-525 Warsaw, Poland
E-mail: z.rymuza@mchtr.pw.edu.pl

This paper reviews the current materials selection criteria for polymeric resists used in Nanoimprint Lithography (NIL).The tribological and adhesive behavior of some selected polymeric NIL resists is presented. The paper concludes with an offer of several solutions to overcoming tribological problems and the problem of adhesion experienced in NIL.

1. Introduction

The nanoimprint lithography (NIL) technology was invented in the USA by Chou and collaborators in 1995.[1-8] Nanoimprint lithography is one of the most promising low-cost, high-throughput technologies for manufacturing nanostructures. As the inventor Chou reported recently during the 20th International Microprocess and Nanotechnology Conference MNC2007 in Kyoto (Japan)[9] that the NIL technology could be considered to be in the 1990's in its childhood in the early 2000's and now entering its adolescence. Many applications are now emerging, for instance, in the fields of optoelectronics and biotechnology. In the future it is expected that NIL will find great utility in organic nano-electronics, nano-optoelectronics and nano-mechatronics technologies.

NIL is primarily a physical deformation process and is different from stamping with self-assembly ink and other lithographical methods. The process is enough simple but unsolved problems have appeared that are connected primarily with the contact, sliding and physical deformation of the polymeric thin films which are most often used as material for formation of nanostructures.

A very important and unique feature of NIL is that it is a 3 dimensional pattern technology. The 3D features manufactured are however connected with the contact on a relatively large tool area and formed material that involves friction and adhesion at the interface. This interfacial contact of two different materials with relative motion on nanoscale is a good example of a nanoscale polymeric tribosystem. The tribological phenomena which occur during the imprint process influence dramatically the imprint quality, the system reliability and the number of high quality imprints which can be performed by a mold.

2. Process of Imprint

The nanoimprint process is relatively simple. Imprinting process is made using a mold with nanostructures on its surface. This mold is used to deform a thin resist film deposited on a substrate (Fig. 1).

The resist can be thermoplastic or UV (or thermal) curable polymer. In the UV process the mold is made of quartz and a laser beam is used to cure the resist under pressure. Then the mold is released. In this case a monomer mix, often in a liquid thin film form on a substrate, is used. The polymer is cured i.e. polymerization and cross-linking processes occur in the liquid and after that the mold is separated from the formed nanostructures.

In the thermal process, with the use of a thermoplastic resist, the resist is heated above its glass temperature. When the resist becomes viscous liquid, the mold is pressed into the resist. The imprinting can be done on a softened resist rather than in the viscous liquid state but time of the process as well as the pressure needed increase, quality of imprints and resolution significantly decrease. In general, such trade-offs need to be addressed to define process windows, which comprise between low cost, high throughput, high yield, good repeatability and reliability of a process. The typical used pressure is 1-10 MPa. The rate of the vertical movement of the mold is in the range of nanometer-micrometer per

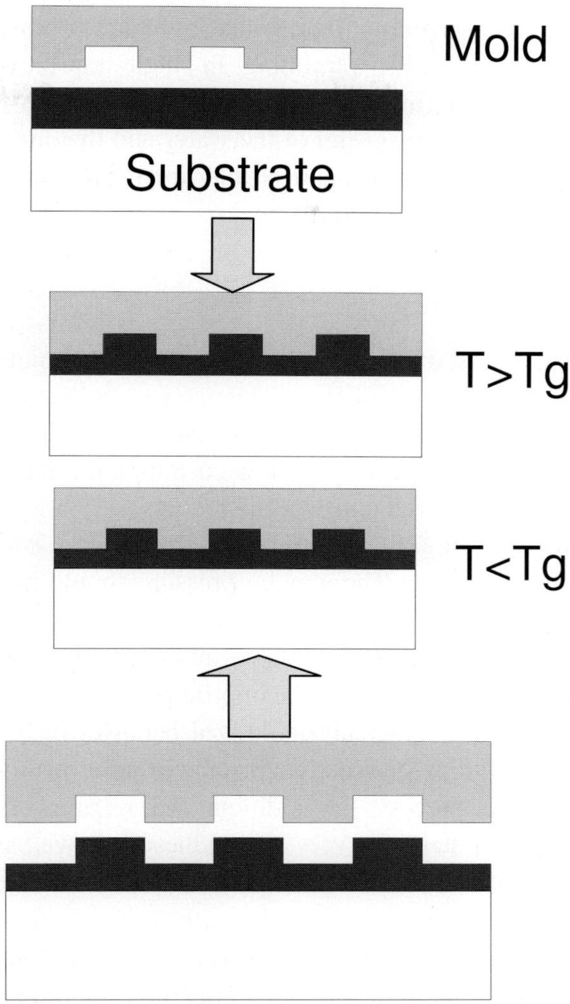

Fig. 1. Model of thermal classic imprint process composed of following steps (from top to bottom): 1-Heat mold and substrate, 2-Press mold and hold, 3-Cool down and hold, 4-Release mold. T_g – glass transition temperature.

second. It describes the relative sliding speed of the mold and resist material during pressing and releasing phases respectively.

The resolution of the nanoimprint lithography is determined by the mechanical strength of the mold and the polymer. Nowadays the resolution below 10 nm is achievable.[10]

The three types of imprint machines are used: single-imprint, step-and-repeat, and roller imprinting. In single-imprint the machine imprints an entire wafer (usually silicon) at once; step-and-repeat NIL machine imprints a small area (die) of the wafer and then moves to a new area of the wafer and imprints again. Roller imprinting has advantages of better uniformity and less imprint force.

3. NIL Problems

The simplicity of the NIL nanofabrication technique is very attractive. However, many difficult problems can arise during imprinting. In general, as was mentioned earlier, the problems are connected with the needed compromise between low cost, high throughput, high yield, good repeatability and reliability of the process.

The accurate filling of cavities in the mold by liquid (melt) polymer is an important issue. The applied pressure should be optimized to obtain complete pattern transfer in the shortest possible time. The knowledge of the visco-elastic properties of ultrathin polymer films used in NIL is a key to the improvement of the imprint process.

The nano-rheological/tribological behavior of ultrathin polymer film during imprint process determines the imprint quality and the throughput of the NIL process. In addition, the release of the mold requires overcoming potentially very large adhesion as well as high friction at the interface mold/resist interface. The detachment force (force needed to release the mold from the resist with its fabricated nanostructures) is a sum of a pull-off force (on horizontal surfaces) and friction force (on vertical surfaces) in the contact. This force should be decreased as much as possible to avoid mechanical damage of formed nanostructures and to increase reliability of the process as well as the lifetime of the mold.

Polymer properties such as viscosity, molecular weight, chain dimensions, glass transition temperature, shrinkage, break limit, adhesion properties against substrate and mold material, etch behavior in dry and wet conditions (for easy removal of residual layer after imprinting) are most important for the successful imprint process. The properties of the used polymeric resist in NIL are completely different from the properties of the resist used in other lithographies (e.g. classic photolithography in microelectronic or Micro Electro Mechanical Systems (MEMS) fabrication[11-14]). The critical properties of the polymer used in NIL are mechanical, tribological and adhesive, whereas in other lithographies the chemical properties are important as radiation sensitivity, resolution and contrast. The best polymer used in NIL should have low viscosity above its glass transition temperature; i.e. over this temperature the chemical forces holding the chain segment of an amorphous polymer together should be as small as possible.

Very important in NIL is an understanding of the behavior of the ultra-thin (below 1000 nm thick) polymer film during both the molding and releasing periods of the imprint process. The relation between the polymer properties and other imprint variables like pressure and imprint time should be determined in developing imprint lithography.

As was mentioned earlier, the critical parameter is adhesion (and friction) of polymer to the mold which can cause damage to the nanoimprint in the form of abrasion and distortion. Also the mold can be mechanically damaged and it is possible that residual polymer structures can contaminate its surface. Covering the mold surface with an anti-adhesive monolayer can reduce adhesion but the lifetime of such a layer is limited.

Nanoimprint lithography is a good candidate for large area nanolithography. Of course it needs multilevel technology to be developed. The main problem is alignment but the progress in this area is enormous.[15] The progress in alignment technology enables the fabrication of sub-30 nm structures using a full wafer mold with wafers up to 8 or even 12 inches in diameter.

Molds (stamps) are manufactured by different methods. One of them is via a metal lift-off process.[1] Such molds are suitable for short and fast use. Si or SiO_2 molds are made by plasma etching process using metal lift-off layer as an etching mask; such molds are good for long-term use. An example of another method is electroplated metal (usually nickel) mold on Si-substrate; such molds are suitable for long-term use and are stable. For fabrication of extremely small nano-cavities in the mold surface electron- and FIB (Focused Ion Beam) lithographies are useful but the cost of the production is very high. Mechanical, tribological (friction ability and wear resistance) and adhesive properties of the mold surface layer are of most importance.

4. Testing of Materials for NIL

4.1. *Introduction*

The evaluation of the properties of materials for NIL application enables the selection of the best material before its application in further real imprint processes. It is an important way to reduce very high costs of testing when several material candidates are tested directly by NIL experiments. As can be concluded from the previous discussion on NIL, the evaluation of the mechanical, tribological/rheological and adhesive/stiction properties of ultrathin polymeric resist films devoted for nanoimprinting is very important task for the preliminary selection of right materials for the imprint fabrication.

A set of such tests was conducted in our laboratory.[16-17] These tests fall into groups: Atomic Force Microscopy (AFM) studies of the surface morphology (roughness), lateral (friction) and adhesion (pull-off force) of the film deposited on silicon substrate; wettability studies of the film and mechanical, tribological and adhesive tests versus temperature.

In the first group of measurements, commercially available AFM cantilevers are used and the tests are performed in air-conditioned laboratory at room temperature. The data on roughness parameters characterizing the surface geometry of the spin-coated ultra-thin (200-300 nm) resist film are collected and compared with the results of friction (lateral) force and pull-off force measurements.

In the second group of tests, the wettability of the resist film is studied. For these tests, a specially constructed computer controlled test rig with CCD cameras is used. Contact angle measurements are performed automatically from the analysis of the droplet's image. The statically deposited series of droplets are imaged after about 50 ms. The diameter of the sessile droplet is between 40-200 μm. Two liquids (water and diiodomethane) with known polar and dispersion components of the surface tension are used. The Owens-Wendt method is applied to estimate polar and dispersion components of the surface free energy of the tested film. The tests are carried out in a special clean-room.

In the third group of tests the AFM-based special microtribometer is used to perform mechanical, tribological/rheological and adhesive/stiction tests on the resist film versus temperature. The specially designed and constructed stiff (200-1000 N/m) AFM cantilever made of beryllium bronze foil with glued ball (0.7 mm in diameter) is used in the tests. Also other commercially available cantilevers with glued micro-balls made of various materials are applied. The heating stage is used to change the temperature of the tested film. The studies are performed as the temperature is both increased and decreased. The Young's modulus is estimated from nanoindentation of the film. A friction loop is than performed to study static (maximum value taken as stiction parameter) and kinetic friction and also preliminary displacement (i.e. the lateral displacement of the surface of the film before transfer to sliding of the probe on the tested film), which enables the nano-rheological behaviour of the film to be charcterized. The pull-off force characterizing adhesive properties of the film is then determined from the Force-Distance Curve (FDC). The AFM-based micro-tribometer is also equipped with laterally oscillating probe. The tests are performed in a special clean-room.

The commercially available cantilevers used in the AFM studies are calibrated by the use of specially constructed micro-springs and the real values of the measured parameters are known.

The material studies are mainly focused on the polymeric resists as their behaviour decides about the final result of the imprint process, however, studies relating directly to the mold material have also been undertaken to estimate, in particular, mechanical properties of the nanometer thick surface layer of molds as well as their wear resistance.

Preliminary experiments have also been carried out to measure the friction coefficient between real mold surfaces and resist films. AFM friction measurements have been initiated to estimate friction coefficient on the real vertical walls of imprinted polymeric nanostructures by the use of special cantilevers.

Recently, we have also constructed a special test rig that enables the experimental modeling/simulation of the NIL process including the measurements of the separation force during detachment of the mold (stamp) from the resist film after imprinting.

4.2. *AFM Studies of Surface Topography, Friction and Adhesion of Ultrathin Resist Films*

The surface topography (roughness) of a resist film after spin-coating and baking, together with its frictional and adhesive (surface free energy) properties are important parameters during molding phase of NIL. The surface free energy is in particular very important in the de-molding phase. The molding process is much more important for the quality of obtained polymer mask in NIL process then de-molding.

The following aspects of roughness of resist film after spin-coating and baking are important for molding phase of NIL process:

- the polymer surface has to be uniform.
- if the applied polymer is rough then the resulting mask (nanoimprint) will effectively be full of defects.
- in case of rough surfaces it is possible that surface won't degas completely during the pumping time associated with the molding process.
- in practice, having a smooth polymer surface increases the probability of obtaining good molding results.
- the softened or melted polymer must have the same floating rate on the entire surface of the substrate which, except for the homogenous heating, depends on the uniformity of the polymer film and on the surface topography of the mold; the polymer during floating should easily wet the mold.

The following aspects of the surface free energy and frictional properties of the mold and resist are important during molding phase of NIL process:

- with the value of surface free energy and its components, interface properties between materials of anti-sticking layer and resist polymer can be described.

- in principle, it is desirable if both materials have a low surface free energy at room temperature and different dominant components of the surface free energy. This will guarantee the probability that both materials will be able to create the same kind of bonding is very small. Also it should have good impact on small features reproduction in the molding process.
- differences in the contact angle measurements in different areas have to be small (homogeneous surface).

The de-molding of mold and resist mask in NIL process is relatively fast and can not be fully controlled. However, in the case where an anti-sticking layer is in good condition and small lateral forces are involved in the disconnection, then the final effect can be very good. In principle, in this process three types of interaction can be described; namely friction in the horizontal direction to surface of the mold, friction in vertical direction to surface of the mold, and adhesive bond disconnection. In this process most important is the lifetime of the anti-sticking coating. If the coating is defective, then it is not possible to have good results from subsequent moldings. These reasons have been taken into consideration in deciding the set of tests to evaluate the resist materials.

The use of AFM has enabled us to obtain good 3D images of topography within sub-nanometer resolution. The roughness parameters were estimated and plots with comparison of the roughness of various polymeric resist films have been used to see the differences between resist films. The AFM technique was sensitive enough to see even the effect of plasma treatment (applied mainly to change adhesive properties) of the film on its roughness (Fig. 2).

The AFM measurements of lateral (friction) force revealed that this technique is very sensitive to find differences in the surface properties of the resist film; 2D maps of the friction force distribution, the homogeneity of the surface layer of the resist film to be quantified, as well as its frictional properties. This test was also very effective to see the difference of the frictional surface properties of the resist film after plasma treatment as compared to the state of the surface layer of an untreated film (Fig. 3).

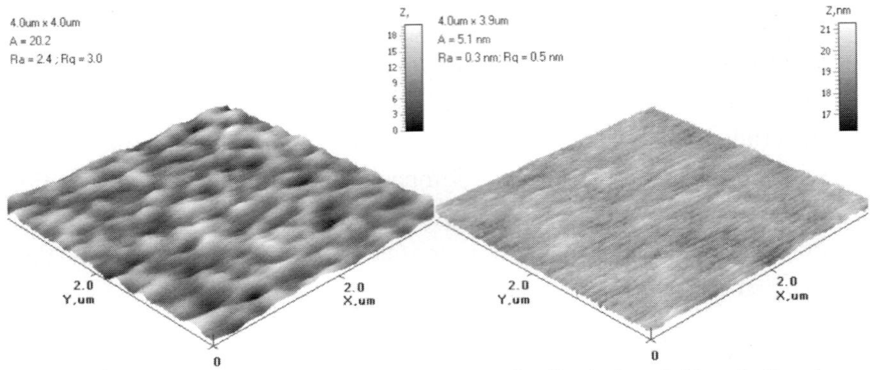

Fig. 2. Surface topography images of MR-I 7030 resist film before (left) and after plasma treatment.

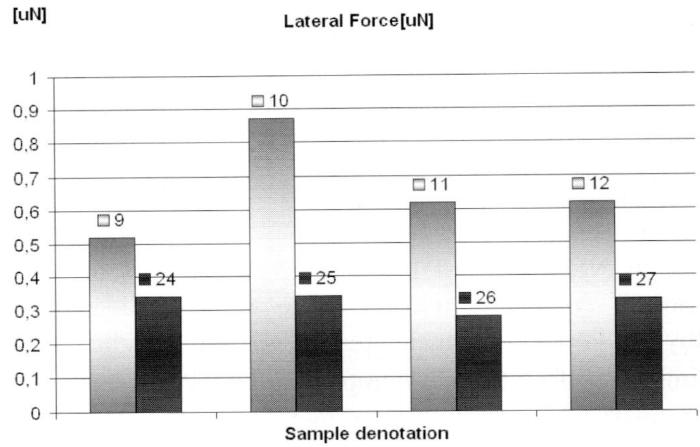

Fig. 3. Effect of SF_6 plasma treatment of NEB22 resist films on lateral (friction) force (left column- before treatment, right – after treatment). Numbers relate to films with various thicknesses: 9-213 nm, 10-303 nm, 11-194 nm, 12-267 nm. Applied load 4.5 μN, scanning speed 9.5 μm/s. AFM cantilever with silicon tip (radius 10 nm) was used.

Adhesive properties have also been evaluated by use of an AFM technique. Force-distance curves (FDC) were performed for various polymeric resist materials.[18-19] It was possible to observe sometimes significant differences in pull-off force for different films (Fig. 4). The value of the pull-off force between the tip of the AFM cantilever and the resist material during detachment of the tip from the film can be used as a measure of sticking ability of the mold and resist film.

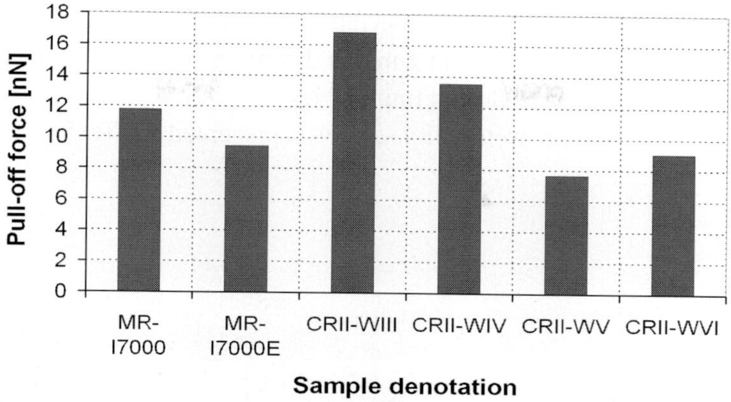

Fig. 4. Comparison of pull-off force between resist film and cantilever's tip measured with AFM for different polymeric resist materials. Normal load applied prior to pull-off was about 5 nN. Nominal imposed detachment velocity was 1 μm/s. AFM cantilever with silicon tip (radius 10 nm) was used.

The AFM evaluation of these three important surface properties of the resist film was effective enough to find differences of many tested materials.

4.3. *Wettability Studies*

The wettability test defines precisely the adhesive properties of the surface.[20-21] The wettability is studied by the measurements of the contact angle.

The contact angle measurement (enabling also estimation of the surface free energy) is very good and simple method to describe surface properties. This method does not need any special shape of the sample or prior preparations. This kind of measurement combined with other standard methods of a surface characterization i.e. AFM is powerful tool for investigation and identification of the surface properties. With the knowledge about the surface properties of the resist film (as well as mold material without or with anti-stiction coating) the whole process of suitable materials selection to decrease adhesion (and also friction) can be much more effective and faster. As it was stated above there is no need, in this important evaluation of material's adhesion, to check every possible material in imprint process itself. The contacted materials can be verified after the tests; those that match desired properties can be applied to further investigations during imprinting.

The setup used for wettability studies is presented in Fig. 5.

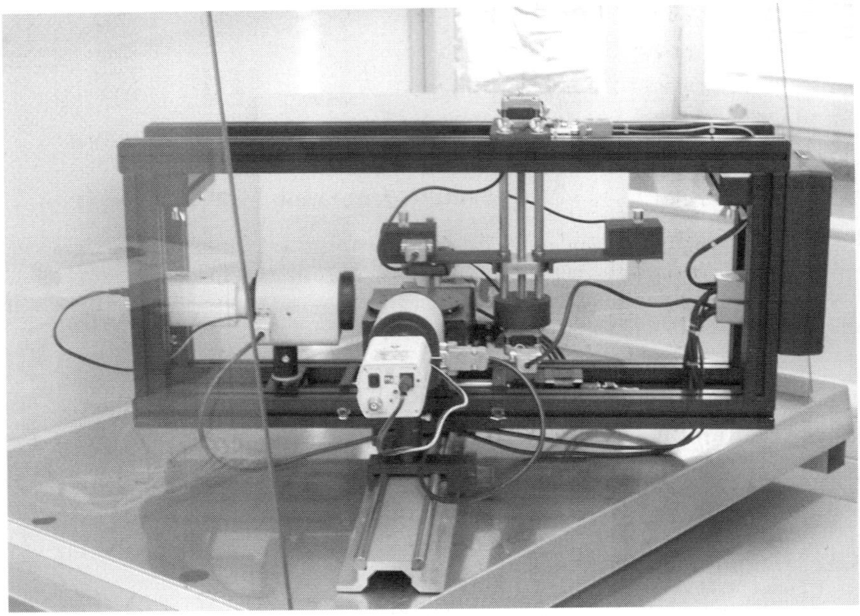

Fig. 5. Test rig used for contact angle measurements.

The droplet is deposited on investigated surface statically, that means without dynamic interaction between the surface and the dosing module (dispenser). Minimum repeatable diameter of the drop is approximately 40 μm for two different testing liquids. Chemically pure water and diiodomethane are needed to determine surface free energy (both polar and dispersion components) of tested material with the application of Owens-Wendt method[22-23]). During about 100 ms after drop deposition the image of the droplet is captured with a CCD camera and PC frame grabber, subsequent image analysis gives the physical shape of the drop. The time when image is captured, because of evaporation, has to be exactly the same each time in order to keep measurement error at a low level. The tests are performed in the ambient of the clean-room lab at the temperature 21°C ± 0.5°C and humidity 50% ± 1%.The accuracy of contact angle determination is about 1%.

Owens-Wendt method is applicable only to materials when the surface free energy is not higher then 60 mJ/m^2. Usual value of surface free energy for modified and not modified surface of polymers is located in area between 20-50 mJ/m^2.[24]

The surface free energy is expressed as follows:

$$\gamma_S = \gamma_S^d + \gamma_S^p \qquad (1)$$

γ_S – surface free energy of solid
γ^p_S – polar component surface free energy;
γ^d_S – dispersion component of surface free energy

Young's equation is a follows

$$\gamma_L \frac{1+\cos\Theta}{2} = (\gamma_S^d \gamma_L^d)^{0,5} + (\gamma_S^d \gamma_L^d)^{0,5} \qquad (2)$$

γ_L – surface tension of liquid; Θ- liquid-solid contact angle.

Two liquids: water γ_L = 72.8 mJ/m^2 with dispersion component γ_d = 21.8 mJ/m^2 and polar component γ_p = 51 mJ/m^2 as well as diiodomethane γ_L = 50.8 mJ/m^2; γ_d = 48.5 mJ/m^2; γ_p = 2.3 mJ/m^2 are used.

After some mathematical operations on afore-mentioned equation (2), the dispersion and polar components of surface free energy of tested polymer can be calculated from the following set of two equations (in which are introduced already values of polar and dispersion components of two used liquids):

$$\begin{aligned}(\gamma_S^d)^{0.5} + 1.53(\gamma_S^p)^{0.5} &= 7.80(1+\cos\Theta_1)\\ (\gamma_S^d)^{0.5} + 0.22(\gamma_S^p)^{0.5} &= 3.65(1+\cos\Theta_2)\end{aligned} \qquad (3)$$

where Θ_1 is water contact angle and Θ_2 is diiodomethane contact angle.

From the set of two last equations it is possible to easily estimate the polar and dispersion components of the surface free energy of the investigated polymeric resist film.

The wettability tests enabled an estimation of the surface free energy of various polymeric resist materials which were found to be in the range of 30-50 mJ/m^2. The dispersion component of the surface free energy of the tested resists is dominant.

The tests were very useful to study the effect of plasma treatment to decrease the surface free energy of the resists as well as the ratio of the dispersion to polar components of the surface free energy. The comparison of the surface free energy of the resist films, before and after plasma treatment, is shown in Fig. 6.

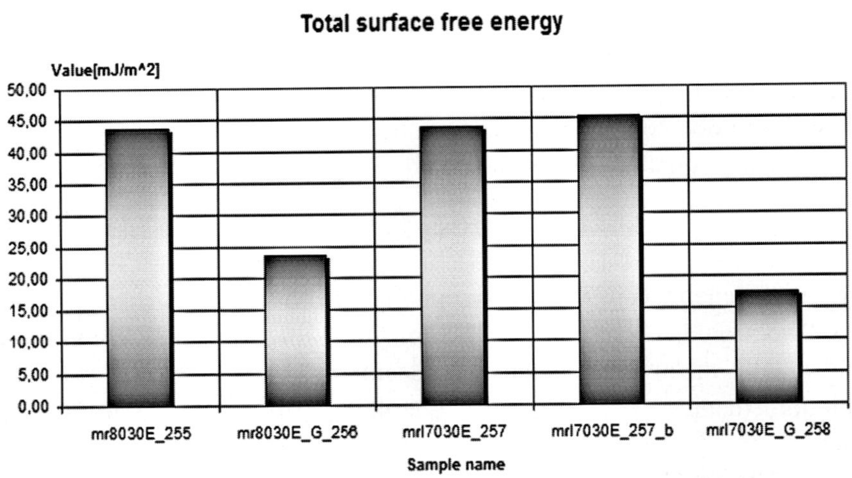

Fig . 6. Surface free energy of polymeric resist films before (first, third and fourth column from left to right) and after plasma treatment (second and last column from left to right respectively).

The plasma treatment was very effective in changing the ratio between the polar and dispersion components of the surface free energy of the tested polymeric resist films (Figs. 7 and 8).

By plasma treatment the surface free energy of testes resists was significantly decreased. The dispersion component of the resist dominated before the plasma treatment. After plasma treatment the polar component was bigger as compared with the dispersion component of the surface free energy of the tested polymeric resist films.

The wettability tests have also been successfully used to estimate the surface energetics of the mold surface layers as well as the surface free energy of anti-sticking coatings deposited on molds.

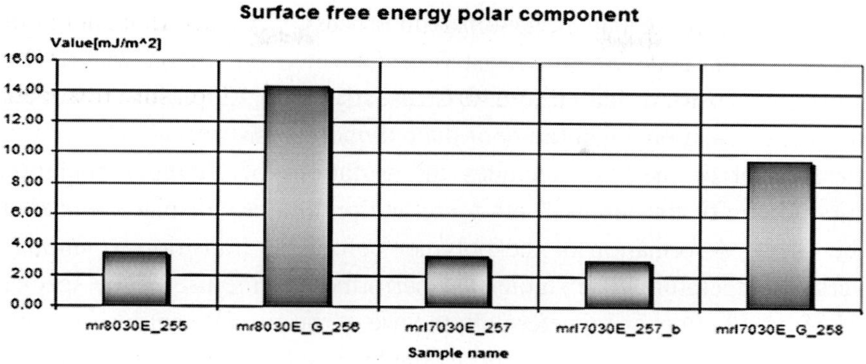

Fig. 7. Polar component of surface free energy of polymeric resist films (before and after plasma treatment) presented in Fig. 6.

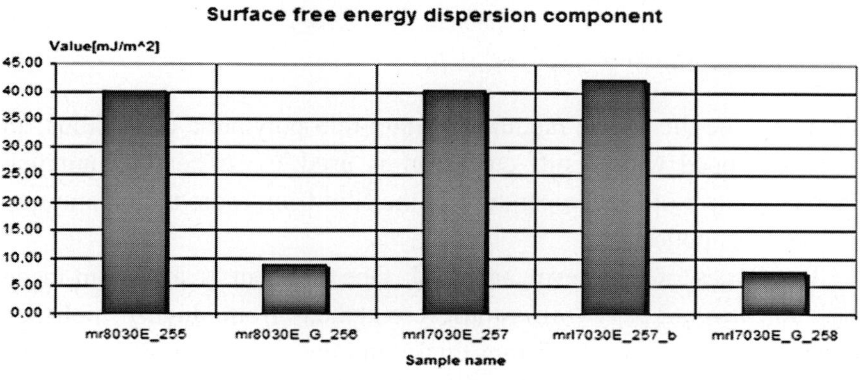

Fig. 8. Dispersion component of surface free energy of polymeric resist films (before and after plasma treatment) presented in Fig. 6.

4.4. *Studies of Properties of Polymeric Resist Films versus Temperature*

4.4.1. *Introduction*

The studies properties versus temperature concerning of the variation of the mechanical, tribological (frictional), adhesive as well as rheological properties are very important to further evaluate the behavior of the polymeric resist during the nanoimprint process. The tests are carried out on ultra-thin (200-300 nm) resist film deposited on silicon wafer. The temperature was varied up to 140°C (usually up to 125°C) to cross over the glass transition temperature of the polymeric resist.

The set of the tests includes the variations of Young's modulus, friction coefficient and pull-off force and preliminary displacement (i.e. the lateral deformation of the polymer before the transition to sliding) versus temperature. The studies are performed by the use of the special AFM-based micro-tribometer with the use of the special cantilevers; both custom made and commercially available. Small balls are added as the tips of these cantilevers. Various ball materials have been used to examine the afore-mentioned properties of the resist films.

The experiments are performed in air (21°C, relative humidity 50%) in clean-room conditions.

4.4.2. *Estimation of Young's Modulus*

To determine the elastic modulus of ultra-thin polymeric resist films, an AFM equipped with a stiff cantilever is used.[25,26] A heating unit that enables control of the temperature of the film from room temperature up to 140°C is applied.

Two types of cantilever are used. One of them is a custom-made high-stiffness (220 N/m) cantilever etched from highly polished beryllium bronze foil (0.07 mm thick) and coated with a 300 nm thick evaporated gold reflection layer. A steel bearing smooth ball with a diameter of 0.691 mm was used as a tip. Another used AFM probe is a rectangular silicon cantilever from MikroMasch (Estonia) modified by Novascan (USA). The modification involved the attachment of a 5 μm diameter SiO_2 particle to the end of the probe. The normal spring constant is k_N=14 N/m.

To evaluate the elastic modulus a series of typical force-distance curve (FDC) are measured. The FDCs are also performed on bare silicon wafers without polymeric resist film. The difference between these two curves represents the deformation of the polymeric film under the normal load applied. The force-displacement curve used in nanoindentation tests[27] was then found. An example of such curve is shown in Fig. 9.[25]

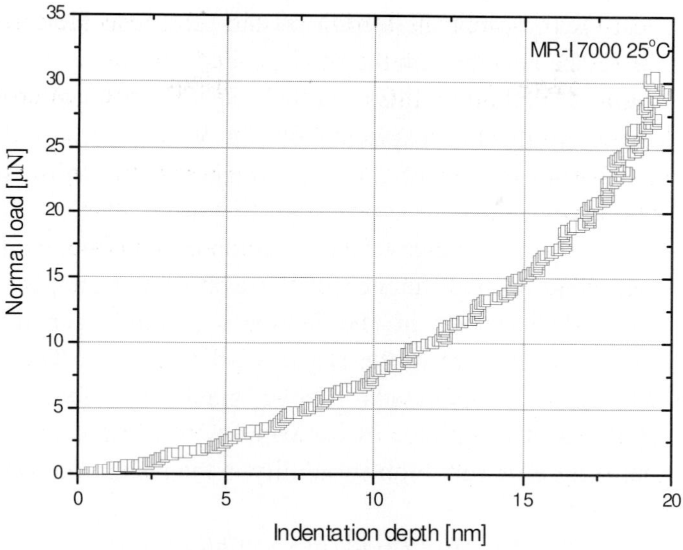

Fig. 9. Example of load versus depth of indentation calculated from approaching FDC.

The nanomechanical properties can be extracted by using Hertz (see below equation (4)) or Johnson-Kendall-Roberts (JKR) (equation (5)) model, derived in ref. 28 and ref. 29 respectively.

$$E = \frac{3}{4} \cdot \left(1 - v^2\right) \cdot R^{-\frac{1}{2}} \cdot k_N \cdot z_{def} \cdot \delta^{-\frac{3}{2}} \tag{4}$$

where $\delta = z_{pos} - z_{def}$

Here v is the Poisson's ratio of the tested material, k_N is the stiffness of the cantilever, z_{pos} is the 'piezostage (scanner) position', z_{def} is the probe deflection corresponding to z_{pos}, R is the radius of the tip and δ is the indentation depth value.

$$E = \frac{9}{4} \cdot \left(1 - v^2\right) \cdot R \cdot k_N \cdot z_{\text{def-PoF}} \cdot \left(\frac{1}{3} \cdot P_1 \cdot R^{-1} \cdot \delta^{-1}\right)^{\frac{3}{2}} \tag{5}$$

where $P_1 = (3 \cdot P_2 - 1) \cdot \left[\frac{1}{9} \cdot (P_2 + 1)\right]^{\frac{1}{3}}$

and $P_2 = \left(z_{def} \cdot z_{\text{def-PoF}}^{-1} + 1\right)^{\frac{1}{2}}$

The $z_{def\text{-}PoF}$ is the probe deflection on the retracting FDC when the cantilever loses the contact with the film.

The Young's modulus values are 0.2 GPa (Hertz) and 0.25 GPa (JKR) according to the data expressed by Fig. 9. Above an indentation depth of 25–30 nm (>10% of the layer thickness), a strong substrate effect was observed.

In the relatively small range of the variation of the temperature of the (25-45°C) no significant difference of the evaluated Young's modulus was observed. The values of the Young's modulus for the tested polymeric resist films were in the range 0.2-0.4 GPa.[25-26] The Young's modulus at higher temperatures (above glass transition) decreased usually 2-3 orders of magnitude as compared with the Young's modulus at room temperature. A very high variability in the results was observed.

4.4.3. *Estimation of Adhesive Properties — Pull-off Force Measurements*

The force-displacement curves (FDCs) are very useful to estimate adhesive properties of the polymeric resist film versus temperature. The pull-off force measurements are carried out with the use of the same cantilevers as used for the evaluation of Young's modulus. The pull-off force was measured during the detachment of the ball from the tested film.

The adhesive force that occurs at the polymer surface is shown in Fig. 10.[30] It can be seen that there is a large increase in the pull-off force above T_g and, in contrast to the case for the friction force (see section 4.4.4), its growth continues even above the melting temperature T_m.

The relatively high values of the standard deviation of the pull-off force can be the result of the oscillation of the cantilever observed at higher temperatures because of the convection of the warm air from the heating stage.

4.4.4. *Tribological (friction) Studies*

To evaluate the tribological properties, the torsion (friction) loop test is used. The loop is obtained by a scan of the tested film. During this loop,

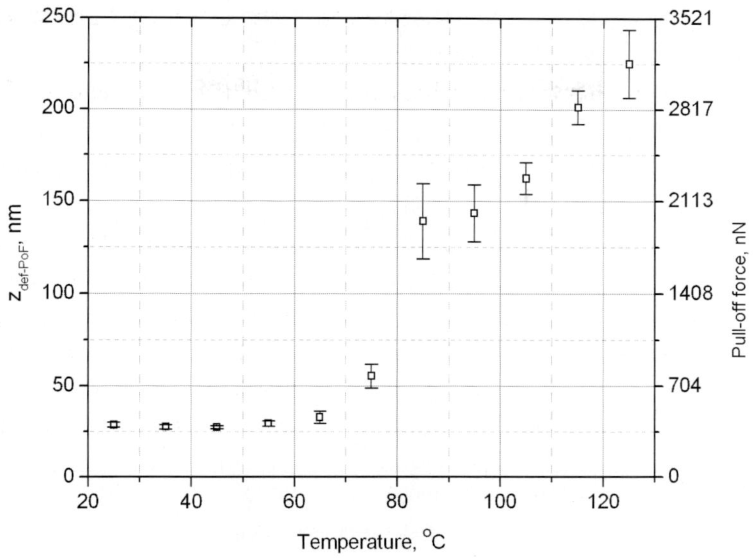

Fig.10. Pull-off force for amorphous thermoplastic polymeric resist mr-I 7000E.

a constant normal load is applied to film by the cantilever while sliding against it. A typical torsion loop showing the measured torsion of the cantilever versus relative displacement (probe-sample) obtained at 25°C is presented in Fig. 11. The horizontal axis indicates the relative distance i.e. horizontal displacement between the film and the base of the cantilever. The vertical axis dU represents the torsion of the cantilever that is used to calculate the lateral (friction) force.

The torsion (friction) characteristic is acquired with both forward and backward directions (of the probe movement) to close the friction loop.

The afore-mentioned two types of cantilevers, one with a glued steel ball and the other commercially available cantilever with a SiO_2 ball (5 µm in diameter) are used for tribological studies. The results presented below are obtained by the use of the cantilever with SiO_2 ball.

The tribological properties of ultrathin (200-300 nm) polymeric resist films were investigated at elevated temperatures starting from 25°C up to 125°C with 10°C steps.[31] The normal load was 280 nN, the sliding speed 38 µm/s, and the total number of scan lines (torsion loops) was 50 each 25.4 µm long (for one sample and one temperature).

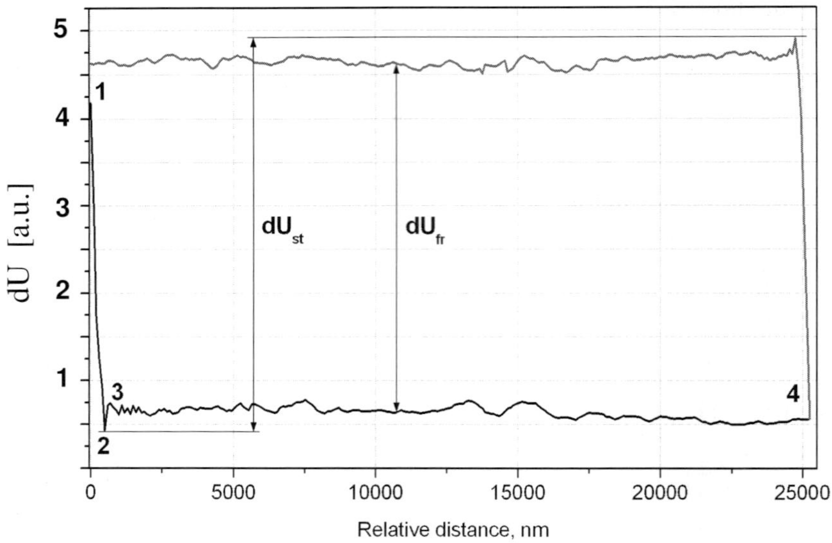

Fig. 11. Example of friction (torsion) loop at 25°C carried out on polymer resist film. 1: initial position of cantilever (in contact), 2: maximum torsion of cantilever, 3: rapid decrease of torsion while start sliding and 4: end of the scan line. dU_{st} is used to calculate static friction, dU_{fr} for kinetic friction.

The following equations are used to calculate the lateral (friction) force:[32]

$$F_L = \frac{3}{2} \cdot k_T \cdot \frac{h}{l} \cdot \frac{dU}{2} \tag{6}$$

where, k_T is the torsional spring constant of the cantilever:

$$k_T = \frac{G \cdot w \cdot t^3}{3 \cdot h^2 \cdot l} \tag{7}$$

where, G is the shear modulus, w is the width, l is the length, t is the thickness of the cantilever and h is the true moment arm:

$$G = \frac{E}{2 \cdot (1+\nu)} \tag{8}$$

where, E is Young's modulus of the cantilever's material and ν is Poisson's ratio.

Note that

$$h = h_{tip} + \frac{t}{2} \quad (9)$$

where h_{tip} is the height of the tip.

Here *dU* is the sensitivity of the photodetector in nm/mV multiplied by the change of an electric signal in mV recorded by the AFM, as mentioned before.

To calculate the preliminary load that was applied by the cantilever to the surface, equation (10) was used:

$$F_{load} = k_N \cdot z_{def} \quad (10)$$

where z_{def} is the constant deflection of the cantilever.

The lateral (friction) force variation versus temperature is shown in Fig. 12. A significant influence of the temperature on the measured lateral force was found. Relatively small effect of temperature on friction was observed in the brittle state of the polymer up to 65°C (T_g). Above this temperature, a significant increase of the lateral force was observed. Above 105°C the friction decreased dramatically; the material of the film started to flow.

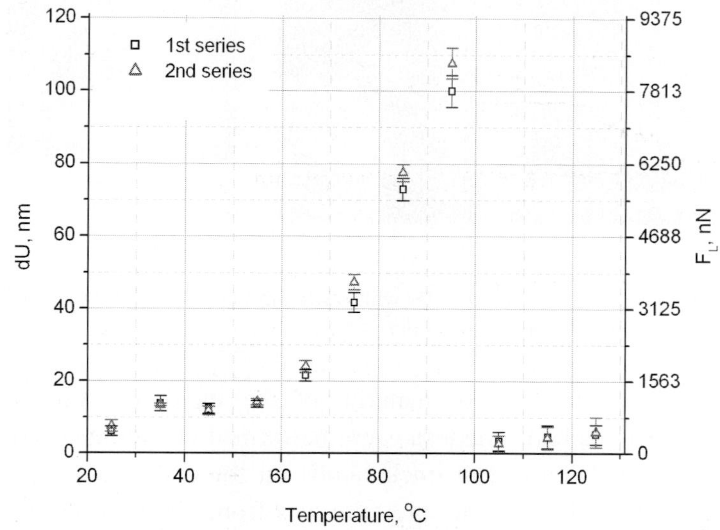

Fig. 12. Lateral (friction) force vs. temperature for polymeric resist mr-I 7000E film for two series of tests (i.e. two samples).

The lateral force that was measured was a few times greater than the normal load that was applied. The main reason for this is probably that there was an increase of contact area between probe and polymeric film as material was softened. The normal load was amplified by the adhesive force that occurs at the polymer surface that is shown in Fig. 11. As it could be seen, the fastest increase in the pull-off force was observed above T_g, and contrary to friction force it continued its growth even above T_m.

5. Studies of Friction between Mold Surface and Resist Film

Preliminary studies of friction between real mold and polymeric resist film have been performed. A specially constructed pin-on-disk microtribometer (Fig. 13a) was used. The polymeric resist film together with the silicon substrate was cut out and glued to the pin (Fig. 13b).

(a)

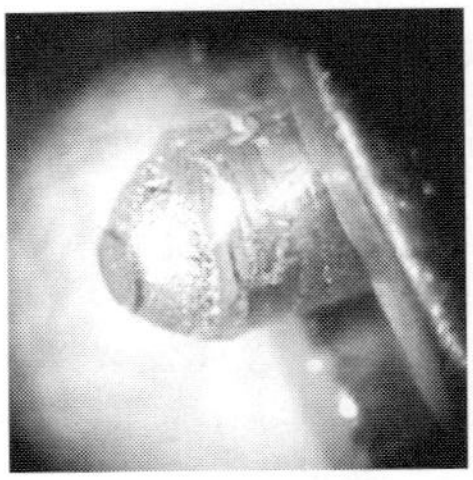

(b)

Fig. 13. (a) Pin-on-disk microtribometer with resist film on silicon substrate. (b) Sample cut out from silicon wafer and glued to pin.

Various combinations of material of the mold as well as resist film have been studied. The tests were performed at sliding speeds from 100 nm/s - 5000 nm/s. The track radius on the disk was varied from 3 mm to 15 mm. The contact pressure was from 200 to 1200 kPa. The sliding distance was from 250 to 1500 μm (at sliding time varied from 50 to 1000 s).

Stick-slip effects were observed for various combinations of mold materials (Si, Si_3N_4, TiN) and resist films. The most smooth sliding was registered for TiN (mold) /MIR 9000 combination. An example of characteristics of the friction coefficient versus sliding speed is presented in Fig. 14.

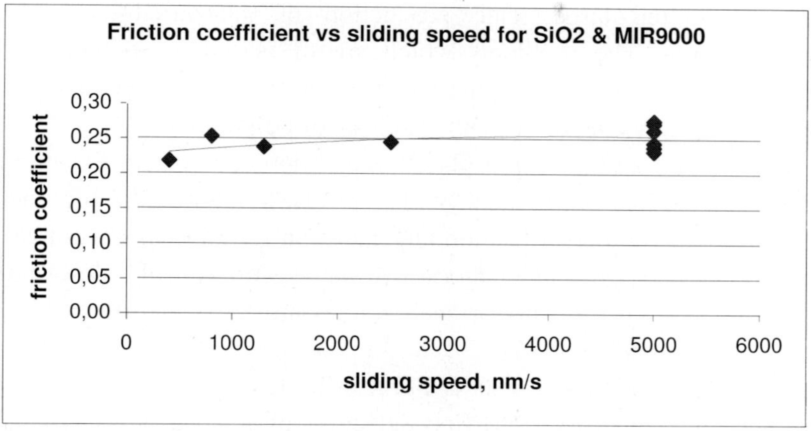

Fig. 14. Friction coefficient vs. sliding speed for rubbing of resist film against mold.

6. Studies of Anti-Stiction Layers

The studies of the anti-stiction layer are similar to the experiments performed on polymeric resist films. Special attention was paid however since the thickness of the layers is very small (usually about 2 nm). The AFM adhesive and friction studies are effective with the use of the special cantilever with glued balls. The contact pressure was kept low as the effect of the substrate can be easily detected.

The preliminary studies of the frictional behavior of anti-stiction layers were however very useful to evaluate the quality of such layers as well as their efficiency in preventing adhesion at the mold/resist interface.[33-34] The level of hydrophobicity (or hydrophilicity) of the anti-stiction layers influence the friction coefficient very much because of the effect of adhesive component of friction (weak or strong respectively).

Wettability allowed the estimation of the polar and dispersive components of the surface free energy when two liquids are used and

Owens –Wendt method is applied. Various studied anti-stiction layers demonstrated very low surface free energy (usually below 10 mJ/m^2).

7. Summary

The nanoimprint lithography (NIL) technique is very simple and effective in the large scale production of nanostructures and in nanopatterning. The problems which are of very big importance for future development and industrial application of this technology is however strongly connected with tribology of polymers.

The friction and adhesion play a very important role in the imprint process and decide on the quality of imprints, reliability and lifetime of the used tools. Studies of friction and adhesion are particularly crucial in the selection of appropriate materials (resist as well as material for mold) for the process. The results of such studies are very important also for modeling of the NIL process.

The elaborated set of tests relating in particular to polymeric resist materials enables reasonable first selection of the various resists and application in further imprinting the resists which demonstrated best properties. The procedure for the evaluation and selection of the polymeric resists for NIL enables testing costs and time to be reduced significantly.

Further studies are needed to estimate the adhesive and frictional properties in particular at the mold/resist interface in the real NIL process. However the studies on physical models (real mold, real imprints) with the use of the AFM technique should enable the prediction of the tribological/adhesive behavior of real interfaces playing most important role in reliability and lifetime of the used tools in NIL as well as quality of nano-imprints.

Acknowledgments

The hard and excellent work on testing NIL materials of my many students is greatly acknowledged. Particular thanks are due to Magdalena Ekwinska, Grzegorz Ekwinski, Andrzej Nowek, Adam Koszewski, Tomasz Gorski, and Dominik Nolbert (who constructed special microtribometer to study friction at real mold/resist interface and performed the tests). The discussions with Dr. David Mendels are acknowledged with thanks.

The materials for testing were obtained from various European labs involved in the EC NaPa project: Micro Resist Technology, Berlin (Germany), LETI in Grenoble (France), Inasmet–Tecnalia in San Sebastian (Spain).

The partial support of the European Commission – funded project NaPa (Contract no. NMP4-CT-2003-500120) is gratefully acknowledged.

References

1. Chou S.Y., Kraus P.R., Renstrom P.J., Appl. Phys. Lett. **67**, 3114 (1995).
2. Sotomayor Torres C.M. (ed): *Alternative Lithography:Unleashing the Potentials of Nanotechnology* Kluwer Academic/Plenum Publishers, New York 2003.
3. Proceedings of Nanoimprint & Nanoimprint Technology NNT'07, Paris 2007.
4. Bhushan B. (ed), *Springer Handbook of Nanotechnology,* Springer, Berlin 2004.
5. Voisin P., Zelsmann M., Gourgon C., Boussey J., Microelectronic Engineering, **84**, Issues 5-8, 916 (2007).
6. Chaix N., Landis S., Gourgon C., Merino S., Lambertini V.G. Durand G., Perret C., Microelectronic Engineering, **84**, Issues 5-8, 880 (2007).
7. Miller M., Schmid G., Doyle G., Thompson E., Resnick D.J., Microelectronic Engineering **84**, Issues 5-8, 885 (2007).
8. Sogo K., Nakajima M., Kawata H., Hirai Y., Microelectronic Engineering **84**, Issues 5-8, 909 (2007).
9. Proceedings of Microprocesses and Nanotechnology 2007 MNC2007, Kyoto 2007.
10. Jeon S., Nam Y-S., Shir D.J.-L., Rogers J.A., Hamza A., Appl. Phys. Lett. **89**, 253101 (2006).
11. Senturia S.D., *Microsystem Design,* Kluwer Academic Publishers, Boston 2001.
12. Madou M., Fundamentals of Microfabrication; The Science of Miniaturization, 2nd ed., CRC Press, Boca Raton 2002.
13. Rebeiz G.M., *RF MEMS*, J.Wiley Interscience, Hoboken 2003.
14. Beeby S.P., Ensell G., Kraft M., White N., *MEMS Mechanical Sensors,* Artech House, Boston 2004.
15. Muhlberger M., Bergmair I., Schwinger W., Gmainer M., Schoftner R., Glinsner T., Hasenfluss Ch., Hingerl K., Vogler M., Schmidt H., Kley E.B., Microelectronic Engineering **84**, Issues 5-8, 925 (2007).
16. Rymuza Z., *Invited Lecture*, Symposium on Progress in Nanoimprint Lithography, Osaka Prefecture University, Osaka, Japan, October 12, 2006.

17. Rymuza Z., Koszewski A., Ekwinska M., Nowek A., in Proceedings of MNC 2007, Kyoto 2007, p. 282.
18. Meyer E., Hug H.J., Bennewitz R., *Scanning Probe Microscopy;The Lab on a Chip*, Springer-Verlag, Berlin 2004.
19. Bhushan B., Fuchs H., Hosaka S. (eds), *Applied Scanning Probe Methods,* Springer-Verlag, Berlin 2004.
20. Berg J. C.(ed), *Wettability,* CRC Press, Boca Raton 1993.
21. Neumann A.W., Spelt J. K. (eds), *Applied Surface Thermodynamics,* Marcel Dekker, New York 1996.
22. Owens D.K., Wendt R.C., J. Appl. Polym. Sci. **13**, 1741 (1969).
23. Żenkiewicz M.: Adhesion and Modification of Surface Layer of High-molecular Materials, WNT Warszawa 2000, in Polish.
24. Van Krevelen D.W., *Properties of Polymers,* Elsevier, Amsterdam 1990.
25. Koszewski A., Gorski T., Reuther F., Rymuza Z., Int. J. Materials Research, **98**, 393 (2007).
26. Koszewski A., Gorski T., Rymuza Z., Int. J. Materials Research, to be published in 2008.
27. Fischer –Cripps A.C., *Nanoindentation*, 2nd ed., Springer Verlag, New York 2004.
28. Kwacz M., Chizhik S.A., Rymuza Z., Kusznierewicz Z.: Polimery No.7/8, 551 (2004) 5, in Polish.
29. Tsukruk V.V.,. Huang Z., S. Chizhik S.A., Gorbunov V.V.: J. Mater. Sci. **33**, 4905 (1998).
30. Koszewski A., *M.Sc. Thesis*, Warsaw University of Technology 2007, in Polish.
31. Koszewski A., Reuther F., Rymuza Z., Proceedings of the STLE/ASME Int. Joint Tribology Conference, San Diego, USA, October 2007, CD disk.
32. Meyer E., Overney R.M., Dransfeld K., Gyalog T., *Nanoscience – Friction and Rheology in Nanometer Scale*, World Scientific, Singapore 2002.
33. Ishihara J., Koszewski A., Kayama M., Kawata H., Rymuza Z., Hirai Y., in NNT'07 Proceedings, Paris 2007, Paper B4.
34. Ishihara J., Koszewski A., Kayama M., Kawata H., Rymuza Z., Hirai Y., in Proceedings of MNC 2007, Kyoto 2007, p. 450.

CHAPTER 20

TRIBOLOGICAL STUDIES OF ULTRA-HIGH MOLECULAR WEIGHT POLYETHYLENE (UHMWPE) THIN FILMS ON SILICON SURFACE

N. Satyanarayana and S. K. Sinha*

Department of Mechanical Engineering, National University of Singapore
9 Engineering Drive 1, Singapore 117576, Singapore
**E-mail: mpesks@nus.edu.sg*

In the present chapter, we first review the literature on the tribological properties of polymer thin films and then present the results on novel ultra-high molecular weight polyethylene (UHMWPE) films formed on Si surface using a simple dip-coating technique. The presently developed UHMWPE films have reduced the coefficient of friction of Si by at least 6 times to 0.09 and increased the wear resistance of the Si by many orders of magnitude with no sign of wear debris formation. Further enhancement of tribological properties of UHMWPE film is described where perfluoropolyether (PFPE) overcoating onto UHMWPE film has increased the wear-life of pristine UHMWPE film by several orders. Finally, the associated wear mechanisms are explained. Extremely wear durable polymer films can find applications in bearing and gear surfaces of machines that are made of Si.

1. Introduction

1.1. *General Background on Polymer Thin Films*

Thin coatings are applied onto objects to dramatically alter their properties so that the interaction of the objects with its surrounding can be controlled. Various thin films include organic coatings such as polymer thin films and self-assembled monolayers (SAMs) and inorganic thin films such as metallic films, ceramic films etc. These coatings alter physical, chemical and mechanical properties such as friction, adhesion, wear, corrosion, wettability, surface energy, surface reactivity towards various elements and species etc. A specific film is selected depending on the desired requirements of a particular application. Thin films on various objects find wide spread applications as paints, varnishes, protective films for bearings, machine elements, magnetic hard-disk, aerospace components, microelectromechanical systems (MEMS) etc.

Polymer coatings are organic thin films formed onto a substrate which are either polymerized prior to coating or polymerized after monomers being coated onto the substrate. A polymer film coated onto a substrate greatly alters its properties by changing the interaction of the substrate with other materials that are in direct contact with it. Polymer films vary in thicknesses from few nanometers (referred as polymer brushes) to few micro-meters (polymer thin films or thick films). The realization of the importance of polymer films dates back to several thousand years. For example, Lacquer generated from tree sap was used as a coating to protect the wooden objects in China some 7000 years ago and around 3000 BC Egyptian ship builders used beeswax, gelatin and clay to produce varnishes and enamels in the form of coatings to waterproof their ships.[1] In the current industrial applications, apart from the simple decoration and protective aspects, polymer films find applications in high-tech areas such as: hard-disks, MEMS/NEMS components, biomedical components, microelectronics and aviation applications.[1] Polymers have superior properties such as low production cost, good toughness coupled with good mechanical properties, lower shear strength, ease in processing to form thin films, excellent corrosion and wear properties etc which makes them attractive as thin films for many applications such as decorations, anti-scratch coatings, corrosion protection, reducing friction, stiction, adhesion, wear and tear, tailored wetting properties, anti-reflection properties etc. Polymer thin films are usually formed onto various substrates that iclude stainless steel, Al, Si etc.

Three applications of polymer films in advanced fields are explained below to emphasize the importance of the developments in this fascinating area of polymer films. First example involves the protection of hard-disk surface by the coating of an ultra-thin polymer layer of

PFPE.[2] By the application of the PFPE film of 2-4 nm thickness the tribological properties are greatly improved, the wear is reduced, and the mean-time-to-failure of the disk is greatly prolonged. The second application involves the use of ultra-thin polymer layers (for example, fluoropolymers)[3] to reduce the stiction (unwanted adhesion) and wear in MEMS components. Procedures have been developed to coat ultra-thin polymers such as fluorocarbon based polymers onto MEMS components which reduce stiction and wear. The third application involves the use of few nanometers thick polymer layer (for example, poly (ethylene oxide))[4] onto blood-exposed implant materials or sensors so that the adhesion of the blood cells onto these surfaces can be mitigated.

The type of interaction between the polymer molecules and substrate surface largely affects the end-properties and hence needs to be properly controlled. Based on the type of interactions between the polymer molecules and solid surfaces, coatings can be either physisorbed or chemisorbed. Generally, the physisorbed film shows inferior wear durability when compared to chemisorbed films; however, the final wear performance depends on many other factors including the molecular structure of the polymer and surface energy.

Below are various processes which are used to form polymer films from solution:

- Solvent casting, painting
- Thermal spray coating of polymers
- Spin coating
- Polymers produced by surface adsorption of monolayers (such as SAMs)
- Floating technique
- Langmuir-Blodgett films, and,
- Dip coating

In the solution-based coating techniques such as spin-coating, dip-coating etc, the polymer molecules are absorbed from the solution, and, the solvent evaporates during and after the coating process. Polymer films with thicknesses ranging from few nanometers to several micro-meters can be obtained in dip-coating technique by appropriate selection of depositional parameters. Primarily, the interaction between the molecules and sample surface are physical and by proper selection of functionalized polymer molecules and/or post thermal treatment, different extent of chemical bonding can be achieved between the polymer film and the substrate.

1.2. *Tribological Properties of Polymer Thin Films*

Table 1 summarizes the tribological results for various polymer thin films obtained from our literature review to date. This table also shows the testing set-up used along with testing parameters, coefficient of friction values and wear life (or wear volume) as well as film thickness etc. This summary will help to understand the present status of the use of polymer thin films for tribological applications and will help in developing new films for future potential applications. Some of the results listed in the Table 1 are explained in detail below.

Liu *et al.*[5] have studied the friction coefficient and wear life of silicone films (polydimethylsiloxane, PDMS) onto modified and unmodified Si surface. Bare Si has shown a coefficient of friction of 0.8 and wear life of 2 cycles in a ball-on-plate tribometer sliding against steel ball at a normal load of 0.5 N and a sliding velocity of 30-100 mm/min. The non-crosslinked PDMS film on hydroxylated Si surface has reduced the coefficient of friction to 0.05-0.08 and increased the wear life to ~1200 cycles under the same loading conditions. This study has also demonstrated that the surface wettability and surface reactivity before PDMS coating greatly alters the adhesion between the PDMS and the Si and affects the final tribological properties.

Pozdnyakov *et al.*[6] have studied the tribological properties of the composite polymer film (polyimide reinforced with C60) on a glass substrate using cylinder-on-plane configuration against steel cylinder at an applied load of 10 N and a sliding velocity of 1 m/sec. The composite polymer film which was imidized at 420°C has shown a coefficient of friction of 0.38 and wear volume of 0.35×10^{-7} mm^3/N-m whereas the pristine PI film without C60 has shown a coefficient of friction of 0.36 and a wear volume of 0.7×10^{-7} mm^3/N-m. Through this study, the

Table 1. Tribological data of various polymer thin films obtained from literature.

S. No.	Substrate	Coating material	Coating technique	Thickness	Testing equipment used, counterface material, load/pressure used, sliding velocity, remarks (if any)	Coefficient of friction	Wear	Reference
1	Si-hydroxylated	PDMS	Spin-coating	180 nm	Ball-on-plate tribometer, steel ball, 0.5 N, 30-100 mm/min	< 0.1	Wear life increased from 2 (of bare Si) cycles to 1200 cycles	[5]
2	Steel	Polyimide-fullerene (C60) composite films	Solution-casting	100 μm	Cylinder-on-plane configuration, steel cylinder, 10 N, 1 m/s	0.37	Wear volume: 2.25×10^{-7} mm^3/N-m	[6]
3	Si	Composite film: APS/PGMA/Polyimide	APS and PGMA are using self-assembly and polyimide using spin-coating	50-80 nm	Ball-on-disk tribometer, steel ball, 0.5 N, 20 mm/sec	0.08	Wear life: more than 25,000 cycles	[7]
4	Glass	Composite film: $PMMA/SiO_2/TiO_2$	Dip-coating	~120 nm	Ball-on-disk tribometer, GCr15 steel ball of 3mm diameter, 3 N, 1.5 mm/s	0.09-0.11	Wear life > 8,000 cycles	[8]
5	Si	PI	Sputtering	20 nm	Reciprocating abrasion test apparatus, sapphire slider of 40 mm diameter, 0.49N, 80 mm/sec	0.21	Wear life = 6,500 cycles	[9]
6	Si	Triplex layer consisting of GPTMS SAM, SEBS polymer and a top hard-layer	Chemical grafting, self-assembly	20-30 nm	Ball-on-disk tribometer, steel ball, 1.2 GPa, 4.43 mm/sec	0.02-0.08	Wear life: ~3500 cycles	[10]

researchers have identified that the addition of C60 helps in improving the wear durability of PI film and moreover, it has been demonstrated that the imidization temperature greatly influencess the coefficient of friction and wear volume of the composite film.

Sun *et al.*[7] have investigated the friction and wear properties of polyimide film of 50-80 nm on aminopropyltrimethoxysilane (APS) SAM modified Si with and without the use of an intermediate adhesion layer of polyglycidyl methacrylate (PGMA). Tribological tests were carried out using ball-on-disk tribometer where the substrate slid against steel ball at an applied load of 0.5 N and a sliding velocity of 20 mm/sec. The composite film with the intermediate adhesive layer has shown a coefficient of friction of 0.08 and a wear life of 25,000 cycles whereas the PI film without the intermediate adhesive layer has shown a coefficient of friction of 0.16 and a wear life of ~1000 cycles. From this study, they have concluded that a good chemical bonding between the polymer film and the substrate is essential for good wear durability.

Gu *et al.*[8] have measured the coefficient of friction and wear life of hydrophobic organic-inorganic composite thin films of PMMA (polymethylmethacrylate)/SiO_2/TiO_2 formed using dip-coating on glass substrate. Friction and wear measurements were done on a ball-on-plate reciprocating friction tester using the steel ball of 3 mm diameter as a counterpart at a normal load of 3 N and a sliding velocity of 1.5 mm/s over a sliding distance of 7 mm. The polymer-inorganic composite film has shown a coefficient of friction of 0.09-0.11 and a wear life of more than 8000 cycles.

Kitoh and Honda[9] have observed that sputtered PI film of 20 nm thickness on Si surface has better abrasion resistance than sputtered PTFE film on Si when tested against sapphire slider of 40 mm in diameter at a normal load of 0.49 N and a sliding speed of 80 mm/sec.

Sidorenko *et al.*[10] have shown that a tri-layer architecture, consisting of poly (methacrylate) layer on SEBS (poly (styrene-b-(ethylene-co-butylene)-b-styrene)) polymer chemisorbed onto Si via glycidoxy-propyltrimethoxysilane (GPTMS) has shown good wear durability. The tri-layer architecture on Si has shown a wear durability of 3300 cycles at a contact pressure of 1.2 GPa and a sliding speed of 4.43 mm/sec when sliding against a 3 mm diameter steel ball using an oscillating friction and wear tester whereas the OTS (octadecyltrichlorosilane) SAM layer has failed within 500 cycles under the same testing conditions.

1.3. *Motivation and Objectives of the Present Study*

Polymers in bulk form have numerous tribological applications ranging from tyre material (high friction) to many bearing materials (low friction). In bulk form, UHMWPE is highly wear resistant compared to many other polymers such as PEEK (polyetheretherketone), LDPE (low density polyethylene), PS (polystyrene), PMMA etc.[11] UHMWPE has shown low coefficient of friction and high wear resistance because of its extremely long and highly entangled molecular chains.[12] An UHMWPE molecule consists of large number of repeating units of ethylene monomers where the backbone structure of the molecule is the C-C covalent bond. Therefore, it is expected that if an ultra-thin film of UHMWPE is formed on the Si surface, it will protect the Si surface from wear and tear and reduce the coefficient of friction. There are no studies in the literature involving the formation of ultra-thin UHMWPE films on Si surface. Si is selected as the substrate because Si is used as structural material for microsystems such as MEMS. A recent study by Bao *et al.*[13] has shown the deposition of UHMWPE coatings on steel substrate using thermal spray deposition for corrosion resistance applications. But such thermal spray processes are very expensive and often not practical to coat on complicated structures. Therefore, we study the feasibility of coating UHMWPE films on Si surface by a simple dip-coating technique. Dip-coating technique is routinely used to obtain ultra-thin lubrication coatings of PFPE on hard-disk surface.[2] Dip-coating was also used for the formation of some polymer thin films,[14,15] for example polystyrene thin films onto micro-textured Si surface.[14] Therefore, one of the novelties of the present work is the establishment of a dip-coating method to apply thin UHMWPE films (sub-micron to few tens of microns) onto Si surface. The challenges in the present study are the selection of a suitable solvent and the optimization of the various parameters such as the solubility of the polymer in the solvent, polymer concentration, dipping time and post-thermal treatment in order to obtain well adhering uniform and defect free UHMWPE films on Si surface.

Further, the wear durability of pristine UHMWPE films on Si surface is further increased by overcoating an ultra-thin film of PFPE layer as top lubricating layer. In this chapter, firstly, the results based on UHMWPE film are presented and discussed followed by the results on the effects of PFPE overcoating onto UHMWPE film.

2. Materials and Sample Preparation

2.1. *Materials*

GUR X143 UHMWPE polymer powder (obtained from Ticona Engineering Polymers, Germany through a local Singapore supplier) was used for the formation of UHMWPE film on Si surface. Table 2 shows the physical properties of UHMWPE, as provided by the supplier.

PFPE Zdol 4000 was used to overcoat onto UHMWPE film. PFPE Zdol 4000 ($OH\text{-}CH_2\text{-}CF_2\text{-}(O\text{-}CF_2\text{-}CF_2)_p\text{-}(O\text{-}CF_2)_q\text{-}O\text{-}CF_2\text{-}CH_2\text{-}OH$, the ratio p/q is 2/3) molecules contain hydroxyl groups located at both ends of its backbone molecular chain.

Table 2. Properties of UHMWPE powder.

Property	Units	Value
Bulk Density	g/cm^3	0.33 ± 0.03
Average particle size d_{50}	μm	20 ± 5
Melting temperature	°C	130-135

After several trials with many solvents, decahydronapthalin (decalin) was selected as a suitable solvent to dissolve UHMWPE.

2.2. *Preparation of UHMWPE Film on Si Surface*

Si substrate was initially ultrasonically cleaned with soapy water and distilled water which was then rinsed with acetone for 10 min. This preliminary cleaning helps to remove any foreign particles/contaminants present on the surface. The above cleaned samples were then subjected to piranha treatment by immersing in a solution of 7:3 (v/v) 98% H_2SO_4 and 30% H_2O_2 at 60-70°C for 50-60 min. The piranha treatment generates plenty of hydroxyl groups on Si surface and further removes any organic/inorganic contaminants present on the surface. The Si surface was very smooth after piranha treatment (rms roughness is 0.3-0.5 nm). After piranha treatment, the samples were thoroughly rinsed with distilled water and acetone respectively and finally dried with N_2 gas.

Firstly, the polymer solution was prepared by dissolving the UHMWPE powder in decalin by heating at a temperature of 150°C for half an hour and at 250°C for the next half an hour. Magnetic stirring was used (along with heating) to enhance the rate of dissolution and to ensure a complete and uniform dissolution of the polymer. A custom built dip-coating machine was used for depositing the polymer thin film which was carried out immediately after ensuring complete dissolution of the polymer powder. The dipping time used during dip-coating was 30 s and the withdrawal speed was fixed as 2.1 mm s^{-1}. The polymer coated samples were briefly dried in air and subjected to post-thermal treatment at a temperature of ~100°C for about 20 h in a clean air furnace. After the thermal treatment, the samples were slowly cooled to room temperature in the furnace itself and then stored in desciccator until further surface or tribological characterization.

2.3. *Experimental Procedures*

Surface characterizations such as water contact angle measurement, AFM (atomic force microscopy) topography, SEM (scanning electron microscopy) morphology, XPS (X-ray photoelectron spectroscopy), FTIR (Fourier Transform Infrared Spectroscopy) are used to measure the properties of the thin films formed. Tribological properties are measured using a micro-tribometer. All these characterization techniques are explained in detail below. The thickness of the polymer was estimated from the cross section observation through SEM. AFM and SEM are used to characterize the wear tracks after sliding tests.

2.3.1. *Water Contact Angle Measurement*

Contact angle is sensitive to the chemical nature of the top molecular layer and the measurement of contact angle is relatively simple, inexpensive and a popular technique to characterize the wettability of surfaces. Contact angle of a liquid on a substrate is the angle between the liquid/solid interface and liquid vapor interface. Water (polar) is the main liquid that is commonly used to measure the contact angle, however, other liquids such as formamide (polar), hexadecane (non-polar), diiodomethane (non-polar) etc are also used to measure their respective contact angles. The contact angle technique is indeed a quick method to qualitatively determine whether or not the target film is being properly coated onto the substrate. A surface is called hydrophobic when the water contact angle is more than or equal to 90° whereas it is called hydrophilic when water contact angle is less than 45°. The surface having a water contact angle between 45° to 90° is defined as the amphoteric hydrophobic/hydrophilic surface.[16] Apart from chemical functionality contact angle also depends on the surface roughness.

In the present work, the static water contact angles on Si before and after each modification were measured with VCA Optima Contact Angle System (AST Products Inc., USA). A water droplet of 0.5 to 1 μl was used for contact angle measurement. At least 5 to 6 replicate measurements were done on at least 3 samples to get an average water contact angle value. The variation in the water contact angle values at various locations of a sample was within $\pm 2^o$ whereas the measurement error was within $\pm 1^o$.

2.3.2. *Topography Measurements with AFM*

In the present work, the surface topography of Si before and after each modification was investigated using Digital Instruments Multimode^{TM} AFM (Veeco, USA). A Si tip was used for imaging in the tapping mode and all images were collected in air. The set point voltage used was 1-2V and the scan rate was 1 Hz.

2.3.3. *FTIR Characterization*

FTIR is an analytical technique to identify organic (and in some cases inorganic) materials. FTIR measures the absorption of infrared radiation by the sample material with respect to the wavelength. The infrared absorption bands identify molecular components and structures. When a material is irradiated with infrared radiation, absorbed IR radiation excites molecules into a higher state of vibration. The wavelengths that are absorbed by the sample are characteristic of its molecular structure.

A Bio-Rad FTIR model 400 spectrophotometer using transmission mode was used to obtain the FTIR spectra for the UHMWPE thin film on Si surface. The piranha treated Si was used for background scan. Each spectrum was collected by accumulating 16 scans at a resolution of 4 cm^{-1}. The measurements were done at about 5 to 6 replicate locations on a sample and the tests were repeated on at least three samples and the spectra were found to be identical at all points of measurement.

2.3.4. *XPS Characterization*

XPS gives information about the atomic composition of the surface and their bonding and is especially sensitive to the local chemical environment of an atom. In the XPS measurement, x-rays are usually generated either by Mg or Al source and directed to the sample surface. The bombarded x-rays cause a core electron to be emitted from the sample surface. The kinetic energy of this electron is detected with an electron multiplier and is equal to the difference between the energy of the x-ray and the binding energy of the electron (1253 eV for Mg and 1486 eV for Al). XPS requires ultra-high vacuum (UHV) conditions. XPS is also known as Electron Spectroscopy for Chemical Analysis (ESCA).

In the present work, XPS measurements were performed on a Kratos Analytical AXIS His spectrometer with a monochromatized Al Kα X-ray source (1486.6 eV photons) at a constant dwell time of 100 ms and pass energy of 40 eV. A photoelectron take-off angle of 90° (with respect to the sample surface) was used to obtain the core-level signals. All binding energies (BE) were referenced to the C1s hydrocarbon peak at 284.6 eV.

2.3.5. *SEM Characterization*

SEM gives very high magnification of the features as high as 100,000x. It can be operated from 10-35 kV. It has to be noted that the higher the voltage the better the resolution for metallic samples whereas for polymers, lower currents are required. Usually the surface has to be conductive for SEM observation but recent environmental SEMs can also image the non-conductive surfaces. Moreover, conductivity of the surfaces can be improved by coating with gold or platinum.

JEOL JSM-5600 LV SEM equipped with EDS (energy dispersive spectroscopy) was used to investigate the surface morphology of the polymer films studied in the present research. Surface analysis was carried out also on worn specimens using SEM/EDS to understand the wear mechanism. Further, the thickness of the polymer films was measured using SEM by observing in the section analysis.

2.3.6. *Tribological Characterization*

For any tribological characterization, friction coefficient and wear volume are the primary measurement quantities. Wear volume is usually measured from the loss of the material. Pin/ball-on-disk and pin/ball-on-plate are the two major test geometries used in tribological characterization. In the case of thin films, the loss of material will be very minute such that it cannot be accurately measured and hence the wear is usually estimated from the number of cycles after which the low friction capability of the film is lost. Depth of the wear track is another way to monitor wear of the film surface.

In the present research, friction and wear durability tests were carried out using a commercial micro-tribometer (UMT-2, Universal microtribometer, Center for Tribology, USA) using the ball-on-disk mode. The Si samples, modified and unmodified, were slid against 4 mm diameter Si_3N_4 ball at a rotational speed of 100-200 rpm (0.021-0.042 ms^{-1}) and at normal loads of 5-7 g which gave a contact pressure of approximately 330-370 MPa (calculated using the Hertzian Contact Model). All of the tests were performed in a class-100 clean booth in ambient atmosphere at room temperature (23°C) and a relative humidity of ~70%. In the present chapter, the initial coefficient of friction reported is the value obtained after 4 s of sliding (i.e. after 6-12 cycles of sliding) i.e. after stabilization of the sliding process. The wear durability in the present research was defined as the number of cycles after which the coefficient of friction exceeded a value of 0.3 or a visible wear scar appeared on the sample, whichever happened earlier. The present wear durability definition is consistent with the literature.[17,18] The average wear life reported in the present chapter was calculated from the data obtained from at least three different samples utilizing at least two different tracks on each sample. The SEM was used to observe the worn surfaces after appropriate number of sliding cycles.

3. Results and Discussion

3.1. *Characterization of the UHMWPE Film on Si Surface*

3.1.1. *Physical Characteristics of the UHMWPE Film*

The water contact angles of bare Si, Si/UHMWPE and bulk UHMWPE polymer samples are shown in Table 3.

Table 3. Water contact angles of bare Si, Si/UHMWPE and bulk UHMWPE samples used in the present study and literature results.

Sample	Water contact angle, deg	Reference
Bare Si	12	Present Study
Si/UHMWPE	127	Present Study
Bulk UHMWPE	107	Present Study
Bulk UHMWPE	87	[19]
Bulk UHMWPE	90	[20]

The coating of UHMWPE film onto Si surface changed the water contact angle from 12° to 127° and made Si surface more hydrophobic (water repellant). Table 3 also shows the water contact angle values of bulk UHMWPE (obtained in the present study and literature) and the variations (10-20 degrees) in the contact angles of various bulk UHMWPE samples (present study and literature) are due to the differences in their grades and surface conditions.[21] The water contact angle of the UHMWPE film on Si surface is ~20° higher than that of bulk UHMWPE solid surface (of present work) and the reasons are explained based on the differences in the surface roughness values, which are explained later. The high water contact angle of UHMWPE film infers that the chemical state of the film is similar to that of the bulk UHMWPE. This means that neither the solvent nor the thermal treatment changed the polymer in any way.

The surface morphology of the UHMWPE film on Si (obtained with SEM) is shown in Fig. 1(a) that depicts uniformly dispersed polymer chains/fibers protruding to the surface and the presence of valleys in between them. Figure 1(b) shows the AFM topography image (along with a typical line profile) of UHMWPE film over a scan area of 40 μm x 40 μm. The AFM image (and also the line scan) shows the presence of islands and valleys of few microns size with an overall rms roughness of 0.556 μm (over 40 μm x 40 μm scan area). Figure 1(c) shows the AFM image and line profile of a bulk UHMWPE solid surface. The bulk UHMWPE surface was subjected to fine polishing before imaging.

The features on bulk UHMWPE are very flat and smooth when compared to those on the UHMWPE film. The rms roughness of the bulk UHMWPE is 0.332 μm, which is less than that of the UHMWPE film (0.556 μm). Therefore, the rough and uneven features of UHMWPE film are believed to be the primary reason for higher water contact angle of the film when compared to the bulk polymer in spite of the same chemical structure/nature of the two. It is observed and well-explained in the literature that the water contact angle value is higher for surfaces with large roughness and those containing needle-like peaks.[22-24]

The thickness of the UHMWPE film formed according to the present experimental conditions is 28 μm which was measured by cross-sectional imaging using SEM.

The mechanical stability of the UHMWPE film on Si surface was studied by subjecting the film coated samples to ultrasonication in decalin (the solvent). The changes in the surface morphology due to ultrasonication for 5, 30 and 100 min were observed using optical microscopy and are shown in Fig. 2. Care was taken to obtain the images at the same spot of the specimen to see the progress of film removal

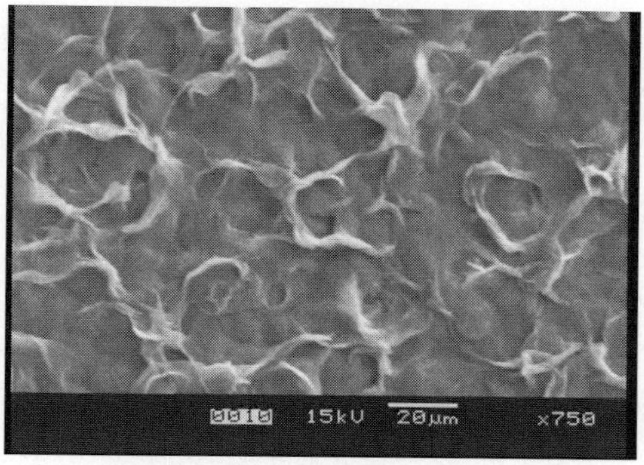

(a)

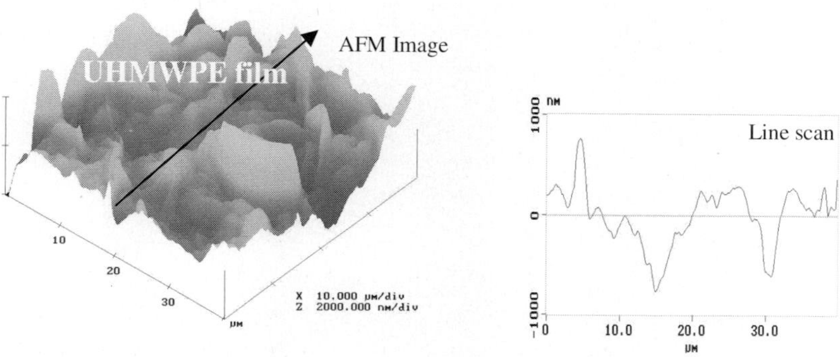

(b)

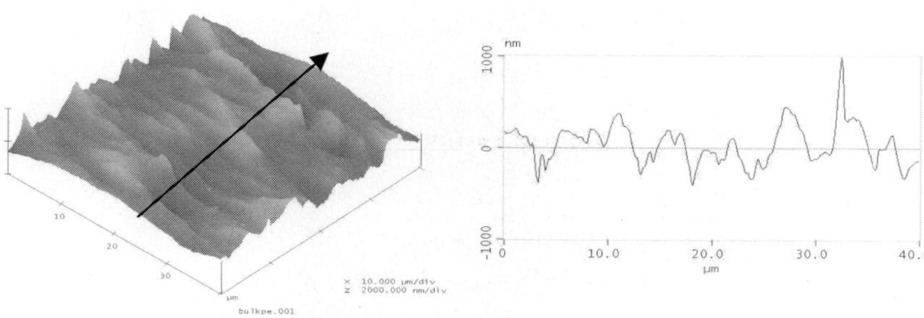

(c)

Fig. 1. (a) SEM morphology of the UHMWPE film on Si surface. It is similar to the structure of bulk UHMWPE. (b) AFM image of the Si/UHMWPE surface with a scan size of 40 μm x 40 μm. The arrow on the 3-dimensional (3D) image shows the location and direction of the line profile shown adjacent to it. (c) AFM image of the bulk UHMWPE with a scan size of 40 μm x 40 μm. A representative line profile on the surface is shown adjacent to it. The vertical scale is 2 μm in both AFM images shown in (b) and (c).

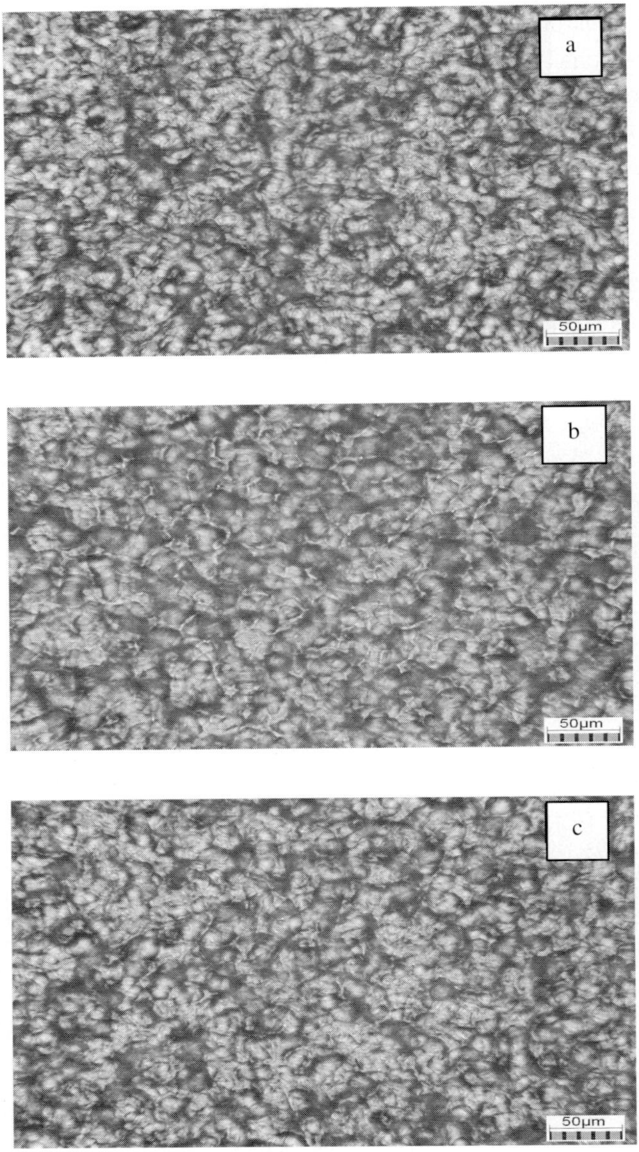

Fig. 2. Optical micrographs of Si/UHMWPE sample after (a) 5, (b) 30 and (c) 100 min of ultrasonication in decalin followed by drying respectively. Note that the scale of the optical images is different from the image shown in Fig. 1.

from the substrate, if any. The optical micrographs did not show any appreciable change in the surface morphology because of ultrasonication for all time durations and this suggests that UHMWPE films are strongly adhered to the Si surface. As there is no covalent bond formation between UHMWPE and Si, the strong adhesion between two must be due to some physical forces and/or mechanical interlocking between the asperities on Si surface and the polymer molecules.

It is believed that, the reactive nature of Si with hydroxyl groups on the surface after piranha treatment and post-thermal treatment at a temperature of 100°C must have helped in good adhesion between the UHMWPE film and the Si surface. But, the exact mechanism responsible for this behavior is not very clear at this moment.

3.1.2. *Chemical Analysis of the Coating*

FTIR and XPS studies are used to investigate the chemical structure of the UHMWPE film.

The transmission spectrum obtained with FTIR is shown in Fig. 3 and represents four fundamental bands of UHMWPE film. The four characteristic bands include two strong CH stretching modes at 2851 and 2919 cm^{-1}, one band at 1462 cm^{-1} corresponding to polyethylene-methylene (CH_2) bending and one more band, the so-called CH_2 rocking mode at 719 cm^{-1}, which are similar to the characteristic features of the bulk PE or UHMWPE polymers.[25-27] The similarity between the peaks of UHMWPE film with those of bulk UHMWPE further proves that the chemical structure of the film is the same as that of the bulk polymer.

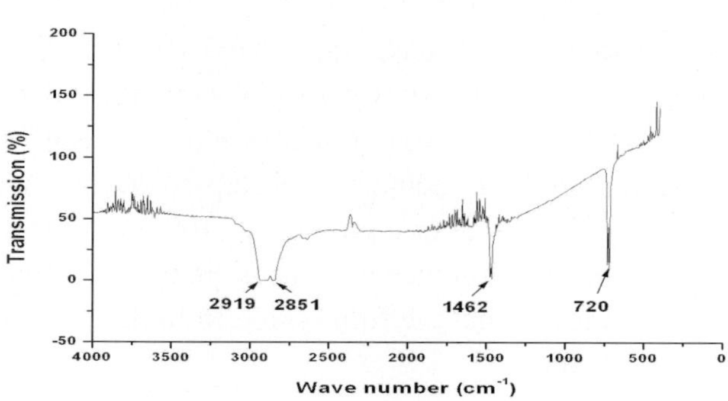

Fig. 3. FTIR spectrum of the UHMWPE coated Si surface.

The wide scan spectra (obtained with XPS) of bare Si and Si/UHMWPE are shown in Fig. 4. The strong intensity O1s peak on bare Si is due to the presence of oxide layer whereas the presence of C1s (with very less intensity) peak on bare Si might come from the contaminants which were not removed during piranha treatment or those adsorbed from the atmosphere during handling before XPS tests. The large intensity C1s peak on UHMWPE film is due to its ethylene groups of backbone polymer chain and the absence of O1s, Si2s and Si2p peaks suggests that the polymer film completely shielded the Si surface and its hydroxyl groups.

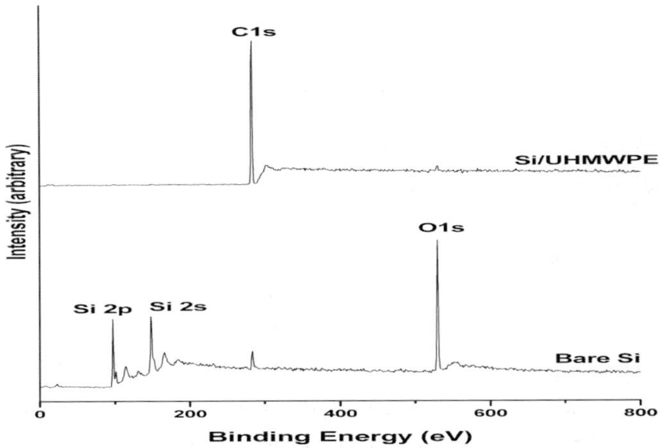

Fig. 4. XPS wide scan spectrum for bare Si and Si/UHMWPE samples.

The surface characterization techniques (physical and chemical) and mechanical stability tests have shown that a well-adhered UHMWPE film (of sub-micron to few microns thickness) onto Si surface can be achieved through the presently demonstrated dip-coating technique. The film is strongly adhered even after a prolonged (100 min) ultrasonication in the solvent (decalin). The absence of Si peak (in the XPS spectrum) on the coated surface suggests that the film is free from defects such as pin-holes or dewetted substrate.

3.1.3. *Tribological Properties*

The coefficients of friction of bare Si and Si/UHMWPE are shown in Table 4. The UHMWPE film has drastically reduced the coefficient of friction of bare Si from 0.6 to 0.09. The low coefficient of friction of UHMWPE film is comparable with that obtained for bulk UHMWPE (~0.12) sliding against Si_3N_4 ball under the same loading conditions. The low coefficient of friction of UHMWPE film is due to its lower shear strength and inherent lubrication property of its molecules because of "smooth molecular profile".[28] Smooth molecular profile is a name given to polymers such as HDPE, PTFE and UHMWPE[28-30] where the polymers contain the linear molecules without any side groups, branching or cross-linking. These polymers have shown very low friction which is attributed to the easiness with which the long chain molecules shear (slip) across each other within a very thin top layer of the material.

Table 4. Coefficient of friction and wear life of bare Si and UHMWPE film modified Si. For comparison, the data for OTS SAM is also included.

Sample	Coefficient of friction	Wear life, number of cycles
Si	0.60	100
Si/UHMWPE	0.09	12000*
Si/OTS#	0.18	1600*

*the lowest and highest wear life data among three tests for UHMWPE film are 8000 and 15000 cycles, respectively, and for OTS SAM are 1000 and 2100 cycles, respectively.
#Data from Satyanarayana and Sinha[31] for comparison purpose.

The variations of the coefficient of friction with respect to the number of sliding cycles for bare Si, Si/OTS and Si/UHMWPE are shown in Fig. 5 and a summary of the wear life data is presented in Table 4 above.

The wear durability data strongly suggests that the UHMWPE film is superior in reducing the friction and wear of the Si surface when compared to the conventional OTS SAM. Bare Si failed within few tens of sliding cycles whereas the UHMWPE film has shown a wear life of ~12,000 cycles, which is much larger when compared to that of conventional OTS SAM (~1600 cycles).

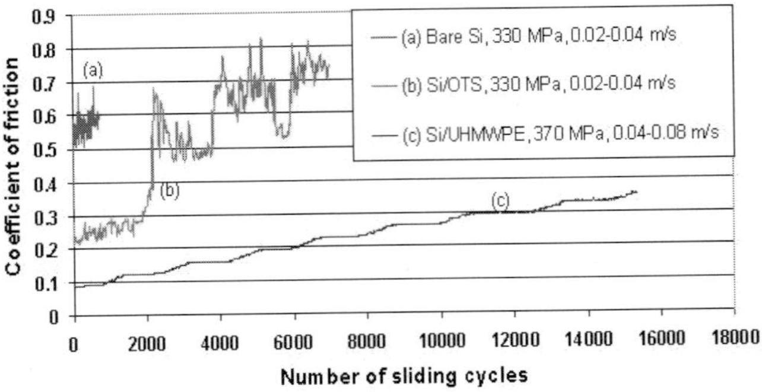

Fig. 5. Coefficient of friction versus number of sliding cycles curves for bare Si and Si/OTS SAM surfaces tested at 330 MPa and 0.02-0.04 ms^{-1} sliding velocities and Si/UHMWPE surface tested at 370 MPa and 0.04-0.08 ms^{-1} sliding velocities.

The SEM micrograph of the wear track of Si/UHMWPE after ~21,000 cycles of sliding is shown in Fig. 6. The wear track does not contain any wear debris and shows the compressed state of the film with localized plastic deformation. The rough asperities and protrusions get ironed or flattened by the ball sliding process. EDS analysis was used to assess the extent of the wear of the film and Si surface and the results are also shown in Fig. 6. The wear track (point A) does not contain Si peak and shows only strong C peak. The EDS spectrum inside the wear track (point A) is identical to that outside the wear track (point B). The strong Si peak obtained (with EDS) on piranha treated Si wafer is shown in Fig. 6(d) for comparison.

The extent of wear of the polymer film cannot be accurately estimated from the intensities of the C peak (inside and outside the wear track) because of the limited accuracy of the EDS as far as quantification of C is concerned. Therefore, EDS results only suggest that the wear of the Si surface has not started even after ~21,000 cycles of sliding in the case of Si/UHMWPE, which is also evident from the SEM morphology inside and outside the wear track (Fig. 6(a)). SEM/AFM analysis of the wear tracks after ~2,000 and ~10,000 cycles of sliding is carried out to understand the reasons for the increase in the coefficient of friction of the UHMWPE film as the sliding progresses with no detectable wear to the polymer film or the Si substrate (Fig. 7).

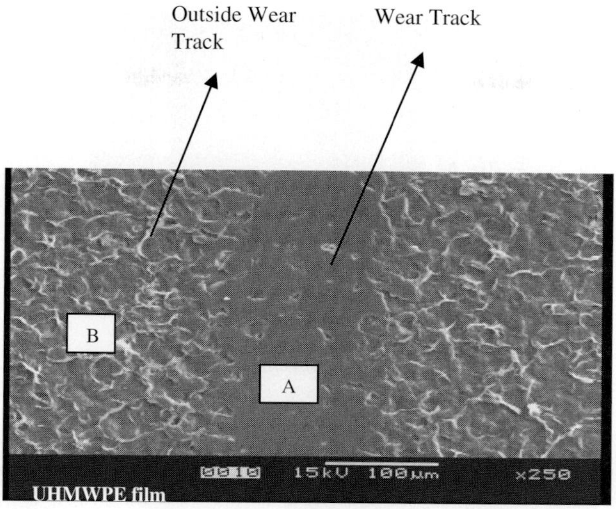

(a)

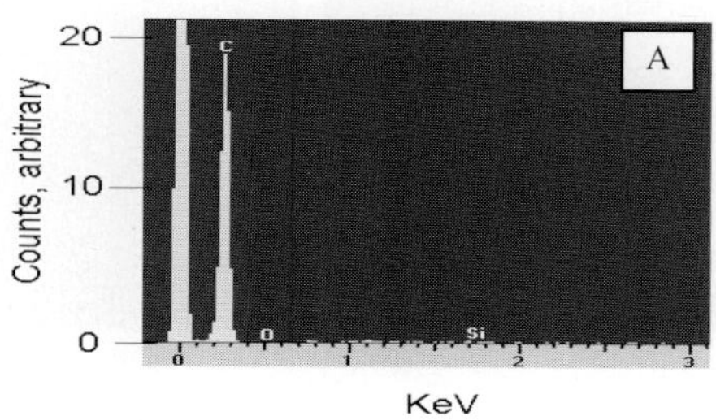

(b)

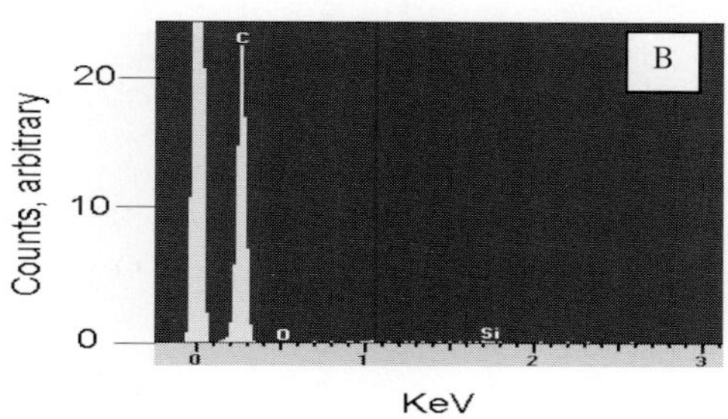

(c)

Fig. 6. (a) SEM image of the wear track of Si/UHMWPE, run up to 21,570 sliding cycles at a contact pressure of 370 MPa. EDS spectrum (b) inside the wear track and (c) outside the wear track for the sample shown in (a). (d) EDS spectrum on piranha treated Si.

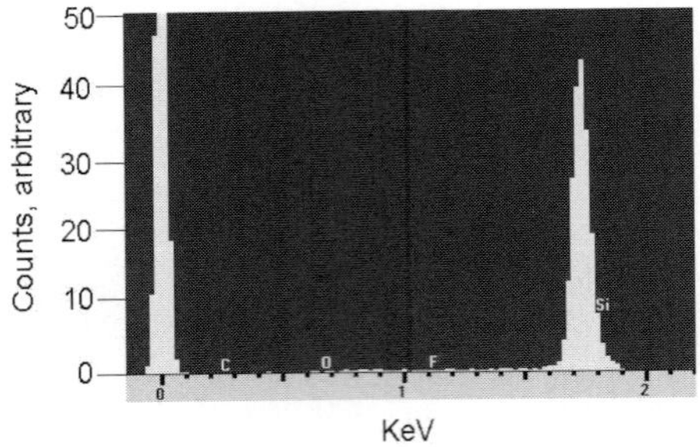

(d)

Fig. 6 (*Continued*)

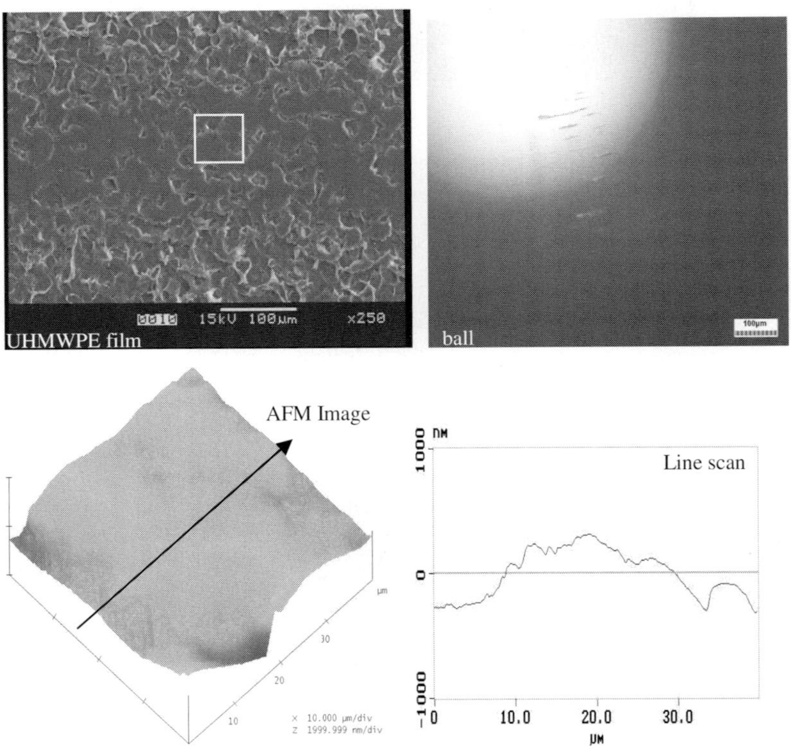

(a)

Fig. 7. (a) Si/UHMWPE sample run up to 2,033 cycles. (b) Si/UHMWPE sample tested up to 10,103 cycles. AFM image of the respective worn surface (the area used for the AFM study is shown with square box on SEM micrograph) together with a line scan is also shown. The magnification of the ball image is 100x.

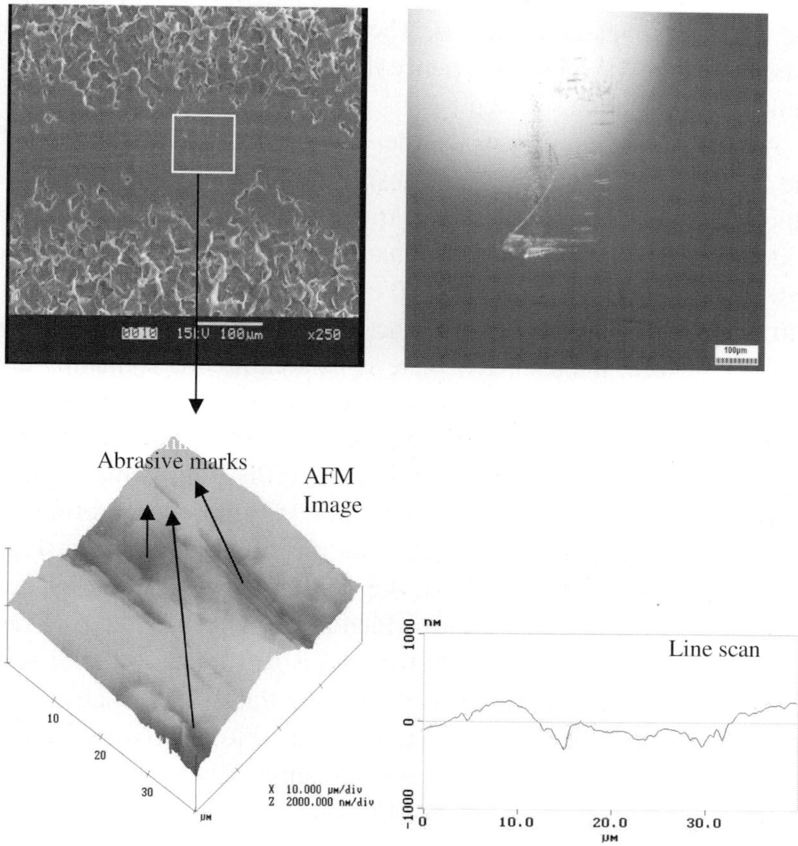

(b)

Fig. 7 (*Continued*)

The wear track after ~2,000 cycles is very smooth (roughness is 0.219 μm) when compared to the un-worn polymer film (roughness is 0.556 μm) showing the compressed state of surface asperities and the corresponding ball surface (after the test) shows negligible material transfer. The AFM image also shows the flattening of the surface features of the polymer film supporting the SEM observation. The AFM image inside the wear track shows higher extent of the compression of the polymer chains when compared to the SEM image of wear track and the difference is attributed to the variation in the scan area used for AFM and SEM morphological studies. The AFM imaging is much localized (when compared to SEM) and hence the surface profile in AFM depends on the image location selected.

As the sliding progresses, further compression of the polymer film and an increase in the material transfer to the ball surface is expected, which is supported on the wear track after ~10,000 cycles (by the evidence of few signs of abrasive marks (Fig. 7(b)) and adhesive plastic flow). The wear track after ~10,000 cycles is even smoother (roughness is 0.196 μm) when compared to that after ~2,000 cycles and un-worn polymer surface. Interestingly, there is no wear debris formation due to the abrasive and adhesive interactions which slightly increased the material transfer to the counterface (ball). It is anticipated that the plastic deformation of the asperities on the UHMWPE film increases the real contact area between the ball and UHMWPE film which results in an increase in the coefficient of friction as the sliding progresses (Fig. 5). Finally, because of this process, the coefficient of friction exceeds a value of 0.3 after ~12,000 cycles which is considered as the film failure in the present study even though the film is still intact to the Si surface and no wear particle is observed on the wear track. The anticipated wear mechanism for UHMWPE films on Si is similar to that observed for bulk UHMWPE sliding against several ceramic counterparts.[32] As the coefficient of friction of UHMWPE film exceeds 0.3 after ~12,000 cycles, further possibilities are explored to reduce the coefficient of friction beyond ~12,000 cycles. One such strategy (PFPE overcoating) is explained below.

3.2. *Effect of PFPE Overcoating onto UHMWPE Film Modified Si Surface on Tribological Properties*

3.2.1. *Background*

Even though UHMWPE film did not show any wear debris until ~20,000 cycles, the coefficient of friction exceeded 0.3 after ~12,000 cycles. Many applications require very low coefficient of friction until the end of the service. Therefore, we looked into the possibilities of improving the wear durability of UHMWPE film even further. Motivated by the improvement in wear durability due to PFPE overcoating onto SAMs (Satyanarayana and Sinha),[31] we overcoated UHMWPE film with an ultra-thin layer of PFPE to enhance the wear durability of UHMWPE film. The results based on the effect of PFPE overcoating onto UHMWPE film are discussed in this section.

PFPE Zdol 4000 was used to coat the UHMWPE films. The PFPE was coated onto UHMWPE film according to the procedure explained in the experimental procedures (Section 2.2).

3.2.2. *Physical Characteristics of the Dual-Layer Film (UHMWPE/PFPE)*

The PFPE overcoating onto UHMWPE film has further increased the water contact angle from 127° to 134° and made the surface even more hydrophobic. SEM observation did not show any further change in the morphology after PFPE overcoating onto UHMWPE film. Also, the SEM measurements did not detect any change in the thickness after PFPE overcoating onto UHMWPE film and suggest that PFPE formed as an ultra-thin film (probably only a few nanometers in thickness). The PFPE film formed on the DLC coated magnetic hard-disk surface under the same depositional conditions used in the present research has shown a thickness of ~2 nm.[33]

3.2.3. *Chemical Analysis of the Dual-Layer Film*

Figure 8 shows the XPS wide scan spectra for UHMWPE film and PFPE overcoated UHMWPE film. The appearance of F1s peak after PFPE overcoating onto UHMWPE film confirms its presence.

3.2.4. *Tribological Properties of the Dual-Layer Film*

The PFPE overcoating onto UHMWPE film has slightly reduced coefficient of friction from 0.09 to 0.08, which is attributed to the low surface energy and lower resistance offered to the sliding by the flexible PFPE molecules. Figure 9 shows the coefficient of friction variation in the sliding test and suggests that PFPE effectively reduced and maintained the coefficient of friction value below 0.2 until 100,000

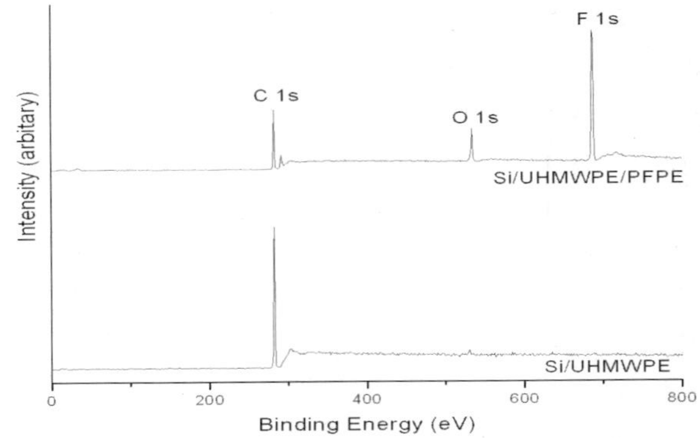

Fig. 8. XPS wide scan spectrum for Si/UHMWPE and Si/UHMWPE/PFPE samples.

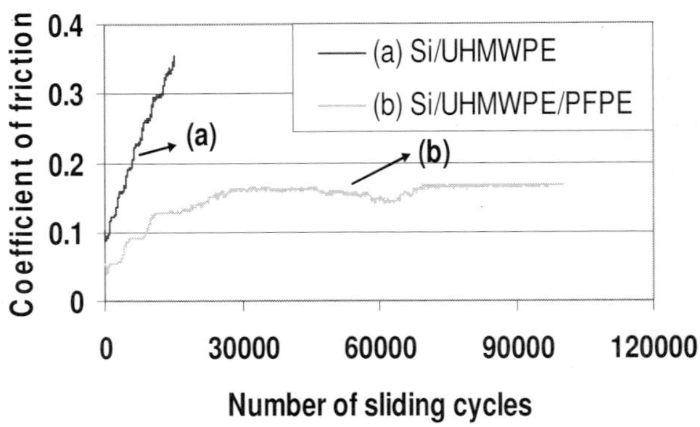

Fig. 9. Coefficient of friction versus number of cycles for UHMWPE film with and without PFPE overcoating at a contact pressure of 370 MPa and a sliding velocity of 0.04-0.08 ms^{-1}.

cycles of sliding. PFPE overcoating onto UHMWPE film has increased the wear durability from 12,000 to > 100,000 cycles (the tests were stopped after 100,000 cycles due to long test duration).

Figure 10 shows the SEM micrograph of the wear track of Si/UHMWPE/PFPE after 100,000 cycles of sliding. The wear track shows no signs of abrasive marks (which is also evident from the AFM image inside the wear track) and is very smooth (roughness is 0.095 μm) with much less plastic deformation when compared to the wear track for Si/UHMWPE after ~10,000 cycles [Fig. 7(b)]. The improved tribological

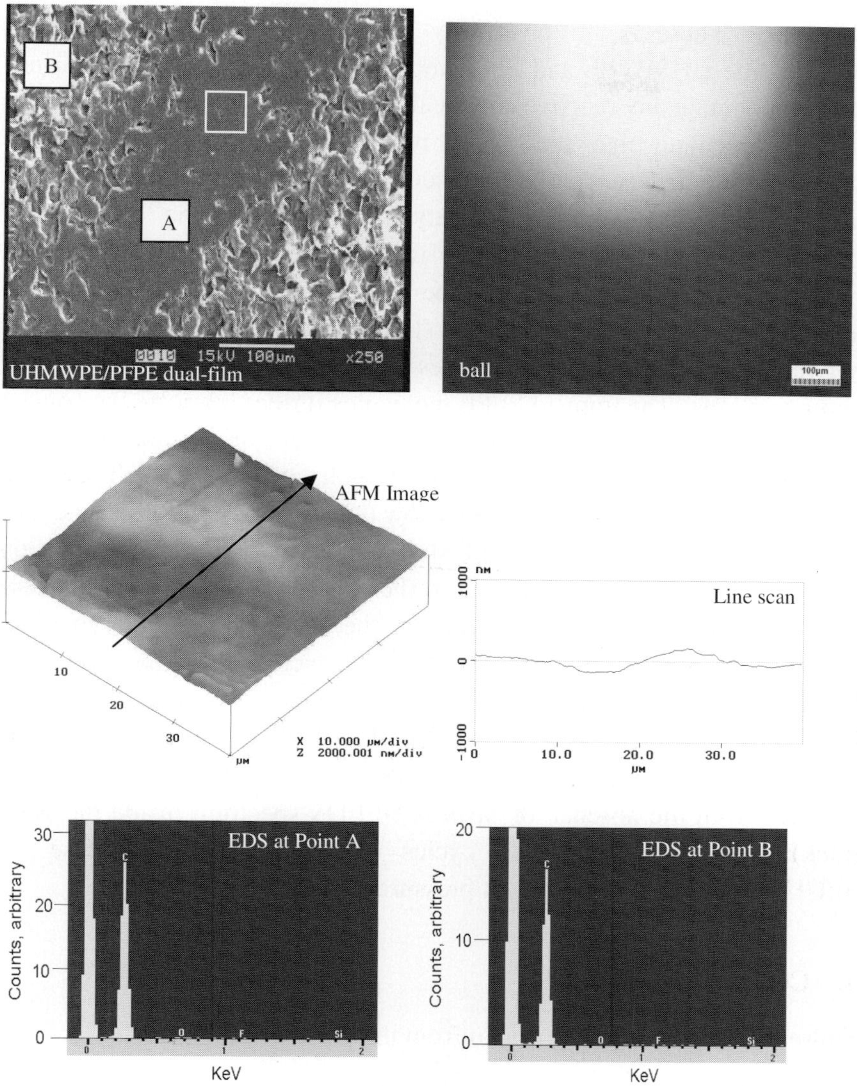

Fig. 10. Wear track of Si/UHMWPE/PFPE, run up to 100,000 cycles. EDS spectrum inside the wear track (Point A) and outside the wear track (Point B) are also shown. AFM image of the respective worn surface together with a line scan is also shown.

properties suggest that the PFPE overcoating greatly influenced the wear mechanism of the UHMWPE film. The role of PFPE has to be understood to explain the reasons for a greater increase in the wear durability. There is no possibility of chemical bonding/interactions between the UHMWPE and PFPE molecules as UHMWPE film surface does not contain any reactive chemical groups.

We speculate that the PFPE molecules are trapped inside the valleys/gaps between the polymer chains/fibers (similar to that anticipated on reactive SAMs in Satyanarayana and Sinha[31]) present on the UHMWPE film surface (Fig. 1(a)). We further speculate that the trapping is only of mechanical type due to plastic deformation of the asperities in the initial few runs of sliding and due to the entanglement between long UHMWPE molecules and PFPE molecules during the sliding process. The trapped PFPE molecules further lubricate the contact (additional to that obtained from UHMWPE film itself) and extend the low friction nature of the film for many thousands of cycles. Therefore, Si/UHMWPE/PFPE has shown very low (~0.15) and stable coefficient of friction up to 100,000 cycles of sliding instead of linearly increasing coefficient of friction with respect to the number of cycles as in the case of UHMWPE film on Si. Moreover, there is no (or little) material transfer to the ball after 100,000 cycles in the case of Si/UHMWPE/PFPE (Fig. 10) whereas some amount of material transfer has been observed in the case of Si/UHMWPE even after only 2000 cycles of sliding (Fig. 7(a)). Further, there is no wear to the Si surface (evident from the absence of Si peak in EDS spectrum inside the wear track) even after 100,000 cycles of sliding in the case of Si/UHMWPE/PFPE at a contact pressure of 370 MPa.

4. Conclusions

Following conclusions are drawn from the present study:

1. A uniformly coated novel UHMWPE film of thickness 28 μm is obtained on Si surface using a simple dip-coating technique. The UHMWPE film is highly hydrophobic with a water contact angle of 127°. Water contact angle and FTIR measurements have shown that the chemical structure of UHMWPE film is identical to that of the bulk UHMWPE.

2. The UHMWPE film has shown a coefficient of friction of 0.09 and a wear life of ~12,000 cycles without the generation of any wear debris even after ~20,000 cycles of sliding. Good adherence between the UHMWPE film and Si surface and self-lubricating properties of UHMWPE are responsible for good tribological properties of the UHMWPE film.
3. PFPE overcoating further increased the hydrophobicity of UHMWPE film by increasing the water contact angle from 127° to 134°.
4. The PFPE overcoating has reduced the coefficient of friction of UHMWPE film from 0.09 to 0.08 and increased the wear life from 12,000 cycles to 100,000 cycles (experiment stopped). The wear track is very smooth even after 100,000 cycles without any wear debris and wear to the Si surface.
5. It has been observed that the large amount of PFPE entrapped into the valleys present on the UHMWPE film lubricates the contact in-situ and hence shows greater improvement in the wear durability.
6. Finally, an optimum combination of good mechanical and tribological properties of UHMWPE and lubrication characteristics of PFPE is greatly beneficial to the wear durability of Si substrate.

Acknowledgments

Authors would like to acknowledge the support by the Faculty of Engineering, NUS research grant #R-265-000-248-112. Authors also wish to acknowledge Mr. K. C. Chung, Ticona Representative Office, Singapore for donating UHMWPE powder for the present study.

References

1. R. C. Advincula, W. J. Brittain, K. C. Caster and J. Ruhe, *Polymer Brushes: Synthesis, Characterization and Applications,* (Wiley-VCH Verlag GmbH & Co., 2004).
2. B. Bhushan, *Tribology and Mechanics of Magnetic Storage Devices.* 2nd Edition, (Springer Verlag, New York, 1996).
3. D. A. Webb. Polymeric Coatings for Micromechanical Devices, *US Patent 5,447,600*, (5 September 1995).

4. K. L. Prime and G. M. Whitesides, *J. Am. Chem. Soc.* **Vol. 115**, 10714 (1993).
5. W. Liu, F. Zhou, L. Yu, M. Chen, B. Li and G. Zhao, *J. Mat. Res.* **Vol. 17**, 2357 (2002).
6. A. O. Pozdnyakov, K. Friedrich and V. V. Kudryavtsev, *J. Mat. Sci. Lett.* **Vol. 20**, 2071 (2001).
7. C. Sun, F. Zhou, L. Shi, B. Yu, P. Gao, J. Zhang and W. Liu, *Appl. Surf. Sci.* **Vol. 253**, 1729 (2006).
8. G. Gu, Z. Zhang and H. Dang, *Appl. Surf. Sci.* **Vol. 221**, 129 (2004).
9. M. Kitoh and Y. Honda, *Thin Solid Films* **Vol. 271**, 92 (1995).
10. A. Sidorenko, H-S. Ahn, D-I. Kim, H. Yang and V. V. Tsukruk, *Wear* **Vol. 252**, 946 (2002).
11. S. K. Sinha, Wear failures of plastics. *ASM International Handbook,* **Vol. 11**, ASM International, Ohio, pp.1019, 2002.
12. A. Wang, A. Essner, V. K. Polineni, C. Stark and J. H. Dumbleton, *Trib. Int.* **Vol. 31**, 17 (1998).
13. Y. Bao, T. Zhang and D. T. Gawne, *J. Mat. Sci.* **Vol. 40**, 77 (2005).
14. H. G. Braun and E. Meyer, *Thin Solid Films.* **Vol. 345**, 222 (1999).
15. J. H. Maas, M. A. Cohen Stuart and G. J. Fleer, *Thin Solid Films.* **Vol. 358**, 234 (2000).
16. H. Yasuda, *Luminous Chemical Vapor Deposition and Interface Engineering*, (CRC Press, 2004), p. 537.
17. K. C. Eapen, S. T. Patton and J. S. Zabinski, *Trib. Lett.* **Vol. 12**, 35 (2002).
18. J. Ruhe, V. Novotny, K. K. Kanazawa, T. Clarke and G. B. Street, *Langmuir.*, **Vol. 9**, 2383 (1993).
19. M. P. Heuberger, M. R. Widmer, E. Zobeley, R. Glockshuber and N. D. Spencer, *Biomaterials* **Vol. 26**, 1165 (2005).
20. J. S. Chen, Z. Sun, P. S. Guo, Z. B. Zhang, D. Z. Zhu, H. J. Xu, *J. Appl. Phys.*, **Vol. 93**, 5103 (2003).
21. L. Torrisi, C. Gentile, A. M. Visco and N. Campo, *Radiat. Eff. Defects. Solids.* **Vol. 158**, 731 (2003).
22. M. Miwa, A. Nakajima, A. Fujishima, K. Hashimoto and T. Watanabe, *Langmuir* **Vol. 16**, 5754 (2000).
23. S. Osawa, M. Yabe, M. Miyamura and K. Mizuno, *Polymer* **Vol. 47**, 3711 (2006).
24. Y. C. Jung and B. Bhushan. *Nanotechnology* **Vol. 17**, 4970 (2006).
25. A. Elliott, *Infra-red Spectra and Structure of Organic Long-chain Polymers*, (Edward Arnold (Pub) Ltd. 1969), p. 48.
26. A. L. S. Alves, L. F. C. Nascimento and J. C. M. Suarez, *Poly. Test.* **Vol. 24**, 104 (2005).
27. P. H. Kang and Y. C. Nho, *Rad. Phys. Chem.* **Vol. 60**, 79 (2001).
28. C. M. Pooley and D. Tabor, *Proc. Roy. Soc. Lond. A.* **Vol. 329**, 251 (1972).
29. Y. Q. Wang and J. Li, *Mat. Sci. and Engg. A.* **Vol. 266**, 155 (1999).

30. Unal and A. Mimaroglu, *Mat. and Design.* **Vol. 24**, 183 (2003).
31. N. Satyanarayana and S. K. Sinha, *J. Phys. D: Appl. Phys.*, **Vol. 38**, 3512 (2005).
32. A. B. Abdallah and D. Treheux, *Wear*, **Vol. 142**, 43 (1991).
33. S. K. Sinha, M. Kawaguchi, T. Kato and F. E. Kennedy, *Trib. Int.* **Vol. 36**, 217 (2003).

CHAPTER 21

TRIBOLOGY OF UHMWPE THIN FILMS ON Si WITH INTERFACIAL LAYERS AND MODIFICATIONS

M. Minn and S. K. Sinha*

Department of Mechanical Engineering, University of Singapore
9 Engineering Drive 1, Singapore 117576, Singapore
**E-mail: mpesks@nus.edu.sg*

This chapter presents the strategies to improve the friction and wear properties of UHMWPE thin film on Si substrate by suitable interfacial modifications. Firstly, hard DLC was used as an intermediate layer in order to increase the load carrying capacity of the polymer film. As a result, the contact area decreased and the wear life of the composite film, Si/DLC/UHMWPE increased significantly. After investigating the effect of UHMWPE thickness on the tribological properties of Si/DLC/UHMWPE, it is found that there is an optimum range of thickness of the polymer film that shows the best friction and wear performance of this composite film. Secondly, Si substrates were modified with different surface energies that could determine the adhesion strength of UHMWPE film to the substrates. The two extreme conditions, the most hydrophilic, 21° and the most hydrophobic, 104° showed much lower wear resistance when compared with moderate surface wettability. It is concluded that the wear performance of UHMWPE thin films on Si substrate could be extended by several orders of magnitude by suitably using a hard interface layer and by modifying the interfaces for different surface energies of the Si substrate.

1. Introduction

Polymers are commonly used as protective coating films in engineering applications because of their self-lubricating properties, low production cost, easy to coat onto complex shapes, excellent wear and corrosion resistance.[1] They can be used as soft coatings due to their low strength and hardness values. However, if the substrate is not properly modified, the soft coating can be easily removed or peeled off from the substrate. There are many parameters to be optimized such as load carrying capacity, thickness and surface energies in order to obtain the long wear life of the coating. In this chapter, we present and discuss some methods to improve the tribological properties of soft polymer coating on Si substrates.

Ultra high molecular weight polyethylene (UHMWPE) has been widely used as a bearing material in highly stressed parts such as total joint replacement for more than forty years.[2] Despite its extensive use, the tribological property of UHMWPE limits the lifespan of the product. Many researchers have been trying to improve the tribological performance of this bulk polymer by various modifications. Enhancing the crosslink density of the polymer is one of the ways to modify the friction and wear properties[3] and it depends on the radiation methods and absorbed radiation dose level.[4-6] Another mechanism widely used to increase the wear resistance is controlling the surface crystallinity of the polymer.[7] Friction and wear properties of UHMWPE is superior to most of the pristine polymers, however, low strength and lower softening temperature of UHMWPE may limit its application in some engineering applications. Use of UHMWPE as thin coating may have some advantage because the substrate can provide strength and thermal conductivity while the film will help in increasing the wear life. Recently UHMWPE has been studied as thin film[8-9] where Bao *et al.*[8] have coated UHMWPE film on steel substrate using thermal spray methods and Satyanarayana *et al.*[9] have used UHMWPE as a protective coating on Si by dip-coating method. These studies have shown that UHMWPE coating can be applied on surface for tribological property enhancement. In this chapter, we have focused on further enhancement in the tribological properties of UHMWPE film and therefore we have selected different intermediated bonding layers between the substrate and the UHMWPE film; in addition to the novel idea of using hard and soft bi-layer coating of DLC/UHMWPE on Si substrate.

The main disadvantage of a soft film is low load carrying capacity that leads to large contact area and eventually high friction force and wear. The contact area can be reduced by introducing a hard layer between the substrate and the soft coating. As a hard layer, diamond like carbon (DLC) films have great attraction for tribological applications due to their high hardness, high wear resistance,[10] chemical inertness, corrosion resistance[11-12] and low surface energy.[13] The range of the coefficient of friction of DLC films is 0.001 ~ 0.7 which is probably the widest among the coated films.[14] The friction and wear behaviours of DLC depend not only on the chemical and structural properties of carbon but also the test conditions such as the substrate, counterface material, rotational speed, applied load, chemistry of the environment etc.[15]

UHMWPE is very difficult to adhere to many substrates since it does not have functional group. The poor adhesion leads to rapid detachment of the polymer film from the substrates. It is necessary to give a pre-treatment to the substrate in order to increase adhesion.[16] Optimized surface wettability is an important criterion to enhance adhesion. Moisture contents can change the surface energy of a substrate that strongly affects the adhesion strength of a coating.[17] Moy and Karasz,[18] Lee and Peppas[19] and Nogueira *et al.*[20] showed that the moisture content on the substrate lowers the glass transition temperature of the coatings that deteriorates film mechanical properties. Wong and Broutman[21] and Xu and Ashbee[22] proved that the moisture contents create cracks or voids in the coatings. As a consequence, the adhesion loss between the coatings and the substrate is inevitable.[23-24] Yoon *et al.*[25] showed that the adhesion force increased with increasing contact angle of the substrate. From the literature, it is obvious that the surface wettability is an important factor in determining the adhesion strength of the polymer film to the substrate.

In this chapter, firstly, we use DLC as an intermediate layer and investigate the tribological advantages of DLC/UHMWPE composite film. The idea of using DLC as an intermediate layer stems from the fact the higher load bearing capacity of DLC would help in reducing contact area between the polymer film and the counterface slider and thus reduce friction and wear. Secondly, we study the effects of UHMWPE thickness on the friction and wear properties for DLC/UHMWPE. Lastly, we explore the role of surface energy on the tribological properties for Si/UHMWPE using different interfaces (different surface wettabilities). Surface energy was varied by applying self-assembled monolayers, heating Si substrate or by hydrogenating the Si surface before polymer film coating.

2. Experimental Procedures

2.1. *Materials*

The polished n-type Si (100) wafers (thickness of 455~575 μm, hardness of 12 GPa and roughness of 0.41 nm) were selected as the substrates. As a hard layer on Si substrates, tetrahedral amorphous carbon, ta-C (non-hydrogenated carbon) was deposited by Filtered Cathodic Vacuum Arc (FCVA) technology. The details of the depositional procedure is mentioned in reference.[26] The thickness, hardness and roughness of DLC layer are 50 nm, 57 GPa and 34.8 nm, respectively. 3-aminopropyltrimethoxysilane (APTMS) and octadecyltrichlorosilane (OTS), obtained from Aldrich Inc., were used for self-assembly monolayers (SAMs) without further purification. The chemical formulae of APTMS and OTS are $H_2N(CH_2)_3Si(OCH_3)_3$ and $CH_3(CH_2)_{17}SiCl_3$ respectively. Toluene (99.5% anhydrous) was used as the solvent for the preparation of SAM solutions. UHMWPE polymer powder (bulk density of 0.33 ± 0.03 g/cm^3 and average particle size of 20 ± 5 μm) was received from Ticona Engineering Polymers, Germany through a local Singapore supplier. Decahydranapthalin (Decalin) was used as the solvent to dissolve UHMWPE for dip-coating purpose.

2.2. *Preparation of Different Layers on Si Substrate*

The cleaning process used to remove contaminants and undesired particles from Si substrate is the most fundamental and necessary part in coating technology. The improper cleaning process prevents optimal bonding of the coating films to the substrate. In our study, we rinsed Si substrates for 1 minute and ultrasonically cleaned for 15 minutes each first with soapy water, second with distilled water and lastly with acetone. Ultrasonic energy removes loose surface materials by alternating compressive and expansive of the cleaning solution. The changes in pressure can produce cavitation bubbles which have the capacity of blasting contaminants away from the surface.[27] The cleaned Si substrates were blow-dried with pure nitrogen gas and immersed into a piranha solution (a solution of 70% H_2SO_4 and 30% of H_2O_2 in volume) at a temperature of 120ºC for an hour to remove any contaminants. The piranha treated substrates were rinsed again with distilled water and acetone for 1 minute each.

The heated Si substrates were prepared by heating the cleaned Si in an oven at a temperature of 100 ºC for an hour. The hydrogen terminated Si (Si-H) substrates were obtained by immersing the cleaned Si substrates into a dilute aqueous solution of hydrofluoric acid (2 vol. %) for 30 seconds. After that, the Si-H substrates were rinsed with distilled water for 1 minute and dried with pure nitrogen gas. The native oxide layer was completely removed from the Si substrate and replaced with hydrogen.

Two different SAMs were formed on cleaned Si substrates by immersing them into APTMS and OTS solutions with a concentration of 3 mM and 5 mM respectively. The immersion time was fixed for 5 hours. After that, the SAMs coated samples were ultrasonically washed with toluene and methanol to remove any physisorbed SAM molecules for seven minutes each. Finally, the samples were dried with pure nitrogen gas.

UHMWPE powder was dissolved in decalin. The dissolving temperatures were 150 ºC for half an hour and increased to 250 ºC for the next half an hour with a magnetic stirrer in order to fasten the dissolution rates. The different modified substrates were dipped in UHMWPE solution for 30 seconds with a constant speed of 2.4 mm/s by simple dip-coating method. The advantages of dip-coating are cost effectiveness, ability to coat recessed or difficult areas (such as corner and edges) and resistance to mechanical damage.[28] The UHMWPE coated samples were given heat treatment in a clean air oven at 100ºC for 15 hours. After the heat treatment, the samples were cooled down to room temperature in the oven. The measured roughness of UHMWPE film was 0.56 μm within a scan area of 10 μm x 10 μm using atomic force microscopy (AFM). Both bulk UHMWPE powder and coated UHMWPE film were tested using differential scanning calorimetry (DSC60, Shimadzu) in order to understand the effect of residue solvent on the coated film. The results show no additional peak of the solvent and thus, it is confirmed that there is no residual effect of the solvent on the final properties of UHMWPE film. A schematic diagram of the composite UHMWPE coating on Si substrate is shown in Fig. 1.

Firstly, we compare the tribological results of Si substrates with and without DLC interface layer where the thickness of top UHMWPE layer was fixed at 28 μm. Secondly, we explore the effect of UHMWPE thickness for Si/DLC/UHMWPE films. Thirdly, we select the best UHMWPE thickness which was coated onto Si substrates with different surface energies in order to investigate the role of surface energy on the top UHMWPE layer.

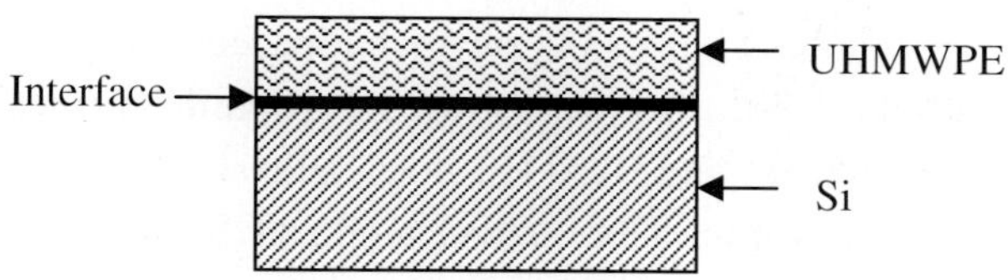

Fig. 1. A schematic (not to scale) diagram of Si/UHMWPE sample with different interfaces.

2.3. *Surface Analysis*

2.3.1. *Water Contact Angle*

One of the easiest ways to estimate the surface properties is to measure the wetting property using water contact angle method. The profile of a standing drop on the plane depends on the free energy (or the surface tension) of the surface. The simplest relationship between the contact angle and the surface free energies was first stated by Thomas Young. The Young equation[29] is written in the form

$$\gamma_{SV} = \gamma_{SL} + \gamma_{LV} \cos\theta \tag{1}$$

where γ is the surface free energy and the subscripts $_{SV}$, $_{SL}$ and $_{LV}$ represent surface-vapour, surface-liquid and liquid-vapour interfaces, respectively. Figure 2 shows an illustration of the Young equation.

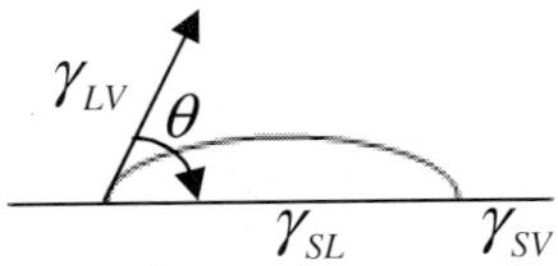

Fig. 2. A schematic diagram of a droplet on a solid surface.

The contact angles were measured using VCA Optima Contact Angle System (AST product, Inc., USA) machine with a water sessile droplet of 0.5 μL. The water contact angles of different surfaces are shown in Table 1. The measurement variation is in the range of ±3°. Bare Si has the most hydrophilic surface with 21°. After the piranha treatment, hydroxyl group is formed on Si which can react with water molecules to form hydrogen bonds. The formation of hydrogen bonds tends to give hydrophilic surface.[30] In other words, the hydrophilic surface can attract more water molecules and lead to low adhesion[17] of the polymer molecules to the substrate. The contact angle of Si after heat treatment (which reduced the amount of water molecules) was 46°. The formation of APTMS and OTS monolayers on Si provided the contact angles of 52° and 104° respectively which were consistent with the work of Satyanarayana *et al.*[31] The termination of hydrogen which removed native oxide from Si (Si-H) had lower tendency to attract water molecules and gave 71° of contact angle.

After coating DLC onto Si, θ increased to 81°. The DLC film used in this study is non-hydrogenated and has only sp^3 hybridized carbons which are less reactive with water molecules. The hydrocarbons in UHMWPE do not form hydrogen bonds and it turns out that the UHMWPE film on Si and Si/DLC surfaces became more hydrophobic with 93° and 91° respectively.

The UHMWPE coatings on modified Si surfaces (heated Si, Si/APTMS, Si-H and Si/OTS) had the same range of contact angle, 91°~93°. This proves that there is no effect of intermediate layers (or substrate surface energies) on the contact angle of top UHMWPE layer.

Table 1. Water contact angles of different surfaces on Si.[32-33]

Surface	Contact Angle, θ (°)
Bare Si	21
Si/DLC	81
Si/UHMWPE	93
Si/DLC/UHMWPE	91
heated Si	46
Si/APTMS	52
Hydrogen terminated (Si-H)	71
Si/OTS	104

2.3.2. *Nanoscratching and Nanoindentation*

The nanoscratching resistance of Si/UHMWPE and Si/DLC/UHMWPE with 28 μm film thickness were compared using MTS Nano Indenter XP machine. The indenter used for nanoscratching was a conical shape diamond tip with 90º cone angle and a tip radius of 5 μm. The scratch tests were conducted using a ramp loading setup from 0~250 mN at a constant velocity of 10 μm/s. The topography of the nano scratches were studied with a JEOL JSM-5600 LV scanning electron microscope (SEM).

Figure 3a shows the nanoscratching results of Si/UHMWPE and Si/DLC/UHMWPE. It is clearly seen that the differences in the penetration depths for the two films increases with increasing loads. That means the load carrying capacity of Si/UHMWPE is much inferior to that of Si/DLC/UHMWPE. Figure 3b shows the SEM images after nanoscratching for the two films in which we can see that Si/UHMWPE film was easier to penetrate and that it peeled off from the substrate as can be seen at the end of the scratch. Furthermore, it shows a clear damage of the polymer film at both sides of the scratch and a pile of polymer at the end of the scratch by a process of plastic deformation and delamination.[34] However, for Si/DLC/UHMWPE, there were no wear debris along the scratch and it resembled plastic deformation of the polymer by ploughing without any visible detachment or delamination of the film.

The modified Si and Si/DLC may influence the hardness and the elastic modulus of the polymer coating that changes the tribological

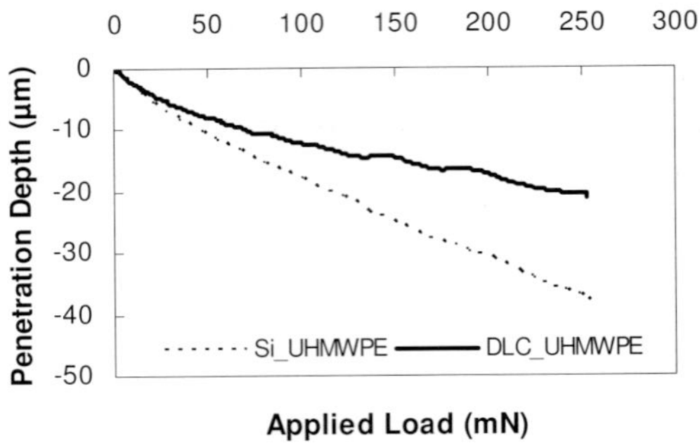

(a)

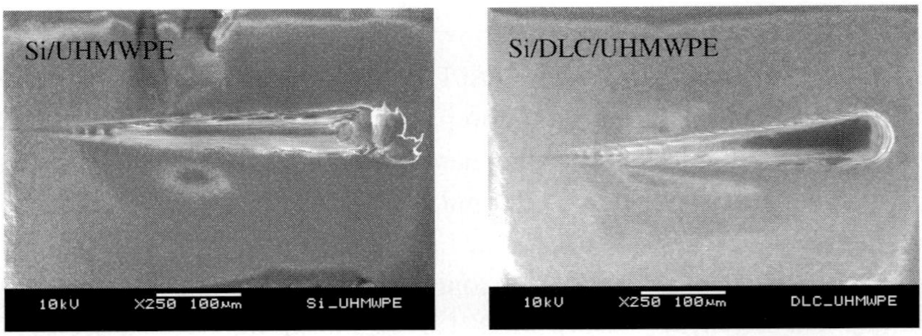

(b)

Fig. 3. (a) Scratch penetration depth as a function of progressively applied normal load and (b) SEM images of the scratch deformation for Si/UHMWPE and Si/DLC/UHMWPE films. Normal load varied from 0 to 250 mN and the scratching direction is from left to right.

properties of the film. Nanoindentation tests were conducted in order to measure the variations in the hardness and the elastic modulus of the polymer film using the same MTS Nano Indenter XP machine. For Si/UHMWPE and Si/DLC/UHMWPE with different UHMWPE thicknesses, we fixed the indentation load at 40 mN with a standard Berkovich diamond tip.

Table 2 provides contact area and contact pressure values for every sample as calculated using Hertzian contact model. Mechanical properties of Si_3N_4 ball (Young's modulus: 310 GPa and Poisson's ratio: 0.22) and UHMWPE (Poisson's ratio: 0.46) were used to calculate the contact area and the contact pressure. As shown in Table 2, the DLC intermediate layer provides higher hardness (high load carrying capacity), higher elastic modulus and shallower penetration depth than those of Si/UHMWPE where UHMWPE thickness is 28 μm. When we reduced the UHMWPE thickness in Si/DLC/UHMWPE film, because of hard DLC, the hardness and the elastic modulus increased whereas the penetration depth decreased. When we measured on the bulk UHMWPE, the hardness and the elastic modulus of the bulk are approximately one order of magnitude lower than those of the 28 μm films indicating that there is significant amount of substrate effect. It is also noted that the contact area decreases gradually but the contact pressure increases remarkably with a decrease in the thickness of the polymer film below 6.2 μm due to the effect of the intermediated hard DLC layer.

For Si substrates with different surface wettabilities (6 ± 2 μm UHMWPE thickness), we used 500 μN of normal load (where the penetration depth is approximately 400 nm) in order to avoid the effect of the substrate. The measured hardness and elastic modulus are 35 MPa and 1 GPa respectively for all samples regardless of the interface condition and the variation is within the measurement errors. The roughness of UHMWPE film on Si with different interfaces was approximately 0.56 μm. The results of nanoindentation have shown that the interface layers have no effect on the hardness and the elastic modulus of the top surface of the polymer film.

Table 2. Mechanical properties and other parameters for different samples.[32]

Sample	Thickness of UHMWPE (μm)	Hardness (GPa)	Elastic Modulus (GPa)	Theoretical		Nanoindentation penetration depth (μm)
				Contact area ($10^{-10}m^2$)	Contact pressure (MPa)	
Bulk UHMWPE	-	0.038	0.993	80	7.7	7.25
Si/UHMWPE	28	0.06	6.51	23.5	26	5.3
Si/DLC/UHMWPE	3.4	11.9	171.87	3.7	165	0.45
	6.2	0.86	43.74	7.3	85	1.52
	12.3	0.12	18.65	12	51.6	3.81
	28	0.09	8.81	19.4	31.8	4.5

2.3.3. *XPS Analysis on Different Interfaces*

Figure 4 shows the wide scan spectrum of XPS results of bare Si, heated Si, Si/APTMS, Si-H and Si/OTS samples before UHMWPE coating. The presence of C1s peaks on bare Si, heated Si and Si-H are due to the organic contaminants which could not be removed during the piranha treatment or were adsorbed from the atmosphere before XPS tests. The O1s peak for bare Si is the highest which indicates the presence of oxide and moisture on the surface. The O1s peak for heated Si is lower than that for bare Si. The possible reason is that the moisture, which can increase O1s peak, is removed during the heat treatment of Si. The presence of hydrogen termination on Si prevents the formation of the oxide (SiO_2) and as a result the O1s peak is the lowest for Si-H specimen. Because of the amine terminal group, we can see N1s peak in Si/APTMS surface. The C1s concentration is the highest in Si/OTS due to its long chain length. The XPS results confirm the successful formation of different layers onto Si substrate.

Any chemical bonding between the modified Si substrates and the polymer films is not expected in such kind of dip-coating method as UHMWPE used in this study were non-functionalized. However, for additional confirmation, we carried out XPS tests to study any possible chemical bonding for all samples. In order to remove chemically unbonded UHMWPE coating, the UHMWPE coated samples were dipped into the solvent (decalin) and heated in an oven at 250 ºC for 15 minutes and ultrasonically cleaned in acetone for 15 minutes. EDS tests were conducted on the cleaned samples and original bare Si in order to compare the Si peaks and the results showed that the intensities of Si peaks were nearly the same for both surfaces. XPS tests were carried out on the cleaned samples and there was no additional or extra peak of any chemical bonding except Si and UHMWPE peaks. The presence of Si peak is due to the uncovered Si substrate and UHMWPE peak is due to the trace of remaining polymer molecules. The XPS tests have confirmed that UHMWPE molecules are not chemically bonded to the substrate but coated onto the substrate by means of physical bonding or mechanical interlocking.

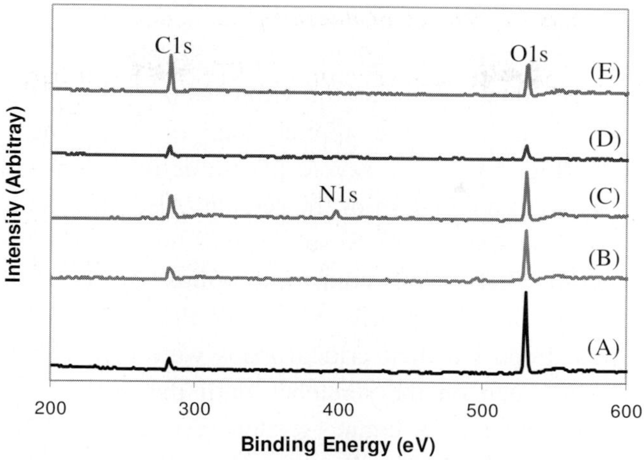

Fig. 4. XPS wide spectrum for (A) bare Si, (B) heated Si, (C) Si/APTMS, (D) Si-H and (E) Si/OTS surfaces.

2.3.4. *Study of Adhesion Strength Using Scratch Tests*

As a comparative study of adhesion strength of UHMWPE on different modified Si substrates, we conducted scratch test with a diamond tip with 2 μm tip radius on every sample. The scratching velocity and the linear scratch distance were fixed at 0.1 mm/s and 1 cm respectively. The applied normal load was varied from 1 g to 7 g with an increment of 1 g. The surface topography of the scratches were studied from the images of field emission scanning electron microscopy (FESEM) using a Hitachi S4300 machine coupled with energy dispersive spectrometer (EDS). Before taking FESEM images, gold coating was performed on the tested polymer films at 10 mA for 40 seconds (JEOL, JFC-1200 Fine Coater).

Table 3 provides the critical loads of samples where critical load refer to the applied normal load when the film fails during scratching. The critical loads were determined by measuring Si peak on the scratches using EDS.

Table 3. The critical load as a function of different interfaces; and the scratching velocity is 0.1 mm/s and the scratch length is 1 cm.[33]

Interface	Bare Si	heated Si	Si/APTMS	Si-H	Si/OTS
Critical load (g)	2	4	4	7	2

It is seen that the critical loads of the scratches of very hydrophilic, Si/UHMWPE and very hydrophobic, Si/OTS/UHMWPE interface are the lowest at 2 g. Figure 5 shows the scratches on Si/UHMWPE and Si-H/UHMWPE films under the applied load of 2 g. The scratch on Si/UHMWPE (Fig. 5a) shows severe plastic deformation with polymer debris (breakage) at both edges of the scratch. Also, there is a clear sign of debris particles coming from Si substrate as they are fragments owing to brittle fracture. EDS results confirm that those brittle particles are Si debris. The scratches on heated Si/UHMWPE and Si/APTMS films showed similar trend but their critical loads were higher. Neither pores nor cracks were seen on the scratches until the applied load of 4 g. Higher critical load means higher scratch resistance and hence better adhesion strength. Although the critical load increases for heated Si/UHMWPE and Si/APTMS/UHMWPE, some pores are clearly seen when the film has failed. The pores show detachment of the polymer from the substrate that is the main reason for film failure in continuous sliding.

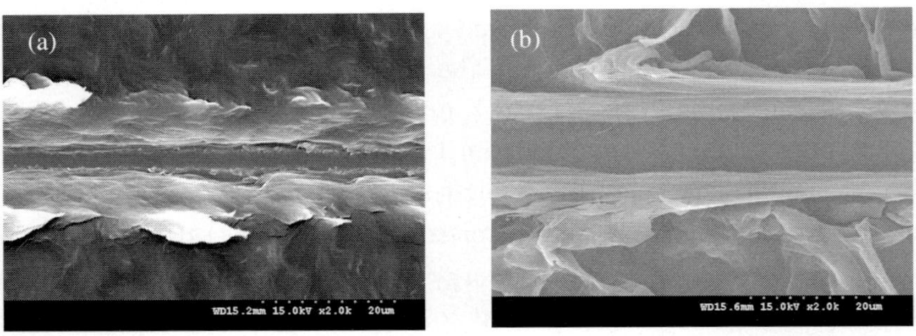

Fig. 5. The FESEM images of the scratches on (a) Si/UHMWPE and (b) Si-H/UHMWPE films where the normal load is 2 g and the scratching velocity is 0.1 mm/s.

In Fig. 5b, the scratch on Si-H/UHMWPE under the applied load of 2 g, shows that the polymer film is very ductile and no wear debris particles are observed along the scratch. For Si-H/UHMWPE film, the high intensity of Si peak was observed when the applied load was 7 g. However, no pores were found at the interface even at the critical load of 7 g confirming that no UHMWPE film was pulled out or torn from the substrate. This suggests that the adhesion strength at the interface (Si-H) is very strong compared to all other interfaces.

2.4. *Friction and Wear Tests*

The friction and wear tests were conducted on a ball-on-disc tribometer (Fig. 6). A 4 mm diameter of silicon nitride (Si_3N_4) ball was chosen as a stationary counterface whereas UHMWPE coated samples acted as the disc. Before the tests, Si_3N_4 balls were cleaned with acetone. The wear track radius and the normal load were 1~2 mm and 4 g respectively. The rotational speed chosen for all tests was 500 rpm.

The normal load was converted from the vertical displacement of the double cantilever that was measured using a laser displacement sensor. The friction force was continuously measured using four strain gauges attached to the same cantilever arms. The sampling rate used to measure the friction data was 5 Hz. The tests were carried out in a clean room environment at a temperature of 25 ± 2°C and a relative humidity of 55 ± 5%. The initial coefficient of friction was taken as an average value of the first 4 seconds of sliding. In this study, we define the wear life of coated sample as the respective number of cycles when the coefficient of friction is above 0.3 or large fluctuations in the coefficient of friction (indicative of film failure) occur continuously, whichever happens first. Energy dispersive spectroscopy (EDS) (Hitachi S4300 FESEM/EDS system) tests were conducted on every wear track in order to confirm the failure of the film by observing the presence of Si peak. Both the coefficient of friction and the wear life data reported are averages of the three repeated tests.

3. Friction and Wear Results

3.1. *Comparison of UHMWPE Film with and without DLC Interface*

As the first part of this chapter, we use DLC as an interface between Si and UHMWPE and then compare the friction and wear properties of Si/UHMWPE and Si/DLC/UHMWPE films.

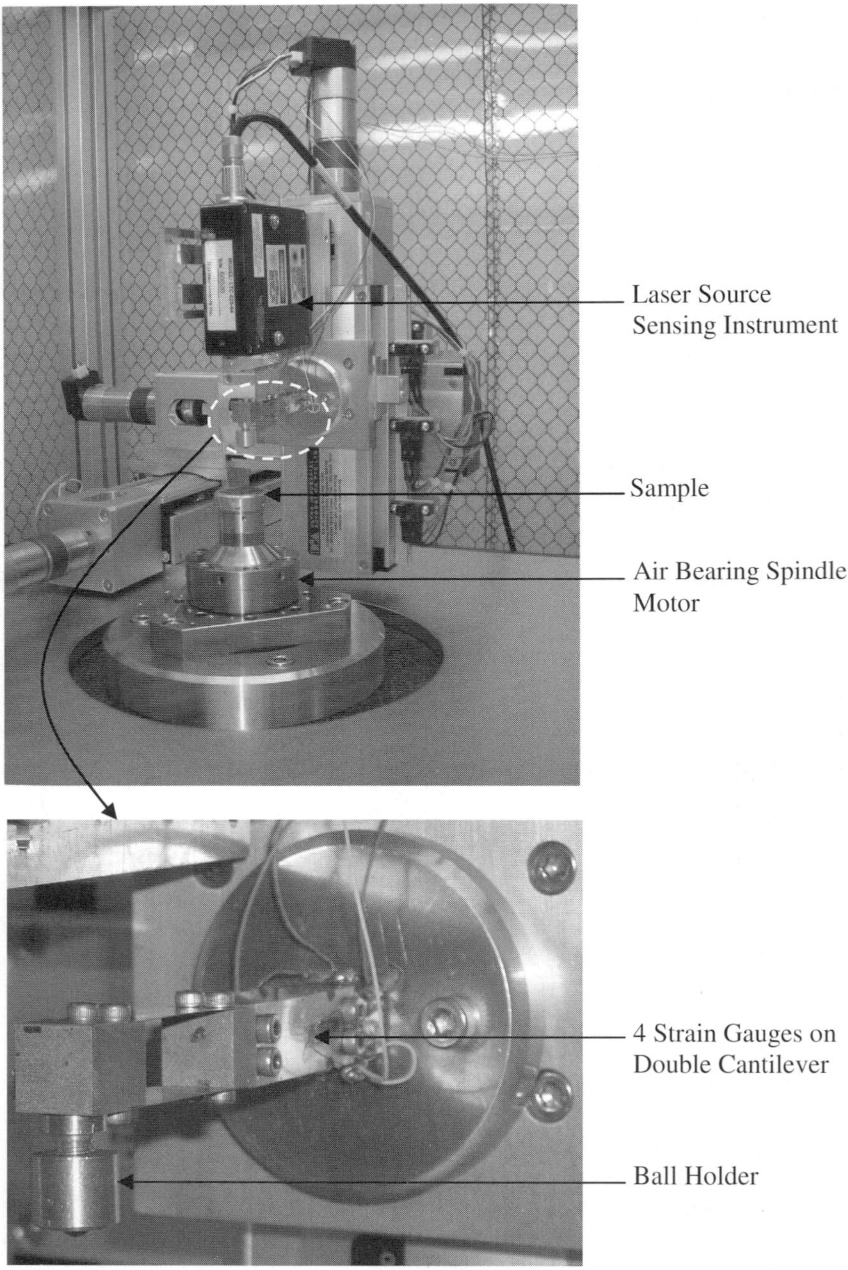

Fig. 6. Ball-on-disc tribometer.

Figure 7 compares the friction and wear life data for samples with and without DLC interface layer; UHMWPE thickness in both cases was 28 μm. Bare Si shows the highest coefficient of friction of 0.65 and the lowest wear life of 5 cycles. Si/DLC reduces the coefficient of friction to 0.25 but its wear life is still low at about 400 cycles. After coating UHMWPE onto bare Si and Si/DLC, the coefficient of friction reduces to 0.18 and 0.13, respectively, due to the self lubricating properties of UHMWPE. Moreover, UHMWPE is soft enough to reduce the shear stress in comparison with bare Si and Si/DLC. It is obvious that the wear life of Si/UHMWPE can improve five times by introducing the hard DLC as an intermediate layer. This indicates that the wear life of composite film, Si/DLC/UHMWPE, has a better tribological property due to the presence of the intermediate DLC film which can provide high load carrying capacity that reduces the contact area.

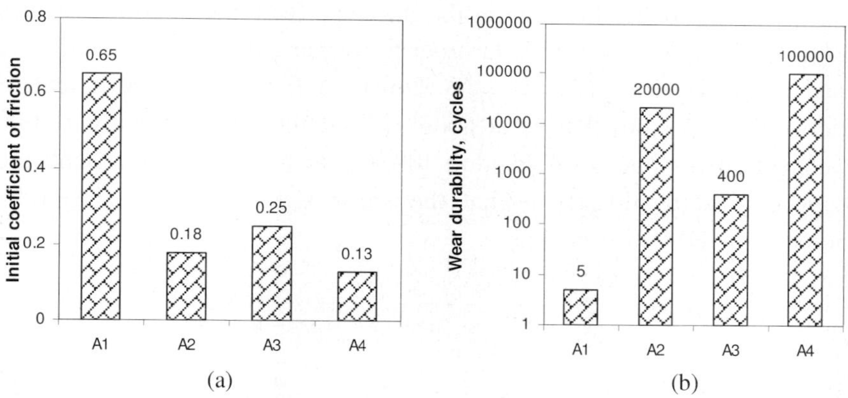

Fig. 7. (a) Initial coefficient of friction and (b) wear life (logarithmic scale) of bare Si and Si coated with different single and composite films where UHMWPE thickness is fixed as 28 μm for all coated samples. (A1 = bare Si, A2 = Si/UHMWPE, A3 = Si/DLC, A4 = Si/DLC/UHMWPE)

3.2. *Effect of UHMWPE Thickness on the Friction and Wear Properties of Si/DLC/UHMWPE*

In the previous section, we have shown that DLC layer can significantly improve the wear life of the composite film, Si/DLC/UHMWPE. In this section, we will further explore the role of soft layer thickness on the tribological properties, in other words, the optimum range of thickness which can provide the best tribological results for Si/DLC/UHMWPE.

We used four different thicknesses of UHMWPE; 3.4 μm, 6.2 μm, 12.3 μm, 28 μm whereas the thickness of the intermediate layer (DLC) was fixed at 50 nm for all cases. As shown in Fig. 8, the coefficient of friction is nearly the same for all thicknesses at the initial stages. When the sliding continues, the effects of thickness are observed on the friction and wear, especially at the higher sliding cycles. For the thinnest film, 3.4 μm, the friction increases with increasing number of cycles. This means that the role of hard DLC is dominant that causes the high shear stress when the top UHMWPE layer is mildly worn out. The film fails at 100,000 cycles. When the film thickness is 6.2 μm, the friction shows the same trend as that of 3.4 μm film, but its wear life extends to 200,000 cycles. The possible reason is that the thicker film can reduce the shear stress, provide better lubricity at the interface for longer period and as a result the wear life increases. In order to further extend the wear life, we used a thicker film, 12.3 μm. As shown in Fig. 8, the coefficient of friction for 12.3 μm film is at about 0.14 ± 0.02 after running 10,000 cycles but that value is stable up to 300,000 cycles when experiments are stopped. The results prove that the shear stress can be reduced by increasing UHMWPE thickness.

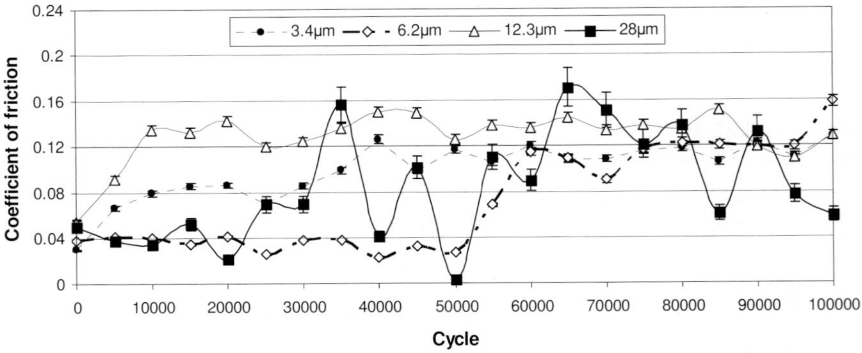

Fig. 8. Coefficient of friction with respect to sliding cycles in typical runs for different thicknesses of UHMWPE in composites films of Si/DLC/UHMWPE.

However, when we use 28 μm thickness, as mentioned in the previous section, Si/DLC/UHMWPE film fails at about 100,000 cycles. Although the friction for 28 μm film is low at the early cycles, the variations become larger beyond 20,000 cycles. The large fluctuations happen due to the different removal rates of the polymer from the sliding track that increases the roughness. Furthermore, the transfer polymer on the ball can also increase the fluctuations.

3.2.1. *Wear Mechanisms for Different UHMWPE Thicknesses on Si/DLC/UHMWPE*

We have shown that there is an optimum range of UHMWPE film thickness that gives the best tribological results. In this section, we will study the wear mechanisms for different thicknesses using the optical images of the tracks during and after the tests.

Figure 9 shows the wear tracks for different UHMWPE film thicknesses for Si/DLC/UHMWPE composite film with respect to the number of sliding cycles. As the sliding continues, the widths of the tracks increase that lead to high friction. It is clearly seen from the optical images of 3.4 μm film that the contact stress is the major factor contributing to friction and wear. It agrees well with the nanoindentation results (Table 2) where the contact pressure of 3.4 μm is the highest and the contact area is the smallest. As a result, the surface image shows some kind of polymer degradation possibly due to high frictional heat generated within a small contact area. The optical image of 6.2 μm film shows a smooth track after 50,000 cycles by removing the asperities and the effect of roughness on friction is less. The smooth track indicates the softening of the sliding surface[35] that can help maintain low coefficient of friction. It is clear from the images of 12.3 μm and 28 μm films that the rates of material removal from the sliding tracks are not uniform. The asperities of the films are removed non-uniformly at an early stage and as a result the surface becomes roughened. For 12.3 μm film, though the polymer is removed unevenly, the surface becomes as smooth as 6.2 μm film with increasing cycles and the effect of surface roughness on friction and wear is not much. For 28 μm film, the early failure of the film is due to the large amount of transfer polymer to the ball (the soft thicker film is easier to penetrate and peel off) that causes large fluctuations.

The optical images of the counterface balls after sliding 100,000 cycles against different UHMWPE thicknesses are shown in Fig. 10. The amount of polymer transferred from the thickest film (Fig. 10d) is remarkably greater than that from film of any other thickness. We can conclude from this thickness effect study that the optimum range of UHMWPE thickness that can give best tribological results is 6.2 μm ~ 12.3 μm.

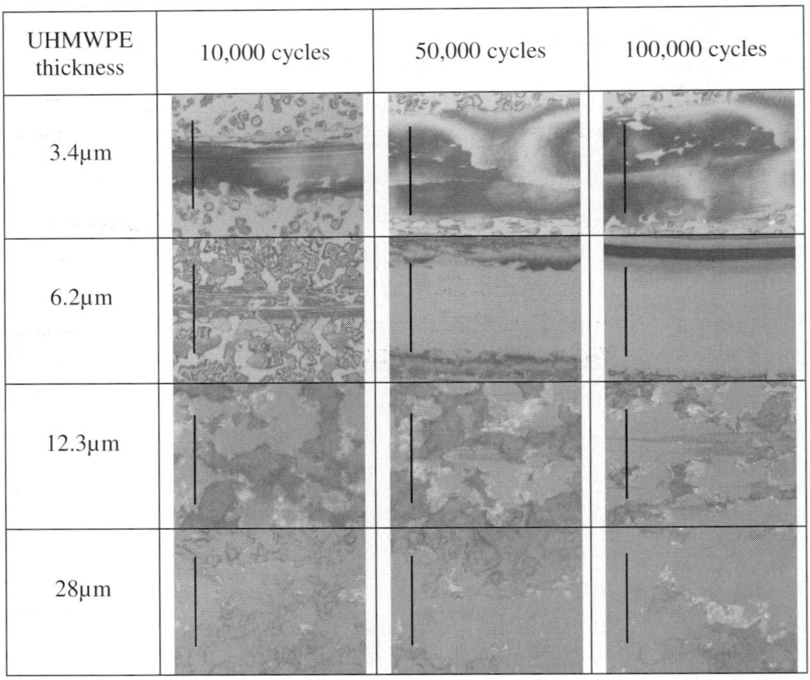

Fig. 9. Wear track optical images of 3.4 μm, 6.2 μm, 12.3 μm and 28 μm UHMWPE thicknesses for Si/DLC/UHMWPE (at a normal load of 40 mN, at a rotational speed of 5.2 cm/s (500rpm) and test radius of 1 mm) against Si_3N_4 counterface ball after 10,000, 50,000 and 100,000 sliding cycles, respectively. The scale bars are 50 μm.

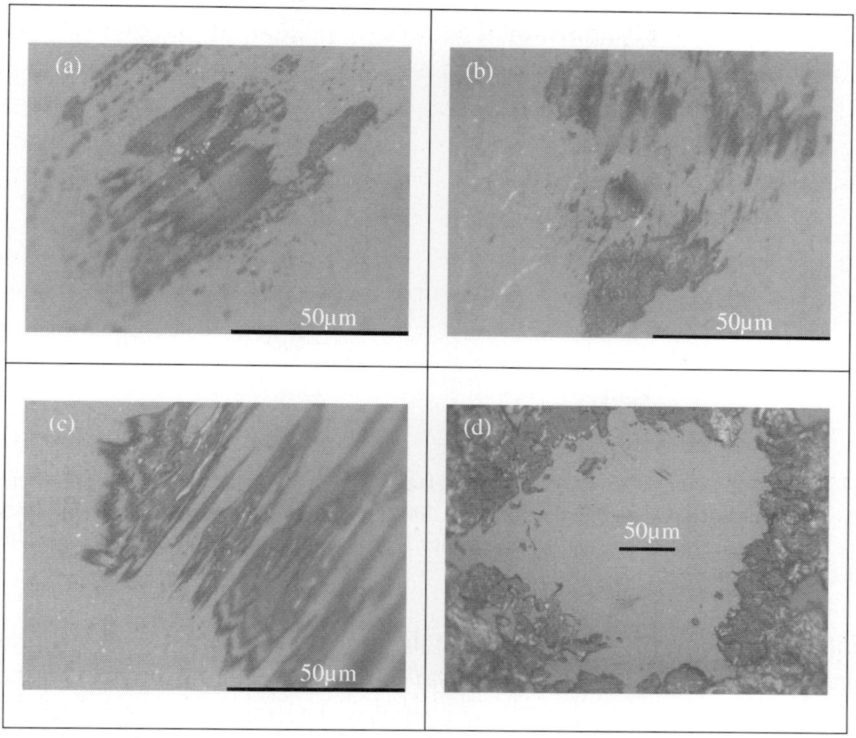

Fig. 10. Optical images of Si_3N_4 counterface ball against Si/DLC/UHMWPE with different polymer film thicknesses (a) 3.4 μm (b) 6.2 μm (c) 12.3 μm and (d) 28 μm after sliding 100,000 cycles. Figs. (a, b, c) are magnified 500 times and Fig. (d) is magnified 200 times.

3.3. *Comparison of UHMWPE Film with Different Interfaces*

As the second part of this chapter, we modify Si substrates in order to change the surface energies (wettabilities) and compare the friction and wear properties of UHMWPE film coated onto differently modified Si substrates.

Figure 11 shows the tribological properties of UHMWPE films coated on Si substrate with different interfaces (surface energies before coating the film). In this study, we have selected UHMWPE thickness as 6 ± 2 μm and three types of interfaces based on surface energies: very hydrophilic (Si), very hydrophobic (Si/OTS) and between these two extremes. It is seen that the initial coefficients of friction for all

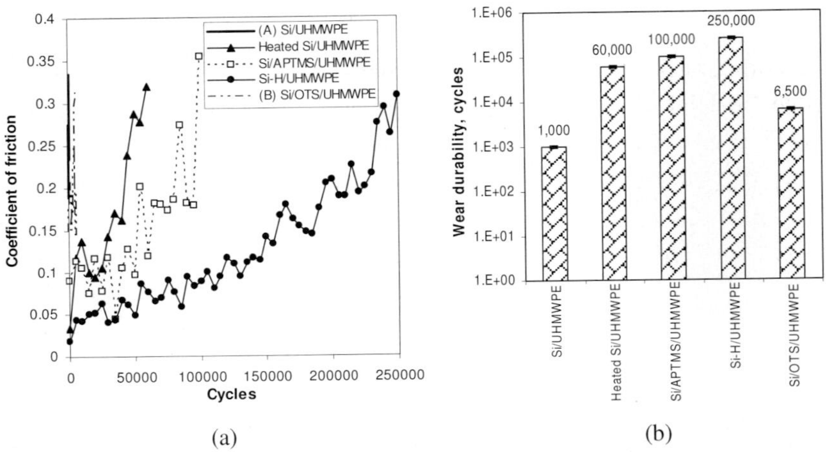

Fig. 11. Friction and wear properties of UHMWPE film with different interfaces where the normal load is 4 g and sliding speed is 500 rpm (0.1 m/s). (a) Typical friction traces as a function of the number of sliding cycles for all samples. (b) Consolidated wear life data for all samples.

interfaces are nearly the same since the uppermost layers for all samples are UHMWPE film. As the top layer is the same, the differences in the trends of friction and wear lives are entirely determined by the properties of the interfaces for long duration sliding.

The coefficient of friction of Si/UHMWPE and Si/OTS/UHMWPE did not increase slowly but abruptly. The sudden increase in the coefficient of friction indicates that within a few sliding cycles, a large amount of polymers are pulled out from the substrates and get transferred to the ball that causes early failure of the film. The possible reason is when the sliding continues, first, the polymer from the ball may get transferred back to the substrate that increases the roughness and second, the lumps of polymer on the ball make the wear track very uneven. As a result, the coefficient of friction increases immediately within a few sliding cycles and wear occurs at 1,000 cycles for the most hydrophilic (Si) and at 6,500 cycles for the most hydrophobic (Si/OTS) films.

Contrary to bare Si and Si/OTS substrates, the coefficients of friction of heated Si/UHMWPE, Si/APTMS/UHMWPE and Si-H/UHMWPE samples increased much slowly with the number of sliding cycles and their average wear lives are 60,000 cycles, 100,000 cycles and 250,000 cycles, respectively.

3.3.1. *Wear Mechanisms for UHMWPE Film with Different Interfaces*

As a comparative study of the wear mechanism, we have selected one sample (Si/UHMWPE) from a group of either very hydrophilic or very hydrophobic and another sample (Si-H/UHMWPE) from a group with interfaces in between these two extremes.

Figures 12a & 12b show the wear tracks on Si/UHMWPE and Si-H/UHMWPE film after 1,000 cycles of sliding. The wear track on Si/UHMWPE shows a clear wear groove and some small debris particles. These small particles either remain on the wear track or get transferred to the ball surface. Continuous sliding over the same track induces further wear of the substrate and high friction.[36] For Si-H/UHMWPE, we do not see any removal of the polymer but only sliding mark (due to ironing of the asperities of the polymer film) on the wear track. That indicates the adhesion strength of UHMWPE on Si-H is much stronger giving higher wear life. Si-H gives intermediate surface energy which provides strongest adhesion between the substrate Si and the polymer film.

Figures 12c & 12d show the optical images of the counterface balls after 1,000 cycles of sliding against Si/UHMWPE and Si-H/UHMWPE. The lumps of polymer transfer on the ball against Si/UHMWPE had taken place that increased the roughness and uneven sliding in the contact zone. The thickness of the transfer polymer near the contact zone (1) is thinner than that away from the contact zone (2). It seems the transfer film is pushed away from the contact zone along the direction of the sliding. When the transfer film is thick enough to block or impose resistance against the sliding, the coefficient of friction sharply increases and the film fails. The same mechanism is applicable to Si/OTS/UHMWPE since the film is patchy or island types which are difficult to withstand the sliding resistance.

Contrary to these two extreme types, heated Si/UHMWPE, Si/APTMS/UHMWPE and Si-H/UHMWPE films have the transfer films which are very thin and of elongated fibrous structures. These structures were formed by the asperities of the film while bulk of the film was firmly attached to the substrate. There was no gross transfer of the film from Si substrate to the counterface ball because of strong bonding between the polymer and the substrate.

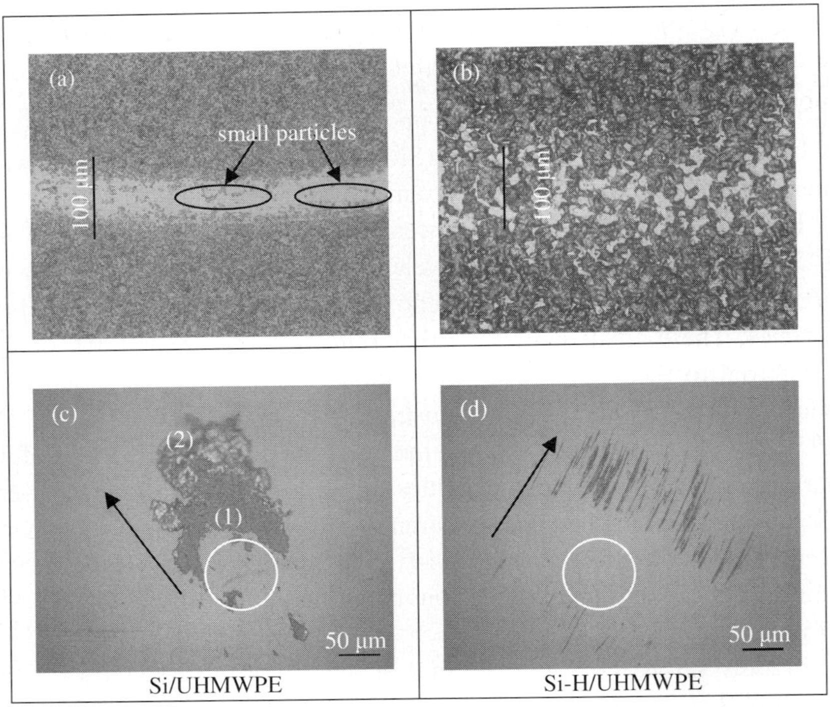

Fig. 12. Optical microscopy images of (a) Si/UHMWPE film and (b) Si-H/UHMWPE film after sliding against Si_3N_4 ball for 1,000 cycles where the normal load is 4 g and the sliding speed is 500 rpm (0.1 ms^{-1}). (c) is the image of the ball after sliding against (a), (d) is image of the ball after sliding against (b). Solid arrows indicate the direction of sliding; white circles indicate the point of contacts.

4. Discussion

Applying DLC on bare Si has two advantages. First, the water contact angle of DLC becomes hydrophobic which repels moisture contents and as a result the adhesion strength is somewhat stronger than it is without DLC as the interface. Second, DLC provides high load carrying capacity that reduces the contact area and decreases the coefficient of friction.

The data from nanoscratching (Table 2) confirm that the penetration resistance of Si/UHMWPE is inferior to that of Si/DLC/UHMWPE where the thickness of UHMWPE is 28 µm. Although the initial penetration depth (5.3 µm for Si/UHMWPE) is shallower than the whole film thickness (28 µm) at the initial stage, as the sliding proceeds, the transfer film on the ball surface is continuously renewed by wearing out of the polymer from the contact point on the film and the ball reaches Si substrate after ~ 20,000 cycles due to weak penetration (low hardness) resistance and large contact area. However, the presence of DLC as an interface provides strong penetration resistance that leads to smaller contact area. As a consequence, we did not see any sign of film failure until 100,000 cycles for Si/DLC/UHMWPE.

Moreover, higher elastic modulus of Si/DLC/UHMWPE provides greater relaxation time which means the elastic modulus does not change much due to interface temperature or creep. This is because, low elastic modulus increases the mobility of the molecules to flow in viscous manner due to less strong inter-molecular bonding. Thus, the relaxation time for larger contact area decreases with decreasing elastic modulus[37] and hence the film is prone to damage because of thermal effect. Thus, the polymer film is easily removed from the sliding track due to thermal and time dependent changes in the elastic modulus of the film and the wear life of Si/UHMWPE is shorter than that of Si/DLC/UHMWPE. Thus, better bonding of the UHMWPE molecules with DLC and better mechanical properties of DLC with self-lubricating property of UHMWPE are the major contributors to the higher tribological performance of the composite film, Si/DLC/UHMWPE.

When the UHMWPE film thickness decreases for Si/DLC/UHMWPE, the load carrying capacity of the film is restrained by the underlying hard DLC substrate and the effect of shear stress or contact pressure is high which leads to the high friction and early film failure. Furthermore, if the film thickness is too low in comparison with the contact radius, the counterface ball will penetrate into DLC film through UHMWPE; hard DLC particles may detach out and serve as third-body abrasive particles at the interface which will increase friction and initiate wear.[38]

When the thickness of UHMWPE film increases, the contact area will increase with deeper penetration depth that leads to high friction and low wear life. At the initial stage of the test, the asperities of the film are removed by microcutting[39] and the removal (wear) rates of the polymer film vary in different part of the sliding tracks (28 μm film in Fig. 8) which happens due to non-uniform contact points. When the transfer polymer on the ball surface is substantial and rough, the friction between the non-uniform film and the roughened polymer adhered ball is highly fluctuating. As a result, the thicker, 28 μm, film fails early at about 100,000 cycles. For 12.3 μm film, though the surface shows non-uniform pattern, the transfer polymer amount is qualitatively low. Moreover, the surface becomes smoother with increasing sliding cycles. Therefore, though the coefficient of friction for 12.3 μm film is slightly high, its value is consistent at 0.14 ± 0.02 for at least 300,000 cycles when the experiments are stopped due to long test duration.

The very hydrophilic, Si surface has higher affinity for water molecules that renders weaker bonding between the Si substrate and UHMWPE. The possibility of the presence of condensed moisture on a surface decreases with increasing surface hydrophobicity.[17] Though the highly hydrophobic Si/OTS substrate has less moisture content, UHMWPE does not adhere well onto Si/OTS due to non-wettability of the substrate during dip-coating. It is evident from the patches of polymer that are clearly seen on Si/OTS/UHMWPE samples. This indicates that very low and very high surface energies (water contact angles) cannot provide sufficient adhesion to UHMWPE film in order to obtain high wear life.

Table 4 provides the samples with water contact angle and wear life. The wear life of UHMWPE increases with increasing the water contact angle up to an optimum value. Beyond the optimum value, the wear life decreases again as shown by Si/OTS/UHMWPE film. The optimum wear life is found when the surface wettability is approximately 71°.

The wear lives of the films are consistent with the critical load data in scratching presented in Sec. 2.3.4. Si/UHMWPE and Si/OTS/UHMWPE films have the lowest critical load and the shortest wear life. The critical loads and wear lives of heated Si/UHMWPE and Si/APTMS/UHMWPE are moderate. Si-H/UHMWPE has the highest critical load and the longest wear life.

Table 4. Water contact angles and wear lives of UHMWPE for different interfacial modifications on Si.[33]

Interface	Bare Si	heated Si	Si/APTMS	Si-H	Si/OTS
Contact angle (°)	21	46	52	71	104
Wear life (cycles)	1,000	60,000	100,000	250,000	3,000

The moisture contents between the substrate and the polymer film are strongly dependent on the surface energies which determine the wear durability and the adhesion strength of the coatings. In general, more hydrophilic surface attracts more water contents that lead to low adhesion. The presence of water contents on the surface can be reduced by increasing the surface hydrophobicity. However, very hydrophobic surface repels not only the water molecules but also the solvent and polymer molecules. It results in poor film formation characterized by uneven or island type on the substrate. A sketch of the interactions between the polymer molecules and the substrates is shown in Fig. 13.

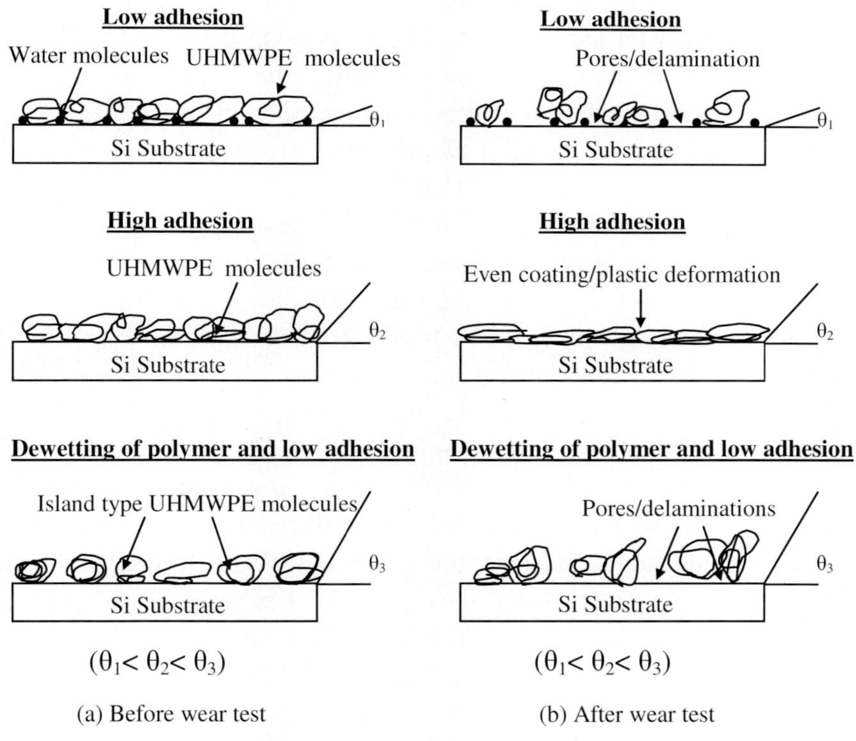

Fig. 13. A diagrammatic model showing the interactions between the polymer molecules and the Si surface with different wettabilities as measured by water contact angle. θ_1, θ_2 and θ_3 represent relative water contact angles of the interfaces before polymer coating. (a) represents the film condition before sliding and (b) is after sliding against a silicon nitride ball.

5. Conclusions

In this chapter, we have modified Si substrates with (1) DLC and (2) different surface energies before UHMWPE film is coated onto them. Then, we have compared the tribological properties of modified films with films on bare Si. The following conclusions can be drawn from our present studies.

The coefficient of friction of Si/UHMWPE is 0.18 and its wear life is approximately 20,000 cycles of sliding when slid against a 4 mm diameter Si_3N_4 ball under a constant applied normal load of 40 mN. The presence of DLC as an intermediate layer between Si and UHMWPE (Si/DLC/UHMWPE) provides higher load carrying capacity (high hardness and elastic modulus) and better adhesion between UHMWPE molecules and hard DLC. As a result, it reduces the coefficient of friction to 0.13 and provides five times improvement in wear life when UHMWPE thickness is fixed at 28 μm. After investigating the effect of UHMWPE thickness on Si/DLC/UHMWPE, we have seen that the wear lives of thin (3.4 μm) and thick films (28 μm) are approximately 100,000 cycles which are shorter than those of moderate (optimum) thickness. The wear lives of 6.2 μm and 12.3 μm films are 200,000 cycles and 300,000 cycles, respectively. The results confirm that the wear life of Si/DLC/UHMWPE can increase several folds by introducing DLC as an interface and carefully controlling the thickness of the top UHMWPE film.

Using the thickness of UHMWPE as 6 ± 2 μm, which is within the optimum range, we have studied the role of different surface energies of modified Si substrates on the tribology of UHMWPE film coated onto them. The results show that the wear life is dependent on the surface energies of the interfaces. The most hydrophilic surface, Si/UHMWPE, and the most hydrophobic surface, Si/OTS/UHMWPE, have the lowest wear lives of 1,000 cycles and 6,500 cycles, respectively. Between these two extremes, the wear life increases with increasing water contact angle.

The final conclusion is that we can increase the tribological properties of UHMWPE film by simply introducing either hard DLC or changing the surface energy of the Si substrate in an optimum range of UHMWPE thickness. Finally, it is necessary to optimize each factor affecting the coefficient of friction and wear life.

Acknowledgments

This study was funded by a research grant from the Faculty of Engineering, NUS (#R-265-000-248-112). The authors would like to acknowledge Ms. Shen Lu of the Institute of Materials Research and Engineering, Singapore, for her assistance in the use of nanoscratching and nanoindentation testers. We also wish to thank Mr. K. C. Chung, Ticona Representative Office, Singapore, for donating UHMWPE powder for this study. One of the authors (M. Minn) would like to acknowledge the Graduate School of Engineering, NUS for the scholarship awarded to support his studies at NUS.

References

1. N. P. Suh, *Tribophysics* (Englewood Cliffs, Prentice-Hall, New Jersey, 1986).
2. S. M. Kurtz, O. K. Muratoglu, M. Evans, and A. A. Edidin, *Biomater.,* 20 (1999).
3. O. K. Muratoglu, C. R. Bragdon, D. O. O'Connor, M. Jasty, W. H. Harris, R. Gul and F. McGarry, *Biomater.,* 20 (1999).
4. H. McKellop, F-W. Shen, B. Lu, P. Campbell and R. J. Salovey. *Orthop. Res.,* 17 (2) (1999).
5. O. K. Muratoglu, C. R. Bragdon, D. O. O'Connor, M. Jasty, W. H. Harris, R. Gul and F. McGarry, *Biomater.,* 20 (16) (1999).
6. O. K. Muratoglu, C. R. Bragdon, D. O. O'Connor, M. Jasty and W. H. Harris, *J. Arthroplasty,* 16 (2) (2001).
7. L. H. Sperling, *Introduction to physical polymer science*, (Hoboken, John Wiley, New Jersey, 2006).
8. Y. Bao *et al. J. Mater. Sci.,* 40 (2005).
9. N. Satyanarayana, S. K. Sinha and B. H. Ong. *Sens. Actuators, A,* 128 (1) (2006).
10. J. Robertson, *Diamond Relat. Mater.*, 1 (1992).
11. Z. H. Liu, P. Lemoine, J. F. Zhao, D. M. Zhou, S. Mailley, E. T. McAdams, P. Maguire, J. McLaughlin, *Diamond Relat. Mater.*, 7 (1998).
12. Z. H. Liu, J.F. Zhao, J.A. McLaughlin, *Diamond Relat. Mater.*, 8 (1999).
13. A. Grill, V. Patel, *Diamond Relat. Mater.*, 2 (1993).
14. A. Erdemir and C. Donnet, *Modern Tribology Handbook*, Ed B. Bhushan, (CRP press, Boca Ranton, 2000).
15. A. Erdemir, C. Donnet, *J. Phys. D: Appl. Phys.*, 39 (2006).

16. R. A. Ryntz, *Prog. Org. Coat.*, 25 (1994).
17. E. P. O'Brien, C. C. White and B. D. Vogt, *Adv. Eng. Mater.*, 8 (1-2) (2006).
18. P. Moy and F. E Karasz, *Polym. Eng. Sci.*, 20 (1980).
19. M. C. Lee and N. A. Peppas, *Prog. Polym. Sci.*, 18 (1993).
20. P. Nogueira, C. Rami'rez, A. Torres, M. J. Abad, J. Cano, J. Lo'pez, I. Lo'pez-Bueno and L. Barral, *J. Appl. Polym. Sci.*, 80 (1) (2001).
21. T. C. Wong and L. J. Broutman, *Polym. Eng. Sci.*, 25 (1985).
22. Z. R. Xu and K. H. G. Ashbee, *J. Compos. Mater.*, 25 (1991).
23. R. D. Armstrong and J. D. Wright, *Electrochim. Acta*, 38 (1993).
24. F. Mansfeld, L. T. Han, C. C. Lee and G. Zhang, *Electrochim.Acta*, 43 (1998).
25. R.-H. Yoon, D. H. Flinn and Ya. I. Rabinovich, *J. Colloid Interface Sci.*, 185 (2) (1997).
26. B. K. Tay, D. Sheeja, Y. S. Choong, S. P. Lau and X. Shi, *Diamond Relat. Mater.* 9 (2000).
27. C. Dunbar, *Surface Preparation for Continuously Applied Coatings*, ASM Handbook, Vol. 5, Surface Engineering, (USA, 1994), p. 1051.
28. *Hot dip Coatings Methods*, ASM Handbook, Vol. 13, *Corrosion,* (USA, 1992), p. 975.
29. T. Young, *Miscellaneous Works*, Ed. G. Peacock, Vol. 1 (Murray, London, 1855), p. 418.
30. R. J. Good, *Contact Angle, Wettability and Adhesion,* Ed. K. L. Mittal (VSP International Science, 1993), p. 3.
31. N. Satyanarayana and S. K. Sinha, *J. Phys. D*: *Appl. Phys.*, **38** (2005).
32. M. Minn and S. K. Sinha, *Surf. Coat. Technol.*, 202 (2008).
33. M. Minn, J. Y. Leong and S. K. Sinha, *J. Phys. D: Appl. Phys.*, 41 (2008).
34. B. J. Briscoe, E. Pelillo and S. K. Sinha, *Polym. Eng. Sci.*, 36 (24) (1996).
35. K. Tanaka, Y. Uchiyama and S. Toyooka, *Wear*, 23 (1973).
36. R. L Fusaro, *NASA Technical Paper*, 2059 (1982).
37. G. A. Gorokhovskii, I. Agulov, *Mech. Compos. Mater.*, 2 (1) (1966).
38. R. D. Arnell, *Surf. Coat. Technol.*, 43-44 (1990).
39. A. Aubert, J.Ph. Nabot, J. Ernoult, Ph. Renaux, *Surf. Coat. Technol.*, 41 (1990).

INDEX